Microwave-Mediated Biofuel Production

Microwave-Mediated Biofuel Production

Veera Gnaneswar Gude
Civil & Environmental Engineering Department
Mississippi State University
Mississippi State, MS
USA

CRC Press is an imprint of the
Taylor & Francis Group, an **informa** business

A SCIENCE PUBLISHERS BOOK

Cover illustrations provided by the author of the book, Veera Gnaneswar Gude

CRC Press
Taylor & Francis Group
6000 Broken Sound Parkway NW, Suite 300
Boca Raton, FL 33487-2742

© 2017 by Taylor & Francis Group, LLC
CRC Press is an imprint of Taylor & Francis Group, an Informa business

No claim to original U.S. Government works

Printed on acid-free paper
Version Date: 20170515

International Standard Book Number-13: 978-1-4987-4515-4 (Hardback)

This book contains information obtained from authentic and highly regarded sources. Reasonable efforts have been made to publish reliable data and information, but the author and publisher cannot assume responsibility for the validity of all materials or the consequences of their use. The authors and publishers have attempted to trace the copyright holders of all material reproduced in this publication and apologize to copyright holders if permission to publish in this form has not been obtained. If any copyright material has not been acknowledged please write and let us know so we may rectify in any future reprint.

Except as permitted under U.S. Copyright Law, no part of this book may be reprinted, reproduced, transmitted, or utilized in any form by any electronic, mechanical, or other means, now known or hereafter invented, including photocopying, microfilming, and recording, or in any information storage or retrieval system, without written permission from the publishers.

For permission to photocopy or use material electronically from this work, please access www.copyright.com (http://www.copyright.com/) or contact the Copyright Clearance Center, Inc. (CCC), 222 Rosewood Drive, Danvers, MA 01923, 978-750-8400. CCC is a not-for-profit organization that provides licenses and registration for a variety of users. For organizations that have been granted a photocopy license by the CCC, a separate system of payment has been arranged.

Trademark Notice: Product or corporate names may be trademarks or registered trademarks, and are used only for identification and explanation without intent to infringe.

Library of Congress Cataloging-in-Publication Data
Names: Gude, Veera Gnaneswar, 1978- author. Title: Microwave-mediated biofuel production / Veera Gnaneswar Gude, Civil & Environmental Engineering Department, Mississippi State University, Mississippi State, MS, USA. Description: Boca Raton, FL : CRC Press/ Taylor & Francis Group, [2017] | "A science publishers book." | Includes bibliographical references and index. Identifiers: LCCN 2016046132| ISBN 9781498745154 (hardback : acid-free paper)| ISBN 9781498745161 (e-book : acid-free paper) Subjects: LCSH: Biomass energy. | Microwave heating. Classification: LCC TP339 .G85 2017 | DDC 662/.88--dc23 LC record available at https://lccn.loc.gov/2016046132

Visit the Taylor & Francis Web site at
http://www.taylorandfrancis.com

and the CRC Press Web site at
http://www.crcpress.com

To my late father Gude Bhaskar Rao
to my mother Gude Fathima
to my sister Kalyani
and to my brothers
Kiran and Karunakar and Preeti Muire

Preface

Increasing environmental pollution issues related to escalating demands for fossil fuels has invigorated the quest for production of renewable fuels such as biofuels in recent years. For biofuel production to be sustainable, energy- and resource-efficient technologies are the starting point. Microwave energy provides a clean and energy-efficient platform for production of various chemicals, fuels and valuable products. Microwave energy based chemical synthesis has several merits and is important from both scientific and engineering standpoints. Microwaves have been applied in numerous inorganic and organic chemical syntheses, perhaps, from the time their ability to work as heat source was discovered. Recent laboratory scale microwave applications in biofuel production proved the potential of the technology to achieve superior results over conventional techniques. Short reaction time, cleaner reaction products, and reduced separation-purification times are the key observations reported by many researchers. Energy utilization and specific energy requirements for microwave based biofuel synthesis are reportedly better than conventional techniques.

Microwaves can be utilized in feedstock preparation, pretreatment, extraction and chemical conversion stages of various biofuel production processes. Microwave technology has advanced in other food, pharmaceutical and polymer chemistry related research and industrial applications. However, it has shown great promise for its beneficial use in biofuel industry for large scale applications based on the laboratory-scale and pilot-scale studies. This book elaborates on the principles and practices of microwave energy technology as applied in biofuels' production. Energy generation, utilization, footprint and economic and environmental analysis of the microwave based biofuel processes are included. Analysis of laboratory scale studies, potential design and operation challenges for developing large scale biodiesel production systems are discussed in detail.

With increasing interest in renewable energy production, microwaves play an important role in improving economic, energy and environmental footprints of the biofuels production. This book focuses on chemical syntheses and processes for biofuel production mediated by microwave energy. Microwaves as an efficient heating mechanism have been evaluated by many researchers around the world. However, this is the first contribution in this area serving as a resource and guidance manual for understanding the principles, mechanisms, design and applications of microwaves in biofuel process chemistry for graduate researchers, professionals and industrial practitioners. The book is a critical resource for researchers from chemical engineering and science, environmental science, mechanical and electrical engineering, and industrial and process engineering and management disciplines including undergraduate and graduate and professional level researchers.

This book is presented in ten chapters. Details of which are presented below.

Chapter No.	Title
1.	Microwaves and green chemistry of biofuels
2.	Microwave heating and interaction with materials
3.	Microwave reactor design and configurations
4.	Microwave mediated biodiesel production
5.	Microwave pretreatment of feedstock for bioethanol production
6.	Microwave enhanced biogas production

7.   Microwave mediated thermochemical conversion of biomass
8.   Heat and mass transfer and reaction kinetics under microwaves
9.   Microwaves and ionic liquids in biofuel production
10.   Microwave hybridization for advanced biorefinery

Chapter 1 presents the evolution, development and history of microwave energy and technological advances. Earlier studies in organic and inorganic chemical syntheses are discussed. Introduction of green chemistry principles and relevance of these principles in microwave enhanced chemistry are discussed. In particular, e-factor, atom economy, mass intensity, reaction mass efficiency are discussed in detail with illustrations for e-factor and atom economy concepts. It provides an overview of the green chemistry applications in microwave enhanced chemistry for biofuel production.

Chapter 2 elaborates on the microwave energy principles, characteristics, and material interactions and heating mechanisms with illustrations for biodiesel production, pyrolysis and other applications. Microwave energy characteristics, energy levels, and heating mechanisms were presented. Principles of microwave heating, microwave-solvent interactions, microwave-solid interactions, interactions with carbon materials, microwave interactions relevant to object size have been discussed. Properties of microwave heating and kinetic rate enhancement are discussed with illustrations.

Chapter 3 discusses the microwave reactor design and configurations. Components of microwave reactors including microwave generators, waveguides, and reactor cavities are described. The current reactor limitations in microwave heating considerations related to process and safety limitations, reproducibility and scalability of reactions have been discussed. Single mode and multi-mode reactors and manufacturer information is provided. Batch, plug-flow and continuous flow reactor configurations are also presented. A few case studies of pilot-scale and large-scale applications of microwave reactor systems have been discussed.

Chapter 4 provides a detailed account of biodiesel production mediated by microwave energy. Biodiesel production methods, niche applications for microwaves in the process reaction and engineering have been discussed. Microwave applications in pre-treatment, lipid extraction, esterification, transesterification reaction are elaborated. Thermodynamic justification for microwave mediated biodiesel production and a comparison with conventional heating method is provided. A comprehensive review of the current state-of-the-art of the biodiesel production and effect of various process parameters is discussed. Glycerol by-product uses, biodiesel quality and process monitoring are discussed. In addition, an evaluation of energy efficiency of microwave application in biofuel production is discussed with examples.

Chapter 5 deals with production of bioethanol under microwave irradiation. Various platforms for bioethanol production from various sources such as starch, lignocellulosic feedstock, pre-treatment methods (chemical, physical and mechanical), microwave applications, interactions with lignocellulosic feedstock, influencing factors in microwave pre-treatment are presented. Energy aspects of microwave enhanced bioethanol production and a summary of recent studies are discussed in detail.

Chapter 6 presents a detailed account of microwave enhanced biogas production. Biogas process chemistry, biomass feedstock pre-treatment methods, and microwave applications are discussed. Factors influencing microwave digestion of biomass, various types of feedstock (activated sludge, agricultural wastes and other microbial and lignocellulosic feedstock) have been discussed. A summary of previous studies involving biogas production and other beneficial product recovery is provided. Concepts such as co-digestion, energy efficiency, and economic analysis are also presented with examples from recent studies.

Chapter 7 provides an overview of microwave mediated thermochemical conversion of biomass into various biofuel products including bio-crude, biochar, syngas and bio-oils. Microwave enhanced pyrolysis process configurations, influencing parameters and a summary of recent studies on pyrolysis and bio-oil production are presented. Reaction products and their yields are compared between the conventional and microwave methods. Process economics and limitations are discussed in detail.

Chapter 8 discusses heat and mass transport phenomena under microwave irradiation. Enhancement of reaction kinetics under microwave irradiation in terms of activation energy and temperature difference are discussed. A comparison of mass and heat transport under conventional and microwave fields is provided. Mathematical models for determining the heat distribution and heat and mass transfer in solid-liquid suspensions are presented. Kinetics related to oil extraction from various feedstock are also discussed. A survey of kinetic models used in the recent research has also been included. Effects of reactants and other process related parameters are discussed for biodiesel, bioethanol and biogas production including pyrolysis and gasification processes.

Chapter 9 provides an overview of benefits and applications of ionic liquids and microwave combination in various biofuel and bio-product processes. Classification and preparation of ionic liquids are discussed. Use of ionic liquids in biodiesel production, their synthesis and recovery after use are also covered. Applications of ionic liquids in bioethanol production using lignocellulosic feedstock, in biogas production and other valuable products such as HMF, furfural, levulinic acid are elaborated with examples from recent literature.

Chapter 10 explores the possibilities of integrating microwaves with various other process intensification techniques to enhance the benefits, especially in biorefinery concepts. Metrics for process intensification in biofuel production are discussed. Microwave hybridization with ultrasound, vacuum, mechanical mixing and ohmic heating is discussed with examples. Process configurations, synergistic benefits in biofuel production are also discussed with case studies. Biorefinery concepts utilizing different feedstocks are presented with examples from recent studies.

Acknowledgments

I would like to express my appreciation to the colleagues in the Civil and Environmental Engineering department, the Bagley College of Engineering, and the Office of Research and Economic Development at Mississippi State University for their support. Collaborations with colleagues in Swalm School of Chemical Engineering (Dr. Todd French; Dr. Rafael Hernandez, now at University of Louisiana; and Dr. Andro Mondala, now at Western Michigan University and Mr. William Holmes, now at University of Louisiana) and department of Chemistry (Dr. David Wipf) and United States Department of Agriculture – Agricultural Research Service (Dr. John Brooks) are gratefully acknowledged. My sincere appreciation and gratitude are extended to my doctoral and post-doctoral advisors Dr. Nagamany Nirmalakhandan, Dr. Ricardo Jacquez, Dr. Shuguang Deng, and Dr. Adrian Hanson of New Mexico State University for their continuous support in my professional career. Prof. Shuguang Deng is gratefully acknowledged for introducing me to biofuels research during my post-doctoral research studies at New Mexico State University. Dr. Prafulla Patil is thanked for sharing his knowledge and some of the wonderful laboratory experiences with me. My graduate research students at Mississippi State University are appreciated for their diligence and energy in work environment and research contributions, especially Dr. Edith Martinez-Guerra for her work on microwave and ultrasound enhanced biodiesel production from microalgae and other waste oil sources. Dr. Bahareh Kokabian and Ms. Sara Fast are acknowledged for their contributions in microbial desalination and advanced oxidation processes research as well. Finally, I would like to thank Ms. Preeti Muire for her continuous support and unconditional love.

Veera Gnaneswar Gude

Contents

Dedication   v

Preface   vii

Acknowledgments   xi

1. Microwaves and Green Chemistry of Biofuels   1
2. Microwave Heating and Interaction with Materials   28
3. Microwave Reactor Design and Configurations   62
4. Microwave Mediated Biodiesel Production   99
5. Microwave Pretreatment of Feedstock for Bioethanol Production   158
6. Microwave Enhanced Biogas Production   206
7. Microwave Mediated Thermochemical Conversion of Biomass   240
8. Heat and Mass Transfer and Reaction Kinetics Under Microwaves   276
9. Microwaves and Ionic Liquids in Biofuel Production   311
10. Microwave Hybridization for Advanced Biorefinery   346

Index   379

1

Microwaves and Green Chemistry of Biofuels

Introduction

The field of microwave technology has a rich history starting from the 19th century. Fundamental discoveries in electromagnetics by Faraday, Maxwell, and Rayleigh have laid the foundations for the working principles while the ingenious experiments of Hertz verified the concepts of electromagnetic-wave propagation (Sobol and Tomiyasu 2002). These foundational efforts combined with those from subsequent investigators in the field have broadened the knowledge in the areas of microwave generation, guidance, detection, and control of short-wavelength electromagnetic-wave propagation. The developments in the microwave technology field over the past 150 years can be characterized, in general, as follows:

  i) Study and applications of electromagnetic-wave propagation, described by Maxwell's equations, in various media with a wide range of boundary conditions;
  ii) Interaction of propagating electromagnetic waves with solids, gases, fluids, and charged particles; and
  iii) Interaction of propagating electromagnetic waves with matter under various states of energy excitation and the reciprocal conversion of microwave energy to other various forms of energy in this matter.

Application of the microwave technology has been a key driver in the development of the field over the past 100 years. Early applications of microwave technology focused on communication technology using electromagnetic propagation in free space and in development of radio, radar, television, long-distance telephony, satellite communication links, and wireless access systems. Accidental discovery of microwave heating reported by Spencer opened the doors for microwave mediated heating in various beneficial applications. Main applications of microwave heating include food processing, drying, polymer synthesis, plastic and rubber treating as well as curing and preheating of ceramics.

In general, microwave radiation is the term associated with any electromagnetic radiation in the microwave frequency range of 300 MHz–300 GHz (Figure 1). Domestic and industrial microwave ovens generally operate at a frequency of 2.45 GHz corresponding to a wavelength of 12.2 cm and energy of 1.02×10^{-5} eV (Jacob et al. 1995). However, not all materials can be heated rapidly by microwaves. Most of the industrial and domestic appliances are authorized to operate only at either 915 MHz or 2.45 GHz (wavelength 32.8 or 12.2 cm, respectively) to avoid interfering with radar and telecommunication frequencies (Harvey 1963). Domestic microwave ovens generally operate at a frequency of 2.45 GHz with few exceptions in some countries (Zhang and Hayward 2006).

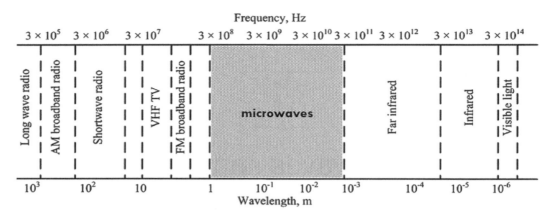

Figure 1. The electromagnetic spectrum and the microwave frequency range.

History of Microwave Technology Development

Microwave technology has become ubiquitous not only for communications and radar applications, but also for the characterization and analysis of materials, food processing and industrial drying, medical diagnosis and treatment, and more importantly in chemical synthesis. Microwaves have transformed the paradigm of modern chemistry with their specific and speedy effects that promote quicker, easier, and energy-efficient reaction schemes. Ever since the discovery of microwave ability to enhance the chemical reactions, they have been used in many inorganic, organic and other process-related chemistries. Difficult reactions that otherwise cannot proceed in conventional heating method have been promoted in a microwave mediated environment proving their special effects. Thus microwaves are often referred to as the new "Bunsen Burner" of the 21st century. A brief history of microwave technology development from the chemistry point of view is presented below.

Brief history of microwave technology development (Kuhnert 2002, Zhang and Hayward 2006)

1831	Faraday discovers electromagnetic induction
1864	J. Maxwell theorized that, if combined, electrical and magnetic energy would be able to travel through space in a wave
1877	Lord Rayleigh presented the theory of sound propagation
1888	H. Hertz first succeeded in showing experimental evidence of radio waves
1940	First generator of microwave power for radar, called magnetron at the University of Birmingham
1946	Percy Spencer engineer at Raytheon Corp., US observed melting of a chocolate bar in his pocket while working with a new MW vacuum tube, magnetron
1947	First commercial MW oven "Radarange", 5.5 feet tall, 750 pounds, over $5,000 was introduced Myths and fear of possible risk of radiation poisoning, going blind and sterile
1954	First commercial household microwave oven was introduced to the market
1975	Widespread domestic use due to technological advances and cheaper prices
1978	First microwave laboratory instrument was developed by CEM Corporation to analyze moisture in solids

1980–82	Microwave radiation was developed to dry organic materials
1983–85	Microwave radiation was used for chemical analysis processes such as ashing, digestion and extraction
1986	First report of MW reaction in organic chemistry, "The use of microwave ovens for rapid organic synthesis"—Hydrolysis of benzamide under acidic condition, Gedye et al. 1986
1990	Microwave chemistry emerged and developed as a field of study for its applications in chemical reactions
1990	Milestone s.r.l. generated the first high pressure vessel (HPV 80) for performing complete digestion of difficult to digest materials like oxides, oils and pharmaceutical compounds
1992–1996	CEM developed a batch system (MDS 200) reactor, and a single mode cavity system (Star 2) that were used for performing chemical synthesis
1997	Milestone s.r.l. and Prof. H.M. (Skip) Kingston of Duquesne University culminated a reference book titled "Microwave-Enhanced Chemistry—Fundamentals, Sample Preparation, and Applications", and edited by H.M. Kingston and S.J. Haswell
2000	First commercial microwave synthesizer was introduced to conduct chemical synthesis
2000–2015	Microwaves have been applied in various chemical reaction schemes involving analytical chemistry, biofuel synthesis, catalysis, drug discovery, environmental remediation, energy production, food processing systems, fine chemicals, high-value chemical products, industrial applications, materials, nanoparticle synthesis, organic metal frameworks, polymers, water and wastewater treatment, waste management, and waste reduction

Microwaves in Organic Synthesis

Microwave mediated organic synthesis is the most applied research area both in academia and industry. The earliest organic reactions were conducted by Richard Gedye and his colleagues, in the hydrolysis of benzamide to benzoic acid under acidic conditions (Gedye et al. 1988). Reaction rate enhancements between 5 and 1000 times in comparison to conventional heating methods were reported. Since then, chemists have successfully conducted a large range of organic reactions. Some of the common microwave mediated organic chemical reactions/synthesis include the following:

- The Diels-Alder reaction
- Rearrangements
- Hydrogenation reactions
- Cycloaddition reactions
- Protection and de-protection reactions
- Olefination reactions
- Oxidation and reduction reactions
- Radical reactions
- Condensation reactions
- Coupling reactions
- Organocatalysis
- Cyclization

Earlier studies concerning microwave mediated organic chemical reactions were conducted on a small scale and under sealed conditions (or in capped Teflon vessels) to create high pressures and temperatures under microwave heating (Bose et al. 1991). Later, explosion proof system incorporating vermiculite (hydrated silicate) were developed to promote high temperature reactions. Design of solid-phase reactions with substrates absorbed on clay, silica, or alumina helped avoid the need for high pressure conditions. Further, domestic microwave ovens were demonstrated as safe and convenient laboratory devices for microwaves mediated organic chemistry.

Rate enhancements

Early demonstrations of rate enhancements were reported by Gedye and co-workers. They reported that organic compounds can be synthesized up to 1240 times faster in sealed Teflon vessels in a microwave oven than by conventional (reflux) techniques (Gedye et al. 1988). Their study also showed that all polar molecules absorb microwave energy rapidly and that the rate of energy absorption varies with the dielectric constant. The rates of reaction of polar molecules in nonpolar solvents are not increased appreciably by the microwave method. The rate enhancement for comparable microwave and conventionally heated reactions, using identical concentrations of reactants, were calculated in the following manner (Gedye et al. 1991):

$$\text{Rate enhancement} = \frac{\text{conventional reaction time}}{\text{microwave reaction time}} \quad [1]$$

where the conventional reaction time and the microwave reaction time are for reactions taken to the same extent of completion (yield or product conversion).

Several different types of organic reactions were carried out both in sealed Teflon vessels in a microwave oven and under traditional reflux conditions. Some representative reactions are given in Table 1. Carrying out reactions in the microwave oven was also found to be very useful for the preparation of derivatives for the identification of organic compounds. This approach was particularly useful for preparing derivatives that require relatively long reflux periods. For example, derivatives that normally required up to 2.5 h to prepare can be synthesized effectively in 2–2.5 min in a microwave oven. This reduces the time required for the preparation of the derivatives by between 24 and 60 times.

The first experiment involved heating 50 mL of several organic compounds in open vessels in a microwave oven for 1 min at 560 W to determine the types of molecules that absorb microwave energy. It was found that all polar compounds absorbed significant amounts of microwave energy. For example, all the polar compounds with boiling points of less than 100°C reached their boiling point within 1 min. While polar molecules such as acids, esters, alkyl halides, primary alcohols and amines, and carbonyl compounds absorb microwave energy readily, the less polar ether, tertiary amine, and higher molecular weight alcohols were poorer absorbers. As expected, little or no microwave energy was absorbed by the non-polar carbon tetrachloride and hydrocarbon solvents.

These studies concluded that homogeneity of the reaction does not affect the rate enhancement. The rate enhancement arises predominantly because the oven superheats the solvent rapidly. Finally, pressure (temperature) measurements have shown that the maximum rate enhancement is achieved when the proper power level and volume of solvent are used. It appears that rate enhancements of approximately 200 are possible for many reactions if the reaction conditions are optimized.

Table 1. Comparison of rate enhancement in conventional and microwave heating.

Compound synthesized	Procedure followed	Reaction time	Average yield (%)	Rate enhancement $\left[\dfrac{k_{MW}}{k_{Class}}\right]$
Hydrolysis of benzamide to benzoic acid in water				
C_6H_5COOH	Classical	1 h	90	6
C_6H_5COOH	Microwave	10 min	99	
Oxidation of toluene to benzoic acid in water				
C_6H_5COOH	Classical	25 min	40	5
C_6H_5COOH	Microwave	5 min	40	
Esterification of benzoic acid with methanol				
$C_6H_5COOCH_3$	Classical	8 h	74	96
$C_6H_5COOCH_3$	Microwave	5 min	76	

Microwaves in inorganic synthesis

Microwave field was applied in inorganic chemical reactions since 1970s (Lidström et al. 2001). High temperature regimes created under microwave heating conditions have favored various inorganic chemical reactions. For example, a fast preparation of zeolite compounds was reported in earlier studies. The reaction times were as short as 10 min in microwave heating conditions which would otherwise be between 10 and 50 h under conventional heating (Arafat et al. 1993). Because of the temperature regime in the microwave oven some zeolites show faster crystalline growth than under conventional conditions, due to the rapid dissolution of the gel. This effect was observed for hydroxy sodalite prepared with sodium or tetramethylammonium salts. In addition, sintering of oxide ceramics (Al_2O_3-MgO and Al_2O_3:ZrO_2-SiO_2 mixtures) in a single mode cylindrical cavity applicator resulted in composites with fine grain size and a high density.

The applicability of dielectric heating for the synthesis of chalcopyrite from the pure elements has also been studied. Calcination of metal oxide (precursors) on activated carbon catalysts, which usually is carried out in a muffled furnace could be carried out by heating the reaction mixture for 10 min in a microwave oven at 2.45 GHz. Specifically for Co_3O_4 on activated carbon the effect of dielectric heating was found to be advantageous. Microwave processing of materials mostly was confined to ceramics, semimetals, inorganic and polymeric materials. It has now been shown that all metallic materials in powder form also absorb microwaves at room temperature and can be very efficiently and effectively sintered providing better quality product (Galema 1997). Table 2 shows the applications and special effects of microwave irradiation with possible explanations for various inorganic synthesis reactions.

Microwaves have been investigated as a possible treatment for many mixed wastes including process sludges, incinerator ash, and miscellaneous wastes. Due to the restrictions on landfilling, it is preferable to reduce the volume of waste being produced. Volumetric reduction is usually achieved through pyrolysis as opposed to incineration to prevent the formation of dioxins, furans and NO_x. Off-gases from the process are sent for further treatment, whilst the residue is usually an inert ash. However, if this remaining material is not inert then further immobilization techniques may be employed. The contaminants may be vitrified within the remaining waste, with the use of glass (network) formers (Jones et al. 2002).

Other microwave applications include the decomposition of H_2S which is acid pollutant, reduction of sulfur dioxide (cause for acid rain) with methane, reforming of methane with carbon dioxide,

Table 2. Microwave mediated gas-solid catalytic reactions: observed effects and possible explanation.

Application	Observed effect	Explanation of observed effects	References
HCN synthesis	Higher selectivity	Selective catalyst heating	Koch et al. 1997
H_2S decomposition	Higher conversion	Hot-spot formation	Zhang et al. 1999
NO_x, SO_2 reduction	Higher conversion	No explanation	Tse et al. 1990, Cha 1994, Cha and Kong 1995, Kong and Cha 1995, Wang et al. 2000, Wei et al. 2011
SO_2 reduction	Enhanced reaction rate Higher conversion Phase transition of catalyst	Hot-spot formation	Zhang et al. 2001
CO oxidation	Higher rate	Imprecise temperature measurement	Perry et al. 1997
Propane oxidation	No effects Negligible effect Higher conversion, reduction in reaction temperature	N/A N/A Hot-spot formation	Silverwood et al. 2006, 2007 Will et al. 2003, 2004, Krech et al. 2008 Beckers et al. 2006
Ethane/propane/n-butane dehydrogenation	No comparison with conventional process	N/A	Cooney and Xi 1996
Ethane dehydrogenation	Higher conversion	Phase transition of catalyst	Sinev et al. 2009

hydrodesulfurization (HDS) of thiophene (an exothermic reaction), and oxidative coupling of methane to higher hydrocarbons. In these applications, the microwave assisted heating resulted in higher product yields (higher than the equilibrium thermodynamic data), higher catalyst performance, and apparent shifts in the equilibrium constants of the reactions (Zhang and Hayward 2006). Microwaves are also utilized in various water and wastewater sample analyses to recover the heavy metals such as mercury, selenium, and arsenic and in some applications to release the nutrients in soluble form and to reduce the solids in the wastewater sludge (Lamble and Hill 1998, Yin et al. 2008, Saidi and Azari 2005).

Explanation of microwave effects in chemical synthesis

Accelerated chemical reactions produced by microwaves were explained by thermal and non-thermal effects using the Arrhenius equation. Table 3 outlines thermal and non-thermal effects by microwaves. Details of these effects and microwave interactions with materials will be discussed in Chapter 2.

Thermal effects are caused by the mechanisms called dipolar polarization and ionic conduction and a combination of the two mechanisms. Thermal effect increases with the polarity of the materials under interaction as described by Gedye et al. (1988). These two mechanisms essentially result in accelerated gains in reaction temperature while non-thermal effects are categorized as effects related to molecular collisions increased by microwave interactions. Thermal effects are reflected through the reaction kinetics. Uniform heat distribution can be obtained with proper mixing. Non-thermal effects are related to the pre-exponential factor of the Arrhenius equation which decreases the activation energy required to promote the chemical reaction. Depending on the material properties and their interaction with the microwaves, hot spots and highly localized temperatures are observed in microwave mediated reactions. The Arrhenius pre-exponential factor depends on the frequency of vibration of the atoms at the reaction interface and it has therefore been proposed that this factor can be affected by a microwave field (de la Hoz et al. 2005). For example, Binner et al. (1995) have reported that synthesis of titanium carbide was accelerated by the microwave field which was purely due to the increase in the pre-exponential factor (by a factor of 3.3) and not by the increase in activation energy. The pre-exponential factor depends on the frequency of vibration of the atoms at the reaction interface and it has therefore been proposed that this factor can be affected by a microwave field. This factor includes the factors such as geometric factor (γ) which includes the number of nearest-neighbor jump sites, the distance (λ) between different adjacent lattice planes (jump distance) and jump frequency (Γ).

Table 3. Possible rationale for special heating effects by microwaves.

Thermal effects (kinetic)	Non-thermal effects (thermodynamic)
Dipolar polarization, ionic conduction	$k = A \exp^{(\Delta E/RT)}$ $A = \gamma \lambda^2 \Gamma$
Uniform heat distribution	Enhancement in collision probabilities Decrease in activation energy
Specificity for polar molecules, effect increases with polarity	Hot spots, high localized microscopic temperatures

Controversies in microwave effects in chemical reactions

Some called microwave heating as the Bunsen burner of the 21st century due to its surprisingly efficient heating which has been witnessed by numerous reports in the organic and inorganic chemical synthesis research. While a few careful investigations incorporating mixing effects (for uniform heat distribution) and fiber optic temperature measuring devices introduced a new revolutionary information, it was strongly argued that non-thermal effects do not exist in microwave mediated reactions. In some studies microwaves proved to be of providing no evidence to non-thermal effects and all the advantages of microwave heating were attributed to thermal effects caused by ionic conduction and dielectric heating. Regardless the debate, microwaves can be considered as mediators for green chemistry as discussed in later sections.

Biofuel research

Research on biofuels has seen an unprecedented growth in recent years due to increasing demands for conventional fossil fuels, their accelerated consumption and associated environmental pollution at global levels. The interest in biofuel research stems from the fact that they help address some of the energy, economic and environmental concerns faced by many countries locally (Satyanarayana et al. 2011). The current rate of petroleum resource consumption will afford the use of global population for another 50 years (Demirbas 2009). It is because the natural replenishment rate for fossil fuels is about 10^5 slower than their consumption rate. In addition, the demand for fossil fuels is expected to rise by 60% or so in the next 25 years (Rittmann 2008). To reduce dependency on the fossil fuel sources and imports from oil-rich countries and maintain environmental sustainability, many countries have committed to renewable energy and fuel production increases and/or greenhouse gas emission reductions at national and international levels (Fabbri et al. 2007). Policy amendments and changes in energy management strategies have also been considered.

Among many renewable energy sources, solar thermal and photovoltaic collectors are still not energy-efficient or economical. For instance, energy conversion efficiency of the photovoltaic modules available in the market is at the maximum 15%. Photovoltaic cells are also referred to as solar energy harvesting factories with an input to output ratios of 1:7. The return energy production rate from the photovoltaic modules is slow over 20–25 years (Pimentel and Patzek 2008). Wind and geothermal sources have limitations such as location, availability, and intensity. Since most of the transportation and industrial sectors need liquid fuels to drive the machinery and engines, more emphasis is needed on alternative fuel sources. Biofuels such as biodiesel, bioethanol, biogas, bio-oils and syngas can be derived from various biomass feedstock which are abundantly available (EIA 2007).

Biofuels offer many benefits:

a) serve as an alternative to petroleum-derived fuel, which implies a lower dependence on crude oil foreign imports;
b) provide favorable energy return on energy invested;
c) reduce greenhouse emissions in line with the Kyoto Protocol agreement;
d) lower harmful gaseous emissions which lead to acid rain and other environmental impacts;
e) are biodegradable and nontoxic to marine life, and other environmentally sensitive areas (Georgogianni et al. 2008, Hoekman 2009, Yang et al. 2011).

Biofuel production can help address challenges related to energy independence, economic prosperity, and environmental sustainability of a nation. Towards this effect, the United States (US) and Europe have encouraged large scale industrial biofuel production. For example, biodiesel production in the US has increased from 75 million gallons in 2005 to 250 million gallons in 2006 and 450 million gallons in 2007, with an expected total capacity of well over 1 billion gallons in the next few years (Kargbo 2010, Haas et al. 2006). Also, the federal government has passed the energy independence and security act (EISA) in 2007 which requires a gradual increase in the production of renewable fuels to reach 36 billion gallons per year by 2022. Furthermore, 28 states have passed their own mandatory renewable energy legislation. For example, Arizona and California will replace 15% and 20% of their electricity sales with renewable energy by 2020, respectively. Texas has a mandate for 5880 MW of renewable electricity capacity by 2015. Other states have mandates to reduce greenhouse gas (GHG) emissions. For instance, Minnesota's strategic goal is to reduce GHG emissions by 80% between 2005 and 2050 (Yang et al. 2011, Hoekman 2009).

Current biofuel production is encircled by two major barriers. They are (i) cost of the feedstock, and (ii) energy-efficiency of conversion processes. Moreover, for biofuels to be viable substitutes for fossil fuels, they should offer superior benefits such as higher energy recovery ratio, lower overall life cycle environmental impacts, and attractive economics (Hill et al. 2006). Biofuels can be produced from various feedstock classified as carbohydrate-rich (sugars and starch), lipid-rich (oils and triglycerides) and lignocellulosic (cellulose and hemicellulose) feedstock. Figure 2 shows various biomass feedstock and possible routes for production of biofuels and bio-products (Huber et al. 2006).

8 Microwave-Mediated Biofuel Production

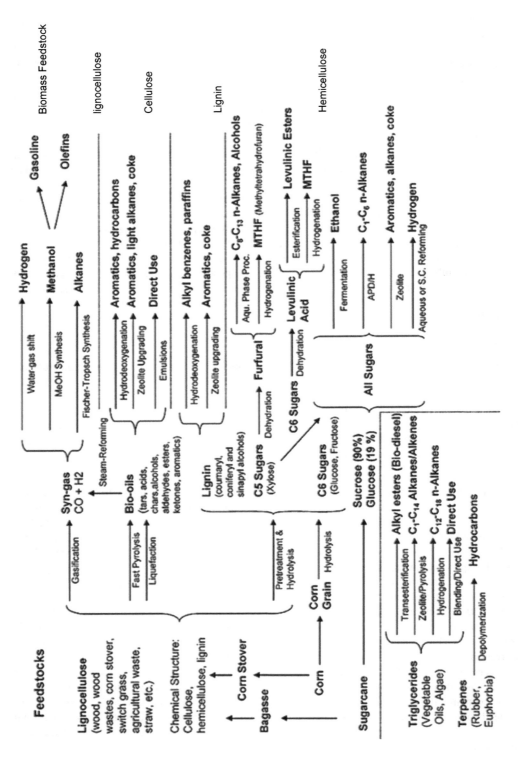

Figure 2. Known routes for production of liquid fuels from biomass.

Starch containing feedstocks are those comprised of glucose polysaccharides joined by aglycosidic linkages, such as amylase and amylopectin, which are easily hydrolyzed into the constituent sugar monomers, making them easy to process such as in first generation bioethanol facilities (Alonso et al. 2010). Triglyceride feedstocks are those comprised of fatty acids and glycerol derived from both plant and animal sources. Sources of triglycerides for the production of biodiesel include various vegetable oils, waste oil products (e.g., yellow grease, trap grease), and algal sources. Lignocellulosic biomass is the most abundant class of biomass. While starch and triglycerides are only present in some crops, lignocellulose contributes structural integrity to plants and is thus always present. In general, most energy crops and waste biomass considered for energy production are lignocellulosic feedstocks, with examples including switchgrass, miscanthus, agricultural residues, municipal wastes, and waste from wood processing. Lignocellulose is comprised of three different fractions: lignin, hemicellulose, and cellulose. As shown in Figure 2, several processes are available for conversion of carbohydrate rich and lipid-rich feedstock. Hydrolysis followed by fermentation can be performed for bioethanol production while liquefaction, pyrolysis and gasification techniques can be used to produce bio-oils and syngas. Utilizing low cost feedstock and/or recycling waste biomass and oils and animal fats can be an alternative to reduce the feedstock costs; process improvements and optimization help reduce the biofuel conversion process costs.

Microwaves in biofuel production

Microwaves are receiving increasing attention due to their ability to complete chemical reactions in very short times. Microwaves have revolutionized the way chemical reactions can be performed with unexplainable results. This amazed the entire scientific and industrial community and resulted in many curious scientists who applied microwaves in different areas of chemistry to benefit from these results. Few advantages with microwave processing can be listed as: rapid heating and cooling; cost savings due to energy, time and work space savings; precise and controlled processing; selective heating; volumetric and uniform heating; reduced processing time; improved quality ("reportedly") and properties; and other effects not achievable by conventional means of heating (Clark and Sutton 1996, Caddick and Fitzmaurice 2009, Ku et al. 2002, de la Hoz et al. 2005, Roberts and Strauss 2005, Varma 1999, Giguere et al. 1986). Microwaves have been used by many researchers around the world in many organic and inorganic syntheses at exploratory levels (Clark and Sutton 1996, Caddick and Fitzmaurice 2009, Ku et al. 2002, de la Hoz et al. 2005, Roberts and Strauss 2005, Varma 1999, Giguere et al. 1986). Recently, many industries have successfully implemented microwave based processes, examples include: ceramic/ceramic matrix composite sintering and powder processing, polymers and polymer-matrix composites processing, microwave plasma processing of materials, and minerals processing (Clark and Sutton 1996). Microwaves have the ability to induce reactions even in solvent-free conditions offering "Green Chemistry" solutions to many environmental problems related to hazardous and toxic contaminants (Varma 1999). Due to these advantages, microwaves provide tremendous opportunities to improve biofuel conversion processes from various feedstock rich in sugars or oils.

Microwaves have been used in production of various biofuels which include biodiesel, bioethanol, and biogas, bio-oils, and syngas and hydrogen. Microwave irradiation is proven to be beneficial in many process stages and from many process related aspects. In biodiesel production, microwaves can be used in feedstock preparation, extraction of oils, and chemical conversion (esterification/transesterification) of oils into biodiesel. Similarly, microwaves have been found to enhance the process yields in bioethanol and biogas production from various sources of biomass especially through a pretreatment step. This chapter will discuss the important steps involved in biofuel production and the role of microwaves in these chemical reactions.

Microwave heating was applied to all processes involving biomass conversion with varying success. The aim of nearly all the studies has been to improve product quality and quantity, reduce processing time, and ultimately increase the overall yields and resource efficiency of the process. Driven by the established benefits of the application of microwaves to organic synthesis, studies on the use of microwave heating in bioethanol, biodiesel and biogas production have been carried out. In the case of bioethanol production, research has been focused on microwave pretreatment. Microwave-assisted hydrolysis of different lignocellulosic biomass have been proven to increase bioethanol yields due to enhanced degradation of lignin, cellulose and hemicellulose, and increased enzymatic susceptibility of the materials. Microwaves have also been studied as pretreatment method of various organic wastes and sludge to be processed by anaerobic digestion, enhancing biomethane production potentials. Moreover, microwave pretreatment favors dewatering, reduces foaming and diminishes the amount of pathogens in the sludge. Microwave-assisted biodiesel production has been widely studied. Microwaves have been successfully applied to biomass pretreatment, lipid extraction and transesterification reactions both in batch and continuous set-ups. Shorter processing times, higher yields and higher product purities have been obtained compared to those achieved under conventional heating. And the use of microwave radiation prior to lipid extraction has been found to enhance the efficiency of the extraction processes. Despite these advantages, some reports have questioned the actual energy savings when using microwave heating in all these processes, and have highlighted the need to optimize each individual process and to carry out a thorough assessment of energy usage.

Esterification and transesterification reactions

Organic reactions such as esterification, etherification, hydrolysis, substitution reactions and Diels-Alder reactions have been studied comprehensively under microwave irradiation. The most common and simple reactions leading to biodiesel production are esterification and transesterification reactions. It is important to note that esterification and transesterification reactions of long chain organic acids mediated by microwaves have received attention in early development of microwave mediated organic synthesis. A few reactions are shown in Table 4 and Table 5a and 5b. Esters can be prepared by following one of the three methods, (i) esterification of carboxylic acids, (ii) transesterification of methyl or ethyl esters, and (iii) alkylation of carboxylate anions. These reactions have been studied widely, but a great need still exists for versatile and simple processes whereby esters may be formed under very mild conditions (3). This is especially so for the two first methods where equilibria are involved (Table 1).

Gedye et al. (1991) evaluated the behavior of a variety of acids and alcohols, including hindered and long-chain acids with heavy alcohols, which are generally difficult cases. Different acids, including KSF montmorillonite, 13X zeolite, and neat p-toluenesulfonic acid (PTSA), were studied. In the first case, the effect of isopropenyl acetate as a water scavenger was studied. The results are given in Table 1. The most favorable reaction conditions involved the use of neat *PTSA,* which led to good yields under exceptionally mild conditions (3–10 min), in a non-modified domestic microwave oven. KSF montmorillonite and 13X zeolite were found to be less efficient.

The benefits of microwaves are mostly manifested by quick and effective synthesis of those special chemicals which otherwise would take significant reaction time under conventional or traditional heating conditions. These reactions represent a large variety, ranging from hydrolysis of nitriles, amides and esters, to the formation of esters and ethers-oxidation and hydrogenations; rearrangements and polymerizations, etc. Esters, amides and nitriles are hydrolyzed very slowly either in basic or acidic mediums under traditional conditions. Hydrolysis of these compounds offers a typical example of application of microwaves. A positive role is played in this case by the presence of strong acids or bases that increase the heating rate under microwave irradiation (Giguere et al. 1986, Gedye et al. 1986, Bose et al. 1991, Mingos 1994).

The Diels–Alder reaction represents a good model to study the effect of microwaves (Galema 1997): the carbonylgroup, besides being important in driving the reaction, acts as a sort of antenna towards the radiation. The reaction is rapid when the starting diene is electron-rich and the dienophile is electron-poor; when dienophiles lack of activating groups the reactions require a high temperature (> 300°C). The reaction between anthracene and maleic anhydride is a classic example of this. This reaction has recently been proposed for undergraduate microwave experiments, because it is simple to carry out and its rate is strongly accelerated in the presence of microwaves.

Esterification is a reaction largely studied under microwaves. The rate of esterification of benzoic acid was found to be increased under microwaves and the increase is a function of the length the hydrocarbon chain of the alcohol. It must be pointed out that the boiling point of the alcohol also increases: higher temperatures of the reaction can therefore also be used in the traditional method: as a consequence, in this case the comparison between the two techniques is no longer homogeneous (Fini and Breccia 1999).

Table 4. Esterification and Transesterification reactions under microwaves reported by Gedye et al. 1991.

Esterification under microwaves (350 W)					
\multicolumn{6}{l}{$RCOOH + R'OH \underset{}{\overset{Acid}{\rightleftharpoons}} RCOOR' + H_2O$ (4 mM) (4 mM)}					
		\multicolumn{4}{c}{NMR (nuclear magnetic resonance) Yield (time, min)}				
R	R'	KSF	KSF + IA	PTSA	13 X
Phenol	nC_8H_{17}	85% (10)	88% (10)	97% (3)	54% (10)
$nC_{17}H_{35}$	nC_8H_{17}	65% (5)	67% (5)	90% (5)	38 (5)
$nC_{17}H_{35}$	$nC_{18}H_{37}$	71% (10)	74% (10)	94% (10)	43% (10)
phenyl	nC_8H_{17}	62% (5)	64% (5)	82% (5)	37% (5)

Table 5a. Acid-catalyzed transesterification under microwaves (600 W).

Acid-catalyzed transesterification under microwaves (600 W)			
\multicolumn{4}{l}{$PhCOOCH_3 + n\ OctOH \underset{}{\overset{Acid}{\rightleftharpoons}} PhCOO_nOct + CH_3OH$ (10 mM) (20 mM)}			
Acid	Amount	GC Yield (time, min)	Final temperature (°C)
Amberlyst 15	0.5 g	18% (4)	170
Amberlyst 15	1.5 g	61% (1)	170
Amberlyst 15	3 g	13% (3)	194
PTSA neat	0.5 mM	5% (2)	146
PTSA neat	2 mM	39% (2)	173
PTSA neat	5 mM	97% (2)	193

Table 5b. Base-catalyzed transesterification under microwaves (600 W).

Base-catalyzed transesterification under microwaves (600 W)			
\multicolumn{4}{l}{$PhCOOCH_3 + n\ OctOH \underset{}{\overset{Base\ (5\ mM)}{\rightleftharpoons}} PhCOO_nOct + CH_3OH$ (10 mM) (20 mM)}			
Base	Relative amount Ester:alcohol	GC yield (time, min)	Final temperature (°C)
KOH/alumina	1:1	60% (3)	140
KOH/alumina	1:2	72% (3)	160
KF/alumina	1:1	55% (3)	135
KF/alumina	1:2	71% (3)	150
K_2CO_3 + aliquat 1.5 mM	1:1	70% (2.5)	184
K_2CO_3 + aliquat 1.5 mM	1:2	90% (2.5)	186

Microwaves in extraction of essential chemicals

Microwaves have been widely used in extraction of various essential chemicals and oils for a variety of uses. As shown in Table 6, some examples include extraction of polyphenols, pigments, soluble proteins, vegetable oils, flavonol glycosides. Oil extraction is an essential step in biodiesel production whereas the protein recovery is a major process step in protein production. The reaction conditions and solvents used in these processes are also shown in Table 6.

Table 6. Microwave extraction with solvents as carrying medium (Stefanidis et al. 2014).

Product	Plant materials	Solvent	Conditions	Microwave power (W)	Time (min)	References
Polyphenols	Green tea	Ethanol/water	1 bar, 85–90°C	700	5	Pan et al. 2003
Pigments	Microalgae	Acetone	Vacuum and atmospheric pressure	25–100	3–13	Pasquet et al. 2011
Soluble proteins	Soybean	Water	60°C, 1 bar	–	10–30	Choi et al. 2006
Vegetable oils	Soybean and rice bran	Ethanol	Continuous system at 73°C, 0.6 and 1.0 l/min	Up to 5 kW	21	Terigar et al. 2011
Flavonol glycosides	*Ginkgo biloba*	Ionic liquid	1 bar	120 W	15	Yao et al. 2012

Green chemistry

A chemistry that is founded by the principles of sustainability can be called green chemistry. Environmental and socio-economic factors need consideration while considering the greenness (environmental friendliness) of a chemical reaction. Thus, green chemistry is not necessarily a new chemistry but it is a chemistry approach based on environmental consciousness. Green chemistry is simply a new environmental priority when accomplishing the science already being performed regardless of the scientific discipline or the techniques applied (Tucker 2006). It is a concept driven by efficiency coupled with environmental responsibility. Green chemistry approach helps develop protocols when developing chemical processes and may play a major role in the production of most essential chemicals, in minimizing the energy demands, creating safer processes, and avoiding hazardous chemical use and production (Cherubini 2010).

Anastas and Warner introduced twelve green chemistry principles in 1998 for safe and environmentally responsible chemistry (Anastas and Warner 1998). The principles are listed as follows: (1) prevention; (2) atom economy; (3) less hazardous chemical synthesis; (4) designing safer chemicals; (5) safer solvents and auxiliaries; (6) design for efficiency; (7) use of renewable feedstock; (8) reductive derivatives; (9) catalysis; (10) design for degradation; (11) real-time analysis for pollution prevention; and (12) inherently safer chemistry for accident prevention. Detailed descriptions of these principles are given in Table 7.

These principles can be implemented in any process that involves a chemical transformation of one compound to another compound. Green chemistry is highly pronounced in industries such as pharmaceuticals, petrochemicals, food, and other chemistry based industrial processes (Beach et al. 2009). However, considering the present energy and environmental crises worldwide, green synthesis of renewable fuels such as biodiesel and bioethanol has become a major and important topic of interest. Efficient synthesis of renewable fuels remains a challenging and important area of research. Embracing the principles of green chemistry might result in a sustainable route for renewable fuel production. Green chemistry provides unique opportunities for innovation via product substitution, new feedstock generation, catalysis in aqueous media, utilization of microwaves and ultrasound, energy and resource efficiency, waste minimization, and scope for alternative or natural solvents (Gude and Martinez-Guerra 2015). The potential of utilizing waste as a new resource, and the development of integrated processes producing multiple products from biomass is highly desirable to improve the economics of the renewable fuels.

Why microwave mediated chemistry is a green chemistry?

Microwave based chemistry is considered green chemistry from several perspectives. This can be discussed in view of the popular green chemistry principles introduced by Paul Anastas and Warner (1998).

1. Prevention

Microwave enhanced reactions promote higher product yields which in turn eliminate the wastes and the need for their cleanup and management.

Table 7. The twelve principles of green chemistry.

No.	Principle	Description
1	*Prevention*	It is better to prevent waste than to treat or clean up waste after it has been created.
2	*Atom economy*	Synthetic methods should be designed to maximize the incorporation of all materials used in the process into the final product.
3	*Less hazardous chemical syntheses*	Wherever practicable, synthetic methods should be designed to use and generate substances that possess little or no toxicity to human health and the environment.
4	*Designing safer chemicals*	Chemical products should be designed to affect their desired function while minimizing their toxicity.
5	*Safer solvents and auxiliaries*	The use of auxiliary substances (e.g., solvents, separation agents, etc.) should be made unnecessary wherever possible and innocuous when used.
6	*Design for energy efficiency*	Energy requirements of chemical processes should be recognized for their environmental and economic impacts and should be minimized. If possible, synthetic methods should be conducted at ambient temperature and pressure.
7	*Use of renewable feedstock*	A raw material or feedstock should be renewable rather than depleting whenever technically and economically practicable.
8	*Reduce derivatives*	Unnecessary derivatization (use of blocking groups, protection/deprotection, and temporary modification of physical/chemical processes) should be minimized or avoided if possible, because such steps require additional reagents and can generate waste.
9	*Catalysis*	Catalytic reagents (as selective as possible) are superior to stoichiometric reagents.
10	*Design for degradation*	Chemical products should be designed so that at the end of their function they break down into innocuous degradation products and do not persist in the environment.
11	*Real-time analysis for pollution prevention*	Analytical methodologies need to be further developed to allow for real-time, in-process monitoring and control prior to the formation of hazardous substances.
12	*Inherently safer chemistry for accident prevention*	Substances and the form of a substance used in a chemical process should be chosen to minimize the potential for chemical accidents, including releases, explosions, and fires.

2. Atom economy

In connection with the above principle, microwaves also enhance the incorporation of all materials into the final product. Hence, utilization of reaction molecules into desired products is increased which in turn increases the atom economy.

3. Less hazardous chemical synthesis

Microwave chemistry is safe and reaction heating rates can be controlled well. Replacement of hazardous chemical reagents may be possible with mild and environmentally benign substances. It has been reported that reactions that are sluggish and require severe reaction conditions have been conducted at much milder conditions involving low temperatures and atmospheric pressures.

4. Designing safer chemicals

Microwaves allow for the production of safer and biodegradable chemicals.

5. Safer solvents and auxiliaries

The use of environmentally benign solvents (e.g., water, and ionic liquids), or of solvent-free conditions has often been applied in combination with microwave heating methods, and in general these conditions have been referred to as green chemistry.

No solvent is a "green solvent"

Chemical synthesis without involving any solvent is an ideal scheme. Microwaves are able to promote solvent-free reactions. Dry media reactions under microwave effect were reported in earlier studies. These studies proved the potential of microwave mediated reactions for biofuel production. Other green solvents are namely ionic liquids with aqueous systems and supercritical fluids. The issue with these solvents is that they are cost prohibitive and toxic.

Water as a green solvent

Water is considered a green solvent especially in microwave mediated chemical synthesis due to its high dielectric properties. It is considered a non-toxic and non-corrosive and abundant solvent available at low costs. Reactions involving water as a solvent have matched the rates, yields and selectivities observed for many reactions in other solvents, or in many cases, surpass those in organic solvents. In contrast to many other solvents, water not only provides a medium for solution chemistry but often participates in elementary chemical events on a molecular scale. Water also offers practical advantages over organic solvents. It is cheap, readily available, non-toxic and non-flammable. In addition to using water for chemistry at ambient pressure in open vessels, there has been a growth of interest in the use of high-temperature water, superheated water and supercritical water. High temperature water is broadly defined as liquid water above 200°C, superheated water as water between 200–300°C and supercritical water as that above 374°C and 218 atm. At these temperatures the water approaches properties more like those of polar organic solvents at room temperature, and can act as an acidic or basic catalyst. The use of water as a solvent for metal-mediated synthesis has attracted considerable research interest since it can be used to implement approaches combining the advantages of homogeneous and heterogeneous catalysis by distributing the reagents, products and catalysts between different phases, pseudo-phases or interface regions. Three distinct types of approaches can be taken based on phase transfer, phase separation or solubilization although most processes involve a combination of two or all three of these. Due to its properties, water is an excellent solvent for microwave-promoted synthesis. Although it has a dielectric loss factor which puts it into the category of only a medium absorber, even in the absence of any additives it heats up rapidly upon microwave irradiation. Using a sealed vessel it is possible to heat water to well above its boiling point. At these elevated temperatures organic substrates become increasingly soluble and, even if they are not, addition of a phase-transfer agent can be used to facilitate solvation. Of course water cannot be used as a replacement for organic solvents for every class of reaction and, at the high temperatures used in microwave reactions, competitive decomposition of starting materials or products can become an issue (Leadbeater 2005). However, the major difficulty in using water as a solvent is its insolubility in most of the organic reactants which makes reactions heterogeneous. The major difficulty with using water as a solvent is the insolubility of most of organic reactants, making reaction mixtures heterogeneous. One way to overcome this is by using phase-transfer catalysts, but their expensive nature means that the resulting method is not economical. Product isolation from aqueous reaction mixture is another critical issue. Usually, evaporation of water is an option, but this is not an energy-efficient technique. Interestingly, these challenges can be tackled successfully by using the microwave heating technique for reactions in aqueous medium.

6. Design for energy efficiency

Due to their dielectric heating technology, microwave heating is often reported as an energy efficient process for heating chemical reactions at elevated temperatures. Since microwave-assisted transformations are typically very fast compared to traditionally heated processes, they should be regarded as energy efficient (less time consumption), and therefore as being green. Although reaction speed is not specifically mentioned as one of the twelve principles of green chemistry, it can be argued that a fast reaction performed at a high temperature regime (Arrhenius equation) is likely to require less energy compared to a transformation that requires significantly longer reaction times at lower temperatures. However, this effect needs to be determined on a case-to-case basis. It should also be emphasized that the extremely short reaction times experienced in microwave chemistry are typically restricted to small scale synthesis. Scaling-up to large scale reactors, both the heating and cooling times in a batch reactor (both requiring energy) will add significantly to the overall processing times. It is therefore not uncommon for a reaction that can be performed in a few minutes in a small scale microwave instrument to require significantly longer processing times in a larger scale reactor.

7. Use of renewable feedstock

Similar to conventional heating processes, microwave mediated processes allow for use of renewable feedstock. For example, the feedstock in biofuel production come from renewable and sustainable sources. These feedstocks are biodegradable and do not contribute to environmental emissions.

8. Reduce derivatives

Higher yields obtained by microwave mediated reactions eventually result in lower derivatives and degradation products (Berlan 1995).

9. Catalysis

If one accepts the provision that "specific" and "non-thermal" microwave effects do not exist for standard synthetic organic chemical transformations then one must assume that the same enhancements (reaction rates, yields and selectivities) seen by applying microwave energy can be reproduced using conventional heating technology if a similar temperature profile can be ensured. Clearly, there may be important synergistic effects in, for example, performing a chemical transformation using microwave heating in a strongly microwave absorbing ionic liquid, or using a microwave transparent inorganic support material under solvent-free conditions. However, it is important to note that the fundamental science will not change when switching from conventional to microwave heating, although the latter method may be significantly more practical to use in many instances. In other words, while these microwave-assisted transformations may well be green, the relative greenness of these procedures is not necessarily linked to the use of microwave energy, but rather is a consequence of the environmentally benign reaction media. Therefore, in our opinion, it is the greenness of the microwave heating process itself that needs to be evaluated.

10. Design for degradation

While not directly related to microwave mediated chemistry, the reactions conducted under microwaves can be controlled to develop products that are biodegradable.

11. Real-time analysis for pollution prevention

Microwave reaction systems can be designed to provide for continuous and real-time monitoring of the process performance in terms of the process conditions such as temperature and pressure, power output and field application. These systems can be operated to facilitate sample collection and analysis for automatic process control to avoid formation of dangerous products since microwave processes can be initiated quickly and stopped faster than conventional processes.

12. Inherently safer chemistry for accident prevention

Microwave reactors are now being developed to closely monitor and selectively control the desired reactions by ensuring optimization of power and temperature distribution. Improved vessel technology and the increased safety features allow for safer chemistry and accident prevention.

Seven green metrics for sustainable chemistry

The development of sustainable chemical processes should consider minimizing negative environmental impacts of the chemical and industrial processes. The chemical processes should be optimized to produce less waste and consume less energy sources. The following metrics are useful when measuring or evaluating how "green" is a chemical process (Constable et al. 2002).

- E-Factor (the environmental factor)
- Atom Economy (utilization) or Atomic Efficiency
- Mass Intensity
- Reaction Mass Efficiency
- Mass Productivity
- Effective Mass Yield
- Carbon Efficiency

The definitions for the green metrics are given below:

1. $E\text{-}factor = \dfrac{Total\ mass\ of\ waste}{Mass\ of\ final\ product}$ [2]

2. $Atom\ Economy = \dfrac{Molecular\ weight\ of\ product \times 100}{\Sigma\ Molecular\ weight\ of\ reactants}$ [3]

3. $Mass\ Intensity = \dfrac{Total\ mass\ in\ process}{Mass\ of\ product}$ [4]

4. $Reaction\ Mass\ Efficiency = \dfrac{Mass\ of\ desired\ product \times 100}{\Sigma\ Mass\ of\ reactants}$ [5]

5. $Mass\ Productivity = \dfrac{Mass\ of\ all\ products}{\Sigma\ Mass\ of\ reactants}$ [6]

6. $Effective\ Mass\ Yield = \dfrac{Mass\ of\ product \times 100}{Mass\ of\ hazardous\ reagents}$ [7]

7. $Carbon\ Efficiency = \dfrac{Carbon\ in\ product \times 100}{Total\ carbon\ in\ reactants}$ [8]

Other important factors that could be considered and not addressed by these metrics are: energy concerns (process—internal and external); utilization of renewable feedstock (starting raw materials); reaction types; catalysts vs. stoichiometric reagents; safety; life cycle analysis; and the environmental quotient. Here, a few illustrations are provided to understand how some of these metrics can be applied in sustainable biodiesel production.

E-Factor

E-factor was proposed by Roger Sheldon (1997, 2010) which considers the amount of total waste generated by a process. It is expressed as:

$E\text{-}factor = \dfrac{Total\ mass\ of\ waste}{Mass\ of\ final\ product}$ [9]

The E-factor represents the actual amount of waste produced in the process, defined as everything but the desired product (Sheldon 2010). It takes the chemical yield into account and includes reagents, solvent losses, process aids, and, in principle, even fuel. Water is generally excluded from the E-factor as the inclusion of all process water could lead to exceptionally high E-factors in many cases and make meaningful comparisons of processes difficult. A higher E-factor means more waste and, consequently, a larger environmental footprint. The ideal E-factor is zero. In simple terms, it is the total mass of raw

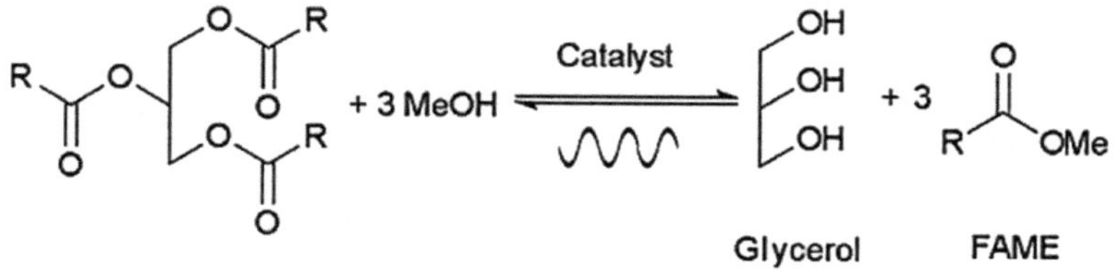

Scheme 1. Transesterification reaction from vegetable oils (triglycerides) to fatty acid methyl esters.

materials minus the total mass of product, all divided by the total mass of product. It can be easily calculated from knowledge of the number of tons of raw materials purchased and the number of tons of product sold, the calculation being for a particular product or a production site or even a whole company. For example, the oil refining process has an E-factor of less than 0.1, compared to fine chemical industry (E-factor between 5 and 50) and the pharmaceutical processes with E-factors between 25 and 100. These numbers suggest that the pharmaceutical chemical processes have the largest scope to implement green chemistry principles to reduce the waste generation.

Illustration I

For biodiesel production, the E-factor could be very small meaning that the process is environmental friendly. For a typical transesterification reaction (reaction scheme shown in Scheme 1), producing biodiesel from virgin oil, the following simple calculations can be made. Biodiesel production essentially requires feedstock such as virgin oils, an alcohol as a reactant, and a catalyst.

A conversion efficiency of higher than 90% is commonly reported (Grant and Gude 2014). The E-factor for biodiesel reaction can be written as follows:

$$\text{Biodiesel E-factor} = \frac{\text{Glycerol (kg)} + \text{Unconverted oil} + \text{Excess methanol}}{\text{kg (Biodiesel)}} \quad [10]$$

In few cases, glycerol is considered a useful product, therefore excluding glycerol, the E-factor can be expressed as:

$$\text{Biodiesel E-factor} = \frac{\text{Unconverted oil} + \text{Excess methanol}}{\text{kg (Biodiesel)}} \quad [11]$$

For example, in theory, at 100% conversion following stoichiometric ratios, for every 100 pounds of oil reacting with 10 pounds of alcohol, 100 pounds of biodiesel (fatty acid alkyl esters, FAAEs) and 10 pounds of glycerol can be produced. The E-factors following the above two equations would then be:

$$\text{Biodiesel E-factor} = \frac{10 \text{ (kg)} + 0 + 0}{100 \text{ kg (Biodiesel)}} = 0.1 \quad [12]$$

$$\text{Biodiesel E-factor} = \frac{0 + 0}{110 \text{ kg (Biodiesel)}} = 0 \quad [13]$$

Since the reactions do not proceed at chemical or thermodynamic equilibrium conditions, excess reactants are required to promote the reactions to completion. The following data are obtained from our laboratory studies, for transesterification reaction of vegetable oil at different ethanol to oil ratios under microwave irradiation (Table 8). The first three columns represent the reactants (feedstock + alcohol + catalyst) and the next three columns show the products obtained for a microwave mediated reaction for two

Table 8. Reactant and product data for E-factor calculations.

Reactants			Products		
Oil (g)	Ethanol (g)	Catalyst (g)	Biodiesel (g)	Glycerol (g)	Unreacted chemicals (g)
20	6.3	0.18	16.5	5.2	3.4
20	8.4	0.18	16.5	5.4	4.9
20	12.5	0.18	17.3	3.5	9.3
20	16.7	0.18	16.0	3.4	14.0

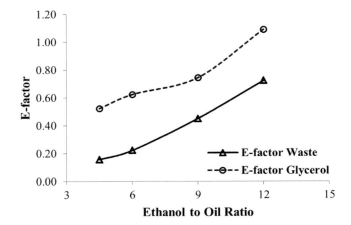

Figure 3. Effect of reactant ratio on the E-factor of the transesterification reaction of waste cooking oil (reaction conditions are summarized in Table 8).

minutes. Unreacted reactants are those that were lost during washing and drying processes. For example, the excess ethanol used in this reaction is lost through evaporation and the catalyst through water washing.

When glycerol was considered a useful product, the E-factors for the above experiments were between 0.16 and 0.72 for different alcohol ratios as shown in Figure 3. When it is considered waste, the E-factors were between 0.52 and 1.1. The optimum ratio of ethanol is very important to reduce the solvent losses and waste generated. The E-factor can vary with size of the production process, smaller systems may have higher values and vice versa. Water is required to grow the crops and clean the biodiesel produced through homogenously catalyzed reactions. If water consumption in the entire life cycle process were to be considered, the E-factor of the biodiesel production can be enormous. While, similar to all other processes, the E-factor considering the raw materials employed in the actual transesterification reaction would seem to provide a benchmark to compare different industrial processes. Biodiesel washing process requires 0.5/1 ratio of water to biodiesel to remove the unreacted methanol and catalyst. Since water is not used as a reactant in the original transesterification reaction, this can be avoided. E-factor should also consider the amount of energy input required (energy footprint) in the process. Energy production involves significant use of resources including water. This can have significant environmental impact on the overall sustainability.

E-factor for bioethanol production

Sheldon (2011) reported the E-factors of fermentation process based on the mass balances. For example, the E-factor for the bulk fermentation product, citric acid, is 1.4 which compares well with the E-factor range of < 1–5 typical of bulk petrochemicals. Roughly 75% of this waste consists of calcium sulfate. During the process, calcium hydroxide is added to control the pH, affording calcium citrate which is reacted with sulfuric acid to produce citric acid and calcium sulfate. Inclusion of water in the calculation afforded an E-factor of 17. It was recently reported that the E-factor of cellulosic ethanol is very high (~ 42). However, if water (36.8 kg/kg ethanol) and carbon dioxide (4.1 kg/kg ethanol) are excluded, the E-factor drops to a more reasonable 1.1. It was further noted that a cellulosic ethanol plant processing 10,000 tons of lignocellulose feedstock per day would produce 870 tons of ethanol and generate 32 million liters of waste water daily, i.e., enough water to supply a town of 300,000 inhabitants. Moreover, this water is contaminated with organic byproducts, thus necessitating a sophisticated industrial waste water treatment in order to decrease their concentrations to the ppm level or below and enable reuse of the water. Fermentation processes for the production of therapeutic proteins (biopharmaceuticals) can have very high E-factors, even compared with those observed for small molecule drugs. The production of recombinant human insulin, for example, involves an E-factor of ca. 6600 (Sheldon 2011). The most important contributors to the waste are urea, acetic acid, formic acid, phosphoric acid, guanidine hydrochloride, glucose, sodium chloride, sodium hydroxide and acetonitrile. If water is included, the E-factor becomes a staggering 50,000.

Atom economy

Conventionally, attaining the highest yield and product selectivity were the governing factors of chemical synthesis. However, knowledge of the stoichiometric equation allows us to predict the theoretical minimum amount of waste that can be expected. This can be done theoretically prior to experimental studies. Atom economy or utilization concept provides a quick assessment of environmental friendliness of alternative processes to produce a particular product (Trost 1991, 1995).

Atom economy is a measure of the proportion of reactant atoms which are incorporated into the desired product of a chemical reaction. Calculation of atom economy therefore also gives an indication of the proportion of reactant atoms forming waste products. Atom economy is defined as the ratio of molecular mass of the desired products divided by the molecular mass of the reactants. For example, the atom economy for the following reaction can be expressed as:

A + B → C + D

$$\text{Atom Economy} = \frac{\text{Molecular weight of (C + D)}}{\text{Molecular weight of (A + B)}} \qquad [14]$$

For transesterification reaction, it can be expressed as:

$$\text{Atom Economy} = \frac{\text{Molecular weight of Biodiesel (FAMEs)}}{\text{Molecular weight of (oil + alcohol)}} \qquad [15]$$

In a case where glycerol is considered as a useful product, the atom economy is defined as:

$$\text{Atom Economy} = \frac{\text{Molecular weight of Biodiesel and Glycerol}}{\text{Molecular weight of oil and alcohol}} \qquad [16]$$

The atom economy of a reaction depends on the reagents used and the type of chemical reaction involved. Most chemical reactions can be classified as rearrangement (migration of an alkyl group), addition, substitution (e.g., chlorination of methane), or elimination (e.g., dehydration). Rearrangement and addition reactions are atom economical by their very nature, since they simply involve reactant atoms being repositioned within the same molecule or incorporated within a second molecule. Substitution reactions, however, involve replacement of one group with another and therefore have intrinsically poor atom economy. Elimination reactions are also inherently atom uneconomical since eliminated atoms are always lost as waste. In developing an atom economical reaction pathway, therefore, the industrial chemist may well prefer rearrangement and addition reactions over less environmental friendly substitution and elimination reactions. Since atom economy provides indicative and theoretical utilization of reactants, the real atom efficiency can be calculated using the following expression:

Atom Efficiency = atom economy × yield [17]

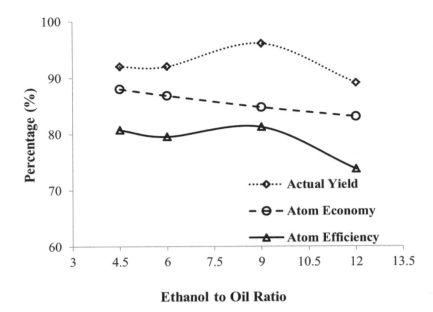

Figure 4. Effect of reactant ratio on atom economy and atom efficiency of the transesterification reaction of waste cooking oil (reaction conditions are summarized in Table 6).

The atom efficiency of the transesterification reaction was calculated using the data from Table 6. The actual yield of the reaction varied between 89% and 96% while the atom economy reduced with increasing use of reactants (from 88% to 83%), in this case alcohol (Figure 4). The atom efficiency was calculated using the actual yield and atom economy, which represents that actual efficiency of the transesterification reaction, which varied between 73.7% and 81.2%.

Mass intensity

Another important metric discussed in these reactions is mass intensity. Mass intensity is expressed as follows (Curzons et al. 2001):

$$\text{Mass intensity, } MI = \frac{\text{Total mass used in a process (kg)}}{\text{Mass of product (kg)}} \qquad [18]$$

Mass intensity (MI) takes into account the yield, stoichiometry, the solvent, and the reagent used in the reaction mixture, and expresses this on a weight/weight basis rather than a percentage. In the ideal situation, MI would approach 1. Total mass includes everything that is used in a process or process step with the exception of water; i.e., reactants, reagents, solvents, catalysts, etc. Total mass also includes all mass used in acid, base, salt and organic solvent washes, and organic solvents used for extractions. Similar to E-factor, water is excluded from mass calculations since it skews mass data in many processes.

It may also be useful to compare MI with E-Factor where:

$E\text{-}Factor = MI - 1$

By expressing mass intensity as its reciprocal and making it a percentage, it is in a form similar to effective mass yield and atom economy. This metric can be called mass productivity.

$$\text{Mass productivity} = \frac{1}{MI} \times 100 = \frac{\text{Mass of product}}{\text{Total mass in process}} \qquad [19]$$

Reaction mass efficiency

When calculating reaction mass efficiency, atom economy (AE), yield and the stoichiometry of reactants are included. RME is the percentage of the mass of the reactants that remain in the product.

For a generic reaction, A + B → C + D, RME is expressed as:

$$Reaction\ mass\ efficiency = \frac{Mass\ of\ product\ (C + D)}{mass\ in\ reactants\ (A + B)} \times 100 \qquad [20]$$

Among the four metrics discussed above, atom economy provides the actual reaction efficiency and utilization of atoms to produce useful products. E-factor allows one to quantify the waste that could be generated from an anticipated process scheme, which is another useful metric. Reaction mass efficiency combines key elements of chemistry and process and represents a simple metric which is very similar to E-factor. Mass intensity may be usefully expressed as mass productivity to understand the efficiency of the overall process schemes. These metrics assist the chemists, process and chemical engineers, and business managers to make meaningful decisions when considering new chemical reaction schemes and processes by providing simple and objective tools for analysis.

Environmental and economic considerations

While the green chemistry principles focus on preventing waste, utilizing renewable materials and reducing the potential environmental effects, these principles can be categorized into environmental and economic benefits/considerations in a broader perspective (see Table 9). All of these expressions apply to the biofuel production process and make meaningful contribution in the process development and sustainable production.

Microwave applications in biofuel production

Applications of microwaves in biofuel production have delivered the following effects as reported in many studies:

- Improved energy efficiency
- Safer solvents or solvent-free reactions
- Material reusability
- Fast catalysis
- Less number of reaction steps
- Economical
- Fast conversion, shorter reaction times, and
- Less hazardous reactions

Table 9. Environmental and economic considerations of biofuel production (after Tucker 2006).

	Environmental	Economic
Atom economy	Minimal byproduct formation	More from less. Incorporate total value of materials
Solvent reduction	Less solvent required, less solvent waste	Reduced capacity requirements, less energy required
Reagent optimization	Catalytic, low stoichiometry, recyclable	Higher efficiency, higher selectivity
Convergence	Related to improved process efficiency	Higher efficiency, fewer operations
Energy reduction	Related to power generation, transport, and use	Increased efficiency, shorter processes, milder conditions
In situ analysis	Reduced potential for exposure or release to the environment	Real time data increases throughout and efficiency, fewer reworks
Safety	Non-hazardous materials and processes reduce risk of exposure, release, explosions and fires	Worker safety and reduced downtime Reduced special control measures

As shown in Figure 5, microwaves provide several benefits promoting green chemistry, enhancing chemical reactions and operational benefits through selective and specific heating.

Examples of microwave mediated reaction benefits over conventional heating as reported in literature are shown in Table 10 (Shekara et al. 2012). Microwaves have been utilized in pretreatment (sugar release in bioethanol and biogas production), extraction (oil extraction in biodiesel production) and other process related applications (bio-oil from pyrolysis) in biofuel production.

Table 10. Microwave mediated reaction benefits over conventional heating in biofuel production as reported in literature (Shekara et al. 2012).

Objective of research	Observation	Reference
Transesterification of *Camelina Sativa* oil using metal oxide catalysts.	In MWH, the reaction rate constants are two orders of magnitude higher than those obtained with CH.	Patil et al. 2011
Investigating production of 5-hydroxy-methylfurfural and furfural from lignocellulosic biomass (Corn stalk, rice straw, and pine wood) in an ionic liquid.	MWH increases product yield and decreases reaction time.	Zhang and Zhao 2010
Preparation of high-grade bio-oils by MWP of wheat straw as compared to the other oil produced by the conventional methods.	The produced oil via MWP contains few impurities and is rich in aromatics pellet form.	Budarin et al. 2009
Crystallization of zeolite T using MWH as well as CH.	MWH has a faster reaction than that in CH.	Zhou et al. 2009
Investigation of hydrothermal carbonization of pine sawdust and cellulose.	MWH increases the carbonization yield.	Guiotoku et al. 2009
Liquefaction of wood with glycols using p-toluenesulfonic acid as a catalyst.	MWH decreases liquefaction time with minimum catalyst use.	Krzan and Zagar 2009
Dissolution of cellulose in N-methylmorpholine-N-oxide.	MWH has a shorter time compared to the traditional methods.	Dogan and Hilmioglu 2009
K-OMS catalyzed oxidation of tetralin.	MWH enhances the conversion compared to CH: (52–88%), and (42–80%), respectively.	Sithambaram et al. 2008
Studying dilution of grass and cellulose in phosphoric acid at different concentrations with water.	MWH produces high yields of glucose in a short time compared to the traditional methods. MWH has a high reaction rate at moderate temperature, which prevents the formation of hot spots.	Orozco et al. 2007
Polycyclic aromatic hydrocarbon extraction from airborne particles.	The extraction time in MWH can be completed in minutes compared to hours as in the traditional methods, with more different chemical components.	Karthikeyan et al. 2006
Pre-treatment of rice straw using microwave/alkali.	Rice straw treatment by microwave/alkali has higher cellulose, lower moisture, lignin, and hemicellulose than that produced by alkali treatment only.	Zhu et al. 2005
Solvent-free microwave extraction of oil from basil, garden mint, and thyme. The energy consumption was 0.25 kWh compared to 4.5 kWh in CH.	First oil droplet was after 5 min of MWH compared to 30 min in the case of CH.	Lucchesi et al. 2004
Pyrolysis of four wet sewage sludges from different treatment plants.	Pyrolysis process is faster than that in CP, MWH has a lower production of non-condensable gases compared to the CH.	Menéndez et al. 2004

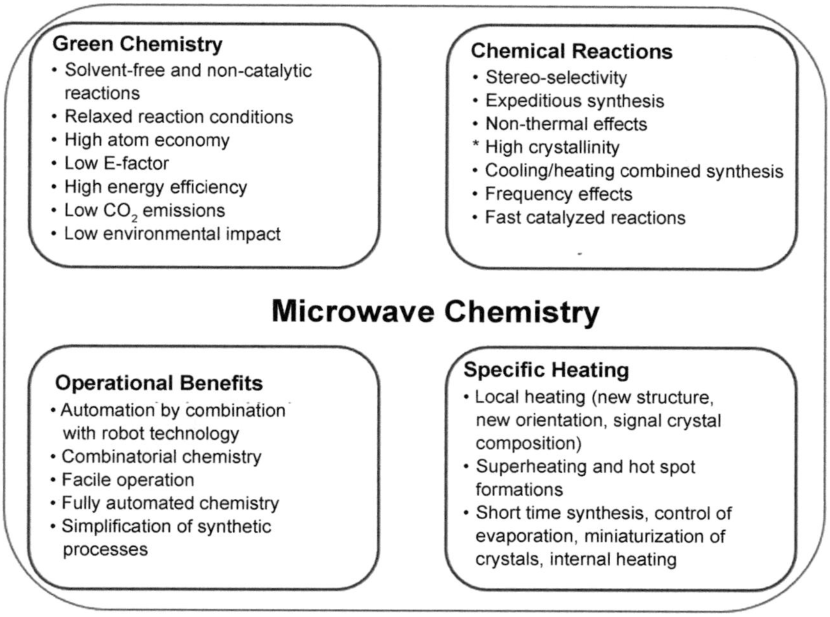

Microwave Chemistry

- **Green Chemistry**
  - Solvent-free and non-catalytic reactions
  - Relaxed reaction conditions
  - High atom economy
  - Low E-factor
  - High energy efficiency
  - Low CO_2 emissions
  - Low environmental impact

- **Chemical Reactions**
  - Stereo-selectivity
  - Expeditious synthesis
  - Non-thermal effects
  - * High crystallinity
  - Cooling/heating combined synthesis
  - Frequency effects
  - Fast catalyzed reactions

- **Operational Benefits**
  - Automation by combination with robot technology
  - Combinatorial chemistry
  - Facile operation
  - Fully automated chemistry
  - Simplification of synthetic processes

- **Specific Heating**
  - Local heating (new structure, new orientation, signal crystal composition)
  - Superheating and hot spot formations
  - Short time synthesis, control of evaporation, miniaturization of crystals, internal heating

Figure 5. Microwave based chemistry and the influencing factors.

Concluding Remarks

Microwave technology holds great potential for advancing the green chemistry and process intensification in biofuel production. The benefits of microwave heating have been confirmed to reflect in enhanced reaction kinetics, increased yields, and increased energy efficiency or energy savings. Other chemistry- and process-specific benefits are lower requirements for solvents, high potential for solvent recovery and reuse and selective heating for recovery of specific high value products. Inherent process intensification characteristics of microwaves embraced with green chemistry principles will undoubtedly offer unlimited opportunities for sustainable biofuel production.

References

Akiya, N. and Savage, P.E. 2002. Roles of water for chemical reactions in high-temperature water. *Chemical Reviews.* 102(8): pp.2725–2750.
Alonso, D.M., Bond, J.Q. and Dumesic, J.A. 2010. Catalytic conversion of biomass to biofuels. *Green Chemistry.* 12(9): pp.1493–1513.
Anastas, P.T. and Warner, J.C. 1998. *Green Chemistry: Theory and Practice.* New York: Oxford University Press.
Arafat, A., Jansen, J.C., Ebaid, A.R. and Van Bekkum, H. 1993. Microwave preparation of zeolite Y and ZSM-5. *Zeolites.* 13(3): pp.162–165.
Beach, E.S., Cui, Z. and Anastas, P.T. 2009. Green Chemistry: A design framework for sustainability. *Energy & Environmental Science.* 2(10): pp.1038–1049.
Beckers, J., van der Zande, L.M. and Rothenberg, G. 2006. Clean diesel power via microwave susceptible oxidation catalysts. *ChemPhysChem.* 7(3): pp.747–755.
Berlan, J. 1995. Microwaves in chemistry: another way of heating reaction mixtures. *Radiation Physics and Chemistry.* 45(4): pp.581–589.
Binner, J.G.P., Hassine, N.A. and Cross, T.E. 1995. The possible role of the pre-exponential factor in explaining the increased reaction rates observed during the microwave synthesis of titanium carbide. *Journal of Materials Science.* 30(21): pp.5389–5393.

Binner, J.G.P., Hassine, N.A. and Cross, T.E. 1995. The possible role of the pre-exponential factor in explaining the increased reaction rates observed during the microwave synthesis of titanium carbide. *Journal of Materials Science*. 30(21): pp.5389–5393.

Bose, A.K., Manhas, M.S., Ghosh, M., Shah, M., Raju, V.S., Bari, S.S., Newaz, S.N., Banik, B.K., Chaudhary, A.G. and Barakat, K.J. 1991. Microwave-induced organic reaction enhancement chemistry. 2. Simplified techniques. *The Journal of Organic Chemistry*. 56(25): pp.6968–6970.

Budarin, V.L., Clark, J.H., Lanigan, B.A., Shuttleworth, P., Breeden, S.W., Wilson, A.J., Macquarrie, D.J., Milkowski, K., Jones, J., Bridgeman, T. and Ross, A. 2009. The preparation of high-grade bio-oils through the controlled, low temperature microwave activation of wheat straw. *Bioresource Technology*. 100(23): pp.6064–6068.

Caddick, S. and Fitzmaurice, R. 2009. Microwave enhanced synthesis. *Tetrahedron*. 65(17): pp.3325–3355.

Cha, C.Y. and Kong, Y. 1995. Enhancement of NOx adsorption capacity and rate of char by microwaves. *Carbon*. 33(8): pp.1141–1146.

Cha, C.Y. 1994. Microwave induced reactions of SO_2 and NOx decomposition in the char-bed. *Research on Chemical Intermediates*. 20(1): pp.13–28.

Cherubini, F. 2010. The biorefinery concept: using biomass instead of oil for producing energy and chemicals. *Energy Conversion and Management*. 51(7): pp.1412–1421.

Choi, I.L., Choi, S.J., Chun, J.K. and Moon, T.W. 2006. Extraction yield of soluble protein and microstructure of soybean affected by microwave heating. *Journal of food processing and preservation*. 30(4): pp.407–419.

Clark, D.E. and Sutton, W.H. 1996. Microwave processing of materials. *Annual Review of Materials Science*. 26(1): pp.299–331.

Constable, D.J., Curzons, A.D. and Cunningham, V.L. 2002. Metrics to "Green" chemistry—which are the best? *Green Chemistry*. 4(6): pp.521–527.

Cooney, D.O. and Xi, Z. 1996. Production of ethylene and propylene from ethane, propane, and n-butane mixed with steam in a microwave-irradiated silicon-carbide loaded reactor. *Fuel Science and Technology International*. 14(10): pp.1315–1336.

Curzons, A.D., Constable, D.J., Mortimer, D.N. and Cunningham, V.L. 2001. So you think your process is green, how do you know?—Using principles of sustainability to determine what is green–a corporate perspective. *Green Chemistry*. 3(1): pp.1–6.

de la Hoz, A., Diaz-Ortiz, A. and Moreno, A. 2005. Microwaves in organic synthesis. Thermal and non-thermal microwave effects. *Chemical Society Reviews*. 34(2): pp.164–178.

Demirbas, A. 2009. Global renewable energy projections. *Energy Sources, Part B*. 4(2): pp.212–224.

Dogan, H. and Hilmioglu, N.D. 2009. Dissolution of cellulose with NMMO by microwave heating. *Carbohydrate Polymers*. 75(1): pp.90–94.

Elizabeth Grant, G. and Gnaneswar Gude, V. 2014. Kinetics of ultrasonic transesterification of waste cooking oil. *Environmental Progress and Sustainable Energy*. 33(3): pp.1051–1058.

Energy Independence and Security Act of 2007: summary of provisions. http://www.eia.gov/oiaf/aeo/otheranalysis/aeo_2008analysispapers/eisa.html.

Fabbri, D., Bevoni, V., Notari, M. and Rivetti, F. 2007. Properties of a potential biofuel obtained from soybean oil by transmethylation with dimethyl carbonate. *Fuel*. 86(5): pp.690–697.

Fini, A. and Breccia, A. 1999. Chemistry by microwaves. *Pure and Applied Chemistry*. 71(4): pp.573–579.

Galema, S.A. 1997. Microwave chemistry. *Chem. Soc. Rev.* 26(3): pp.233–238.

Gedye, R., Smith, F., Westaway, K., Ali, H., Baldisera, L., Laberge, L. and Rousell, J. 1986. The use of microwave ovens for rapid organic synthesis. *Tetrahedron Letters*. 27(3): pp.279–282.

Gedye, R.N., Rank, W. and Westaway, K.C. 1991. The rapid synthesis of organic compounds in microwave ovens. II. *Canadian Journal of Chemistry*. 69(4): pp.706–711.

Gedye, R.N., Smith, F.E. and Westaway, K.C. 1988. The rapid synthesis of organic compounds in microwave ovens. *Canadian Journal of Chemistry*. 66(1): pp.17–26.

Georgogianni, K.G., Kontominas, M.G., Pomonis, P.J., Avlonitis, D. and Gergis, V. 2008. Conventional and *in situ* transesterification of sunflower seed oil for the production of biodiesel. *Fuel Processing Technology*. 89(5): pp.503–509.

Giguere, R.J., Bray, T.L., Duncan, S.M. and Majetich, G. 1986. Application of commercial microwave ovens to organic synthesis. *Tetrahedron Letters*. 27(41): pp.4945–4948.

Giguere, R.J., Bray, T.L., Duncan, S.M. and Majetich, G. 1986. Application of commercial microwave ovens to organic synthesis. *Tetrahedron Letters*. 27(41): pp.4945–4948.

Grant, G.E. and Gude, V.G. 2014. Kinetics of ultrasonic transesterification of waste cooking oil. *Environmental Progress and Sustainable Energy*. 33(3): pp.1051–1058.

Gude, V.G. and Martinez-Guerra, E. 2015. Green chemistry of microwave-enhanced biodiesel production. In: *Production of Biofuels and Chemicals with Microwave* (pp.225–250). Springer Netherlands.

Guiotoku, M., Rambo, C.R., Hansel, F.A., Magalhães, W.L.E. and Hotza, D. 2009. Microwave-assisted hydrothermal carbonization of lignocellulosic materials. *Materials Letters*. 63(30): pp.2707–2709.

Haas, M.J., McAloon, A.J., Yee, W.C. and Foglia, T.A. 2006. A process model to estimate biodiesel production costs. *Bioresource Technology*. 97(4): pp.671–678.

Harvey, A.F. 1963. Microwave engineering. Academic Press.

Hayward, D. 1999. Apparent equilibrium shifts and hot-spot formation for catalytic reactions induced by microwave dielectric heating. *Chemical Communications*. (11): pp.975–976.

He, H., Wang, T. and Zhu, S. 2007. Continuous production of biodiesel fuel from vegetable oil using supercritical methanol process. *Fuel.* 86(3): pp.442–447.

Hill, J., Nelson, E., Tilman, D., Polasky, S. and Tiffany, D. 2006. Environmental, economic, and energetic costs and benefits of biodiesel and ethanol biofuels. *Proceedings of the National Academy of Sciences.* 103(30): pp.11206–11210.

Hoekman, S.K., 2009. Biofuels in the US–challenges and opportunities. *Renewable Energy.* 34(1): pp.14–22.

Huber, G.W., Iborra, S. and Corma, A. 2006. Synthesis of transportation fuels from biomass: chemistry, catalysts, and engineering. *Chemical Reviews.* 106(9): pp.4044–4098.

Jacob, J., Chia, L.H.L. and Boey, F.Y.C. 1995. Thermal and non-thermal interaction of microwave radiation with materials. *Journal of Materials Science.* 30(21): pp.5321–5327.

Jon, C.J.G. and Tai, H.S. 1998. Application of granulated activated carbon packed-bed reactor in microwave radiation field to treat BTX. *Chemosphere.* 37(4): pp.685–698.

Jones, D.A., Lelyveld, T.P., Mavrofidis, S.D., Kingman, S.W. and Miles, N.J. 2002. Microwave heating applications in environmental engineering—a review. *Resources, Conservation and Recycling.* 34(2): pp.75–90.

Kargbo, D.M., 2010. Biodiesel production from municipal sewage sludges. *Energy & Fuels.* 24(5): pp.2791–2794.

Karthikeyan, S., Balasubramanian, R. and See, S.W. 2006. Optimization and validation of a low temperature microwave-assisted extraction method for analysis of polycyclic aromatic hydrocarbons in airborne particulate matter. *Talanta.* 69(1): pp.79–86.

Koch, T.A., Krause, K.R. and Mehdizadeh, M. 1997. Improved safety through distributed manufacturing of hazardous chemicals. *Process Safety Progress.* 16(1): pp.23–24.

Kong, Y. and Cha, C.Y. 1995. NOx abatement with carbon adsorbents and microwave energy. *Energy and Fuels.* 9(6): pp.971–975.

Krech, T., Möser, C., Emmerich, R., Scholz, P., Ondruschka, B. and Cihlar, J. 2008. Catalytic and heating behavior of nanoscaled perovskites under microwave radiation. *Chemical Engineering and Technology.* 31(7): pp.1000–1006.

Kržan, A. and Žagar, E. 2009. Microwave driven wood liquefaction with glycols. *Bioresource Technology.* 100(12): pp.3143–3146.

Ku, H.S., Siores, E., Taube, A. and Ball, J.A. 2002. Productivity improvement through the use of industrial microwave technologies. *Computers & Industrial Engineering.* 42(2): pp.281–290.

Kuhnert, N. 2002. Microwave-assisted reactions in organic synthesis—Are there any nonthermal microwave effects? *Angewandte Chemie International Edition.* 41(11): pp.1863–1866.

Kusdiana, D. and Saka, S. 2001. Kinetics of transesterification in rapeseed oil to biodiesel fuel as treated in supercritical methanol. *Fuel.* 80(5): pp.693–698.

Kusdiana, D. and Saka, S. 2001. Methyl esterification of free fatty acids of rapeseed oil as treated in supercritical methanol. *Journal of Chemical Engineering of Japan.* 34(3): pp.383–387.

Kusdiana, D. and Saka, S. 2004. Effects of water on biodiesel fuel production by supercritical methanol treatment. *Bioresource Technology.* 91(3): pp.289–295.

Landry, C.C. and Barron, A.R. 1993. The Synthesis of Polycrystalline Chalcopyrite Semiconductors by Microwave Irradiation. *MRS Proceedings* (Vol. 327, p.89). Cambridge University Press.

Landry, C.C., Lockwood, J. and Barron, A.R. 1995. Synthesis of chalcopyrite semiconductors and their solid solutions by microwave irradiation. *Chemistry of Materials.* 7(4): pp.699–706.

Leadbeater, N.E. 2005. Fast, easy, clean chemistry by using water as a solvent and microwave heating: the Suzuki coupling as an illustration. *Chemical Communications.* (23): pp.2881–2902.

Li, C.J. and Trost, B.M. 2008. Green chemistry for chemical synthesis. *Proceedings of the National Academy of Sciences.* 105(36): pp.13197–13202.

Lidström, P., Tierney, J., Wathey, B. and Westman, J. 2001. Microwave assisted organic synthesis—a review. *Tetrahedron.* 57(45): pp.9225–9283.

Loupy, A. 2004. Solvent-free microwave organic synthesis as an efficient procedure for green chemistry. *Comptes Rendus Chimie.* 7(2): pp.103–112.

Loupy, A., Petit, A., Ramdani, M., Yvanaeff, C., Majdoub, M., Labiad, B. and Villemin, D. 1993. The synthesis of esters under microwave irradiation using dry-media conditions. *Canadian Journal of Chemistry.* 71(1): pp.90–95.

Lucchesi, M.E., Chemat, F. and Smadja, J. 2004. Solvent-free microwave extraction of essential oil from aromatic herbs: comparison with conventional hydro-distillation. *Journal of Chromatography A.* 1043(2): pp.323–327.

Ma, F. and Hanna, M.A. 1999. Biodiesel production: a review. *Bioresource Technology.* 70(1): pp.1–15.

Melo-Júnior, C.A., Albuquerque, C.E., Fortuny, M., Dariva, C., Egues, S., Santos, A.F. and Ramos, A.L. 2009. Use of microwave irradiation in the noncatalytic esterification of C18 fatty acids. *Energy and Fuels.* 23(1): pp.580–585.

Menéndez, J.A., Domínguez, A., Inguanzo, M. and Pis, J.J. 2004. Microwave pyrolysis of sewage sludge: analysis of the gas fraction. *Journal of Analytical and Applied Pyrolysis.* 71(2): pp.657–667.

Mingos, D.M.P. 1994. The applications of microwaves in chemical syntheses. *Research on Chemical Intermediates.* 20(1): pp.85–91.

Orozco, A., Ahmad, M., Rooney, D. and Walker, G. 2007. Dilute acid hydrolysis of cellulose and cellulosic bio-waste using a microwave reactor system. *Process Safety and Environmental Protection.* 85(5): pp.446–449.

Pan, X., Niu, G. and Liu, H. 2003. Microwave-assisted extraction of tea polyphenols and tea caffeine from green tea leaves. *Chemical Engineering and Processing: Process Intensification.* 42(2): pp.129–133.

Pasquet, V., Chérouvrier, J.R., Farhat, F., Thiéry, V., Piot, J.M., Bérard, J.B., Kaas, R., Serive, B., Patrice, T., Cadoret, J.P. and Picot, L., 2011. Study on the microalgal pigments extraction process: Performance of microwave assisted extraction. *Process Biochemistry.* 46(1): pp.59–67.

Patil, P., Gude, V.G., Pinappu, S. and Deng, S. 2011. Transesterification kinetics of Camelina sativa oil on metal oxide catalysts under conventional and microwave heating conditions. *Chemical Engineering Journal.* 168(3): pp.1296–1300.

Perry, W.L., Katz, J.D., Rees, D., Paffet, M.T. and Datye, A.K. 1997. Kinetics of the microwave-heated CO oxidation reaction over alumina-supported Pd and Pt catalysts. *Journal of Catalysis.* 171(2): pp.431–438.

Pimentel, D. and Patzek, T.W. 2008. Biofuels, solar and wind as renewable energy systems. Benefits and risks. New York: Springer.

Saidi, A. and Azari, K. 2005. Carbothermic reduction of zinc oxide concentrate by microwave. *Journal of Materials Science and Technology-Shenyang.* 21(5): p.724.

Saka, S. and Kusdiana, D. 2001. Biodiesel fuel from rapeseed oil as prepared in supercritical methanol. *Fuel.* 80(2): pp.225–231.

Satyanarayana, K.G., Mariano, A.B. and Vargas, J.V.C. 2011. A review on microalgae, a versatile source for sustainable energy and materials. *International Journal of Energy Research.* 35(4): pp.291–311.

Shekara, B.C., Prakash, B.J. and Bhat, Y.S. 2012. Microwave-induced deactivation-free catalytic activity of BEA zeolite in acylation reactions. *Journal of Catalysis.* 290: pp.101–107.

Sheldon, R.A. 1992. Industrial Environmental Chemistry, ed. DT Sawyer and A. E. Martell. pp.99–119.

Sheldon, R.A. 2000. Atom efficiency and catalysis in organic synthesis. *Pure and Applied Chemistry.* 72(7): pp.1233–1246.

Sheldon, R.A. 2008. E factors, green chemistry and catalysis: an odyssey. *Chemical Communications.* (29): pp.3352–3365.

Sheldon, R. 2010. Introduction to green chemistry, organic synthesis and pharmaceuticals. *Green Chemistry in the Pharmaceutical Industry.* pp.1–20.

Sheldon, R.A. 2011. Utilisation of biomass for sustainable fuels and chemicals: Molecules, methods and metrics. *Catalysis Today.* 167(1): pp.3–13.

Shiying, Y., Ping, W., Xin, Y., Guang, W.E.I., Zhang, W. and Liang, S.H.A.N. 2009. A novel advanced oxidation process to degrade organic pollutants in wastewater: Microwave-activated persulfate oxidation. *Journal of Environmental Sciences.* 21(9): pp.1175–1180.

Silverwood, I.P., McDougall, G.S. and Whittaker, A.G. 2006. A microwave-heated infrared reaction cell for the *in situ* study of heterogeneous catalysts. *Physical Chemistry Chemical Physics.* 8(46): pp.5412–5416.

Silverwood, I., McDougall, G. and Whittaker, G. 2007. Comparison of conventional versus microwave heating of the platinum catalysed oxidation of carbon monoxide over EUROPT-1 in a novel infrared microreactor cell. *Journal of Molecular Catalysis A: Chemical.* 269(1): pp.1–4.

Sinev, I., Kardash, T., Kramareva, N., Sinev, M., Tkachenko, O., Kucherov, A. and Kustov, L.M. 2009. Interaction of vanadium containing catalysts with microwaves and their activation in oxidative dehydrogenation of ethane. *Catalysis Today.* 141(3): pp.300–305.

Siskin, M. and Katritzky, A.R. 2001. Reactivity of organic compounds in superheated water: general background. *Chemical Reviews.* 101(4): pp.825–836.

Sithambaram, S., Nyutu, E.K. and Suib, S.L. 2008. OMS-2 catalyzed oxidation of tetralin: A comparative study of microwave and conventional heating under open vessel conditions. *Applied Catalysis A: General.* 348(2): pp.214–220.

Sobol, H. and Tomiyasu, K. 2002. Milestones of microwaves. *Microwave Theory and Techniques, IEEE Transactions on.* 50(3): pp.594–611.

Stefanidis, G.D., Munoz, A.N., Sturm, G.S. and Stankiewicz, A. 2014. A helicopter view of microwave application to chemical processes: reactions, separations, and equipment concepts. *Reviews in Chemical Engineering.* 30(3): pp.233–259.

Rittmann, B.E. 2008. Opportunities for renewable bioenergy using microorganisms. *Biotechnology and Bioengineering.* 100(2): pp.203–212.

Roberts, B.A. and Strauss, C.R. 2005. Toward rapid, "green", predictable microwave-assisted synthesis. *Accounts of Chemical Research.* 38(8): pp.653–661.

Roger, A. Sheldon. 1997. Catalysis: The Key to Waste Minimization. *J. Chem. Tech. Biotechnol.* 68: pp.381–388.

Tang, S.L., Smith, R.L. and Poliakoff, M. 2005. Principles of green chemistry: Productively. *Green Chemistry.* 7(11): pp.761–762.

Terigar, B.G., Balasubramanian, S., Sabliov, C.M., Lima, M. and Boldor, D. 2011. Soybean and rice bran oil extraction in a continuous microwave system: From laboratory-to pilot-scale. *Journal of Food Engineering.* 104(2): pp.208–217.

Trost, B.M. 1991. The Atom Economy—A Search for Synthetic Efficiency. *Science.* 254(5037): pp.1471–1477.

Trost, B.M. 1995. Atom economy—a challenge for organic synthesis: homogeneous catalysis leads the way. *Angewandte Chemie International Edition in English.* 34(3): pp.259–281.

Tse, M.Y., Depew, M.C. and Wan, J.K. 1990. Applications of high power microwave catalysis in chemistry. *Research on Chemical Intermediates.* 13(3): pp.221–236.

Tucker, J.L. 2006. Green chemistry, a pharmaceutical perspective. *Organic Process Research and Development.* 10(2): pp.315–319.

Varma, R.S. 1999. Solvent-free organic syntheses. Using supported reagents and microwave irradiation. *Green Chemistry.* 1(1): pp.43–55.

Wei, Z.S., Zeng, G.H., Xie, Z.R., Ma, C.Y., Liu, X.H., Sun, J.L. and Liu, L.H. 2011. Microwave catalytic NOx and SO_2 removal using FeCu/zeolite as catalyst. *Fuel.* 90(4): pp.1599–1603.

Will, H., Scholz, P., Ondruschka, B. and Burckhardt, W. 2003. Multimode microwave reactor for heterogeneous gas-phase catalysis. *Chemical Engineering and Technology*. 26(11): pp.1146–1149.

Will, H., Scholz, P. and Ondruschka, B. 2004. Heterogeneous gas-phase catalysis under microwave irradiation—a new multimode microwave applicator. *Topics in Catalysis*. 29(3-4): pp.175–182.

Yao, H., Du, X., Yang, L., Wang, W., Yang, F., Zhao, C., Meng, X., Zhang, L. and Zu, Y. 2012. Microwave-assisted method for simultaneous extraction and hydrolysis for determination of flavonol glycosides in Ginkgo foliage using Brönsted acidic ionic-liquid [HO3S (CH2) 4mim] HSO4 aqueous solutions. *International Journal of Molecular Sciences*. 13(7): pp.8775–8788.

Yang, J., Xu, M., Zhang, X., Hu, Q., Sommerfeld, M. and Chen, Y. 2011. Life-cycle analysis on biodiesel production from microalgae: Water footprint and nutrients balance. *Bioresource Technology*. 102(1): pp.159–165.

Yin, S., Liu, B. and Sato, T. 2008. Microwave-assisted hydrothermal synthesis of nitrogen-doped titania nanoparticles. *Functional Materials Letters*. 1(03): pp.173–176.

Zhang, X., Hayward, D.O., Lee, C. and Mingos, D.M.P. 2001. Microwave assisted catalytic reduction of sulfur dioxide with methane over MoS 2 catalysts. *Applied Catalysis B: Environmental*. 33(2): pp.137–148.

Zhang, X. and Hayward, D.O. 2006. Applications of microwave dielectric heating in environment-related heterogeneous gas-phase catalytic systems. *Inorganica Chimica Acta*. 359(11): pp.3421–3433.

Zhang, Z. and Zhao, Z.K. 2010. Microwave-assisted conversion of lignocellulosic biomass into furans in ionic liquid. *Bioresource Technology*. 101(3): pp.1111–1114.

Zhou, R., Zhong, S., Lin, X. and Xu, N. 2009. Synthesis of zeolite T by microwave and conventional heating. *Microporous and Mesoporous Materials*. 124(1): pp.117–122.

Zhu, S., Wu, Y., Yu, Z., Liao, J. and Zhang, Y. 2005. Pretreatment by microwave/alkali of rice straw and its enzymic hydrolysis. *Process Biochemistry*. 40(9): pp.3082–3086.

2

Microwave Heating and Interaction with Materials

Introduction to Material Interactions with an Electric Field

Materials and solvents interact with an electric field in different ways. The interaction depends on the frequency of the electric field and the dielectric properties namely dielectric constant (ε'), dielectric loss factor (ε'') and tangent loss (tan δ). These properties are important to understand the material interactions as well as heating characteristics and behavior of materials in an electric field. These properties also depend on the microwave frequency, temperature, types of materials, etc. Selection of right frequency is important either to increase the penetration depth of the microwaves or reduce the energy requirement. Overall, the complex dielectric constant (ε^*) indicates the charge storing capacity of the material irrespective of the sample dimension (Gabriel et al. 1998).

A material may have several dielectric mechanisms or polarization effects that contribute to its overall permittivity of an electric field (Figure 1). A dielectric material has an arrangement of electric charge carriers that can be displaced by an electric field. The charges become polarized to compensate for the electric field such that the positive and negative charges move in opposite directions. At the microscopic level, several dielectric mechanisms can contribute to dielectric behavior. Among which, dipole orientation and ionic conduction interact strongly at microwave frequencies. For example, water molecules are permanent dipoles, which rotate to follow an alternating electric field. These mechanisms are quite lossy leading to

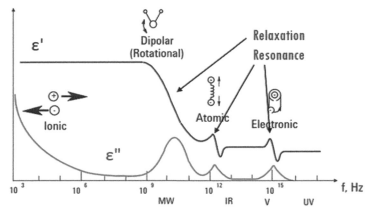

Figure 1. Broadband permittivity variation of materials.

rapid heat generation. Atomic and electronic mechanisms are relatively weak and usually constant over the microwave region. Each dielectric mechanism has a characteristic "cut-off frequency." As the frequency increases, the slow mechanisms drop out in turn, leaving the faster ones to contribute to ε'. The loss factor (ε'') will correspondingly peak at each critical frequency. The magnitude and "cut-off frequency" of each mechanism is unique for different materials. Water has a strong dipolar effect at low frequencies—but its dielectric constant rolls off dramatically around 22 GHz. PTFE (Polytetrafluoroethylene), on the other hand, has no dipolar mechanisms and its permittivity is remarkably constant well into the millimeter-wave region (Komarov et al. 2005). A resonant effect is usually associated with electronic or atomic polarization. A relaxation effect is usually associated with orientation polarization.

Microwave Characteristics

Microwave irradiation is the electromagnetic spectrum between infrared waves and radio waves with frequency range of 0.3–300 GHz and wavelengths between 0.01 and 1 m. Commercial microwave ovens approved for domestic applications operate at a frequency of 2.45 GHz to avoid interference with telecommunication and cellular phone frequencies. Typical bands approved for industrial applications are 915 and 2.45 GHz. Most of the reported microwave chemistry experiments are conducted at 2.45 GHz (the corresponding wavelength is 12.24 cm) since this frequency is approved worldwide and used in currently available commercial microwave chemistry equipment. At this frequency, the microwave energy is absorbed most by liquid water. Interaction of dielectric materials with microwaves leads to what is generally described as dielectric heating due to a net polarization of the substance (Stuerga and Delmotte 2002, Mingos and Whittaker 1997, Mingos and Baghurst 1991, Gabriel et al. 1998). There are several mechanisms which are responsible for this, including electronic, ionic, molecular (dipole), and interfacial (space-charge) polarization which will be discussed further (Evaluserve 2005).

A material can be heated by applying energy to it in the form of microwaves which are high frequency electromagnetic waves. The electric field component of microwaves exerts a force on the charged particles found in the compound. If the charged particles are able to move freely through the electrical field, a current is induced. However, if the charged particles are bound in the compound, restricted in their movements, they merely reorient themselves in phase with the electric field. This is termed as dielectric polarization. The dielectric polarization can be made up of four components as shown in Equation [1] based on the different types of the charged particles in matter: electrons, nuclei, permanent dipoles and charges at interfaces (Jacob et al. 1995).

$$\alpha_t = \alpha_e + \alpha_a + \alpha_d + \alpha_i \qquad [1]$$

where α_t = total dielectric polarization;

α_e = electronic polarization due to polarization of electrons surrounding the nuclei;

α_a = atomic polarization due to polarization of the nuclei;

α_d = dipolar polarization due to polarization of permanent dipoles in the material; and

α_i = interfacial polarization due to polarization of charges at interfaces.

Microwave Energy

Energy associated with microwaves is lower than the energy of Brownian motion which is not strong enough to even break chemical bonds. As such, microwaves cannot induce chemical reactions (see Table 1). The influence of microwave energy on chemical or biochemical reactions is both thermal and non-thermal (Perreux and Loupy 2002). The microwave energy quantum is given by the well-known equation, $W = h\nu$. Within the frequency domain of microwaves and hyper-frequencies (300 MHz–300 GHz), the corresponding energies are $1.24 \times 10^{-6} - 1.24 \times 10^{-3}$ eV, respectively. These energies are much lower than ionization energies of biological compounds (13.6 eV), of covalent bond energies such as OH^- (5 eV), hydrogen bonds (2 eV), van der Waals intermolecular interactions (less than 2 eV) and even lower than the energy associated

Table 1. Energies of chemical bonds in comparison to different microwave energies (Nüchter et al. 2004).

		Energy (eV)	Energy (kJ mol^{-1})
1	CC single bond	3.61	347
2	CC double bond	6.35	613
3	CO singe bond	3.74	361
4	CO double bond	7.71	744
5	CH bond	4.28	413
6	OH bond	4.80	463
7	Hydrogen bond	0.04–0.44	4–42
8	MW 0.3 GHz	1.2×10^{-6}	0.00011
9	MW 2.45 GHz	1×10^{-5}	0.00096 ~ 1 J/mol
10	MW 30 GHz	1.2×10^{-3}	0.11

1 eV = 1.602177×10^{-19} J

with Brownian motion at 37°C (2.7×10^{-3} eV) (Metaxas 1993, Chemat-Djenni et al. 2007). Microwaves, as an energy source, produce heat by their interaction with the materials at molecular level without altering the molecular structure (Varma 2001, Refaat 2010).

Microwave heating offers several advantages over conventional heating in the manner that they provide non-contact heating (reduction of overheating of material surfaces), energy transfer instead of heat transfer (penetrative radiation), reduced thermal gradients, material selective and volumetric heating, fast start-up and stop and reverse thermal effect (i.e., heat starts from the interior of material body). For biofuel production, this means more effective heating, fast heating of catalysts, reduced equipment size, faster response to process heating control, faster start-up, increased production, and elimination of process steps (Chemat-Djenni et al. 2007).

Microwave Heating Mechanisms

Microwave heating mechanism is complex. The microwave method of heating can be illustrated as shown in Figure 2. A comparison with conventional heating method would provide a suitable platform to compare the differences in heating mechanisms and further realize the advantages associated with microwave heating.

In conventional heating, heat transferred to the sample volume is first utilized to increase the temperature of the surface of the vessel followed by the internal materials. This is also called "wall heating". Therefore, a large portion of energy supplied through conventional energy source is lost to the environment through conduction of materials and convection currents. Heating effect in the conventional method is heterogeneous and dependent on thermal conductivity, specific heat, and density of materials, which result in higher surface temperatures causing heat transfer from the outer surface to the internal

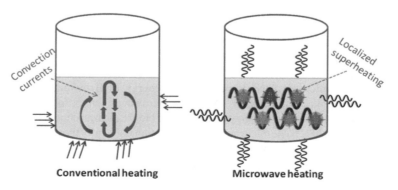

Figure 2. Conventional and microwave heating methods.

sample volume as seen in Figure 3. As a result, non-uniform sample temperatures and higher thermal gradients are observed in conventional heating (Bogdal 2005, Groisman and Gedanken 2008).

Figure 3a shows the temperature profiles for a 5 mL sample of ethanol boiled at 160°C in a single mode closed vessel microwave irradiation and open vessel oil bath heating conditions (Kappe 2006). The temperature profiles show that microwave heating method allows for rapid increase of solvent temperature and quick cooling as well, whereas in conventional heating (oil bath), the pace of heating and cooling are very slow. Figure 3b shows thermal behavior of microwave versus oil bath heating. Temperature gradients shown in Figure 3b suggest that microwave irradiation rises the temperature of the whole volume evenly and simultaneously whereas in oil bath heating the reaction mixture in contact with the vessel wall is heated first. Inverted thermal gradient differences can be observed between the two heating methods (Li and Yang 2008, Kappe 2004). The advantages of this enabling technology have more recently also been exploited in the context of multistep total synthesis and medicinal chemistry/drug discovery and have additionally penetrated fields such as polymer synthesis, material sciences, nanotechnology, and biochemical processes (Herrero et al. 2008).

Microwaves transfer energy into materials by dipolar polarization, ionic conduction and interfacial polarization mechanisms to cause localized and rapid superheating of the reaction materials. If a molecule possesses a dipole moment, when it is exposed to microwave irradiation, the dipole tries to align with the

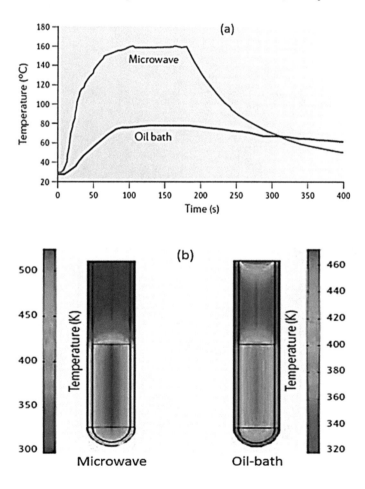

Figure 3. (a) Temperature profiles for a 5 mL sample of ethanol boiled at 160°C in a single mode closed vessel microwave irradiation and open vessel oil bath heating conditions; (b) Comparison of thermal behavior and conduction of conventional and microwave heating methods (Kappe and Dallinger 2006).

applied electric field. Since the electric field is oscillating, the dipoles constantly try to realign to follow this movement. At 2.45 GHz, molecules have time to align with the electric field but not to follow the oscillating field exactly (Figure 4). The angle δ represents the phase lag between the polarization of the material and the applied electric field. This is therefore an extremely useful quantity in determining how efficiently microwave heating will take place (Figure 4a). This continual reorientation of the molecules results in friction and thus heat. If a molecule is charged, then the electric field component of the microwave irradiation moves the ions back and forth through the sample while also colliding them into each other. This movement again generates heat (Figure 4b). In addition, because the energy is interacting with the molecules at a very fast rate, the molecules do not have time to relax and the heat generated can be, for short times, much greater than the overall recorded temperature of the bulk reaction mixture. In essence, there will be instantaneous localized superheating. Therefore, the bulk temperature may not be an accurate measure of the temperature at which the actual reaction is taking place. The interfacial polarization method can be considered as a combination of the conduction and dipolar polarization mechanisms. It is important for heating systems that comprise a conducting material dispersed in a non-conducting material such as metal oxides in polar solvents (Chemat-Djenni et al. 2007, Refaat 2010).

The introduction of microwave energy into a chemical reaction that has at least one component which is capable of coupling with microwaves can lead to much higher heating rates than those that can be achieved conventionally. Using very cheap and readily available microwave cavities, heating rates of 2–4°C/s may be readily achieved even for common organic solvents. Such heating rates are more difficult

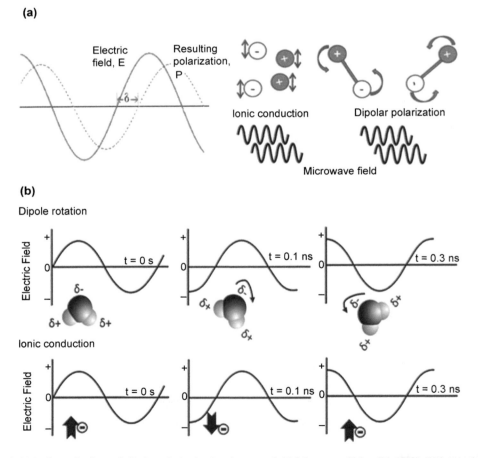

Figure 4. (a) Ionic conduction and dipolar polarization in microwave field (Metaxas and Meredith 1983); (b) interaction of a water molecule with the electromagnetic field involving dipole rotation and ionic conduction (Bilecka and Niederberger 2010).

to achieve using conventional heating although of course dropping sealed tubes into heated furnaces at > 1000°C could result in comparable heating rates.

The microwave energy is introduced into the reactor remotely and therefore there is no direct contact between the energy source and the sample undergoing heating. This combined with the above phenomenon may lead to very different temperature-time profiles for the reaction and as a consequence may lead to an alternative distribution of chemical products in the reaction. The remote nature of the interaction results in much higher heating and cooling rates for the microwave technique as well as higher reaction temperances that are normally achieved under conventional heating. Therefore, microwave dielectric heating resembles a flash heating process whereby the energy is generated much more rapidly and the sample cools more rapidly at the end of the reaction. The different profiles may therefore lead to significantly different products, particularly if the reaction product distribution is controlled by complex and temperature-dependent kinetic profiles.

Chemicals and the containment materials for chemical reaction do not interact equally with the commonly used microwave frequencies for dielectric heating and consequently selective heating may be achieved. Specifically it is possible to cool the outside of the vessel with a coolant that is transparent to microwaves (solid CO_2 or liquid N_2) and thereby have cold walls that still allow the microwave energy to penetrate and heat the reactants which are microwave active in the vessel. Also for solid state reactions, contamination from the crucible walls may be minimized.

The degree of selective heating should not be exaggerated for solvent mixtures. For example, if a mixture of methanol which has a loss factor and benzene which is transparent to microwaves is exposed to a microwave field, the whole mixture heats up rapidly. The microwave process involves translation and rotation and although the effect may have its origins in the vicinity of methanol molecules, the rate of energy transport is so fast that benzene molecules are also heated up rapidly. Therefore, it is not possible to store the microwave energy selectively either within parts of a molecule or in active molecules in two component mixtures.

The boiling phenomenon is a kinetic as well as thermodynamic process and therefore solvents heated under microwave conditions often boil at elevated temperatures even though they remain contained under one atmosphere. The precise elevation of this nucleation-limited boiling point depends on the power input, the occurrence of effective stirring and the limitation of the number of nucleation sites for example by having smooth surfaces and no boiling chips. This effect can be enhanced by allowing the pressure to rise above atmospheric level. It is possible to increase the temperature of a reaction in a common organic solvent by up to 100°C above the conventional boiling point. Ethanol which has a conventional boiling point of 79°C when microwave dielectrically heated boils at 164°C at a pressure of 12 atm. This temperature rise when translated into the Arrhenius expression leads to an enhancement of ~ 10^3 in the reaction rate.

In solid samples, the rate of energy transport is less and consequently the development of hot spots is more significant. A careful analysis of heterogeneous catalysis suggests that hot spot formation around the catalyst not only enhances the reaction rate but may also contribute to shifts in the equilibrium constant.

Figure 5 shows the range of microwave frequency and the variations of ionic conduction and dipolar polarization within the microwave frequency range. The effective loss factor as a function of the microwave frequency due to dipolar relaxation and Maxwell-Wagner or ionic conduction mechanisms are shown in this figure. Ionic conduction dominates in certain frequencies while the dipolar rotation becomes dominant at some other high frequencies.

Principles of Microwave Heating

Microwave chemistry generally relies on the ability of the reaction mixture to efficiently absorb microwave energy, taking advantage of "microwave dielectric heating" phenomena involving dipolar polarization or ionic conduction mechanisms (Dallinger and Kappe 2007). In most cases this means that the material used for a particular transformation must be microwave absorbing. The ability of a specific solvent to convert microwave energy into heat at a given frequency and temperature is determined by the so called loss tangent (tan δ), expressed as the quotient, tan $\delta = \varepsilon''/\varepsilon'$, where ε'' is the dielectric loss, indicative of

34 *Microwave-Mediated Biofuel Production*

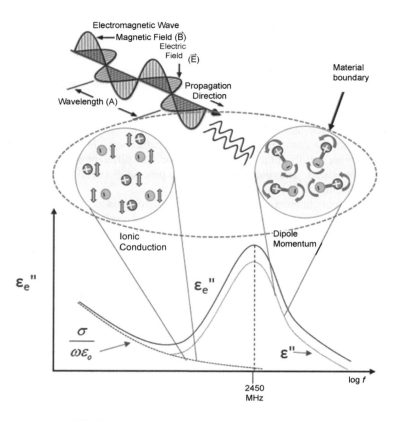

Figure 5. Ionic conduction and dipolar reorientation variation with microwave frequency (Meridith and Metaxas 1983).

the efficiency with which electromagnetic radiation is converted into heat, and ε' is the dielectric constant, describing the ability of molecules to be polarized by the electric field. A reaction medium with a high tan δ at the standard operating frequency of a microwave synthesis reactor (2.45 GHz) is required for good absorption and, consequently, efficient heating (Kitchen et al. 2013).

Dipolar polarization, P_d, occurs on a time scale of the order of those associated with microwaves. When a dielectric material is subjected to an external electric field of strength E, the polarization is related to the intrinsic properties of the material which is given by the relation shown in Equation 2.

$$P_d = \varepsilon_0(\varepsilon_r - 1)E \qquad [2]$$

where ε_0 is the permittivity of free space and ε_r is the relative permittivity of the material. Given that in reality permittivity is a complex quantity, ε^* (equal to the product ε_0 and ε_r which is expressed as $\varepsilon^* = \varepsilon' + i\varepsilon''$), where ε' represents the time-independent polarizability of a material in the presence of an external electric field and ε'', the time-dependent component of the permittivity, quantifies the efficiency with which electromagnetic energy is converted to heat. By contrast, the interaction of charge carriers with the applied electric field in solids leads to ohmic heating. In such cases the complex permittivity is modified to take account of losses by including a separate conduction term (Equation 3)

$$\varepsilon^* = \varepsilon' - i\varepsilon'' - \left(\frac{i\sigma_i}{\omega\varepsilon_0}\right) \qquad [3]$$

Where σ_i is the conductivity of the material and ω is the frequency of the microwave field. The losses in this case arise not from a phase lag, as the relaxation time for this process is of the order of that of a microwave phase period, but from the intrinsic resistance of the material to an electric current. This term is extremely significant in metallic conductors or semiconductors such as carbon. The dielectric properties of a material can also be used to determine the amount of power absorbed, P, by a sample as shown in Equation 4.

$$P = \sigma |E|^2 = 2\pi\varepsilon_0 \varepsilon'' |E|^2 = 2\pi f \varepsilon_0 \varepsilon' \tan\delta |E|^2 \qquad [4]$$

where σ is the total effective conductivity, E is the magnitude of the internal electric field, ε_0 is the permittivity of free space, and f is the microwave frequency. Finally, a key aspect in practical design of microwave solid state syntheses and processes is consideration of how the efficiency of MW heating is mediated by sample volume. As the size of the sample being irradiated is increased, an absorbance loss factor becomes progressively more significant. The penetration depth, D_p, is the distance into the sample at which the electric field is attenuated to 1/e of its surface value as shown in Equation 5 (Stefanidis et al. 2014).

$$D_p = \frac{\lambda_0 \sqrt{\varepsilon'}}{2\pi\varepsilon''} \qquad [5]$$

where λ_0 is the wavelength of the microwave radiation. For materials at microwave heating frequencies, D_p ranges between several micrometers for metals and several tens of meters for some low-loss polymers. For many materials at microwave frequencies, the penetration depth is of the same order of magnitude as the dimensions of the sample, a fact that often has important implications for heating uniformity. From the above it is apparent, for example, that high dielectric loss materials, such as carbon, are more suited for small batch processes or a continuous feed system. If the volume of the sample is too large then insufficient microwave energy reaches the center of the sample, resulting in non-uniform heating and inhomogeneous, impure products (Kitchen et al. 2013).

Microwave Interaction with Materials

Materials can be classified into three categories based on their interaction with microwaves: (1) materials that reflect microwaves, which are bulk metals and alloys, e.g., copper; (2) materials that are transparent to microwaves, such as fused quartz, glasses made of borosilicate, ceramics, Teflon, etc.; and (3) materials that absorb microwaves which constitute the most important class of materials for microwave synthesis, e.g., aqueous solutions, polar solvent, etc. Dissipation factor (often called the loss tangent, tan δ), a ratio of the dielectric loss (loss factor) to the dielectric constant, is used to predict material's behavior in a microwave field. The microwave absorption ability of a material is directly proportional to its dissipation factor (Li and Yang 2008). Figure 6 shows a general representation of the relation between dielectric loss factor of a material and a representative power absorption capacity under microwave field. Microwave power absorbed per unit volume of the material depends on the interaction of the material with microwave field which is expressed as dielectric loss factor.

According to their microwave absorption characteristics, materials may be classified as conductors, insulators, and absorbers (Church 1993).

Conductors possess a property to reflect the incident waves off the material surface. Metals such as copper, brass, silver, and aluminum are good conductors, and waveguides in a microwave oven are often made of brass or aluminum.

Insulators such as Teflon and polypropylene (PP) are substantially transparent to the microwaves and possess a property to partially reflect and transmit the incident waves traveling through the material, but not store microwave energy in the form of heat. Insulators are often used in microwave ovens to support the material to be heated.

Absorbers possess a property to absorb microwave radiation, direct energy transfer, and thus effectively get heated at room temperature. Their ability to absorb microwave energy is related to the loss tangent of the materials, which depends on the functional groups and the volume of the molecules (Gabriel et al. 1998).

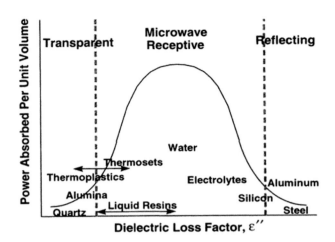

Figure 6. Relationship between the dielectric loss factor and ability to absorb microwave power for some common materials (adapted from Thostenson and Chou 1999).

Several factors affect the dielectric properties of materials which includes the frequency of the microwave field, mode of applications, temperature of the material, density, composition and structure of the material (Venkatesh and Raghavan 2004). More interestingly, dielectric properties are not specific to a substance but they change with the microwave field frequency and temperature. For example, water has different dielectric loss factors at different temperatures at different microwave frequencies as shown in Figure 7 (data taken from Hasted 1973). Even though the frequency is usually fixed, the loss tangent is important as the temperature of a reaction has been shown to vastly affect the dielectric properties of a material. This observation is important, particularly in the case of solids that do not absorb microwave energy at room temperature; often such materials will begin to absorb when the temperature is increased. In these cases, a susceptor material is often used to raise the reaction temperature to a point where the dielectric properties are more favorable for the reactant (Peng et al. 2002, Kitchen et al. 2013). An example of this is in the MW heating of alumina which becomes 3000% more efficient when the temperature is raised from 200 to 1200°C (Sutton 1989). However, this increase in heating efficiency can cause problems such as thermal runaway (where an increase in temperature results in a change in reaction rate which leads to further increase in temperature), and without careful monitoring the temperature can become uncontrollable.

Variations in dielectric permittivity and dielectric loss of water between 0°C and 100°C at different wavelengths are shown in Figure 8. The arrows show the effect of increasing temperature or increasing water activity (Figure 8). The wavelength range 0.01–100 cm is equivalent to 3 THz–0.3 GHz respectively. As the temperature increases, the strength and extent of the hydrogen bonding, both decrease. This lowers both the static and optical dielectric permissivities, reduces the difficulty for the movement of dipole and so allows the water molecule to oscillate at higher frequencies, and reduces the drag to the rotation of the water molecules, so reducing the friction and hence the dielectric loss. Most of the dielectric loss is within the microwave range of electromagnetic radiation ($\sim$ 1–300 GHz, with wave number 0.033 cm^{-1}–10 cm^{-1}, and wavelength 0.3 m–1.0 mm respectively). The frequency for maximum dielectric loss lies higher than the 2.45 GHz (wave number 0.0817 cm^{-1}, wavelength 12.24 cm) produced by most microwave ovens. This is so that the radiation is not totally absorbed by the first layer of water it encounters and may penetrate further into the foodstuff, heating it more evenly; unabsorbed radiation passing through is mostly reflected back, due to the design of the microwave oven, absorbed in subsequent passes.

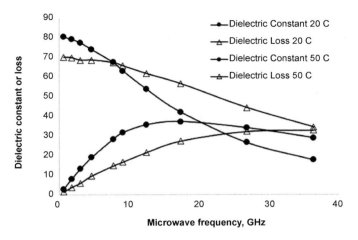

Figure 7. Variations in dielectric properties of water at different temperatures under microwave irradiation.

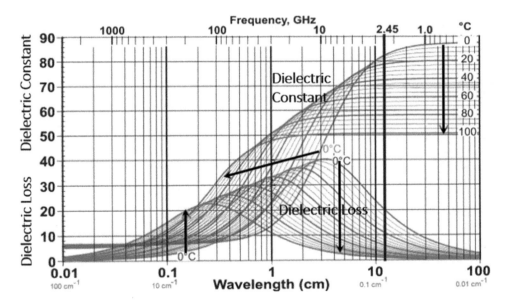

Figure 8. Variations in dielectric properties of water at different temperatures under microwave irradiation corresponding to different wavelengths.

Microwave-solvent Interactions

Organic solvents are generally categorized in three different groups according to their high, medium, and low absorbing properties. High microwave absorbing solvents typically have a tan $\delta > 0.5$, whereas medium and low absorbing solvents have values of 0.1–0.5 and < 0.1, respectively (Dallinger and Kappe 2007, Bilecka and Niederberger 2010). Table 2 provides a list of commonly used solvents in chemical synthesis and grouping according to their microwave field absorbance level. Some of the solvents possess high dielectric constants. However, their lower dielectric loss factor reduces the loss tangent factor which is reflected in reduced microwave absorbance level.

Table 2. Most commonly used solvents separated by microwave absorbance level.

Absorbance level	Solvents
High	DMSO, ethanol, methanol, propanol, nitrobenzene, formic acid, ethylene glycol
Medium	Water, DMF, NMP, butanol, acetonitrile, HMMPA, methyl ethyl ketone, acetone, and other ketones, nitromethane, o-dichlorobenzene, 1,2-dichloroethane, 2-methoxyethanol, acetic acid, trifluoroacetic acid
Low	Chloroform, dichloromethane, carbon tetrachloride, 1,4-dioxane, THF, ethers, ethyl acetate, pyridine, trimethylamine, toluene, benzene, chlorobenzene, xylenes, pentane, hexane, and other hydrocarbons

Some other common solvents without a permanent dipole moment such as carbon tetrachloride, benzene, and dioxane are more or less microwave transparent. Therefore, microwave synthesis in low absorbing or microwave transparent solvents is often not feasible unless either the substrates or some of the reagents/catalysts are strongly polar and therefore microwave absorbing, raising the overall dielectric properties of the reaction medium to a level that allows sufficient heating by microwaves. Water can be considered only a medium microwave absorbing solvent with a loss tan δ of 0.123 (Table 3).

Table 3. Dielectric properties of different materials under microwave field.

Material or solvent	Dielectric constant (ε')	Dielectric loss (ε'')	Loss tangent ($tan\ \delta$)
Vacuum	1.00	0	0
Air	1.0006	0	0
Water	80.4	9.89	0.123 (2.45 GHz)
Methanol	32.6	21.48	0.659 (2.45 GHz)
Ethanol	24.3	22.86	0.941 (2.45 GHz
Glass (Pyrex)	4.82	0.026	0.0054 (3 GHz)
Hexane	1.9	0.038	0.020
Benzene	2.3		
Carbon tetrachloride	2.2		
Chloroform	4.8	0.4368	0.091
Acetic acid	6.1	1.0614	0.174
Ethyl acetate	6.2	0.3658	0.059
THF (tetrahydrofuran)	7.6	0.3572	0.047
Methyl chloride	9.1	0.3822	0.042
Acetone	20.6	1.1124	0.054
Acetonitrile	36.0	2.232	0.062
Dimethyl formamide	36.7	5.9087	0.161
Dimethyl sulfoxide	47.0	37.885	0.825
Formic acid	58.0	41.876	0.722
2-propanol			0.799
1-butanol			0.571
2-butanol			0.447
1,2-dichlorobenzene			0.280
1,2-dichloroethane			0.127
Chlorobenzene			0.101
Ethyl acetate	6.2	0.3658	0.059
Dichloromethane	8.9	0.3738	0.042
Toluene			0.040
Styrofoam	1.03	0.0001	0.0001 (3 GHz)
PTFE	2.08	0.0008	0.0004 (10 GHz)
Titanium oxide	50	0.25	0.005
Zirconia	20	2	0.1
Zinc oxide	3	3	1
Magnesium oxide	9	0.0045	0.0005
Aluminum oxide	9	0.0063	0.0007

Microwave-solid Material Interactions

Microwave heating effects result from the interaction of the electric component of the microwave field with charged particles in a material (or more rarely, the magnetic component may interact with magnetic dipoles). The precise nature of the interaction depends upon the mobility of charged particles and may give rise to one or both of the two major microwave heating processes. In substances where the charges are bound as dipoles, the electric field induces motion until it is balanced by electrostatic interactions; this is known as dipolar polarization (Pd) and is most significant in the liquid phase. For materials in which the charge carriers are mobile, as with electrons and fast ion conductors, the alternating microwave field gives rise to a current traveling in phase with the field and causing resistive heating in the sample. Generally, this mechanism is the dominant effect in solid materials and is referred to as conduction heating. Despite these generalizations, solids with bound solvent molecules, for example, may display dipolar polarization effects, and ionic solutions may equally well display the effects of conduction heating. It is important to note that microwave heating in condensed phases is quite distinct from the quantized energy absorption observed in microwave spectroscopy. While absorption of microwaves in solid and liquid samples is frequency dependent, it is, in effect, non-quantized. Instead, the material behaves as though reacting to a high frequency electric field.

Since the influence of a dielectric material depends on the amount of mass interacting with the electromagnetic fields, the mass per unit volume, or density, will have an effect on the dielectric properties (Nelson 1992). This is especially notable with particulate dielectrics such as pulverized or granular materials. In understanding the nature of the density dependence of the dielectric properties of particulate materials, relationship between the dielectric properties of solid materials and those of air–particle mixtures, such as granular or pulverized samples of such solids, are useful. In some instances, the dielectric properties of a solid may be needed when particulate samples are the only available form of the material. This was true for cereal grains, where kernels were too small for the dielectric sample holders used for measurements (Nelson and You 1989) and in the case of pure minerals that had to be pulverized for purification. For some materials, fabrication of samples to exact dimensions required for dielectric properties measurement is difficult, and measurements on pulverized materials are more easily performed. In such instances, proven relationships for converting dielectric properties of particulate samples to those for the solid material are important. Several well-known dielectric mixture equations have been considered for this purpose (Venkatesh and Raghavan 2004).

Microwave Interaction with Carbon Materials

Carbon derived compounds are often used in biofuel synthesis as heat absorbers or catalysts or heat dissipating media (in lieu of boiling stones). For instance in biodiesel production, these materials interact with microwaves to provide superheating and thus act as catalysts for heat enhancement. In high temperature conversion processes such as pyrolysis and gasification, microwaves may present a hazard of thermal runaway. Carbon materials can be used as media for heat absorption to avoid thermal runaway. The dielectric properties of different forms of carbon materials are shown in Table 4.

Carbon materials were introduced in microwave pyrolysis process as well, especially to increase the syngas concentration. Microwaves can generate microplasmas, which promote heterogeneous catalytic reactions, but not all materials can be heated by means of microwave irradiation, since some materials are transparent to microwaves. Addition of carbon-rich materials has been proposed to absorb microwaves (Menéndez et al. 2010, 2011). The material to be pyrolyzed is then heated by conduction. Use of the char obtained from MSW pyrolysis process, as microwave absorbent is an attractive solution since it avoids the addition of materials that might increase the cost of the process.

Microwave Interaction Relevant to Object Size

Microwaves possess wavelengths in vacuum ranging from 1 m to 1 mm. Whether the wave-like nature of electromagnetic fields is relevant to specific applications depends largely on the geometrical sizes of objects

Table 4. Different types of carbon materials and their dielectric properties.

Carbon material	tan δ = $\varepsilon''/\varepsilon'$	Reference
Coal	0.02–0.08	Yang and Wu 1987, Marland et al. 2001
Carbon foam	0.05–0.20	Fang et al. 2007
Charcoal	0.11–0.29	Wu et al. 2008, Challa et al. 1994
Carbon black	0.35–0.83	Atwater and Wheeler 2004a, Ma et al. 1997
Activated carbon	0.57–0.80	Challa et al. 1994, Atwater and Wheeler 2003, 2004b
Activated carbon	0.22–2.95	Dawson et al. 2008
Carbon nanotube	0.25–1.14	Lin et al. 2008, Zhang and Zhu 2009
CSi nanofibers	0.58–1.00	Yao et al. 2008

or equipment that are interacting with the field (Sturm et al. 2013, 2014). More specifically, it depends on their size relative to the field's wavelength. Figure 9a–c illustrate the differences between the characteristics of these interactions. Figure 9a illustrates the case in which the wavelength is (very) large with respect to a dielectric object that is exposed to the electromagnetic field. The arrows indicate the direction of the electric field and the grayscale of the object indicates the amplitude, or peak field strength, of the field in the object—darker zones have higher intensity. For a field with a frequency of zero and an infinitely long wavelength, the electric field would be static, i.e., unchanging in time. For non-zero frequencies, the field would exhibit a sinusoidal oscillation with two reversals of the field direction per oscillation period. There are some variations in the field intensity in the object. Inside and in the vicinity of the object, the

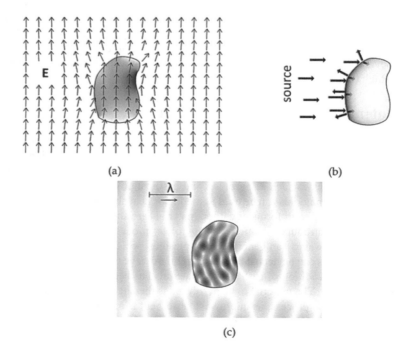

Figure 9. The variability of the interactions of an electromagnetic wave field with an object. Figure (a) represents the case in which the wavelength is (very) large relative to the object, (b) illustrates the case in which the wavelength is (very) short relative to the object, and (c) represents the case in which the wavelength is of the same order of magnitude as the object geometry (Sturm et al. 2013, 2014).

field is affected by the presence of the object, but at some distance there is no influence of its presence. Under the above conditions—electrically non-conductive, non-magnetic media, the analysis of this case reduces to quasi-static analysis via potential theory, reducing the complexity of the problem considerably.

Figure 9b illustrates the case in which the wavelength is (very) small with respect to the object. Unlike in Figure 9a, the arrows in Figure 9b indicate the direction in which the field propagates, not the direction of the electric field vector. The rightward arrows represent the field originating from a source and incident onto the object. The field is partially transmitted into the object where it dissipates in an exponential profile along its trajectory, assuming dissipative—electromagnetically absorbing—medium properties. Hence the dark zone on the right side of the object that represents the time-average intensity of the field. The zone of highest intensity in the object is located to the right, facing the source. On the interface (or boundary) between the object and the surroundings, the field is partially reflected away from the object. Like the previous case, the analysis of this case is relatively simple, as it can be conducted via optical analysis, for example by ray-tracing accounting for reflection, refraction and attenuation of optical beams. On a side note, in case the wavelength is relatively short and in the absence of medium properties variations, energy is locally traveling in straight lines through space. This is a characteristic of radiation and, therefore, the electromagnetic interactions can be considered as such. For cases with longer wavelengths relative to the objects involved, and in which an object is present in the direct vicinity, its presence will affect the electromagnetic field and energy cannot a-priory be expected to travel in straight lines. Hence in those cases the electromagnetic interactions cannot be considered radiation. More appropriate terminology in such case would be "field" or "wave field"; these terms will therefore be maintained throughout this paper. The case in which the size of the object is of the same order of magnitude as the wavelength is illustrated in Figure 9c. Note that the wavelength is approximately equal to the width of the object in this case. Similarly in Figure 9a and b, the grayscale represents the time average intensity of the electric field component, which is darker for higher intensities. Around the object, the instantaneous field strength (irrespective of field vector direction) of the wave pattern that travels rightward past the object is represented by the lighter gray coloring. It is evident that the intensity distribution inside the object is much more complicated than in the former two cases. Furthermore, outside of the object, the field is affected considerably; four bands of low wave intensity stretch far outward from the object. In all, the interactions are much more complicated than in the other two cases, which is due to the wave fields that are simultaneously scattered and redirected by the object, traveling through and around it, and constructively and destructively combining, thus forming spatially complex interference patterns of high and low intensities. The latter type of interaction—with object or equipment sized at the same order of magnitude as the wavelength—is the type relevant to microwave assisted chemistry applications; at 2.45 GHz for example, the wavelength is 122 mm in vacuum and 13 mm in water, which are length scales that are commonly encountered in processing systems.

Small parametric variations can cause large effects on modal field patterns; they cause some modes to become stronger while other modes become weaker. Hence, the electromagnetic field is very sensitive to variations in geometries and in the medium properties of objects and fluids. As the modal patterns stretch over the entire spatial domain that is involved with the electromagnetic field, local variations cause global effects. Therefore, the introduction of objects in the field as well as variations in geometries and in medium properties change the field pattern globally. This adds to the complexity of the analysis, because, in the context of chemical processing, there are many phenomena that are likely to affect the spatial distribution of the electromagnetic medium properties. Examples in this respect are: progressing reaction coordinates, limited manufacturing tolerances of glassware, and hydrodynamic variations of fluids. Although mode stirrers have been suggested to improve microwave field uniformity (Meredith 1998), effectively these stirrers only add to the complexity of the problem, because they introduce yet another varying parameter that affects the field.

Experimental and modeling results for a typical domestic microwave oven (Sharp R-2S57) are presented in Figure 10. This oven has a 285 mm by 290 mm by 190 mm applicator cavity. For an empty rectangular cavity it is possible to analytically calculate the free resonance modes (Pozar 2005); within a 30 MHz frequency band—typical for magnetron tubes—around a 2.45 GHz center frequency, this cavity supports six free resonance modes. The system is therefore a multimode system. Inserting a load does not change this; while this may cause some resonance modes to shift out of the generated frequency band, it will

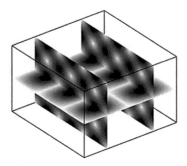 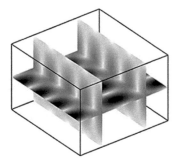

Figure 10. Modal patterns inside a 285 mm by 290 mm by 190 mm applicator cavity at a 2.449724 GHz resonance frequency (left) and at a 2.45032 GHz resonance frequency (right). The figures present slice plots of the electric field amplitude; darker zones concur with higher field intensity (Sturm et al. 2013, 2014).

also cause other resonance modes to shift into this spectral band. By means of modal analysis in COMSOL Multiphysics 3.5 with the RF module (COMSOL 2008), two spatial patterns of the distribution of the electric field strength were calculated for two free resonance modes that this cavity has around 2.45 GHz. Figure 10 presents these spatial field strength distributions; their spatial distributions differ considerably even though only a very small 0.6 MHz frequency variation would cause a shift from one to the other.

Further, the results show that there is an interference pattern of alternating high and low electromagnetic dissipation inside the water volume. Figure 11, a multiple-slice plot in a different projection to show the three-dimensional distribution, supplements this. This interference pattern does not even remotely follow an exponentially decaying trend as would be expected if in the analysis the microwave field would be performed using the ray (or optical) approximation. On occasions such modeling approximation has been proposed but more regularly, studies mention the existence of penetration depth limitations. In both these discussions the microwave field is described using the ray (or optical) approximation. This simulation

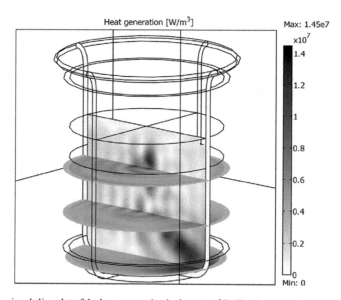

Figure 11. Three-dimensional slice plot of the heat generation in the water filled beaker heated in a multimode cavity described in Figure 10 at a frequency of 2.45 GHz. A complex three-dimensional interference pattern appears in the water volume. The spatial distribution of heat generation does not at all resemble an exponentially decaying trend that would be expected if one would predict the field intensity based on the penetration depth of a microwave field at this frequency in water (13 mm) (Sturm et al. 2013, 2014).

suggests that for this water filled beaker and for all systems similar to it—which essentially includes the whole field of microwave assisted chemistry—wave field physics would be more appropriate on this geometrical scale relative to the wavelength.

Properties of Microwave Heating

MW heating is fundamentally different from conventional heating methods, both in theory and in practice. Unique factors associated with microwaves make the process more rapid and energy-efficient than the corresponding conventional heating techniques. They also result in the potential for different reaction pathways to exist and can lead to structural and/or chemical differences in the reaction product.

Microwave heating may provide the following advantages in comparison to conventional heating for chemical synthesis (Bilecka and Niederberger 2010).

  i) high heating rates, thus increasing the reaction rates;
  ii) no direct contact between the heating source and the reactants and/or solvents;
  iii) excellent control of the reaction parameters, which is not only important with respect to the quality of the product, but also addresses a serious safety issue;
  iv) selective heating, if the reaction mixture contains compounds with different microwave absorbing properties;
  v) higher yields;
  vi) better selectivity due to reduced side reactions;
  vii) improved reproducibility; and
  viii) automatization and high throughput synthesis.

Direct heating

Microwave irradiation interacts directly with the reaction components which minimize the energy expended in heating furnaces, containment materials, and the sample environment (Sutton 1989). In theory, this requires at least one of the reactants to be capable of converting microwave energy into heat, but the efficiency of modern microwave applicator designs effectively allows almost all materials to be heated in this way. As a consequence of the direct heating of samples, the temperature profile of a microwave-heated material is the inverse of that seen in conventionally heated examples (Figure 3), resulting in the surface being cooler than the interior (Clark and Sutton 1996).

Volumetric heating

Direct heating of the reaction components also promotes volumetric heating (Metaxas and Meredith). In a homogeneous material, this means that heating is uniform throughout the sample. In most real samples, the thermal homogeneity is reduced by MW field and sample inhomogeneities and the temperature dependence of the interaction of the sample with a MW field. In contrast with conventional heating, volumetric heating results in a reduced requirement for heat transfer via thermal conduction within the sample, and this allows relatively large samples to be heated much more efficiently and with a much more uniform thermal history throughout the sample.

Instantaneous heating

The direct and volumetric nature of microwave heating results in a very fast transfer of microwave energy into heat, which can lead to extremely rapid temperature rises, far beyond that which can be achieved in a conventional furnace, which results in very rapid reaction times (Sutton 1989). Similarly, when the application of microwave power is terminated at the end of a reaction, heating stops immediately (Clark and Sutton 1996), which often results in the reaction essentially being quenched and can lead to metastable reaction products that are inaccessible using conventional methods.

Selective heating

The direct heating process allows specific reactants that interact more strongly with the microwave field to be heated selectively. As a result, microwave heating is uniquely capable of generating extremely high temperatures in specific regions of the sample while maintaining lower temperatures in others (Sutton 1989). This principle is widely used in the specific heating of active sites in supported metal catalysts (Zhang et al. 2003). The energy transfer at a molecular level can have additional advantages of selective handling of materials. The molecular structure affects the ability of the microwaves to interact with materials and transfer energy. When materials in contact have different dielectric properties, microwaves will selectively couple with the higher lossy material. This phenomenon of selective heating can be used for a number of purposes. In conventional joining of ceramics or polymers, considerable time and energy is wasted in heating up the interface by conduction through the substrates. With microwaves, the joint interface can be heated *in situ* by incorporating a higher lossy material at the interface. In multiple phase materials, some phases may couple more readily with microwaves. Thus, it may be possible to process materials with new or unique microstructures by selectively heating distinct phases. Microwaves may also be able to initiate chemical reactions not possible in conventional processing through selective heating of reactants. Thus, new materials may be created in this process (Venkatesh and Raghavan 2004).

The effect of microwave selective heating in organic synthesis was evaluated by Chemat and Esveld (1998). They conducted homogeneous and heterogeneous catalytic esterification process under direct and indirect exposure of microwaves to the samples. A continuous flow microwave reactor was developed to study the selective heating effect of microwave heating by using heterogeneous catalysts. Three different methods were considered as shown in Figure 12. The first configuration involves direct heating of the reaction mixture with catalyst under microwaves while the second configuration involved indirect heating of the heterogeneous catalyst and the third configuration included heating of the catalyst and reaction mixture by conventional heating. The first configuration increased the yield by 150% while the other two configurations produced similar yields. It was reported that the catalyst surface interacted with microwave field and produced hot spots at surface sites.

The kinetics for homogeneous catalytic reactions were equal to batch or conventional methods. This configuration has an advantage over other continuous microwave reactors since the heterogeneous catalyst remains contained in the microwave cavity. An increase of reaction rate of 50–150% for heterogeneous esterification was observed when they were conducted with microwave heating as compared to classical heating. It was proven crucial for the rate enhancement to have direct heating of the catalyst by microwaves. This may be because of the selective heating of the catalyst resulting in an increased reaction rate at the hot spot. The temperature of the reaction was calculated to be 9–18°C above the bulk temperature. The combination of a continuous reactor with selective heating of the catalyst promises to be an industrial alternative for reactions where the bulk temperature of the reactants is bound to certain limits.

Superheating Under Microwave Effect

Microwaves cause higher boiling points due to the superheating effect. Significant differences in boiling points were observed when compared under conventional and microwave heating methods. Table 5 presents a set of data for different solvents including relaxation time and boiling points under conventional and microwave heating conditions. The ability to operate at higher boiling points could be beneficial in many organic and inorganic chemical syntheses in which boiling is an undesirable phenomenon.

The boiling point of a liquid undergoing microwave irradiation was investigated. Several works have established that the enthalpy of vaporization is the same under both microwave and conventional heating (Roussy et al. 1986). These studies have also shown that the rate of evaporation, as well as the temperature of both vapor and liquid at the interface, strongly depends on experimental conditions and particularly on the absorbed microwave power. The hydrodynamic behavior of organic liquids during microwave evaporation stems directly from the field distribution and the thermal dependency of the dielectric loss of the concerned solvents (Stuerga et al. 1994). Studies of boiling phenomena related to microwave chemistry appeared only in 1992, with the pioneering paper of Baghurst and Mingos (1992). Microwave

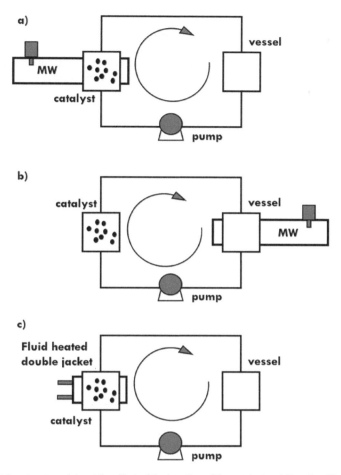

Figure 12. Experimental system to point out the effect of the location of the catalyst and the role of hot spots in microwave activation (Chemat and Esveld 1998).

Table 5. Boiling point elevations under microwave heating.

Material or solvent	Dielectric constant	Loss tangent	Relaxation time (ps)	Boiling point -normal conditions	MW conditions	Difference
Water	80.4	0.123 (2.45 GHz)	9.04	100	104	4
Methanol	32.6	0.659 (2.45 GHz)	51.5	65	84	19
Ethanol	24.3	0.941 (2.45 GHz)	170	79	103	24
Ethyl acetate	6.2	0.059	4.41	78	95	17
THF (tetrahydrofuran)	7.6	0.047	3.49	66	81	15
Methyl chloride	9.1	0.042				
Acetone	20.6	0.054	3.64	56	81	25
Acetonitrile	36.0	0.062	4.47	81	107	26
Dimethyl formamide	36.7	0.161	13.05	153	170	17
1-propanol		0.757	332			
2-propanol		0.799		82	100	18
Dichloromethane		0.042	3.12	40	55	15

heated liquids boil at temperatures above the equilibrium boiling point at atmospheric pressure. For many solvents, the super-boiling temperature under microwave irradiation can be 10°C to nearly 40°C above the boiling point under conventional heating.

Chemat and Esveld (2001) studied the super-boiling of the methanol under microwave and conventional heating methods. They evaluated the effect of microwave power and the solvent volumes. It was shown that the boiling point elevations under microwave heating were higher at higher microwave power applications. The sample volume has also shown positive correlation with the boiling point elevation. Table 6 shows the comparison between boiling point elevations for different alcohols, acids, alkanes, and ethers commonly used in organic synthesis under conventional and microwave heating methods. The number of carbon atoms seems to play an important role in the boiling point elevation.

Table 6. Super-boiling under microwave heating (Chemat and Esveld 2001).

	Compound	nr. of C	$BP_{convention}$ (°C)	BP_{mw} (°C)	ΔT (K)
Alcohol	1-methyl alcohol	1	65	79	14
	1-ethyl alcohol	2	78	89	11
	1-propyl alcohol	3	97	105	8
	1-butyl alcohol	4	117	122	5
	1-pentyl alcohol	5	137	143	6
	1-hexyl alcohol	6	156	173	17
	1-octyl alcohol	8	196	216	20
Acid	formic acid	1	100	110	10
	acetic acid	2	118	126	8
	propionic acid	3	141	147	6
	valeric acid	5	186	200	14
	hexanoic acid	6	202	218	16
Bromoalkane	1-bromoethane	2	37	40	3
	1-bromopropane	3	71	78	7
	1-bromobutane	4	103	115	12
	1-bromopentane	5	130	145	15
	1-bromoheptane	6	180	189	9
	1-bromooctane	8	201	208	7
Ether	di-ethyl	4	34	40	6
	di-propyl	6	68	75	7
	di-butyl	8	141	146	5
Nitrile	acetonitrile	1	81	86	5
	valeronitrile	5	140	150	10
	hexanenitrile	6	163	176	13
	heptanenitrile	8	200	206	6

Rate Enhancement by Microwaves

Thermal effects of microwaves

The rate of a reaction is determined by the Arrhenius equation ($k = Ae^{-Ea/RT}$), where T is the absolute temperature that controls the kinetics of the reaction. In conventionally heated reactions, this temperature is a bulk temperature (T_B). Microwave-assisted reactions are different (Hayes 2004). Microwave irradiation will directly activate most molecules that possess a dipole or are ionic (Lidstrom et al. 2001). Since energy transfer occurs in less than a nanosecond (10^{-9} s), the molecules are unable to completely relax (~ 10^{-5} s) or reach equilibrium (Hayes 2002). This creates a state of non-equilibrium that results in a high instantaneous temperature (T_i) of the molecules and is a function of microwave power input. The instantaneous temperature is not directly measurable, but it is much greater than the measured T_B ($T_i \gg T_B$). Thus, the greater is the intensity of microwave power supplied to a chemical reaction, the higher and more consistent T_i will be. The concept of instantaneous temperatures can be used to explain reactions occurring at a lower bulk temperature than expected, while using microwave irradiation (de Pomerai et al. 2003). The instantaneous temperature (T_i), not T_B, ultimately determines the kinetics of microwave reactions. After numerous studies performed over the past two decades, chemists have found that microwave-enhanced chemical reaction rates can be faster than those of conventional heating methods by as much as 1,000-fold (Lidstrom et al. 2001).

Assuming a standard first-order rate law (rate = k[A]), the Arrhenius rate equation can be used to calculate the instantaneous temperatures required to get three different reaction enhancements (10-, 100-, and 1,000-fold). The assumption was based on a desired reaction bulk temperature of 150°C and activation energy of 50 kcal/mol for the transformation. For a 10-fold rate increase, it can be shown that a temperature enhancement of only 17°C would be needed, relative to a bulk temperature of 150°C. Microwave energy can provide that temperature increase instantly. Likewise, for a 100-fold rate increase, the instantaneous temperature would have to reach 185°C, approximately a 35°C increase over the bulk temperature. A 1000-fold enhancement would need a 56°C increase over T_B. These instantaneous temperatures are very consistent with the temperatures that would be expected in a microwave system and are directly responsible for the enhancements in reaction rates and yields (Hayes 2004).

Illustrations

Example 1

Superheating by microwaves can have unique advantages. For instance Chemat and Esveld (2001) studied the reaction rate of esterification under both microwave and conventional heating. Theoretically, there is no limit to the microwave super-heating temperature at atmospheric pressure except for safety conditions. For example: methyl alcohol with a boiling point of 65°C has a superheated temperature of 79°C when 75 W of microwave power are used, and 108°C at 300 W.

Rate constants were determined at different temperatures, in the range 60°C to 110°C, for the esterification reaction of benzoic acid in excess of methyl alcohol catalyzed by Sulfuric acid (Figure 12). Using the Arrhenius plot, the activation energy was determined for esterification reaction by both conventional heating [E_a (conventional) = 45 ± 0.5 kJ mol^{-1}] and microwave heating [E_a (microwave) = 46 ± 0.5 kJ mol^{-1}]. It can be seen that the kinetics are similar under microwave or conventional heating, and that no non-thermal microwave effects were detected. However, a significant expansion of the thermal

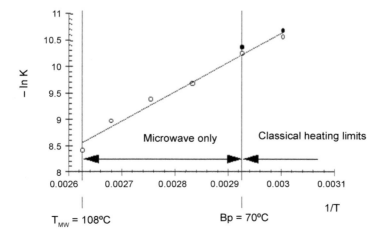

Figure 13. Kinetics of thermal cyclization (○) microwave heating; (•) classical heating (Chemat and Esveld 2001a).

region normally possible at atmospheric pressure was observed. Microwave super-heating allows for conducting reactions over boiling temperature without boiling. The esterification reaction was conducted at 40°C higher than the boiling point of the reaction mixture. Reaction times were reduced from days and/or hours under classical heating to minutes under microwave heating.

Example 2

In general, microwave processing under sealed-vessel conditions (taking advantage of increased reaction rates at higher temperatures) will be significantly more energy-efficient than conventional heating in open vessels at the solvent reflux temperature. This is clearly seen for the Diels–Alder reaction presented in Figure 14 (Razzaq and Kappe 2008). While the thermal reflux processes at 110°C (Diels–Alder in toluene, Table 7) require several hours to reach completion, the same transformations can be completed within less than 10 min by using sealed-vessel microwave heating at 200°C.

Although the faster reactions are a consequence of the higher reaction temperatures, sealed vessel microwave heating in dedicated instruments for a few minutes requires only a fraction of the energy used as compared to conventional heating in an oil bath or heating mantle for several hours. It is important to note that the energy savings in these cases are mainly the result of the significantly shortened reaction times and are not directly connected to the heating mode. This is made evident by comparing the consumed energy from open-vessel microwave heating experiments.

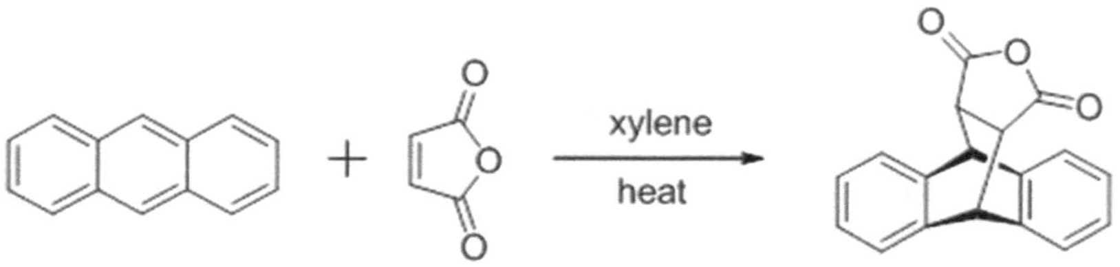

Figure 14. Cycloaddition between anthracene and maleic anhydride.

Table 7. Comparison of the energy consumed by microwave (MW) and oil-bath (Δ) heating for the cycloaddition between anthracene and maleic anhydride (Razzaq and Kappe 2008).

Entry	Heating method	Scale [mmol]	Solvent [V(mL)]	T_R [°C]	t_R [min][a]	Yield [%][b]	E [kWh][c]	$t_{\Delta total}$ [min][c]	E [kWhmol^{-1}]
1	oil bath	10	toluene (8)	111	180	93	0.165	200	17.74
2	oil bath	10	DCE (8)	83	300	85	0.274	315	32.23
3	MW[d]	3.7	toluene (3)	200	1	87	0.014	3	4.34
4	MW[d]	10	toluene (8)	200	2	90	0.015	4	1.67
5	MW[d]	3.7	DCE (3)	200	1	85	0.006	2.5	1.91

[a] Reaction time (t_R) is defined as the time period the reaction mixture is kept at the specified maximum reaction temperature ("fixed hold time" in the case of microwave heating).
[b] Isolated yield of pure product after filtration.
[c] Energy consumption as measured by the Wattmeter for the total heating time ($t_{\Delta total}$) specified. For oil-bath experiments, this includes the time required for heating the bath to the desired bath temperature. For microwave experiments, this includes a 30-s pre-stirring period at room temperature, the ramp time and reaction time at the specified temperature.
[d] Single-mode microwave reactor (Initiator Eight EXP 2.0, 400 W maximum magnetron output power).

Non-thermal (or extra-temperature) effects of microwaves

Kinetics observed under conventional heating are purely thermal whereas microwave heating combines both kinetic and thermodynamic heating patterns (Hayes 2004). It is often difficult to identify the individual effects, i.e., reaction rate enhancements, caused by thermal and non-thermal impacts especially for microwave mediated reactions. This is mainly because of the limitation of measuring the exact reaction temperature. Bulk temperature measured by using the fiber optics do not quite represent the actual temperatures responsible for the rate enhancements. These effects are sometimes referred as "extra-temperature" effects of microwaves.

Illustration

Rosana et al. (2014) proposed three different hypotheses for understanding the special effects caused by a microwave field in a chemical reaction. Figure 15 shows the thermodynamic parameters for benzylation of p-xylene reaction in which a 35% conversion after 5 min at an integrated average temperature of 99.8°C was observed.

Quantifying MW-specific thermal effects in terms of an imaginary impact on E_a may have some merit. If one takes the position that the measured solution temperature accurately reflects the thermal energy available to the substrate, then one can use the measured temperature and solve the Arrhenius equation for imaginary impacts of MW energy on either the effective activation energy or the pre-exponential term.

Option 1

It can be interpreted that MW-specific thermal effects disturb the thermal equilibrium between solute and solvent, such that the measured solution temperature does not accurately reflect the solute kinetic molecular energy. The effective temperature of the reactant was thus calculated to be T_{eff} = 393 K (120°C). However, one can alternatively quantify the observed MW rate enhancement in terms of impacts on the pre-exponential term (A) or the activation energy (E_a), which are all part of the same equation.

Option 2

If the measured temperature was used in the Arrhenius equation, the pre-exponential factor can be calculated as follows. The actual Arrhenius parameters for this reaction are E_a = 30.2 kcal/mol and A = 9.64 × 10^{13} s^{-1}. However, one can take the observed reaction conversion of 35% corresponding to a rate constant k = 14.4 × 10^{-4} s^{-1} and the integrated average (measured) temperature of 99.8°C and solve for "imaginary" or "effective" MW impacts on these parameters. Assuming that E_a is unchanged, the new pre-exponential under MW heating would be A = 1.70 × 10^{15} s^{-1}.

Option 3

Alternatively assuming an unaltered pre-exponential value, the imaginary impact on activation energy would be ΔE_a = −1.5 kcal/mol; the observed rate acceleration is as if the activation energy were lowered to E_a = 28.7 kcal/mol. One could also postulate partial impacts on both A and E_a simultaneously, or even spread across all three factors (A, E_a, and T), to rationalize the observed rate enhancement. Our current thinking is that the most reasonable explanation is selective MW heating of the solute to a higher effective temperature than the rest of the solution. However, these effects cannot be rigorously quantified in Arrhenius terms because the system is not thermally homogeneous, and Arrhenius calculations do not account for thermal gradients in the reaction mixture.

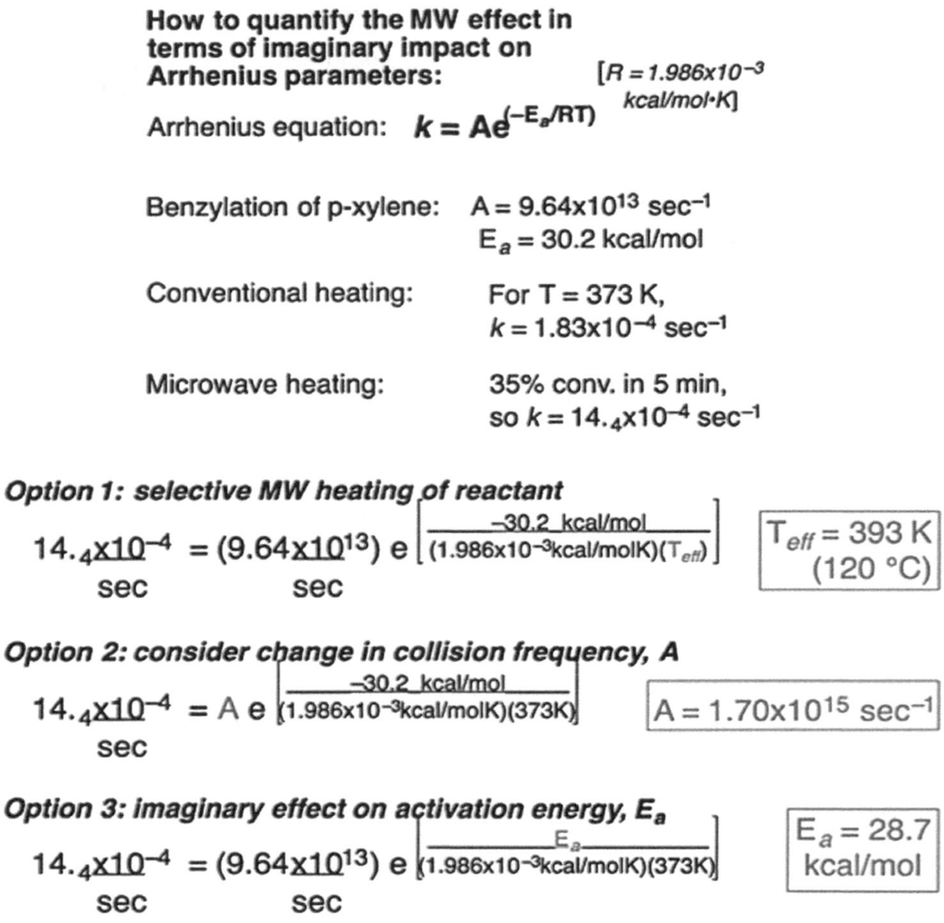

How to quantify the MW effect in terms of imaginary impact on Arrhenius parameters: $[R = 1.986 \times 10^{-3}$ kcal/mol·K]

Arrhenius equation: $k = A e^{(-E_a/RT)}$

Benzylation of p-xylene: $A = 9.64 \times 10^{13}$ sec^{-1}
$E_a = 30.2$ kcal/mol

Conventional heating: For T = 373 K, $k = 1.83 \times 10^{-4}$ sec^{-1}

Microwave heating: 35% conv. in 5 min, so $k = 14.4 \times 10^{-4}$ sec^{-1}

Option 1: selective MW heating of reactant

$$\frac{14.4 \times 10^{-4}}{\sec} = \frac{(9.64 \times 10^{13})}{\sec} e^{\left[\frac{-30.2 \text{ kcal/mol}}{(1.986 \times 10^{-3} \text{kcal/molK})(T_{eff})}\right]} \quad \boxed{T_{eff} = 393 \text{ K} \ (120\,°C)}$$

Option 2: consider change in collision frequency, A

$$\frac{14.4 \times 10^{-4}}{\sec} = A\, e^{\left[\frac{-30.2 \text{ kcal/mol}}{(1.986 \times 10^{-3} \text{kcal/molK})(373K)}\right]} \quad \boxed{A = 1.70 \times 10^{15} \text{ sec}^{-1}}$$

Option 3: imaginary effect on activation energy, E_a

$$\frac{14.4 \times 10^{-4}}{\sec} = \frac{(9.64 \times 10^{13})}{\sec} e^{\left[\frac{-E_a}{(1.986 \times 10^{-3} \text{kcal/molK})(373K)}\right]} \quad \boxed{E_a = 28.7 \text{ kcal/mol}}$$

Figure 15. Mathematical treatments of the Arrhenius equation to put observed MW-specific thermal effects in terms of Arrhenius parameters. The MW reaction rate constant (k = 14.4 × 10^{-4} s^{-1}) and integrated average temperature (T = 99.8°C) do not equate for the Arrhenius parameters of this reaction (A = 9.64 × 10^{13} s^{-1} and E_a = 30.2 kcal/mol).

Microwave Heating in Biodiesel Production

Modified domestic microwave ovens have been widely used for conducting many organic synthesis reactions. A typical microwave reactor with a reflux system is shown in Figure 16 (Zhang et al. 2012). As shown in Figure 17, FFAs conversion of conventional heating esterification (THE) is about 98.4% under the optimized conditions for 8 h reaction time; however, FFAs conversion of microwave assisted esterification (MAE) could reach 97.4% under the optimal conditions of reaction temperature 60°C, methanol/acidified oil mass ratio 2.0:1, 120°C annealed catalytic membrane loading 3 g, microwave power 360 W and reaction time 90 min (1.5 h). The reaction temperature, catalyst loading and methanol consumed in MAE are slightly lower and the reaction time is much lower than that in THE, which could lead to lower cost in biodiesel production. These results strongly demonstrates that the application of microwave condition could offer a fast, economical and easy route to produce biodiesel and MAE catalyzed by CERP/PES catalytic membrane is an appropriate option for biodiesel production.

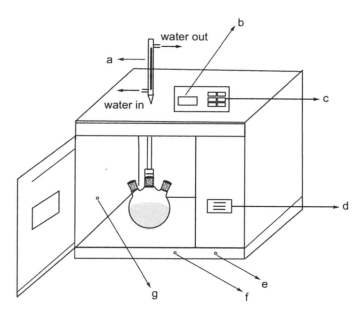

Figure 16. Schematic diagram of the microwave accelerated reaction system. (a) condenser pipe; (b) display panel; (c) control panel; (d) reaction chamber switch; (e) mains switch; (f) magnetic stirring knob; (g) infrared temperature measurement.

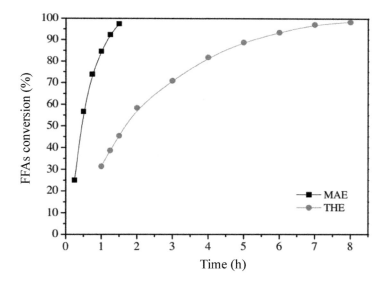

Figure 17. Comparison of conventional heating method and microwave assisted esterification catalyzed by CERP/PES membrane under optimal conditions respectively.

Compared with traditional heating esterification, microwave assisted esterification requires less reaction time, lower reaction temperature, less energy and lower methanol additive. Thus, manufacturing biodiesel using microwave represents a fast, easy and effective route with advantages of a short reaction time, a low methanol/acidified oil mass ratio, an ease of operation, reduced energy consumption and all with lower production cost (Zhang et al. 2012).

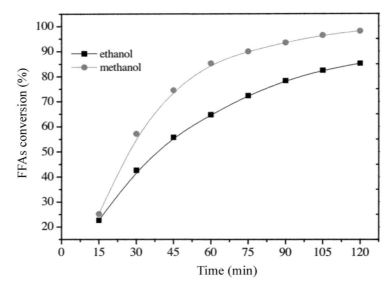

Figure 18. Effect of different alcohols on esterification. Reaction conditions: acidified oil 20 g, reaction temperature boiling point of each alcohol, alcohol/acidified oil molar ratio 23.5:1, microwave power 240 W and reaction time 90 min.

Alcohols used in esterification and transesterification are mainly methanol, ethanol, propanol and butanol. Methanol and ethanol are used most frequently in both laboratory research and the biodiesel industry. Methanol is more favorable due to its low cost while ethanol is derived from agricultural products (renewable sources) and is more environmental friendly than methanol. Thus, ethanol is the ideal candidate for the synthesis of any renewable fuel. To study the esterification of methanol and ethanol with acidified oil under microwave condition, different experiments were performed and the results are shown in Figure 18. It is observed that methanol has better FFAs conversion than ethanol. This is associated with the fact that methanol has stronger microwave absorption capacity and it can easily react with acidified oil under the microwave condition (Yuan et al. 2008).

Microwave Interactions with Biodiesel Feedstock

Biodiesel reactions are mostly performed at 2.45 GHz and occasionally at 0.915 GHz frequencies. It is critical to understand the dielectric properties of the reaction elements to optimize the microwave biodiesel production process. Optimization of microwave energy, frequency and application may have significant effects on process benefits. Muley and Boldor (2013) have reported on the dielectric properties of biodiesel precursors. They were measured using a vector network analyzer and a slim probe in an open ended coaxial probe method at four different temperatures (30, 45, 60 and 75°C) and in the frequency range of 280 MHz to 4.5 GHz. The dielectric properties of oil feedstock (soybean oil), ethanol, methanol, and their mixtures of sodium hydroxide catalyst are shown in Table 8 for microwave frequencies of 0.915 GHz and 2.45 GHz. It can be noted that the dielectric constant (ε') and the dielectric loss (ε'') of the different compounds vary in different patterns at different microwave frequencies and reaction temperatures. It can be observed from this table that the dielectric properties of ethanol and methanol were significantly affected by the presence of catalyst (Muley and Bolder 2013).

Reaction environment and preparation methods can have significant effects as well. Two different scenarios were considered to study the impact of preparation method on the dielectric properties under microwave irradiation during biodiesel production.

Table 8. Values for dielectric constant and dielectric loss of pure components and alcohol-catalyst mixture at 915 and 2450 MHz (Muley and Bolder 2013).

Component	Temperature (°C)	0.915 GHz			2.450 GHz		
		ε'	ε''	Tan δ	ε'	ε''	Tan δ
Soybean Oil	30	2.90 ± 0.01	0.23 ± 0.0001	0.079	2.86 ± 0.03	0.10 ± 0.003	0.035
	45	2.97 ± 0.01	0.28 ± 0.006	0.096	2.85 ± 0.02	0.14 ± 0.003	0.052
	60	2.99 ± 0.03	0.28 ± 0.007	0.094	2.95 ± 0.04	0.14 ± 0.007	0.049
	75	2.85 ± 0.21	0.21 ± 0.01	0.075	2.81 ± 0.05	0.11 ± 0.009	0.04
Ethanol	30	15.85 ± 0.18	9.54 ± 0.11	0.602	8.21 ± 0.11	6.83 ± 0.08	0.831
	45	17.68 ± 0.14	7.57 ± 0.25	0.428	10.02 ± 0.02	7.63 ± 0.03	0.762
	60	18.22 ± 0.04	5.06 ± 0.06	0.278	13.01 ± 0.05	7.32 ± 0.02	0.562
	75	18.35 ± 0.05	3.63 ± 0.07	0.198	14.98 ± 0.04	6.20 ± 0.08	0.413
Methanol	30	29.51 ± 0.14	6.55 ± 0.16	0.222	23.03 ± 0.04	11.56 ± 0.13	0.502
	45	28.34 ± 0.26	4.63 ± 0.22	0.163	24.33 ± 0.03	9.13 ± 0.02	0.375
	60	26.84 ± 0.39	3.30 ± 0.10	0.123	24.25 ± 0.04	6.95 ± 0.02	0.286
	75	-NA-	-NA-	-NA-	-NA-	-NA-	-NA-
Ethanol + NaOH	30	13.57 ± 0.28	11.04 ± 0.09	0.814	7.03 ± 0.08	6.95 ± 0.11	0.988
	45	15.76 ± 0.27	10.253 ± 0.25	0.650	8.97 ± 0.03	7.96 ± 0.27	0.888
	60	16.50 ± 0.01	8.41 ± 0.06	0509	11.39 ± 0.03	8.06 ± 0.02	0.707
	75	16.32 ± 0.08	6.92 ± 0.0048	0.424	13.04 ± 0.06	6.89 ± 0.05	0.528
Methanol + NaOH	30	23.34 ± 0.15	27.12 ± 0.32	1.162	17.99 ± 0.05	16.24 ± 0.002	0.903
	45	21.84 ± 0.03	30.42 ± 0.48	1.392	19.10 ± 0.12	16.32 ± 0.07	0.854
	60	20.80 ± 0.03	34.18 ± 0.21	1.643	19.22 ± 0.01	16.74 ± 0.02	0.871
	75	-NA-	-NA-	-NA-	-NA-	-NA-	-NA-

- In Case 1, oil and the alcohol-catalyst mixture were heated separately and were mixed only when the desired temperature was attained. The dielectric properties were measured immediately after adding oil and ensuring sample homogeneity via stirring.
- In Case 2, oil was mixed with the solvent-catalyst mixture at room temperature, followed by heating in a controlled temperature water bath at different temperatures with continuous stirring. Measurements were collected as the temperature reached each of the values prescribed above. This methodology provided the dielectric properties at different temperatures for a sample undergoing transesterification.

The two types of heating methods (while undergoing transesterification and prior to it) will help in understanding the microwave dielectric heating for the biodiesel precursor components and will be useful in designing and optimizing a microwave-based transesterification process.

Results (shown in Table 9 and Figure 19) indicate that the microwave dielectric properties depend significantly on both temperature and frequency. Addition of catalyst significantly affects the dielectric properties. Dielectric properties behave differently when oil, alcohol and catalyst were mixed at room temperature before heating and when the oil and the alcohol catalyst mixture were heated separately to a pre-determined temperature before mixing. These results can be used in designing microwave based transesterification system (Muley and Bolder 2013).

Case 1—Oil methanol catalyst heated separately

Dielectric constant for oil and methanol + catalyst mixture was relatively constant at lower frequency but decreased gradually as the frequency increased (Figure 19a) as well as with change in the temperature of methanol-oil mixture. This difference was apparent at 60°C as with the increasing temperature, the

Table 9. Values of dielectric constant and dielectric loss at 915 and 2450 MHz for oil alcohol catalyst mixture (Case 1 & 2) (Muley and Bolder 2013).

Component	Temperature (°C)	0.915 GHz		2.450 GHz	
		ε'	ε''	ε'	ε''
Oil+EtOH-NaOH (Case 1)	30	15.78 ± 0.13	11.55 ± 0.09	8.38 ± 0.07	7.65 ± 0.09
	45	12.23 ± 1.64	6.90 ± 1.56	6.98 ± 0.07	5.70 ± 0.1
	60	11.19 ± 0.01	3.38 ± 0.03	8.47 ± 0.04	3.93 ± 0.03
	75	8.13 ± 0.33	1.58 ± 0.18	7.94 ± 0.18	2.77 ± 0.02
Oil+EtOH-NaOH (Case 2)	30	9.13 ± 0.17	5.00 ± 0.23	5.72 ± 0.08	3.46 ± 0.01
	45	5.61 ± 0.09	1.97 ± 0.05	4.47 ± 0.06	1.75 ± 0.02
	60	5.88 ± 0.05	0.94 ± 0.01	5.25 ± 0.03	1.08 ± 0.01
	75	6.20 ± 0.007	0.72 ± 0.02	5.61 ± 0.02	0.85 ± 0.01
Oil+MeOH-NaOH (Case 1)	30	18.61 ± 0.08	21.03 ± 0.18	21.03 ± 0.05	12.38 ± 0.03
	45	19.84 ± 2.3	26.40 ± 4.32	26.40 ± 0.24	14.28 ± 0.24
	60	10.55 ± 0.56	19.43 ± 2.0	19.43 ± 0.75	9.43 ± 0.99
Oil+MeOH-NaOH (Case 2)	30	11.11 ± 0.16	12.05 ± 0.39	8.81 ± 0.17	6.75 ± 0.1
	45	15.64 ± 0.19	22.43 ± 0.91	13.34 ± 0.33	11.69 ± 0.41
	60	14.38 ± 0.09	22.75 ± 0.02	12.65 ± 0.9	10.97 ± 0.1

reaction proceeded towards completion. For oil-methanol mixture in Case 1, a steep decrease in the ε'' (dielectric loss) values from a very large value was observed at lower frequencies (until the frequency reached 1500–1600 MHz) after which the decrease was much less pronounced (Figure 19c). This behavior is observed for any given temperature. Similar trends were observed for methanol-catalyst solution but the overall values were lower with the addition of oil. Separate heating of oil and methanol-catalyst solution reduces the probability of the mixtures undergoing transesterification reaction. Nevertheless, at high temperature there could be a possibility of reactions taking place hence, a decrease in ε'' was noticed as the temperature was increased.

Case 2—Oil, methanol catalyst heated together

The dielectric constant changed with temperature for oil–alcohol + catalyst mixture (Figure 19b). This was true for ethanol–oil at all the temperatures and it also stands true for methanol–oil mixture except at 30°C. This exception could be because at 30°C, ion solvation takes place, reducing the ε' of the mixture. As the temperature rises to 45°C, the catalyst ions participate in the reaction reducing the effect of ion solvation; hence the rise in ε' is observed. These results can be compared with the results obtained by other researchers (Sorichetti 2011) for fatty acid methyl esters (FAME) before and after washing. They noted that as the temperature increases, more and more methanol reacts with the oil molecules; the product obtained is a non-polar product with lower ε'. Thus, the concentration polar impurity in the form of methanol reduces with temperature and the dielectric value decreases. It should be noted that the frequency range (400–10,000 Hz) at which the ε' values were measured were much lower for the cited work than in present study.

Dielectric loss for the oil–methanol-catalyst mixture decreased asymptotically with increasing frequency at all the temperatures (Figure 19d). The ε'' is lower at 30°C compared to 45 and 60°C (which were very close to each other). These results can be compared to literature data for FAME (Sorichetti 2011). For an unwashed FAME product, the ε'' decreased with an increase in temperature. Unwashed FAME product contained unreacted methanol which increases the polarity and ultimately the ε''. However, with an increase in temperature, more and more methanol undergoes reaction and the amount of unreacted methanol decreases, decreasing the ε''. When the product was washed and was free of any polar impurities, the ε'' increased with temperature as expected.

Microwave Interactions with Pyrolysis Feedstock

Another example can be discussed to understand the dielectric constants of various feedstock under high pressure and temperature conditions. Figure 20 shows the variations of dielectric properties with respect to two different microwave frequencies (Motasemi et al. 2014). It can be shown that the dielectric properties of switch grass are independent of the microwave frequency within the temperature range evaluated for drying and pyrolysis. During the drying stage, the dielectric properties decreased sharply from room temperature to 75°C and then remained roughly constant until ~ 200°C, reflecting the release of free and lightly bound water which are strong absorbers. In the pyrolysis region, the dielectric properties drop between 200°C

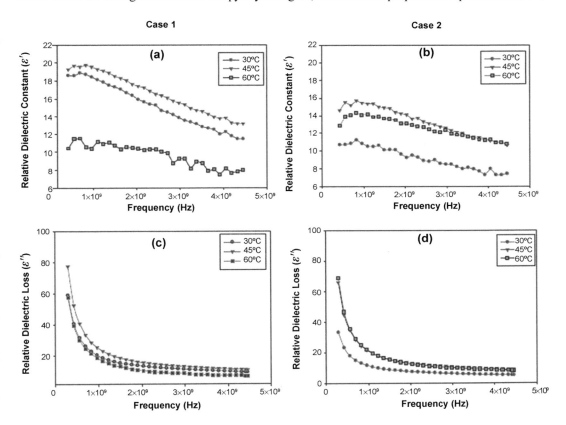

Figure 19. Dielectric properties of oil-methanol-catalyst mixture (a) Dielectric constant for Case 1, (b) Dielectric constant for Case 2, (c) Dielectric losses for Case 1, and (d) Dielectric loss for Case 2 (Muley and Bolder 2013).

and 350°C (reflecting the breaking of bonds and the release of volatile matter) after which the constant values up to 450°C indicate little reaction. The weight loss and the resulting densities may have reduced the polarizability of the feedstock.

The penetration depth also decreases dramatically as the temperature increases in the char region as shown in Figure 21. The low penetration depths above 550°C at both frequencies indicated the strong microwave absorption capability of the switchgrass char. The penetration depth was 0.5 cm and 0.3 cm at 915 and 2450 MHz, respectively at 689°C. Hence, the microwave penetration depth at this temperature for 2450 MHz frequency was less than the size of the switchgrass pellet which confirms that the microwaves

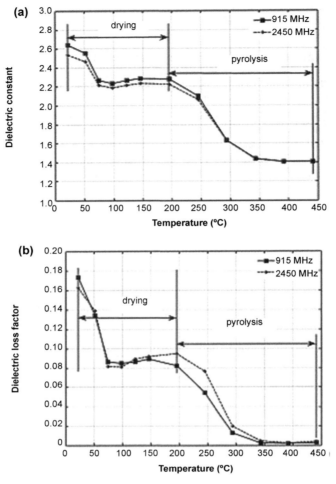

Figure 20. Dielectric properties of switchgrass vs. temperature under nitrogen environment at 915 MHz and 2450 MHz with initial density of 0.94 ± 0.050 g/cc; (a) relative dielectric constant, and (b) relative loss factor (Motasemi et al. 2014).

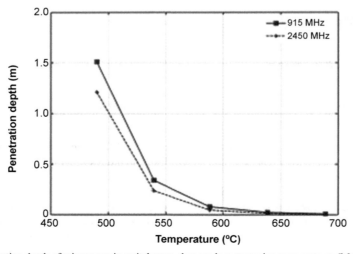

Figure 21. Penetration depth of microwave in switchgrass char product at varying temperatures (Motasemi et al. 2014).

are absorbed by the char material. The sudden change in the penetration depth at high temperatures might be associated with the complete loss of volatile matters, as confirmed by the weight loss in the sample. Furthermore, this will result in increase in the density of delocalized electrons in the char. This relates to the structure transformation of the biomass material during pyrolysis.

Salema et al. (2013) studied the dielectric properties of oil palm biomass (oil palm fiber and oil palm shell) and biochar at varying frequency in the range 0.2–10 GHz using a coaxial probe attached to a network analyzer. The dielectric constant was found to be inversely proportional to the frequency and the loss factor was directly proportional to the frequency. The dielectric properties of oil palm shell and its biochar were found to be almost similar and higher than oil palm fiber.

Methods of dielectric property measurement

There are various techniques available for measuring the dielectric properties which are shown in Table 10 (Venkatesh and Raghavan 2005, Tereshchenko et al. 2011, Rattanadecho and Makul 2016). These techniques include the transmission/reflection line, perturbation, free-space, and open-ended methods. Selection of the appropriate method depends on the operating frequency of the microwave system. For example, at the commonly utilized frequency of 2.45–0.05 GHz, the dielectric property should be measured by transmission line. This method allows for the measurement of specimens with a slab-shaped geometry to be placed across the cavity. However, this method is limited by the air-gap effect. The open-ended probe method could also be used, which involves measuring the phase and amplitude of the reflected signal at the end of an open-ended coaxial line inserted in a specimen. This method, too, has limitations due to the reflection effect from the specimen.

Choices of the measurement equipment and design of the sample holder depend on several factors, including the dielectric material to be characterized and the extent of information, as well as the equipment and resources that are available. Two types of dielectric analyzers have recently been used: the scalar network analyzer and the vector network analyzer. Vector network analyzers are quite popular, very versatile, and useful for extensive studies; however, they are also very expensive. Although relatively inexpensive, scalar network analyzers and impedance analyzers are generally still too costly for many applications. The HP 8510 can measure the magnitude and phase characteristics of linear networks, such as filters, amplifiers, attenuators, and antennae. Regarding all network analyzers, the HP 8510 apparatus measures the reflection and transmission of electromagnetic waves at an interface. An incident signal generated by a radio frequency source is compared with the signal transmission through the analyzer or reflected from the wave input when passing through the waveguide. However, this method is strongly affected by the air-gap content, with high air content tending to decrease the dielectric property of the material. It is also influenced by the consistency of dielectric materials. In particular, when materials behave like liquids, the settling of the material under its own weight can result in the non-uniform behavior of the whole specimen. Consequently, the measurement results are not representative of the whole body of the specimen. To perform the transmission/reflection-line technique, a sample is placed in a section of the waveguide or coaxial line, and the complex scattering parameters are measured at the two ports with a vector network analyzer. Note that the system must be calibrated before the measurements can be made. Both the reflected signal (S11) and the transmitted signal (S21) are measured. Various equations are used to relate the scattering parameters to the complex permittivity and permeability of the material. The S-parameters are converted to the complex dielectric parameters by solving the equations with the Nicholson-Ross-Weir (NRW) technique. An open-probe technique can be applied to measure the dielectric properties of samples before MWD processing. A portable dielectric measurement kit can be used to measure the complex permittivity values of a wide range of solid, semisolid, granular, and liquid materials. This system performs all of the necessary control functions, microwave signal treatments, calculations, and data-processing steps, and provides a representation of the results. The software controls the microwave reflectometer to measure the complex reflection coefficient of the tested material. It detects the cavity resonant frequency and quality factor, which it converts into the complex permittivity of the tested material. Finally, the measurement results are displayed in a graphical format or saved to disk (Rattanadecho and Makul 2016).

Table 10. Summary of techniques for measuring dielectric properties (Venkatesh and Raghavan 2005, Tereshchenko et al. 2011, Rattanadecho and Makul 2016).

Technique	Characteristics	Feature	Dielectric properties
Perturbation (Cavity) (Resonant)	The measurement is made by placing a sample completely through the center of a waveguide that has been made into a cavity. *Advantages*: Simple and easy data reduction; Accuracy and high temperature capability; Suited to low dielectric loss materials. *Disadvantages*: Need high frequency resolution vector network analysis (VNA); Limited to narrow band of frequency band only	Vector Network Analyzer with Source, Receiver, Processing unit, and Cavity containing MUT	ε_r, μ_r
Transmission/ reflection line	The material under test (MUT) must be made into slab geometry. *Advantages*: Common coaxial lines and waveguide; Used to determine both ε_r, μ_r of the MUT. *Disadvantages*: Limited by the air-gap effect; Limited to low accuracy	Vector Network Analyzer with Source, Receiver, Processing unit, MUT with Reflected and Transmitted signals	ε_r, μ_r
Open-ended probe	The technique calculates the dielectric properties from the phase and amplitude of the reflected signal at the end of an open-ended coaxial line inserted into a sample to be measured. *Advantages*: After calibration, can be routinely measured in a short time. *Disadvantages*: Available for reflection measurement	Vector Network Analyzer with Source, Receiver, Processing unit, Probe and MUT	ε_r
Free-space	The sample is placed between a transmitting antenna and a receiving antenna, and the attenuation and phase shift of the signal are measured. er, μr. *Advantages*: Used for high frequency and allow non-destructive measurement; Measure MUT in hostile environment. *Disadvantages*: Need large and flat MUT and diffraction effects at the edge of sample; Limited to low accuracy	Vector Network Analyzer with Source, Receiver, Processing unit, MUT and Sample holder	ε_r, μ_r

Concluding Remarks

This chapter discussed the characteristics of microwaves, dielectric properties of microwaves, microwave interaction with materials, and elucidated the special effects of microwave heating in various applications. The properties and thus the benefits of microwave heating depend on several factors. These factors were separated into microwave related such as frequency and material related such the conductivity, transparency and insulating properties of the materials. Several examples were discussed from the literature to emphasize the selective and special effects of microwave heating such as the dielectric properties of biodiesel components, and microwave pyrolysis of biomass. The illustrations discussed here suggest that microwave effects can be determined as a prior step to optimize a chemical process. This optimization process based on the understanding of special microwave effects and their interactions with materials will lead to better process design, energy efficiency and resource utilization.

References

Atwater, J.E. and Wheeler, R.R. 2003. Complex permittivities and dielectric relaxation of granular activated carbons at microwave frequencies between 0.2 and 26 GHz. *Carbon*. 41(9): pp.1801–1807.
Atwater, J.E. and Wheeler Jr, R.R. 2004a. Microwave permittivity and dielectric relaxation of a high surface area activated carbon. *Applied Physics A*. 79(1): pp.125–129.
Atwater, J.E. and Wheeler Jr, R.R. 2004b. Temperature dependent complex permittivities of graphitized carbon blacks at microwave frequencies between 0.2 and 26 GHz. *Journal of Materials Science*. 39(1): pp.151–157.
Baghurst, D.R. and Mingos, D.M.P. 1992. Superheating effects associated with microwave dielectric heating. *J. Chem. Soc. Chem. Commun.* (9): pp.674–677.
Bilecka, I. and Niederberger, M. 2010. Microwave chemistry for inorganic nanomaterials synthesis. *Nanoscale*. 2(8): pp.1358–1374.
Bogdal, D. 2005. *Microwave-Assisted Organic Synthesis: One Hundred Reaction Procedures* (Vol. 25). Elsevier.
Challa, S., Little, W.E. and Cha, C.Y. 1994. Measurement of the dielectric properties of char at 2.45 GHz. *Journal of Microwave Power and Electromagnetic Energy*. 29(3): pp.131–137.
Chemat, F. and Esveld, M.P. 1998. Activation of heterogeneous catalysis reactions using a continuous microwave reactor. *Journal of Microwave Power and Electromagnetic Energy*. 33(2).
Chemat, F. and Esveld, E. 2001a. Microwave assisted heterogeneous and homogeneous reactions. In: *Fifth International Electronic Conference on Synthetic Organic Chemistry (ECSOC-5)*.
Chemat, F. and Esveld, E. 2001b. Microwave super-heated boiling of organic liquids: Origin, effect and application. *Chemical Engineering and Technology*. 24(7): pp.735–744.
Chemat-Djenni, Z., Hamada, B. and Chemat, F. 2007. Atmospheric pressure microwave assisted heterogeneous catalytic reactions. *Molecules*. 12(7): pp.1399–1409.
Church, R.H. 1993. USBOM Report of Investigations. p.9194.
Clark, D.E. and Sutton, W.H. 1996. Microwave processing of materials. *Annual Review of Materials Science*. 26(1): pp.299–331.
Dallinger, D. and Kappe, C.O. 2007. Microwave-assisted synthesis in water as solvent. *Chemical Reviews*. 107(6): pp.2563–2591.
Dawson, E.A., Parkes, G.M.B., Barnes, P.A., Bond, G. and Mao, R. 2008. The generation of microwave-induced plasma in granular active carbons under fluidised bed conditions. *Carbon*. 46(2): pp.220–228.
de Pomerai, D.I., Smith, B., Dawe, A., North, K., Smith, T., Archer, D.B., Duce, I.R., Jones, D. and Candido, E.P.M. 2003. Microwave radiation can alter protein conformation without bulk heating. *FEBS Letters*. 543(1): pp.93–97.
Evalueserve, I.P. 2005. Developments in Microwave Chemistry.
Fang, Z., Li, C., Sun, J., Zhang, H. and Zhang, J. 2007. The electromagnetic characteristics of carbon foams. *Carbon*. 45(15): pp.2873–2879.
Gabriel, C., Gabriel, S., Grant, E.H., Halstead, B.S. and Mingos, D.M.P. 1998. Dielectric parameters relevant to microwave dielectric heating. *Chem. Soc. Rev.* 27: pp.213–223.
Groisman, Y. and Gedanken, A. 2008. Continuous flow, circulating microwave system and its application in nanoparticle fabrication and biodiesel synthesis. *The Journal of Physical Chemistry C*. 112(24): pp.8802–8808.
Hasted, J.B. 1973. Aqueous Dielectric. Chapman and Hall, London.
Hayes, B.L. 2002. Microwave synthesis: chemistry at the speed of light.
Hayes, B.L. 2004. Recent advances in microwave-assisted synthesis. *Aldrichimica Acta*. 37(2): pp.66–77.
Herrero, M.A., Kremsner, J.M. and Kappe, C.O. 2008. Nonthermal microwave effects revisited: on the importance of internal temperature monitoring and agitation in microwave chemistry. *The Journal of Organic Chemistry*. 73(1): pp.36–47.
Jacob, J., Chia, L.H.L. and Boey, F.Y.C. 1995. Thermal and non-thermal interaction of microwave radiation with materials. *Journal of Materials Science*. 30(21): pp.5321–5327.
Kappe, C.O. 2004. Controlled microwave heating in modern organic synthesis. *Angewandte Chemie International Edition*. 43(46): pp.6250–6284.
Kappe, C.O. and Dallinger, D. 2006. The impact of microwave synthesis on drug discovery. *Nature Reviews Drug Discovery*. 5(1): pp.51–63.
Kitchen, H.J., Vallance, S.R., Kennedy, J.L., Tapia-Ruiz, N., Carassiti, L., Harrison, A., Whittaker, A.G., Drysdale, T.D., Kingman, S.W. and Gregory, D.H. 2013. Modern microwave methods in solid-state inorganic materials chemistry: from fundamentals to manufacturing. *Chemical Reviews*. 114(2): pp.1170–1206.
Komarov, V., Wang, S. and Tang, J. 2005. Permittivity and measurements. *Encyclopedia of RF and Microwave Engineering*.
Li, Y. and Yang, W. 2008. Microwave synthesis of zeolite membranes: a review. *Journal of Membrane Science*. 316(1): pp.3–17.
Lidström, P., Tierney, J., Wathey, B. and Westman, J. 2001. Microwave assisted organic synthesis—a review. *Tetrahedron*. 57(45): pp.9225–9283.

Lin, H., Zhu, H., Guo, H. and Yu, L. 2008. Microwave-absorbing properties of Co-filled carbon nanotubes. *Materials Research Bulletin.* 43(10): pp.2697–2702.

Ma, J., Fang, M., Li, P., Zhu, B., Lu, X. and Lau, N.T. 1997. Microwave-assisted catalytic combustion of diesel soot. *Applied Catalysis A: General.* 159(1): pp.211–228.

Marland, S., Merchant, A. and Rowson, N. 2001. Dielectric properties of coal. *Fuel.* 80(13): pp.1839–1849.

Menéndez, J.A., Arenillas, A., Fidalgo, B., Fernández, Y., Zubizarreta, L., Calvo, E.G. and Bermúdez, J.M. 2010. Microwave heating processes involving carbon materials. *Fuel Processing Technology.* 91(1): pp.1–8.

Menéndez, J.A., Juárez-Pérez, E.J., Ruisánchez, E., Bermúdez, J.M. and Arenillas, A. 2011. Ball lightning plasma and plasma arc formation during the microwave heating of carbons. *Carbon.* 49(1): pp.346–349.

Meredith, R.J. 1998. *Engineers' Handbook of Industrial Microwave Heating* (No. 25). IET.

Metaxas, A.A. and Meredith, R.J. 1983. *Industrial microwave heating* (No. 4). IET.

Metaxas, R.J. 1993. Meredith. Industrial Microwave Heating, Peter Peregrinus Ltd.

Mingos, D.M.P. and Baghurst, D.R. 1991. Tilden Lecture. Applications of microwave dielectric heating effects to synthetic problems in chemistry. *Chem. Soc. Rev.* 20(1): pp.1–47.

Mingos, D.M.P. and Whittaker, A.G. 1997. Microwave dielectric heating effects in chemical synthesis. John Wiley and Sons, and Spectrum Akademischer Verlag: New York, Heidelberg. pp.479–514.

Motasemi, F., Afzal, M.T., Salema, A.A., Mouris, J. and Hutcheon, R.M. 2014. Microwave dielectric characterization of switch grass for bioenergy and biofuel. *Fuel.* 124: pp.151–157.

Muley, P.D. and Boldor, D. 2013. Investigation of microwave dielectric properties of biodiesel components. *Bioresource Technology.* 127: pp.165–174.

Nelson, S.O. and You, T.S. 1989. Microwave dielectric properties of corn and wheat kernels and soybeans. *Transactions of the ASAE.* 32(1): pp.242–0249.

Nelson, S.O. 1992. Correlating dielectric properties of solids and particulate samples through mixture relationships. *Transactions of the ASAE.* 35(2), pp.625–629.

Nüchter, M., Ondruschka, B., Bonrath, W. and Gum, A. 2004. Microwave assisted synthesis–a critical technology overview. *Green Chemistry.* 6(3): pp.128–141.

Peng, J.H., Binner, J. and Bradshaw, S. 2002. Microwave initiated self-propagating high temperature synthesis of materials. *Materials Science and Technology.* 18(12): pp.1419–1427.

Perreux, L. and Loupy, A. 2002. Nonthermal effects of microwaves in organic synthesis. *Microwaves in Organic Synthesis.* pp.61–114.

Pozar, D.M. 2005. Microwave Engineering, 3rd ed. New York: Wiley.

Rattanadecho, P. and Makul, N. 2016. Microwave-assisted drying: A review of the state-of-the-art. *Drying Technology.* 34(1): pp.1–38.

Razzaq, T. and Kappe, C.O. 2008. On the energy efficiency of microwave-assisted organic reactions. *ChemSusChem.* 1(1-2): pp.123–132.

Refaat, A.A. 2010. Different techniques for the production of biodiesel from waste vegetable oil. *International Journal of Environmental Science and Technology.* 7(1): pp.183–213.

Rosana, M.R., Hunt, J., Ferrari, A., Southworth, T.A., Tao, Y., Stiegman, A.E. and Dudley, G.B. 2014. Microwave-specific acceleration of a Friedel–Crafts reaction: evidence for selective heating in homogeneous solution. *The Journal of Organic Chemistry.* 79(16): pp.7437–7450.

Roussy, G., Thiebaut, J.M. and Colin, P. 1986. Latent heat of evaporation of a microwave irradiated polar liquid. *Thermochimica Acta.* 98: pp.57–62.

Salema, A.A., Yeow, Y.K., Ishaque, K., Ani, F.N., Afzal, M.T. and Hassan, A. 2013. Dielectric properties and microwave heating of oil palm biomass and biochar. *Industrial Crops and Products.* 50: pp.366–374.

Sorichetti, S.D.R.P.A. 2011. Dielectric Spectroscopy in Biodiesel Production and Characterization. Springer-Verlag London Limited.

Stefanidis, G.D., Munoz, A.N., Sturm, G.S. and Stankiewicz, A. 2014. A helicopter view of microwave application to chemical processes: reactions, separations, and equipment concepts. *Reviews in Chemical Engineering.* 30(3): pp.233–259.

Stuerga, D., Steichen-Sanfeld, A. and Lallemant, M. 1994. An orginal way to select and control hydrodynamic instabilities: microwave heating. III: linear stability analysis. *Journal of Microwave Power and Electromagnetic Energy.* 29(1): pp.3–19.

Stuerga, D. and Delmotte, M. 2002. Wave-material interactions, microwave technology and equipment. *Microwaves in organic synthesis.* pp.1–33.

Sturm, G.S., Verweij, M.D., Van Gerven, T., Stankiewicz, A.I. and Stefanidis, G.D. 2013. On the parametric sensitivity of heat generation by resonant microwave fields in process fluids. *International Journal of Heat and Mass Transfer.* 57(1): pp.375–388.

Sturm, G.S., Verweij, M.D., Stankiewicz, A.I. and Stefanidis, G.D. 2014. Microwaves and microreactors: Design challenges and remedies. *Chemical Engineering Journal.* 243: pp.147–158.

Sutton, W.H. 1989. Microwave processing of ceramic materials. *American Ceramic Society Bulletin.* 68(2): pp. 376–386.

Tereshchenko, O.V., Buesink, F.J.K. and Leferink, F.B.J. 2011. An overview of the techniques for measuring the dielectric properties of materials. *In: Proceedings of the General Assembly and Scientific Symposium, Istambul, Turquia* (Vol. 1320, p.14).

Thostenson, E.T. and Chou, T.W. 1999. Microwave processing: fundamentals and applications. *Composites Part A: Applied Science and Manufacturing*. 30(9): pp.1055–1071.

Varma, R.S. 2001. Solvent-free accelerated organic syntheses using microwaves. *Pure and Applied Chemistry*. 73(1): pp.193–198.

Venkatesh, M.S. and Raghavan, G.S.V. 2004. An overview of microwave processing and dielectric properties of agri-food materials. *Biosystems Engineering*. 88(1): pp.1–18.

Venkatesh, M.S. and Raghavan, G.S.V. 2005. An overview of dielectric properties measuring techniques. *Canadian Biosystems Engineering*. 47(7): pp.15–30.

Wu, K.H., Ting, T.H., Wang, G.P., Yang, C.C. and Tsai, C.W. 2008. Synthesis and microwave electromagnetic characteristics of bamboo charcoal/polyaniline composites in 2–40 GHz. *Synthetic Metals*. 158(17): pp.688–694.

Yang, J.K. and Wu, Y.M. 1987. Relation between dielectric property and desulphurization of coal by microwaves. *Fuel*. 66(12): pp.1745–1747.

Yao, Y., Jänis, A. and Klement, U. 2008. Characterization and dielectric properties of β-SiC nanofibres. *Journal of Materials Science*. 43(3): pp.1094–1101.

Yuan, H., Yang, B.L. and Zhu, G.L. 2008. Synthesis of biodiesel using microwave absorption catalysts. *Energy & Fuels*. 23(1): pp.548–552.

Zhang, L. and Zhu, H. 2009. Dielectric, magnetic, and microwave absorbing properties of multi-walled carbon nanotubes filled with Sm_2O_3 nanoparticles. *Materials Letters*. 63(2): pp.272–274.

Zhang, H., Ding, J. and Zhao, Z. 2012. Microwave assisted esterification of acidified oil from waste cooking oil by CERP/PES catalytic membrane for biodiesel production. *Bioresource Technology*. 123: pp.72–77.

Zhang, X., Lee, C.S.M., Mingos, D.M.P. and Hayward, D.O. 2003. Carbon dioxide reforming of methane with Pt catalysts using microwave dielectric heating. *Catalysis Letters*. 88(3-4): pp.129–139.

3

Microwave Reactor Design and Configurations

Microwave reactors are designed to function in single mode (or monomode) and multimode configurations. They are also designed to facilitate batch and continuous operations and processing of chemical reactions. This chapter will describe the essential parts of the microwave reactors, various configurations, design details, effect of wave guides, large scale applications, commercial and laboratory scale microwave reactor models and reliability and reproducibility issues in microwave reaction chemistry. In addition, the challenges and limitations in the microwave reactor design, processing and recommendations for future design and development are also presented.

Components of a Microwave Reactor

The major components of a microwave reactor consist of the power supply and microwave generator, the applicator, and the control circuitry. Typical microwave heating system consists of three basic components: (i) source (magnetrons, klystrons, gyrotrons and travelling wave tubes) to generate microwave power, (ii) waveguide components to transmit microwave energy into, (iii) a cavity which accommodates the load.

Microwave source

Microwave generator produces microwaves at appropriate frequency. The microwave source can be provided through numerous devices, most of which are vacuum tube based. Common generators include crossed-field amplifiers, gyrotrons, klystrodes, klystrons, magnetrons, power grid tubes and traveling wave tubes. They all have different characteristics with regard to power, frequency, gain, linearity, noise, phase and amplitude stability, coherence, size, weight and cost (Stein 1994, Metaxas and Meredith 1983). Magnetron is more commonly used due to its robustness, energy efficiency, and stability in frequency and ready availability. Klystrons are expensive and available in higher powers with longer operating life. Magnetrons vary in power supply ranging between 1 and 6 kW for laboratory applications to greater than 70 kW for high power industrial applications. High power applications use a frequency of 915 MHz while low power laboratory and small scale applications operate at a more common 2.45 GHz frequency. Klystrons have a somewhat longer operating life of 10000 h when compared with magnetrons which have an operating life of 80000 h.

The magnetron consists of a cylindrical cathode surrounded by a coaxial anode (shown in Figure 1). Between these, an electric field is created by an applied direct current (dc) potential. A longitudinal magnetic field is applied by an external magnet. Thus the electric and magnetic field are perpendicular to one another. Electrons, traveling from the anode to the cathode through this field, interact to cause a net energy transfer from the applied dc voltage to the microwave field. The cavity magnetron is the most efficient of the microwave tubes. Efficiencies of 90% have been achieved and 70–80% are common. The microwave frequency is determined by the size of the magnetron: the bigger the size; the lower will be the frequency. The microwave power is determined by the voltage of the direct current applied to magnetron. However cooling presents a practical limit for each magnetron. At 2450 MHz and 900 MHz, 20 kW and 100 kW, respectively, magnetrons are available, and 1000 kW magnetrons are available for even lower frequencies (Stein 1994, Metaxas and Meredith 1983). As magnetrons are very common, especially at 2450 MHz, they are relatively inexpensive. The frequency generated by a magnetron varies significantly compared to, e.g., a klystron. However this is of little significance for heating applications.

A typical magnetron unit is shown in Figure 2. Power output from low power permanent magnet magnetrons can be achieved by thyristor control of the anode voltage, while high power systems usually use an electromagnet to vary the anode current. It is usual to insert a circulator between the magnetron and the load. This is a 3-port structure that couples power clockwise between adjacent ports. As microwave power can be reflected, the circulator protects the magnetron from excessive reflected power by diverting reflected power to a water-cooled matched load, which may also be equipped with a power meter. The circulator is particularly important in preliminary investigations where the dielectric properties of the load are not well known and where a general purpose applicator is being used which may result in significant amounts of reflected power.

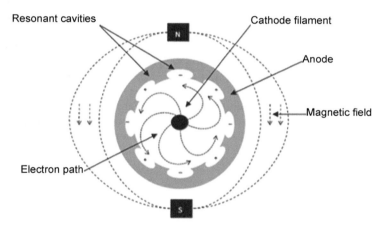

Figure 1. Schematic of a magnetron in microwave systems based on information from Driggers (2003).

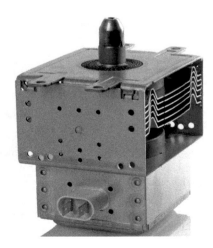

Figure 2. Photo image of a typical magnetron in small scale microwaves.

Transmission waveguides

The various microwave components are connected by waveguides (generally rectangular section brass) of appropriate dimensions for the operating frequency (86.36 × 43.18 mm at 2450 MHz). A stub tuner is often placed between the load and the magnetron to allow impedance matching. This device can readily be adjusted by hand to minimize reflected power, while automatic tuning is also possible, using stepper motors. A dual directional coupler allows measurement of forward and reflected power. The directional coupler is a passive microwave device that couples a fraction of the power transmitted in a given direction to a detector port equipped with a crystal diode detector. Nominal couplings of 30 and 60 dB are common. In the latter case, the ratio of measured to actual power is 10^{-6} (for a forward power of 1 kW, 1 mW would be measured at the detection port). Calibration of these devices requires considerable expertise and sophisticated equipment. Control circuitry will usually allow temperature control by power manipulation, and sometimes automatic impedance matching.

Transmission waveguides are rectangular hollow metal cylinders used to transmit microwaves in a controlled fashion as shown in Figure 3. These channels, being dimensioned according to the frequency of the microwaves passing through, allows transmission of power with almost no loss of power. Thus the microwave generator can be physically separated from the applicator. This has design advantages which improves flexibility and safety. The heat losses in the waveguide are mostly due to resistance losses in the metals. With its low losses, aluminum is the preferred choice of material. Other metals are also in use. Notably stainless steel is used in corrosive environments or under high temperatures. Due to border phenomena with the metal encasing as the microwaves travel through, the wavelength of the microwaves are increased. Microwaves of 2450 MHz increase from 122 mm to 174 mm. This should be taken into consideration when constructing waveguides to avoid flanges at whole multiples of the wavelength. For lower power systems, such as consumer (domestic) microwave ovens, the microwave generator and applicator is manufactured as one unit. The waveguide will then be very short, and integrated into the unit. Additional equipment such as multi-stub-tuners, circulators and terminating loads are adapted as part of the waveguide to prevent reflective power and ensure the safe operation of the magnetron on low and varying loads. Waveguides can also be used as specialized applicators. These topics, and a comprehensive coverage of the briefly mentioned topics can be found in (Meredith 1998).

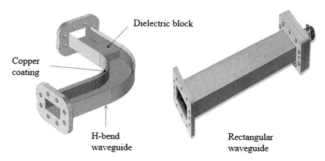

Figure 3. Examples of waveguides in microwaves reactors.

Chokes

Circular waveguides of diameter less than the cut-off diameter of a frequency have a high attenuation rate. This makes them useful as microwave chokes, preventing microwaves from passing through while not allowing other things like feedstock, stirrers or measuring equipments to pass through. The cut-off diameter is dependent on frequency and dielectric properties of the medium filling up the center of the waveguide. The cutoff diameter of air is approximately 150% of polytetrafluoroethylene (PTFE), twice that of quartz and thrice that of alumina. Thus the attenuation of the choke rate will decrease if a quartz glass rod is inserted into a previously empty choke. Since some generators, such as magnetrons, generate a broader spectrum of frequencies, the lowest frequency should be used for these considerations. As a rough approximation the waveguide diameter should be less than half of the microwave wavelength and the waveguide length should be longer than the microwave wavelength.

Microwave applicators or cavities

Microwave applicators are metallic enclosures that contain the material to be heated, and their design depends on the processing requirements. These are also called cavities which are generally classified into two categories: (a) single mode (mono-mode) and (b) multimode. Single mode microwave cavity generates a homogeneous energy field of high density around a small cavity. Single mode cavities are useful for processing small quantities of material, particularly those with low effective loss factors. On the other hand, multimode cavity is larger with reflective walls to prevent leakage of radiation and to increase the efficiency of the oven (Toukoniitty et al. 2005). Multimode cavity is basically a large box, with at least one dimension somewhat larger than the free space wavelength of the radiation (122 mm at 2450 MHz). The domestic microwave oven is an example of multimode cavity; it is an affordable device for preliminary laboratory scale investigations. Common domestic microwave ovens are unable to maintain a constant microwave power and temperature and therefore overheating may occur.

Microwave reactors are usually made of either aluminum or stainless steel. Aluminum is a better choice considering the costs and high electrical conductivity. This property helps reduce power loss especially when small loads are heated in large reactor cavities. Doors can be made with 1/4 wave chokes to contain microwave radiation, or more simply by ensuring good electrical contact around the entire sealing interface. This can easily be achieved by using copper braid as a gasket. Feed ports can be provided into the microwave cavity; generally holes of less than 10 mm diameter will act as chokes at 2.45 GHz and prevent leakage of radiation. Larger apertures can be provided but require electromagnetic design and possibly need to be provided with absorbing materials and water-cooling. Details on choking can be found in Metaxas and Meredith (1983).

Microwave radiation entering a multimode cavity undergoes multiple reflections to form a complex standing wave pattern, governed by the dimensions of the cavity and the nature of the load. The multimode cavity is versatile and suited to heating large loads and can be adapted for continuous processing. However, its convenience is offset by problems of poor electromagnetic uniformity and difficulties in modelling and design. Specialized features such as mode stirrers and slotted waveguide feeds can overcome the former (Kashyap and Wyslouzil 1977). The mode stirrer is the most common of these devices, consisting of a rotating vaned metallic fan. Rotation speeds are typically 1–10 rev/s (Metaxas and Meredith 1983). The domestic microwave oven is an example of the multimode cavity, and is often used as a general-purpose cavity for initial laboratory scale investigations. However, domestic ovens are not designed for high temperature operation, have no provision for temperature control and generally have extremely rudimentary power control (by variation of the duty cycle). Although they are useful for initial screening, unless used with care, seriously misleading results can result from such experimentation. Generally, the performance of a multimode cavity will improve as the filling factor increases. Conversely, a small load in a large cavity will result in a large Q-factor meaning that only a small fraction of the energy applied to the cavity is dissipated in the load.

Because materials heated in a microwave oven lose heat from the surface to the relatively cool interior of the oven, careful attention has to be given to insulation of the workpiece, especially for high temperature applications. In general the insulation material must be microwave transparent at the operating temperature while possessing a sufficiently low thermal conductivity. High alumina, zirconia and silica-alumina refractories are commonly used. If the insulation couples appreciably with microwaves, microwave power will be attenuated and efficiency will be reduced. The alternative is to place the insulation outside the cavity, but if the cavity is large relative to the size of load, this will also be thermally inefficient. Hybrid insulation systems using materials which couple with microwaves at low temperatures (susceptors) have been developed to insulate materials that are difficult to heat with microwaves at low temperatures (Janney et al. 1992). An alternative to the use of susceptors, and one that is much easier to control, is the use of an additional heat source, such as gas firing or resistive heating, to provide an increased ambient temperature. This approach has the added benefit that it is usually much more efficient than using either microwave or conventional heating alone.

Reactor Cavities

Single mode

Maxwell equation can be used to describe the response of microwaves with appropriate boundary conditions for geometry (Thostenson and Chou 1999). Single mode applicators are designed based on solution of the Maxwell equations to support one resonant mode. Thus, the size of single mode applicators is of the order of approximately one wavelength, and to maintain the resonant mode, these cavities require a microwave source that has little variation in the frequency output. Because the electromagnetic field can be determined using analytical or numerical techniques, the areas of high and low electromagnetic field can be estimated, and single mode applicators have non-uniform, but predictable, electromagnetic field distributions. In general, single mode cavities have one "hot spot" where the microwave field strength is high. This ability to design an applicator where the locations of high and low field strengths are known can offer some distinct advantages. Through proper design, single mode applicators can be used to focus the microwave field at a given location. This technique has been exploited in specific applications such as joining ceramics (Palaith et al. 1988). An additional advantage of single mode cavities is the ability to monitor the dielectric properties during processing (Wei et al. 1994). Figure 4 shows the schematic representation of a single mode microwave reactor.

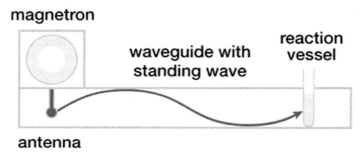

Figure 4. Schematic representation of a single mode microwave reactor (anton-paar.com).

Multimode

Unlike the design of single mode applicators, which are designed based on solutions of the electromagnetic field equations for a given applicator geometry, the design of multimode applicators are often based on trial and error, experience, and intuition (Thostenson, Chou 1999). As the size of the microwave cavity increases, the number of possible resonant modes also increase. Consequently, multimode applicators are usually much larger than one wavelength. The presence of different modes results in multiple hot spots within the microwave cavity. Like single mode cavities, local fluctuations in the electromagnetic field result in localized overheating. To reduce the effect of hot spots, several techniques are used to improve the field uniformity. The uniformity of the microwave field can be improved by increasing the size of the cavity. Because the number of modes within a multimode applicator increases rapidly as the dimensions of the cavity increase, the heating patterns associated with the different resonance modes begin to overlap. The rule of thumb to achieve uniformity within an applicator is to have the longest dimension be 100 times greater than the wavelength of the operating frequency (Kimrey and Janney 1988). Since this is not possible, inclusion of a turntable improves the heating patterns. The purpose of the turntable is to reduce the effect of multiple hot spots by passing the food through areas of high and low power and, therefore, achieve time-averaged uniformity. Another technique for improving the field uniformity is through mode stirring. Mode stirrers are reflectors, which resemble fans that rotate within the cavity near the waveguide input. The mode stirrers "mix up" the modes by reflecting waves off the irregularly shaped blades and continuously redistribute the electromagnetic field. Like turntables, mode stirring creates time-averaged uniformity. In addition, adding multiple microwave inputs within a multimode cavity can further enhance the uniformity. Most techniques for creating uniformity depend on modifying the electromagnetic field within the microwave cavity. Another method developed to achieve more uniform heating is hybrid heating. Hybrid heating can be achieved through combining microwave heating with conventional heat transfer through radiation, convection, or conduction. Figure 5 shows the schematics of multimode microwave reactors suitable for parallel synthesis and single-batch reactor (top view).

The variable frequency microwave furnace could be another solution to overcome the problems of power non-uniformity within multimode cavities. The system makes use of the travelling wave tubes source to sweep the frequency of the microwave field. The cavity used in this type of system is a multimode cavity. The result is time-averaged power uniformity within the microwave cavity. The ability to excite many different resonant modes by sweeping the frequency allows for uniform heating in a small cavity.

The distinctive features of single mode and multimode microwave reactors are presented in Table 1.

Microwave Reactor Considerations

Microwave ovens operating at 2450 MHz are common appliances in the households of USA and around the world. Hundreds of 2450 and 915 MHz systems between 10 to 200 kW heating capacities are used in the food industry for precooking bacons, tempering deep frozen meats when making meat patties, and precooking many other food products. When evaluating an extraction process it is important to consider the various factors affecting it during a scale up to commercial operations. In microwave processing this usually means a change in frequency from 2450 MHz to 915 MHz. Microwaves at 915 MHz (used industrially) have much higher penetration depths into the material as compared to the higher frequency of

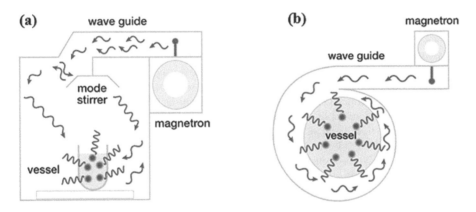

Figure 5. Schematic representation of a multimode microwave reactor: (a) microwave field distribution in a common multimode parallel synthesis reactor; and (b) microwave field distribution in a multimode single-batch reactor (anton-paar.com).

Table 1. Distinct features of single and multimode microwave reactors (Stein 1994).

Single mode	Multimode
High fields are possible	Suitability for bulk processing applications
The applicators can operate in the standing or traveling wave configurations	Oven dimensions that are often determined by product dimensions
Fields are well defined	Moderate to high efficiency
Fields can be matched to product geometry	Adaptability to batch or continuous product flow
The applicators are useful for heating both low-loss and high-loss materials	Performance that is less sensitive to product position or geometry
The applicators are compatible with continuous product flow	Good uniformity that may require motion of product or hybrid heating
High efficiency is possible	

2450 MHz, commonly used in laboratory sized equipment. The higher penetration depths allow for much larger diameter tubes and processing flow rates, and microwave generators can be built for significantly higher power and efficiencies when compared to smaller generators.

Proper application of microwave energy may result in greater benefits in terms of energy efficiency and reaction product quality. Understanding the characteristics of the reactants and nature of the desired reactions is critical in many applications. In certain polymerization reactions where the reaction temperatures change with nature of the reactions (endothermic versus exothermic), a better control of the microwave power dissipation is desired. In these applications, pulsed type microwave heating rather than continuous heating might result in improved energy efficiencies without affecting the quality of the reaction products due to too high or too low reaction temperatures (Bogdal 2005). In some applications, the organic chemicals under study may not have the capacity to absorb the microwave energy. In such cases, it is beneficial to introduce materials that have strong microwave absorption capability. This helps initiate the desired chemical reactions using organic chemicals. Here, the materials introduced, whether it is a solvent or metal particle acts both as a chemical catalyst as well as an energy converter. In addition, by using a proper microwave pulse train, it is further possible to control the desired selectivity in the products.

Microwave Reactor Limitations

As discussed earlier, microwave reactor design requires consideration of limitations related to penetration depths, hot spot formation, thermal runaway, scale-up and safety concerns.

Penetration depth limitation

One of the main limitations of the microwave technology reported by many experts is its inability to penetrate through large sample volumes. This limitation challenges the scalability of microwave applications from laboratory small-scale synthesis (milli-molar level) to industrial multi-kilogram production (kilo-molar level). For this reason, the replacement of conventional processes by microwave has several limitations. Measurement and control of temperature are difficult and temperature distribution is non-uniform in large batch reactors, it may indeed simulate thermal currents similar to conventional heating. Microwaves generally have a few centimeters depth of penetration capacity into the absorbing materials depending on their dielectric properties. As such, in large batch type reactors, the microwave power density varies greatly from outside surface to inside sample material. Therefore, materials in the center of the reaction vessel are heated only by convection and not by microwave dielectric heating. When trying to heat large quantities of materials, additional problems arise. As the volume of the mixture increases, the energy required for heating it also increases and higher radiation intensity is needed. Safety of the pressurized vessel with large quantities of batch operation needs to be considered as well.

The dissipation factor or penetration capacity of the microwave radiation depends on factors such as ion concentration, size of the ions, the dielectric constant and viscosity of the reacting medium and the microwave frequency. The dissipation factor of water and most organic solvents decreases with increasing temperature, i.e., the absorption of microwave radiation in water decreases at higher temperatures. In turn, the penetration depth of microwaves increases (Nüchter et al. 2004). For some heterogeneous reactions, the microwaves may not be able to penetrate through large sample volumes.

Hot spots (local heating)

An important characteristic of microwave heating is the phenomenon of 'hot spot' formation, whereby regions of very high temperature form due to non-uniform heating (Hill and Marchant 1996). This thermal instability arises because of the non-linear dependence of the electromagnetic and thermal properties of the material on temperature (Reimbert et al. 1996). The formation of standing waves within the microwave cavity results in some regions being exposed to higher energy than others. This results in an increased rate of heating in these higher energy areas due to the non-linear dependence. Cavity design is an important factor in the control, or the utilization of this hot spot phenomenon. Considering high production flow rates, it is beneficial to design the reactor in a fashion that simulates the plug flow reactor. In this case, the sample volume exposed to microwave field can be sized to the power dissipation capacity of microwave heat source. Plug flow reactors or small quantities of batch reactions in a continuous chain type operation mode can be designed to enhance the utilization of microwave energy (Kappe et al. 2012, Boldor et al. 2008).

Super heating has been the cause for the exploding reactions in laboratory scale reactors. Many microwave absorbing solvents can achieve temperatures well in excess of their boiling point during microwave irradiation, which can result in sudden, rapid and significant increase in vapor pressure. As indicated previously, the loss tangent is a measure of how well a substance absorbs microwave energy. The relaxation time, t, is defined as the time it takes one molecule to return to 36.8% (1/ε) of its original state when the electric field is switched off. The relaxation time is temperature dependent and decreases with increasing temperature. For microwave frequency of 2.45 GHz, if an organic solvent has a relaxation time of 0.65 picoseconds, then the loss tangent will increase with temperature and the majority of solvents fall within this category. Under microwave irradiation, as the temperature increases the solvent converts more microwave energy into thermal heat energy which increases the rate of temperature rise and self-heating occurs which can lead to thermal runaway. For example water and methanol have a relaxation time, 65 picoseconds and would not be expected to lead to thermal runaway with continued irradiation but ethanol and higher alcohols, which have t values 0.65 picoseconds would. Although this is not a chemical reaction hazard, appropriate choice of solvent, reactor configuration and reaction conditions are important to avoid high pressures as a result of solvent super-heating or thermal runaway. For a continuous reactor, a knowledge of the heat generated by microwave absorbance at the reaction temperature is necessary (along with the reaction power output) to determine the reactor cooling requirements.

Thermal runaway and voltage breakdown

Thermal runaway occurs when the rate of heat generation in the material exceeds its rate of heat dissipation which results in localized heating and subsequently voltage breakdown. These factors become very important for high throughput pilot scale reactors and are the primary safety considerations during a scale-up operation. Numerous experimental reports of thermal runaway phenomena have been published (Wu et al. 2002). When input power level is below a critical input power, the steady state temperature of the material increases continuously with increasing input power (Wu et al. 2002). Yet the temperature of the processed material suddenly rises to a very high value as the input power level exceeds the critical input power. The temperature corresponding to the critical input power is called the critical temperature. The temperature of the material cannot be stabilized between the critical temperature and some higher value unless some control mechanisms are introduced. Thermal runaway may be explained by the energy balance equation:

$$\rho C_p \frac{dT}{dt} = q_{abs} - q_{loss} \qquad [1]$$

Where ρ, C_p are the density and specific heat of the material, respectively, and q_{loss} is the rate of heat loss, which normally includes losses due to convection and radiation. Both q_{loss} and q_{abs} are functions of temperature. If when starting from some critical temperature, q_{abs} continues to exceed q_{loss}, thermal runaway will occur. Since q_{loss} increases with T, the increase of ε'' with increasing T provides a possible means for thermal runaway. If the required processing temperature exceeds the maximum stable temperature, a control system must be employed, or the temperature will continue increasing until the material is damaged, which occurs when sintering ceramic powder. In early experiments, when the desired temperature was reached, the power source was adjusted manually. These attempts sometimes led to success but, when materials are in a thermal runaway state, the temperature rise rate is rapid and temperature is highly sensitive to the incident power, making control highly problematic.

If the reactants have a low loss tangent, high external electric field stress E_{ext} is applied to convert it into AC. E_i is the internal electric field stress generated when a dielectric material is heated in a MW field. The power dissipation density p can be calculated by the internal electric field stress by $p = 2\pi f \varepsilon_o \varepsilon'' E_i^2$, where f is MW frequency, ε_o is the permittivity of free space, and ε'' is the loss factor.

Voltage breakdown occurs if any of the internal or external electric field stress reaches a critical value which results in ionization of the gases present. Ionized gases form an arc, establishing a low resistance electric path within the reactor which results in high power loss. As a result the sample can be burnt or denatured, and the equipment may undergo a meltdown crisis. Moreover, breakdown voltage is likely to happen at low pressure or at high temperature, leading to low density or low pressure of the gas inside the reactor. Therefore, care must be taken while using a low loss tangent reactant (requiring high external field stress) and experimenting at elevated temperatures (causing lower density in the reactor) as the conditions are conducive to arching.

Microwave safety

Although microwave radiation is a non-ionizing radiation, it still poses a safety hazard. It is important to ensure integrity of all microwave joints and doors. Microwave leakage detectors are readily available for this purpose. A radiation flux of 5–10 mW/cm^2 at a distance 50 mm from the equipment is a generally accepted limit. Doors should preferably be fitted with interlocks and water flow monitors should also be installed. When choking systems requiring resistive choking are used, it is essential that safety cut-outs are provided. Compatibility with other communication systems is important, particularly as cellular telephones operate close to the 915 MHz band (Bradshaw et al. 1998).

For all electric radiation of frequency f, the energy E of its individual photons is given by $E = hf$ where h is Planck's constant. As the frequency, and consequently photon energy, is less than visible light, microwaves are non-ionizing radiation (Matthes et al. 1999). Concerns over possible non-thermal effects of microwaves has sparked some debate, but no convincing occurrences of such effects have been observed. The human nervous system does not sense temperature changes inside the body as easily as on the skin surface. Thus a subject could be exposed to deep penetrating radiation, such as microwaves, for a relatively longer time than the equivalent dose of surface absorbed radiation, such as infrared, before noticing. Tissue temperature significantly in excess of body temperature is harmful. The temperature increase in the tissue is dependent on the heat dissipation within the body. Consequently, organs with poor circulation, such as the eye, is especially susceptible to harm (Metaxas and Meredith 1983). The generally accepted limit on safe stray microwave leakage is 100 W/m^2 measured at 50 mm from the reactor. These limits are significantly less strict than those for interference with telecommunication, and are generally easily satisfied (Meredith 1998).

Microwave Reactor Configurations

Most instrument companies offer a variety of diverse reactor platforms with different degrees of sophistication with respect to automation, database capabilities, safety features, temperature and pressure monitoring, and vessel design. Importantly, single mode reactors processing comparatively small volumes also have a built-in cooling feature that allows for rapid cooling of the reaction mixture with compressed air after completion of the irradiation period. The dedicated single mode instruments available today can process volumes ranging from 0.2 to about 50 mL under sealed vessel conditions (250°C, ca. 20 bar), and somewhat higher volumes (ca. 150 mL) under open-vessel reflux conditions. In large multimode instruments, several liters can be processed under both open- and closed-vessel conditions. Continuous-flow reactors are currently available for both single- and multimode cavities that allow the preparation of kilograms of materials by using microwave technology.

For extraction of oils, two basic designs of microwave reactors are available. The first one is a scientific/industrial/laboratory level multimode cavity which in principle is similar to the domestic microwave unit. In multimode configuration, microwaves reflect of the walls and generate a standing wave pattern in which waves intersect at specific points in the cavity. The second design involves a waveguide in which microwaves are focused and reflected at a specific location (Boldor et al. 2008, Glasnov and Kappe 2007). In comparison, the design with a single mode applicator (as appeared to multimode commonly used in household microwaves) focuses the microwaves in the center of the applicator, where the material flows in a processing tube. This resonance mode allows for very high electric field values which increase the heating rate. This focusing creates an electrical field distribution with the highest values in the center of the applicator tube which decreases as it nears the walls of the tube. Therefore, if the flow in the tube is laminar, the fluid with highest velocity in the center receives the highest amount of microwave energy. The fluid with the lowest velocity near the wall receives lower amounts of energy, therefore creating a more uniform temperature distribution when exiting the microwave applicator (Baxendale et al. 2007, Salvi et al. 2010). While this difference in electric field distribution may not play a significant role in small diameter tubes, when scaling up to higher flow rates and consequently larger diameter tubes, temperature uniformity becomes more important. Continuous processes using a 5 kW, 915 MHz microwave have been successfully applied so far for beverage and vegetable purees' sterilizations, for aseptic processing (Coronel et al. 2008) and for ballast water treatments (Boldor et al. 2008, Terigar 2009).

Batch operations are energy-efficient when conducted at micromole or millimole level scale. Large volumes of batch reactors may not be energy efficient with microwaves and a continuous flow process is more favorable. In a continuous flow, the mixture is continuously pumped and heated in a microwave cavity. This process is more complex due to the addition of momentum transfer to the heat generation from microwave heat transfer in the solvent/solid matrix, mass transfer through the solid/solvent (Bhattacharya and Basak 2006). The drawbacks of a continuous-flow microwave apparatus are that it can be difficult to process solids, highly viscous liquids, or heterogeneous reaction mixtures. In addition, adaptation of conditions from simple small scale reactions to the continuous-flow cell could end up being time-consuming (Bowman et al. 2007). Continuous flow systems will allow for large-scale production with reduced costs. A continuous flow system was tested by Groisman for transesterification of canola and sunflower oils. Very high FAMEs yield of 92% and 89% were obtained for both the oils respectively. For comparison, a batch reaction with a volume of 500 mL of oil was conducted and this test resulted in 64% yield. Again, the batch test with a one-tenth volume (50 mL) of the oil resulted in 97% yield. This confirmed the inability of the microwaves to influence large scale reactions (Groisman and Gedanken 2008).

Microwave based continuous flow biodiesel production has not been developed to date. Many other industries, including food, rubber, ceramics, and mining, successfully use microwave heating on a large scale, often using different frequencies that can increase penetration depths through solvents and solids. The option to change the frequency may provide an alternative solution to the problem of microwave scale-up. However, these applications do not require the controlled heating of delicate organic molecules while suspended in relatively low boiling and often flammable solvents (the context for pharmaceutical chemistry) (Moseley and Woodman 2007, Moseley et al. 2007).

Types of Reactors and Configurations

Microwave heating technology has some inherent physical limitations (e.g., magnetron power, penetration depth) and two different approaches for microwave-assisted synthesis on a larger scale (~ 100-mL volume) have emerged. Several reactor configurations have been developed over the past two decades to address these issues. Whereas companies like Anton Paar8 (Graz, Austria) have developed a batch-type multimode reactor, others like Milestone (Sorisole, Italy) are offering continuous flow systems. CEM10 (Matthews, USA) in contrast tried to overcome the scale-up problems by developing a stop-flow concept, which can be regarded as a combination of both. The Voyager SF is a monomode reactor with an 80 mL glass vessel, which is automatically filled and drained by a pump system. Reactions are carried out on a 50-mL scale sequentially and fully automated. Therefore, it was of great interest to us to evaluate the potential of commercially available microwave reactors on a large scale (with at least 300–500 mL reaction volume), to speed up the processing time for reactions that require long reaction time under conventional conditions (Lehmann and LaVecchia 2005).

Microwave vendors

Microwave reactors are manufactured and marketed by Anton Paar, Biotage, CEM, Milestone and few other vendors. Different types of reactors used in microwave based organic chemical synthesis are as follows:

Anton Paar (synthos 3000, with XQ80); Biotage (Advancer); CEM (voyager, MARS-open, MARS-a/c); Milestone (flow SYNTH, microSYNTH-open, microSYNTH-a/c, autoclave). The functional, operational and process parametric data are provided in Tables 2a and 2b. These tables were reproduced from the work presented by Moseley et al. (2007), and Bowman et al. (2007). For more information on these reactor specifications and performances, readers are referred to these references (Bowman et al. 2007, Moseley et al. 2007). Bowman and co-researchers have worked on different types of reactors in batch and continuous flow based configurations with the objective of identifying the concerns for scale-up needs as well as to compare their performances. For batch technologies, they found that open reactor vessels offer operational advantages while still giving good yields of desired products. In cases where volatile or toxic reagents are used, closed vessel reactors are better.

CEM microwave reactors

CEM manufactures both batch and continuous flow reactors suitable for various chemical syntheses. Discover™ and ExplorerPLS systems are both batch type reactors (Figures 6a and 6b). Discover can handle reaction vessels from 5 to 125 mL in size with working volumes as low as 0.25 mL with ambient and elevated pressure reaction environments. The Explorer system allows multiple reactions to be run unattended within a single cavity. The Navigator™ System provides for high-throughput robotics with the capability of the Discover microwave reactor (Figure 6c). The Voyager™ system offers two separate flow through modules designed to handle the entire range of scale-up applications (Figure 6d). Each module incorporates the Discover reactor to satisfy the need to scale-up reactions with microwave energy in a safe, controlled, and consistent manner. This technology provides a microwave solution to the need for scaling-up reactions in process development. The Voyager family includes a *Continuous Flow* module, which is the ideal choice for scaling up solution-based chemistries, and a *Stopped Flow* module, which is the optimum choice for scaling up chemistries with solid reagents, highly viscous materials, or chemistries that form insoluble products. Similar to *Continuous Flow* option, the *Stopped Flow* module automates

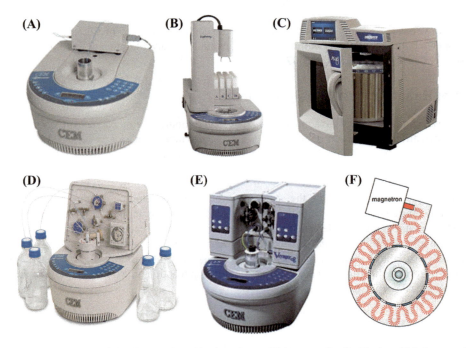

Figure 6. Reactor designs by CEM Corporation: (a) The Discover™ System; (b) The ExplorerPLS System; (c) The Navigator System; (d) The Voyager System; (e) MARS Xpress™ System with 40 place parallel synthesis set; (f) Self-tuning, single mode cavity (CEM Corporation—patented).

the necessary pumping and valve components to route the process stream through a flow cell that serves as the reactor. Reactions developed on the Discover or Explorer Systems may be easily transferred to Voyager for scale-up. The system is also a viable alternative for the production of valuable intermediates or the timely creation of starting materials, which means that medicinal chemists can now better manage their development projects.

The MARS 5™ platform provides a multimode system with all of the necessary features to allow its successful use in synthetic chemistry, when properly employed (Figure 6e). Multimode cavities provide their maximum benefit when heating large quantities of material or multiple containers of similarly absorbing material. This means that multimode systems are often employed in traditional parallel synthesis schemes. The reported issues with uneven heating of these multiple samples are often more a result of the varying absorptive characteristics of the samples encountered than an issue of positional sensitivity resulting from multimode energy distribution (Ferguson 2003). Multimode systems present a field that consists of multiple modes of energy with multiple nodes that vary in intensity. While samples are often circulated through the field, it is impossible to completely reproduce the exact field conditions in replicate experiments, reducing the reproducibility of experiments in multimode systems. Due to the relatively low field energy intensity of multimode systems (from 25–35 watts per liter), these systems are not well suited to heating individual, small sample quantities or poor absorbing materials. This is where single mode microwave systems present a significant advantage. Single mode systems, by definition, propagate only one mode of microwave energy, effectively eliminating the positional sensitivity issues encountered with multimode systems.

The recent advance in single mode cavity technology developed by CEM Corporation circumvents problems associated with tuning single mode cavities with small samples. The technical breakthrough also eliminates the reflected-power issues associated with single mode cavities. The cavity design features multiple slots in the interior wall to provide optimum coupling. No mechanical device is required to maintain coupling and there are no moving parts in the system. When the sample changes its coupling characteristics to the microwave field, the cavity constantly and without any intervention allows the field to instantaneously retune to the sample encountered, a first in microwave cavity designs. As shown in Figure 6f, the self-tuning slots simply let the sample dictate the field tuning. A considerable secondary advantage of the self-tuning circular cavity is that larger samples heat indirect proportion to the small samples. Thus, reactions can be designed on a small scale and applied directly to larger scale or upscale reactions. This allows great ease in scaling up reactions in a single system and affords the largest cavity capacity of any single mode system available to chemists today. This cavity technology is present in all of the Focused Microwave Systems from CEM.

Milestone

Milestone produces both batch and continuous type microwave reactors at very small batch/flow capacities to high levels. MicroSYNTH provides precision reaction monitoring and control with highly effective stirring. The FlowSYNTH system is a continuous flow microwave reactor (Figure 7a). This system resembles industrial-scale production system which allows to scale-up reactions from grams to kilograms. Once the chemistry has been optimized with the batch reactors, it can be easily transferred to the FlowSYNTH, which uses a flow-through reaction scheme with full control of all reaction parameters. The salient features of this system are easy scale-up from grams to kilograms, high purity of reaction products, fast heating and cooling rates which minimize reaction by-products, capability to fulfill large productivity needs, flow rate adjustable from 12 to 100 mL/minute and advanced temperature and pressure control functions. Reactions can be conducted at temperatures up to 200°C, and pressures up to 30 bar with less safety issues since small flow quantities will be used.

The Milestone innovative RotoSYNTH combines all the "traditional" advantages of microwave-enhanced synthesis, with the unique capability of physically rotating the reaction vessel, to achieve very homogenous bulk heating of heterogeneous reaction mixtures (Figure 7b). The advantage with this unit is highly homogeneous heating of solid materials can be achieved which allows efficient processing of slurries and dense reaction media. Fast removal of volatile side-products and full control of reaction parameters

Table 2a. Microwave reactors: process parametric data.

Make & model	General description	Power (W)	Mode	Reaction volume (per cycle)	Vessel size (mL)	Max temp (°C)	Max pressure (bar)	Overall size (kg)
Anton Paar Synthos 3000	autoclave, multiple (16)	1400	multi	1000	16 × 100	240	40	M (s74)
with XQ80	autoclave, multiple (8)	1400	multi	400	8 × 80	300	80	M (74)
Biotage Advancer	autoclave, single vessel	1200	multi	250	350	250	20	L 450
CEM Voyager	autoclave, stop-flow	300	mono	50	80	250	20	S (29)
MARS (open)	cavity for lab glassware	1600	multi	3000	5000	solvent bp	1	M (54)
MARS (a/c)	autoclave, various	1600	multi	700	14 × 75	200	20	M (54)
Milestone FlowSYNTH	continuous flow	1000	multi	unlimited	200	200	30	L (110)
MicroSYNTH (open)	cavity for lab glassware	1000	multi	1000	2000	Solvent bp	1	M (90)
MicroSYNTH (a/c)b	autoclave, various	1000	multi	1200	6 × 300	200	20	M (90)
Ultraclave	autoclave, various	1000	vari	2000	3500	300	200	L (400)

Table 2b. Microwave reactors: functional parameters.

Make & model	General description	Reaction volume (per cycle)	Agitation	Automated charging	Continuous addition/sampling	Hetero generous	Active cooling
Anton Paar Synthos 3000	autoclave, multiple (16)	1000	Magnetic	No	No	Yes	Air
with XQ80	autoclave, multiple (8)	400	Magnetic	No	No	Yes	Air
Biotage Advancer	autoclave, single vessel	250	Mechanical	No	Yes	Yes	Adiabatic flash cooling
CEM Voyager	autoclave, stop-flow	50	Magnetic	Yes	No	No	Compressed air
MARS (open)	cavity for lab glassware	3000	Both	No	Yes	Yes	Air
MARS (a/c)	autoclave, various	700	Magnetic	No	No	Yes	Air
Milestone FlowSYNTH	continuous flow	unlimited	Mechanical	yes	Yes	No	Water jacket
MicroSYNTH (open)	cavity for lab glassware	1000	Magnetic	No	No	Yes	Air jacket
MicroSYNTH (a/c)b	autoclave, various	1200	Both	No	Yes	Yes	Air jacket
Ultraclave	autoclave, various	2000	magentic	No	No	Yes	Water jacket

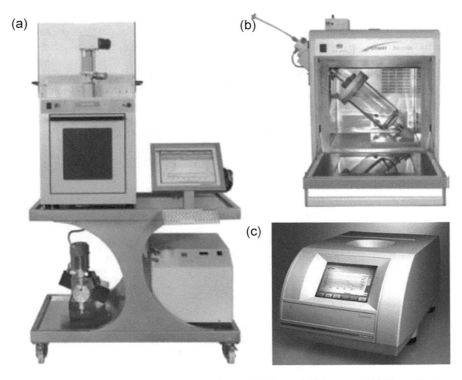

Figure 7. (a) FlowSynth and (b) RotoSynth (Milestone/MLS) and (c) Monowave 300 (Anton Paar).

with high flexibility of operation is another major benefit. The RotoSYNTH system represents an excellent platform for method optimization and for scale-up, as it can host various other accessories for sealed or reflux microwave synthesis. Therefore, it offers an extremely high flexibility of operation in open vessel, closed vessel or under vacuum; with full control of all reaction parameters (time, microwave power, heating and cooling rate, temperature, pressure, and vacuum), in a very safe environment.

Anton-Paar

Monowave 400 and Monowave 200 are high performance microwave reactors specially designed for small scale microwave synthesis applications in research and development laboratories (Figure 7c). An 850 W unpulsed microwave output power with extremely high field density and simultaneous temperature measurement with IR and fiber-optic sensor provides the highest accessible temperature/pressure (300°C/30 bar) conditions. The reaction times up to 100 h can be performed with high temperature homogeneity. However, the maximum sample volume is 30 mL. This limits the applications to academic and laboratory level syntheses.

Biotage

The fourth generation systems manufactured by Biotage provide for controlled temperatures and pressures up to 300°C and 30 bar which open new possibilities to complete difficult reactions. Even low boiling point solvents can be run at higher temperatures. The system automatically senses and performs reactions at their highest possible temperatures. The Initiator[+] supports vials from 0.2 to 20 mL, delivering greater flexibility and direct scale-up from milligrams to grams (Figure 8). The four different vial sizes can be used in any order or combination without system modifications. The single mode applicator and the Dynamic

Field Tuning™ features offer faster and more powerful heating (400 W) of a broader range of solvents. The setting for low microwave absorbing solvents enhances the heating for, e.g., toluene and 1, 4-dioxane.

The Initiator+ can be upgraded from a single-sample manual format to an automated 8- or 60-position system. The modular design allows a user to add on different automated sample processors dependent on throughput requirements. The 8-position sample bed gives the user a compact automation solution to start scale-up process and library build-up. The 8-position system is useful in a multi-user environment or for queuing multiple reactions. Flexible operation enables the use of both large and small vials in combination at any time and in any order without manual intervention. The 60-position sample bed supports the production of focused libraries, multi-user environments and scale-out, and use of both large and small vials in any order without manual intervention.

Reproducibility and Scalability

Reproducibility among the microwave reactors is a major concern. The reliability of experimental results can be enhanced by repetitive verifications. This may vary from single mode systems to multimode systems and with reactor design and microwave propagation. Reproducibility can vary from a batch process to a continuous process as well as between reactors from different manufacturers.

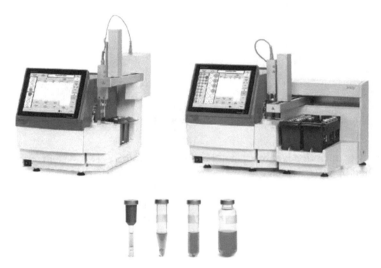

Figure 8. Biotage® Initiator Robot Eight and Robot Sixty microwave systems with different vial sizes (0.2–0.5 mL; 0.5–2.0 mL; 2.0–5.0 mL and 10–20 mL).

Reproducibility of alkylation reactions in commercially available single mode and multimode microwave reactors was first reported by Alcazar et al. (2004). Alkylation of four secondary amines at 150°C for 10 minutes was evaluated. They concluded that the alkylation reactions are reproducible in microwave reactors. Microwave chemistry approach was first optimized and then, it was used for comparison of results between single mode and multimode microwave reactors. It was found that small differences in yield and conversion could be due to small differences in the temperature inside each vessel. The maximum difference of temperature ranged ± 10°C of the set temperature on the control vessel. Using fully optimized conditions, four different compounds were prepared in 0.3 mmol scale using a single mode instrument and successfully scaled up by a factor of ten in a multimode unit in a single experiment in parallel. No major yield differences were observed in multimode instrument as shown in Table 3. The combination of single mode and multimode instrumentation was demonstrated to be a valuable tool for the organic chemists to quickly optimize reactions and subsequent scale up in parallel.

Solvent-free reactions are much different from conventional reaction schemes. They are known to be non-reproducible. Diaz-Ortiz and Alcczar showed for the first time that solvent-free reactions performed

Table 3. Alkylation of four secondary amines at 150°C for 10 minutes in single mode and multimode microwave reactors.

Compound	Amine	Single mode Conversion	Yield	Multimode Conversion	Yield
1-Benzyl-4-(2-phenoxyethyl) piperazine		100	93	100	93
Benzylmethyl-(2-phenoxyethyl) amine		99	94	99	94
2-(2-Phenoxyethyl)-1,2,3,4-tetrahydroisoquinoline		99	84	99	81
1-(2-Phenoxyethyl)-4-phenylpiperidin-4-ol		99	99	100	99

in domestic ovens could be reproduced, scaled and parallelized in controlled microwave monomode and multimode reactors (Díaz-Ortiz et al. 2007). The 1, 3-dipolar cyclo addition of nitrile N-oxides with nitriles was selected as a model reaction, and this was previously performed by the same authors using a domestic oven. In another work (Díaz-Ortiz et al. 2011), several solvent-free reactions previously carried out in domestic ovens have been translated into a single mode microwave reactor and then scaled up in a multimode oven. The results showed that most of these reactions, although not considered as reproducible, can be easily updated and applied in microwave reactors using temperature-controlled conditions.

The lack of reproducibility limits the general applicability and the scale up of these reactions. Computational calculations can assist to explain and/or predict whether a reaction will be reproducible or not. Computational calculations have been carried out on selected reactions in order to determine energy and physical parameters required to understand and predict the origin of improvements and selectivities observed in organic reactions performed under microwave irradiation (Rodríguez et al. 2015). For this purpose, the influence of enthalpy and polarity in the reactions was studied. It was reported that exothermic reactions with low activation energies proceed well under conventional heating, and improvements should not be expected under microwave irradiation. However, the corresponding endothermic reactions can be improved slightly under microwave irradiation under specific conditions such as controlled temperatures and pressures. Endothermic and exothermic reactions with moderate or high activation energies can be improved under microwave irradiation in polar media (reagents, solvent, and catalyst). Reactions with very high activation energies can be performed under microwave irradiation when two requirements are fulfilled: the reaction is exothermic and a polar medium is present. Endothermic reactions do not take place under either conventional heating or microwave irradiation.

Slower reacting schemes, in general, tend to show better effects under microwave irradiation. Since radiation is selectively absorbed by polar systems, a characteristic that leads to selective heating profiles may be observed. The presence of a polar solvent, reagent or support in the reaction media leads to strong coupling with the radiation. This fact is particularly important in heterogeneous systems where it could also generate microscopic hot spots or selective heating. The presence of small amounts of a strongly microwave absorbing 'doping agent' or 'susceptor', such as an ionic liquid, leads to very efficient interactions with microwaves through the ionic conduction mechanism (Rodriguez et al. 2015).

Reproducibility in Small vs. Large Scale Applications

An investigation on the scalability of the microwave technology was reported by Lehmann and LaVecchia (2005). Microwave Emrys was used for small scale reactions. Two different microwave systems designed for large-scale operation (Multiwave 3000 and CEM Voyager SF) were evaluated to characterize strengths and weaknesses of each instrument for special use in a kilolab with focus on temperature/pressure limits, handling of suspensions, ability for rapid heating and cooling, robustness, and overall processing time. An aromatic substitution of aryl halides with nucleophiles like phenolates or amines was considered to study the effects on reaction time, selectivity, work-up procedure, and overall processing time. Results of scheme 1 are summarized in Table 4.

In theory, conditions in a microwave reactor, with regard to temperature and pressure, should also be attainable by using an autoclave. Therefore, some reactions were done in an autoclave for comparison. The main difference was the longer processing time, mainly caused by the long heating and cooling period when an autoclave was used. This leads, as shown in example 2 of our study, to a different selectivity and to significantly higher amount of by-products, which as a consequence had to be separated by additional purification steps. In such a case, using microwave heating is obviously more advantageous than using an autoclave.

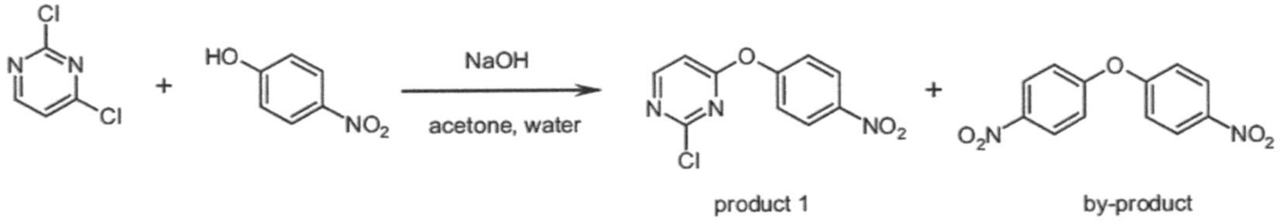

Scheme 1.

Table 4. Results of reaction 1 under different conditions.

Entry	Method or unit	Batch size (mL)	Reaction temp., time	Processing time	Isol. yield (%)	Purity[a] (%)	Remarks
1	Conv. heating	20	62°C, 12 h	12 h 30 min	66	80	5% educt left, 20% by-product
2	Microwave, Emrys Optimizer	15	120°C 5 min	10 min	83	> 96	100% conversion of educts, < 5% by-product
3	Microwave, Multiwave 3000	8 × 50	120°C, 10 min	30 min	76	90	10% by-product
4	Microwave, Voyager SF	400	120°C, 15 min	8 × 30 min	81	95	5% by-product, precipitation of product blocked the line
5	Autoclave	400	120°C, 10 min	275 min	74	85	> 10% by-product

[a] measured by area % HPLC.

On a small scale under reflux in acetone/water, 5% of starting material remained after 12 h reaction time and about 20% of the by-product was formed (entry 1). Performing the reaction at the same concentration in a lab-scale microwave device (Emrys Optimizer, entry 2) at 120°C, the reaction was complete after 5 min and gave a product of significantly higher purity and in higher yield. In the next step, 400 mL of reaction mixture was reacted in an eight-vessel rotor batch microwave (entry 3) at the same temperature for 10 min. After a cooling down period of 18–20 minutes, the precipitated product was isolated by simple filtration in 76% yield but also 10% of the by-product was formed under these conditions. The overall processing time, which includes heating, maintaining the temperature for 10 minutes, and cooling was only 30 minutes.

In contrast to this, conversion of a reaction volume of 400 mL in the Voyager SF took about 4 h (entry 4) because eight cycles on a 50 mL scale were necessary to convert this volume. Each cycle consists of adding the starting material, heating, maintaining the temperature for 15 minutes, cooling, and removal of the mixture from the reaction vessel, which required 30 minutes. Furthermore, when the mixture was pumped out of the vessel after the reaction, precipitation of the product blocked some lines and valves, so that the process had to be interrupted manually. For comparison, the reaction was also carried out in an autoclave (1 L volume, entry 5). Due to the long heating and cooling process, processing time in total was 275 min, leading to a product yield of 74% that contained more than 10% of the byproduct.

Another homogenous reaction was evaluated to compare the influence of different solvents under conventional heating and microwave heating. Results are shown in Table 5.

Under reflux conditions on a small scale in acetonitrile, the reaction was complete only after 24 h (entry 2). In a microwave reactor at 150°C, 86% of isolated yield was achieved after 20 minutes of reaction time (entry 3). Running the same reaction in DMSO or DMF at 150°C gave complete conversion of starting material but due to losses during work up, isolated yield was lower (entry 4 and 5). For scale-up trials, the Voyager SF was used, and the reaction was done at 180°C for 20 min in each cycle. To process a 250 mL batch, five cycles of 50 mL each had to be performed, leading to a processing time of 2 h 40 min (entry 6). The temperature rises very rapidly to 180°C and is held for 20 minutes, and the pressure during the reaction is about 11 bar. Then, the vessel is cooled by air ventilation to about 60°C within 8–10 minutes, which results in a cycle time of about 32 minutes.

The Multiwave 3000 as a batchmode reactor provides a relatively large reaction volume (16–70 mL), allows high temperature and pressure (250°C/40 bar), and has proved to be very robust. Handling of suspensions revealed no problems, and the system was shown to have a very good temperature control. In various experiments it was shown that the use of microwave technology leads to a significant decrease of the reaction time and in some cases also to less by-products and a higher yield. This technology provides the opportunity to optimize the reaction with focus on work up and purification, independent of reaction temperature and boiling point of the solvent. In most cases, the reaction conditions applied on a 15 mL scale in the Emrys Optimizer could be transferred without further optimization to the tested microwave reactors and lead to comparable results. Additional optimization in a few cases was limited to a moderate adjustment of reaction temperature or reaction time.

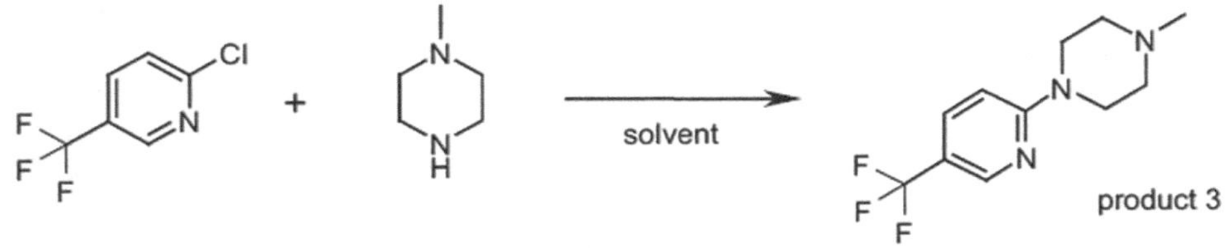

Scheme 2.

Table 5. Results of reaction 2 under different conditions.

Entry	Method or unit	Batch size (mL)	Solvent, temp., time	Processing time	Isol. yield (%)	Purity[a] (%)	Remarks
1	Conv. heating	20	CH$_3$CN, reflux	3 h	69	> 97	Reaction not completed, work-up A
2	Conv. heating	20	CH$_3$CN, reflux	24 h	97	> 97	Reaction completed, work-up A
3	Microwave, Emrys	15	CH$_3$CN, 150°C, 20 min	28 min	86	> 97	Reaction not completed, work-up A colorless crystals
4	Conv. heating	20	DMF, 150°C, 20 min	50 min	60	> 99	Work-up B brownish crystals
5	Conv. heating	20	DMSO, 150°C, 20 min	50 min	63	> 99	Work-up B beige crystals
6	Microwave, Voyager SF	5 × 50	CH$_3$CN, 180°C, 20 min	2 h 40 min	72	95	Work-up B colorless crystals

Work-up A: reaction mixture was poured into water and extracted with DCM. The organic phase was evaporated and crude product was purified by flash-chromatography.
Work-up B: reaction mixture was poured into water; precipitated product was filtered, washed, and dried.
[a]Measured by area % HPLC.

Scale-up Issues

Development of continuous processing is an important consideration when discussing the commercial advancement of microwave techniques for materials production. Despite being of paramount importance from an industrial perspective, relatively little work has been carried out on development of continuous-flow processes. Significant process development and scale up is difficult to consider without significant improvements in control and reproducibility, which is ultimately obtained from *in situ* analysis and monitoring. In terms of industrial processing, some of the potential advantages offered by MW methods include rapid and uniform energy transfer, volumetric and selective heating, environmental compatibility, increased throughput, fast on and off switching, compact equipment–space savings, clean environment at the point of use-enhanced worker safety, and unique characteristics of the products.

The scale-up of MW processing is a complex process and will require a combination of computational simulation, expert system design, and comprehensive cost and benefits analysis. It has already been established that a MW system can efficiently deliver energy to a reaction vessel; however, there are several considerations which must be taken into account:

- ensuring a homogeneous electric field profile,
- optimum reactor design which takes into account penetration depths,
- control of temperature and pressure within the reactor,
- cost of the reactor and spare parts, and
- safety issues and MW leakage.

Reactor design—Scale-up of microwave systems can be realized by working with the wavelength and equipment dimensions to the same factor. If, for example, one would lower the frequency to the 915 MHz band, which is the nearest ISM frequency band under the 2.45 GHz band, then all dimensions are scaled to a factor of 2.68 with respect to a system operating at 2.45 GHz. This would result in a 19-fold volumetric increase with the same, but scaled-up, distribution of the microwave field (Stefanidis et al. 2014).

The size of the microwave reactor and the microwave power required is calculated by the throughput capacity using microwave studies for dielectric materials. Once the reactor size and the number of magnetrons required to deliver the given power are known, the reactor configurations are subjective, and self-inventive configurations are used (Li et al. 2009, Lin et al. 2012). The usual configuration is a continuously agitated cavity fitted with magnetron assembly and instrumentation (Waheed ul Hasan and Ani 2014).

Determination of dielectric parameters—The varying trends of ε'' with temperature indicate the necessity of an on-line computer control of microwave input power to prevent overheating and damage to the reactant and product. Temperature can be used as feedback parameter to the on-line power to maintain constant rate of heating, switch the input power off at a set reaction temperature and regulate power in order to maintain the reactant at a set temperature until the reaction terminates. The dynamic change in dielectric properties also suggest that the microwave-assisted system particularly in the continuous mode of processing needs for a dynamic impedance matcher in order to minimize the reflected power and increase the efficiency of energy consumption (Zaini and Kamaruddin 2013).

Thermal runaway—In high temperature processes such as pyrolysis, the main challenge in scaling-up the microwave heating process is thermal runaway, in which the reactant temperature cannot be stabilized and regularly deteriorates into arcing (Meredith 1998). Heat absorbed in microwave heating is proportional to the dielectric loss, ε''. Rapid increase in ε'' with increasing temperature will provide a potential for thermal runaway to occur (Wu et al. 2002). It has been reported that the dielectric loss of most carbon materials increase with increasing temperature (Menéndez et al. 2010, Atwater et al. 2003) which cause thermal runaway in microwave-assisted activation. This is more critical if the reactant is heterogeneous (i.e., impregnated with chemical agents) because the value of ε'' of the dielectric property is not constant and varies not only with frequency, but also with temperature, moisture content, physical state (solid or liquid) and composition (Meredith 1993, 1998).

Reaction phase—Most of the reports on microwave process scale-up related to liquid-phase organic chemistry are available. Few studies report on the considerations and challenges applicable in solid systems, i.e., solid-liquid phase such as biomass processing in biofuel production. Microwaves are currently used in food industry in a variety of applications. For example, the tempering of blocks of food from −20 to −2°C can be done using a 60 kW system which occupies 1/6th of the space of conventional equipment. Use of microwave hybrid baking and cooking allows retention of distinctive flavor, color, and texture of oven baking with the increased throughput associated with microwaves. In the rubber industry MW heating of blocks of rubber of up to several hundred kilograms in weight has replaced conventional vulcanization (Leonelli and Mason 2010). Other applications include various areas of waste processing: processing of automotive tires (Appleton et al. 2005), treatment of hospital and municipal wastes (Lee et al. 2004) treatment of toxic substances (Abramovitch et al. 2003), recovery of waste plastics, and hospital waste sterilization (Kitchen et al. 2013).

Current Status of Scale-up and Large-scale Applications

Large scale industrial production through microwave processing has several limitations. The first and major limitation is the volume of reactor that can be processed without any safety issues. Manufacturers make vessels of an optimum size that enables the strength components to work effectively and withstand high pressures, and design to fit a reasonable number of them inside the cavity. The throughput from these systems is limited. Because most high-pressure vessels have an internal volume of approximately 100 mL, the sample size in a difficult organic matrix, such as heavy fuel oil or a polymeric material like polypropylene, is limited due to the fact that it will generate around 600–800 psi of gaseous digestion products and reagent vapor pressure at the temperatures required for adequate digestion (200–220°C). Adding more sample increases the pressure level to the limit of the Teflon materials used for most vessel designs, and although quartz vessels can be used, the seals are still made of Teflon. In addition, rigid materials like quartz have proved to be problematic and can fail catastrophically, unlike non-rigid polymeric materials (Barclay 2004).

A continuous flow system could be a viable alternative for efficient microwave utilization. Since the microwave field varies in the multimode cavity microwave reactors, a serpentine or coil type continuous flow pattern would allow for an optimized microwave field exposure for reactants. Large organic samples can benefit from vent-and-reseal technology, used as one of the earliest commercial vessel designs. However, the true advantages of this technique were not noted until the development of high-temperature systems with vessels that could maintain a set pressure limit and bleed off gas formed by the digestion in excess of this pressure. These vessels are designed to maintain a temperature at which a "good" digestion will result and vent after the oxidation is completed. Essentially, the vessel bleeds off the excess pressure during the rapid exothermic portion of the digestion and then reseals once the rapid gas evolution has ceased. However, such technology should be approached with care, because some manufacturers set the vent pressure above 500 psi. At these pressures, the force of the vent can be sufficient to aerosol the sample and eject a portion of the vessel contents, losing sample and analytes of interest. Before venting, ideally below 500 psi, the bulk of the sample and reagent digestion has to be complete, thus forming analytes of interest from solvated ionic species, which have an oxidation state that has no volatile species.

Focused Microwave Reactors

For pretreatment and extraction purposes, focused microwave reactors can be instrumental. These can be used for reactions involving severe conditions such as high pressures and high temperatures. A single mode focused microwave reactor was used for algal lipid extraction under continuous flow conditions (Balasubramanian et al. 2011). The microwave system had an output rate of 1.2 kW at 2.45 GHz frequency. This unit is comprised of a power generation unit, connecting wave guide, circulator, 3-stub manual tuning coupler, a cooling system, and an elliptical focusing cavity which holds the PTFE applicator tube through which the process fluid flows. The microwave system components were purchased from commercial vendors (Richardson Electronics, IL, USA and IMS LLC, Morrisville, NC) and assembled in-house. The

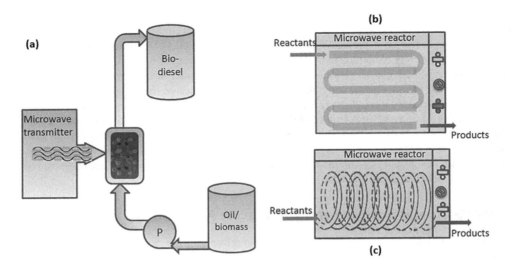

Figure 9. Potential industrial scale microwave reactor for biodiesel production; (a) focused microwave reactor; (b) serpentine plug-type continuous flow reactor; (c) helical coil-type continuous flow reactor.

entire processing system consisted of a feed tank, microwave processing unit, a constant temperature water bath, and peristaltic pump to circulate the feedstock through the entire system (Figure 10).

The tuning stubs were adjusted at the start of the study using algal-water solution (1:1 weight basis) and a network analyzer (HP8753 C, Hewlett Packard, Palo Alto, CA) to maximize the power absorbed by the load (product) flowing through the focusing cavity. The ability of single mode microwave applicators to focus and match the generated energy into the process medium resulted in effective heating at center of the tube. Effective coupling of the generated microwave energy with the process medium is carried out by "internal cavity" matching accomplished by the focusing cavity and tuning of the system based on the dielectric properties (Asmussen et al. 1987). By effective electromagnetic coupling of the generated microwave energy onto the process medium in the applicator tube, the focusing cavity helped increase power utilization efficiency and reduced the amount of microwaves reflected back (Boldor et al. 2008). Teflon tubes of inner diameter of 0.953 cm (3/8") were used to convey the process mixture from the feed tank to the microwave unit and convey the processed fluid to 50 mL beakers placed in the water bath.

A maximum oil yield of 76–77% of total recoverable oil was extracted at 95°C and 30 min. The control system (water bath) extracted only 43–47% of lipids. Extraction time and temperature had significant influence ($p < 0.0001$) on extraction yield. Oil analysis indicated that microwaves extracted oil containing higher percentages of unsaturated and essential fatty acids (indicating higher quality). Higher oil yields, faster extraction rates and superior oil quality demonstrate this system's feasibility for oil extraction from a variety of feedstock.

Terigar et al. (2011) reported a continuous flow microwave system similar to Figure 9a for extraction of oils from soy flour and rice bran. A custom-designed, multimode 2450 MHz laboratory-scale, batch type-converted to continuous microwave extraction system was investigated and the optimization results obtained were used to develop and test a pilot-scale 5 kW, 915 MHz (to allow for higher penetration depths) focused cavity microwave extraction system. Higher penetration depths allow for larger scale equipment to be used, e.g., larger diameter tubes, which allow for higher processing flow rates. Oil was extracted from soy flour and rice bran at various time–temperature combinations with an ethanol:feedstock ratio of 3:1. Both processes were optimized to minimize extraction time while maximizing the quantity of oil extracted. Extraction yields were compared to conventional and Soxhlet extraction. Yields in the laboratory-scale system for soybean oil were highest at the highest temperature (73°C) and longest exposure time (21 min), while for rice bran oil, the highest yields were obtained at 73°C and 17 min. Using the pilot-scale extraction system, greater than 93.0% of total recoverable oil was extracted from both feedstock at all

Microwave Reactor Design and Configurations 83

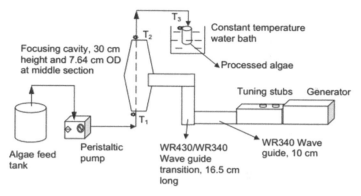

Figure 10. Focused microwave reactor for oil extraction from *Scenedesmus obliquus* algal species (Balasubramaniam et al. 2011).

tested flow rates and extraction times, more than the oil recovered by the laboratory-scale unit. Time of extraction and flow rate did not have a significant influence on extraction yield at tested temperatures for the pilot-scale process for soybeans. For rice bran, extraction yield decreased slightly with increasing flow rate (from 0.6 to 1.0 l/min) and extraction time also influenced oil yields (Terigar et al. 2011). Overall, the results indicated that oil yields obtained from both laboratory- and pilot-scale systems were better than the yields obtained by conventional extraction. The quality of the extracted oils was sufficient to meet biodiesel feedstock standard specifications, and a MW-based extraction method was considered a viable alternative to the current state of the art (Kitchen et al. 2013).

Focused microwave reactors are also widely used in food and dairy extraction processes. Kumar et al. (2008) developed a 5-kW microwave system, shown in Figure 11, consisting of a 5-kW microwave generator (Industrial Microwave Systems, Morrisville, NC) operating at 915 MHz, a waveguide of rectangular cross-section, and a specially designed focused applicator (Drozd and Joines 2001). A tube of 1.5 in nominal diameter (0.038 m ID) made of Polytetra-fluoroethylene (PTFE or Teflon) was placed at the center of the applicator through which the product was pumped using a positive displacement pump (Model MD012, Seepex GmbH+ Co., Bottrop, Germany). The length of applicator tube exposed to microwaves was 0.28 m. The average residence time of the product in the applicator for a flow rate of 0.9 L/min was 21.1 s.

Two vegetable purees were processed for a run time of 8 h (a typical duration of plant operation in industry) in the industrial-scale system based on the procedures developed. Temperature measurement revealed significant temperature non-uniformity in the pilot-scale 5 kW microwave reactor process. The temperature difference between the center and the wall of the applicator tube at the outlet became smaller with increasing reaction temperature. It was probably due to the enhanced absorption of MWs at higher temperatures or the reduced viscosity of the purees at higher temperatures facilitating faster thermal dissipation. Temperature non-uniformity was to be overcome by the use of static mixers installed at the exit of each microwave applicator, in which the purees were channeled through a geometric arrangement of mixing elements. For further scale-up, the purees were processed in a 60-kW microwave reactor process (Figure 12) with the objective of successful operation for at least 8 h. Processing was performed for 8 h in the 60-kW microwave reactor process with an output power of 30–40 kW and a flow rate of 3.78 L min^{-1}. These results illustrate the effectiveness of mixers to reduce temperature differences during reactions. One of the key challenges in solid-state temperature non-uniformity was successfully avoided by using static mixers to decrease temperature differences and with careful engineering control these effects can be minimized (Kumar et al. 2008).

Processing in the 5-kW microwave system prior to processing in the 60-kW microwave systems offered several advantages. It can be used as a compatibility test for continuous flow microwave heating. If a product does not reach the required temperatures in a desired application, it is concluded that continuous flow microwave heating will not be suitable for that particular product in its present form. Processing in the 5-kW system also helps in understanding the behavior of the desired products (Kumar et al. 2008).

Continuous flow Systems

The majority of research to date on biofuels using microwave irradiation has employed domestic microwave ovens and the reported data should be rather considered as a part of an initial screening (Brasoveanu and Nemtanu 2014). Since microwave processing requires very precise temperature control, it is advisable to use laboratory microwave appliance designed and equipped with thermocouple temperature probe

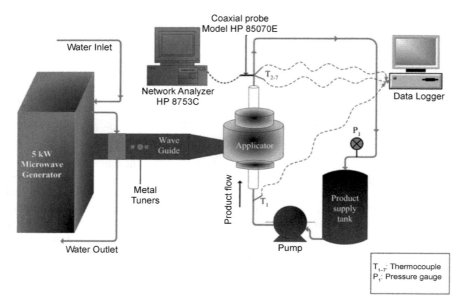

Figure 11. Schematic representation of a 5-kW focused continuous flow microwave reactor.

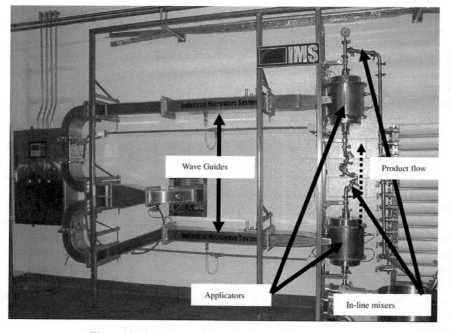

Figure 12. 60-kW focused continuous flow microwave reactor.

and temperature controller for research accuracy. Several dedicated microwave reactors equipped with highly automated and precise control systems of temperature and pressure are currently available. These appliances enable controlled and reproducible operating conditions, optimal for relatively easy scale-up and technology transfer (Kappe 2004, Toukoniitty et al. 2005).

For continuous flow processing, homogeneity of the reaction mixture is very important. When the mixture is homogeneous, it is possible to move from small scale sealed-vessel conditions to the continuous-flow apparatus without any modification of reaction conditions or loss in product yield. When either the starting materials or the product mixture contains particulate matter, continuous processing can prove to be a challenge, but optimization of reaction conditions as well as reduction of the concentration may allow these difficulties to be overcome (Bowman et al. 2007, Moseley et al. 2007). For biodiesel applications, assuming the reactants are well mixed and homogeneous, a focused continuous flow microwave reactor or serpentine plug-type reactor or helical reactor may be considered to monitor and control the reactants temperature and detention times as shown in Figure 9 (Terigar 2009, Cleophax et al. 2000). This type of reactor was used in solvent-free organic reactions and extraction of polycyclic aromatic hydrocarbons from sediment, soil, and air samples.

There are many reports available in literature on continuous microwave processing of materials in a variety of process industries. A few examples can be discussed here. The use of a commercial FlowSynth microwave reactor (shown in Figure 7a) for the continuous production of biodiesel from palm oil was demonstrated (Choedkiatsakul et al. 2015). A high ester content of 99.4% was obtained in only 1.75 min residence time at a methanol to oil molar ratio of 12, a microwave heating power of 400 W, reaction temperature of 70°C and NaOH catalyst loading of 1% wt of oil. Figure 13 presents the effect of microwave heating power (200 W, 400 W, 600 W, 800 W) on biodiesel yield. A long total reaction time of 10 min was required to produce high biodiesel yields of at least 96.50% at 200 W, while MW heating powers of 400 and 800 W provided high biodiesel yields in accordance with ASTM standard in the same reaction time (4 min).

It was reported that high microwave power produces high biodiesel yields due to enhanced dielectric effects and the destruction of the boundary layer between methanol and oil, hence reducing the dielectric constant and polarity of methanol. The phenomena results in methanol and oil homogeneity. Saifuddin and Chua (2004) have reported that irradiation power must be controlled to avoid overheating which can destroy some organic molecules. This may be why a slight increase in biodiesel yield can be observed at the high MW heating power of 800 W. The optimal MW heating power was therefore chosen to be 400 W.

The feed flow rate can have significant effect as shown in Figure 14. Three feed flow rate values of 100, 150 and 200 mL/min, which correspond to residence times of 1.75, 1.17 and 0.875 min, respectively, were investigated. Results indicate that the reaction time needed to reach the steady state condition of biodiesel yield only depends on feed flow rate. It can be seen that a long reaction time of 8 min was required for a feed flow rate of 200 mL/min while only 4 min was needed at 100 mL/min (Choedkiatsakul et al. 2015).

The process energy consumption analysis proved the advantages of using microwave irradiation for biodiesel production. The energy consumption of palm oil transesterification as low as 0.1167 kWh/L of biodiesel was required, proving that microwave reactors need less energy consumption as compared to conventional processes (Choedkiatsakul et al. 2015).

Increasing energy efficiency in continuous flow systems

Two major issues in conventional single mode continuous microwave processing are energy efficiency and the residence time of the reaction medium in the microwave heating zone (Figure 15). These issues can be solved by the short circuit transmission line resonator concept (Morschhäuser et al. 2012). Microwaves coupled into the stainless steel transmission line resonator at the front end propagate rotationally symmetric with the flow of the reaction medium inside the microwave transparent reaction tube. On the way to the short circuit, microwave energy will be absorbed by the reaction mixture and transferred into heat. Consequently, the field strength of the individual modes exhibits an exponential decay along the tube. Any remaining energy at the short circuit will be reflected back along the tube but now in the opposite direction. Compared to a conventional single mode batch reactor, the energy released from the magnetron will be able to interact with the reaction medium over several modes of the standing wave. Dipoles inside

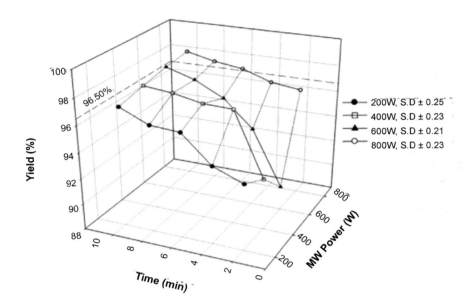

Figure 13. Effect of microwave power heating on biodiesel yield (methanol/oil molar/oil molar ratio of 12, catalyst loading 1 wt%. of oil, reaction temperature of 70°C and feed flow rate of 100 mL/min).

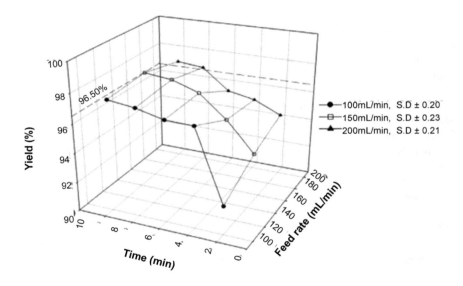

Figure 14. Effect of feed flow rate on biodiesel yield (methanol/oil molar/oil molar ratio of 12, catalyst loading 1 wt%. of oil, reaction temperature of 70°C and microwave heating power of 400 W).

the tube do have more time to convert the provided microwave energy into heat via dielectric heating mechanisms. The wave will be reflected and sent back; on the way back to its origin, microwave energy transfer into heat occurs. To prevent damage to the microwave generator by reflected microwave radiation, a ferrite isolator is used to eliminate reflected microwave power, which can be easily measured using a watt-meter. The difference between forwarded energy released from a magnetron and reflected energy

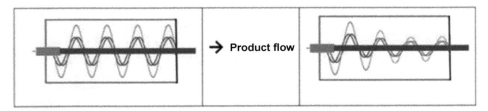

Figure 15. Left: Reaction tube centered in the monomode resonator microwave field without medium (Creation of a standing wave in a monomode resonator). Right: Exponential decay of field strength along the reaction tube in presence of reaction medium with dielectric loss.

coming back from the applicator unit provides direct access to the efficiency of the system as represented by equation 2 below.

$$Q_{eff} = \frac{(Q_{generated} - Q_{reflected})}{Q_{generated}} \qquad [2]$$

In this system, microwave power is generated with a standard magnetron with variable power output from 0.6 to 6 kW. This power range also represents the adjustable power released from the magnetron measured by a watt-meter. The energy provided to the dielectrically active medium will be absorbed instantaneously to almost 100%. The microwave power generated by a magnetron is guided via a rectangular waveguide to a launching section that couples the microwave power into the stainless steel transmission line resonator. A standing wave is created along the length (approx. 70 cm) of the transmission line waveguide (see Figure 15) which will be absorbed by the medium.

Figure 16 displays the general set-up of the continuous flow microwave processing unit using the standing wave concept for energy transfer. This unit is capable of processing 20 L/h. The throughput of the system can be readily increased by increasing the diameter of the tube, increasing the flow rate and employing more powerful magnetrons (e.g., 30 kW), which can ensure rapid and efficient volumetric heating of the reactor zone. An important additional variable is the microwave frequency, as the use of, e.g., 915 MHz technology allows a higher penetration depth coupled with an even further increase in energy efficiency. This reactor allows the safe and highly energy efficient processing of organic reaction mixtures (with or without solvent) under high-temperature/high pressure conditions.

Increasing microwave field uniformity in continuous flow systems

The major challenges associated with microwave systems are the introduction of microwave field and its uniformity in the cavity to improve the reaction control. Placing a reactor inside a microwave requires serious considerations for the material of the reactor and the reaction conditions. To address these issues, the internal transmission line (INTLI) concept (Franssen et al. 2014), in which the microwave is brought into the reactor by a specially shaped waveguide can be used. In this concept, the reactor vessel can be made of any material that can withstand severe process conditions such as high pressures and high temperatures. In addition, the controllable energy transfer to the medium along the length of the lossy transmission line is not only beneficial in terms of uniformity, but it also reduces the total power consumption by minimization of the reflected power. A study on polyester preparation in the INTLI reactor showed that the reflected power was almost fourfold lower compared to a conventional multimode cavity (Komorowska-Durka et al. 2013).

On the other hand, in typical multimode microwave equipment, the field and the intensity of irradiation are highly non-uniform, especially at large scale applications. Such non-uniformity may be acceptable in operations such as drying, thawing, or pasteurization, but it is not acceptable in the case of chemical reactions. The non-uniformity of the field as well as the resultant orientation of the flowing reactants with respect to the field plays critical role in these reactions (Sturm 2013). These issues can be successfully addressed using a traveling wave reactor (TWR) as shown in Figure 17a (Sturm 2013). Similarly to a

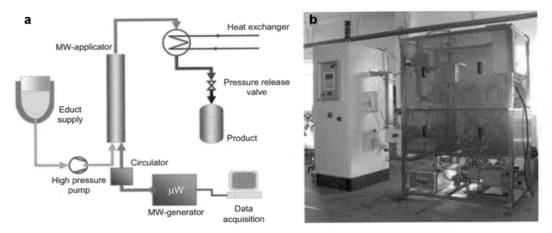

Figure 16. (a) General design of the continuous microwave synthesis unit; (b) photo-image of the unit (Morschhauser et al. 2012).

coaxial cable, a TWR consists of outer and inner conductors with filler between them. A monolithic support with catalytic micro-channels/milli-channels is placed on the filler, and the microwave is allowed to travel along those channels. With a full cylindrical symmetry, the channels placed at the same radius are evenly irradiated with microwaves, as shown in Figure 17b. Additionally, one can manipulate the diameter of the inner conductor to compensate for attenuation.

For large-scale applications, bundles of TWRs (similar to multi-tubular reactor designs) are envisaged. On the other hand, in heterogeneous catalytic systems, guidelines for the rational design of microwave-receptive catalysts that can attain the optimum heating rate and pattern for a given process do not exist. The key dielectric properties (real and complex permittivity for the prediction of the interaction of the material with the microwave field and the induced heating rate) are seldom measured. Dielectric characterization, when done, is done at room temperature and under nonreactive conditions. Consequently, the dielectric property values obtained may not be representative of those during the process.

There is a need to develop instrumentation that enables the measurements of the dielectric properties of microwave-irradiated catalytic reactors under real process conditions over a wide frequency range. The process conditions concern the actual reaction temperature, which may be several hundred degrees Celsius, and the exposure of catalyst to reactants and products in gas or liquid form. In the next step, the results should be used to guide the development of new models based on effective medium theories (Sihvola 1999) which will correlate the dielectric properties of catalyst composites as function of the catalyst size, shape, composition, and temperature.

Industrial Scale Microwave Systems

In industrial applications, microwave field can be distributed along the process line using a slotted waveguide to ensure the uniform distribution and to overcome overheating issues. These allow for step input of microwaves to optimize and increase the exposure to microwaves. A general schematic of an industrial level conveyor type microwave system is shown in Figure 18.

Chemat-Djenni et al. (2007) developed a 6-kw conveyor type microwave system to study heterogeneous catalytic reactions at atmospheric pressure (Figure 19). Microwave heterogeneous catalytic reactions as an "environmentally friendly" synthesis method suitable for preparation of xylene, nitriles, oxides, etc. was demonstrated. They report that the microwave process avoids the use of large quantities of solvent and voluminous reactors for large scale applications. It could also be used to produce larger quantities by using existing large scale microwave reactors suitable for the heterogeneous catalytic reactions of 10, 20 or 100 kg of products at a time. A 20-fold reduction in energy demands and rapid reaction and reduced environmental impacts were the benefits from this process.

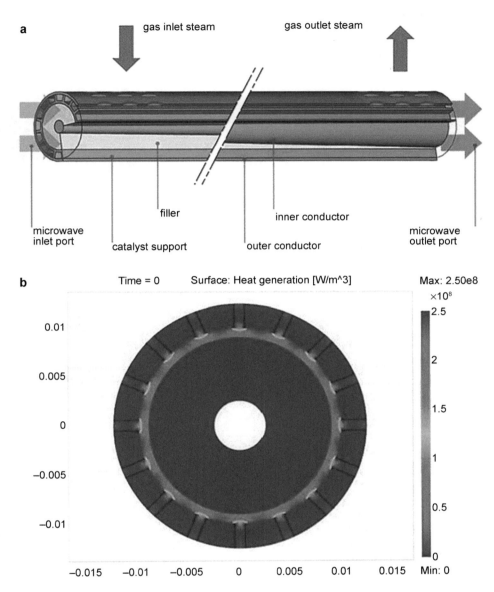

Figure 17. (a) TWR for gas-phase heterogeneous catalysis. (b) Heating rate distribution over the channels of the reactor.

Laboratory/Pilot Microwave System

Thermex manufactures both pilot scale and industrial scale microwave systems for various process applications including laboratory-pilot scale batch systems (Figure 20). The Thermex Laboratory/Pilot Microwave System is designed with a self-contained microwave source to bridge the gap between small bench top ovens and production systems. The ability to establish consistent and reproducible test data provides the confidence that is important for establishing the groundwork for a successful manufacturing process. The unique features of this system coupled with the PLC controls yield the ultimate in material testing, data acquisition, trending and process development.

An optional load cell is available that when incorporated into the turntable and programmed, the microwave power will turn off at a predetermined weight. An IR Pyrometer to monitor sample temperature

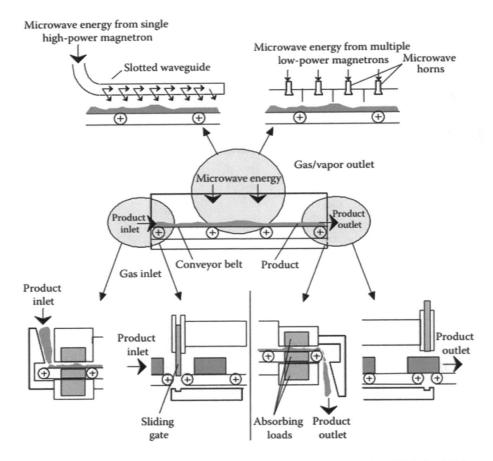

Figure 18. Typical industrial conveyor belt microwave oven (adapted from Regier and Schubert 2001).

and/or shut off the microwave power at a predetermined temperature is included. Additional options include: PLC and Control Upgrades; Applicator Light; Exhaust Fan; Fire Suppression System; and Power Density Control. Applications include Ceramics, Catalysts, Pyrolysis, Vulcanizing, Composites, Bulk fibrous components, Sand cores, General drying & heating of electrically non-conductive materials, and research and development. The details of design for different reactors are shown in Table 6.

Conveyor Type Modular Industrial Microwave Systems

Thermex Thermatron's Conveyorized Modular Industrial Microwave Systems are built in a large variety of shapes and sizes according to the requirements of the continuous heating process. Each module can apply up to 100 kW of power to the product being heated and operates at 915 MHz. Based on the power requirements, two or more modules can be installed on the same conveyor as shown in Figure 21. To assure uniform heat distribution in a large variety of load configurations, multimode cavity is provided with a waveguide splitter with dual microwave feed points and mode stirrers. The product being heated is carried through the oven by a unique conveyor belt. The belt material and configuration are selected based on the nature of the product being heated. Each end of the conveyor is provided with a special vestibule to suppress any microwave energy leakage. Air intake and exhaust ports are provided for circulating air to be used in cases where vapors or fumes are developed during the heating process.

Microwave Reactor Design and Configurations 91

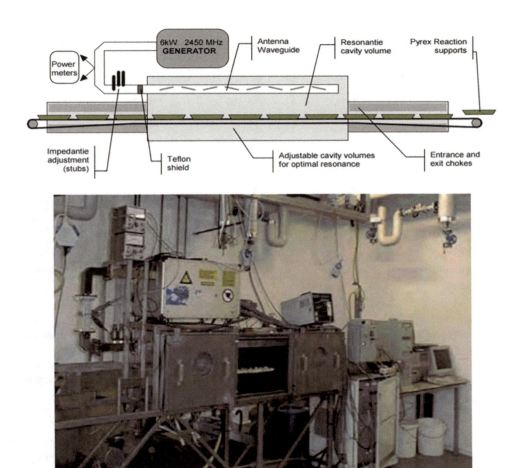

Figure 19. Schematic of a 6 kW continuous microwave heterogeneous-media reactor. Total length 4 m (Chemat-Djenni et al. 2007).

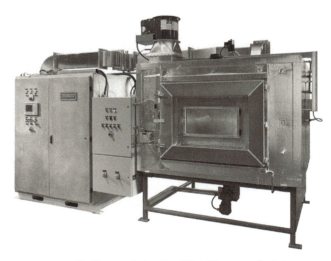

Figure 20. Thermex Laboratory/Pilot Microwave System.

Figure 22 shows a schematic representation of the unit. This system is equipped with a programmable logic controller which provides automatic operation, programmable to operate at maximum efficiency,

Table 6. Dimensions of Thermex Laboratory/Pilot Microwave System.

Model Number	TM6BP	TM15BP	TM75BP	TM100BP
Power Output (Adjustable)	6 kW	15 kW	75 kW	100 kW
Phases	3	3	3	3
Frequency	2450 MHz	2450 MHz	915 MHz	915 MHz
Generator Dimensions	64" High 18" Deep 36" Wide	76" High 24" Deep 62" Wide	78" High 48" Deep 72" Wide	78" High 48" Deep 72" Wide
Applicator Dimensions	38" High 48" Deep 48" Wide	38" High 48" Deep 48" Wide	60" High 72" Deep 72" Wide	60" High 72" Deep 72" Wide
Optional Turntable	36" diameter	36" diameter	42" diameter	42" diameter

recording and storage of up to 32 operating parameter variables, auxiliary hot air system uses electric heat to preheat the air circulating throughout the cavities, on screen diagnostics, help screens and alarm messaging to assist with troubleshooting, designed for ease of inspection, and installation and maintenance. Applications include ceramics, catalysts, pyrolysis, vulcanizing, composites, bulk fibrous components, sand cores, drying and heating of electrically non-conductive materials, and research and development.

Batch and Continuous Flow Reactors

SAIREM, a European company collaboration with École Nationale de Chimie in Montpellier, installed a pilot plant at 915 MHz at a semi-industrial scale (Figure 23). The main parameters of the pilot-scale installation are: installed maximum microwave power of 30 kW, 915 MHz; fast control of forward and reflected power and high attainable microwave power densities; integration between microwave generator and reactor ensures internal compatibility and control of all system components; continuous flow reactor up to 5 L/min; batch reactor integral with INTLI, with variable speed mechanical stirring, with maximum volume of 100 L adapted particularly for vegetal-type extraction in aqueous phase or solvents; possible recirculation of the extracted product back in to the reaction mixture (loop operation); process could be carried out under an inert atmosphere (N_2, Ar, CO_2, etc.); possibility of on-line filtration or distillation of the products; reactor external cooling via a cooling jacket with automated temperature control; and *in situ* temperature measurement.

Microwave pyrolysis reactors

A Scandinavian biofuel company has developed a continuous flow microwave pyrolysis reactor (Figure 24). This process is more flexible and easy to operate when compared with conventional incineration. Incineration is complicated and expensive to control, and will usually create harmful or toxic components that have to be removed from the flue gas. The pyrolysis is a process with no oxygen. For the same reason dioxins cannot occur, as the formation of dioxins is dependent on the presence of oxygen. The process is completely enclosed, and all products are collected and duly treated without any emissions to the

Figure 21. Photo-image of Thermex Thermatron's Conveyor type Modular Industrial Microwave Systems.

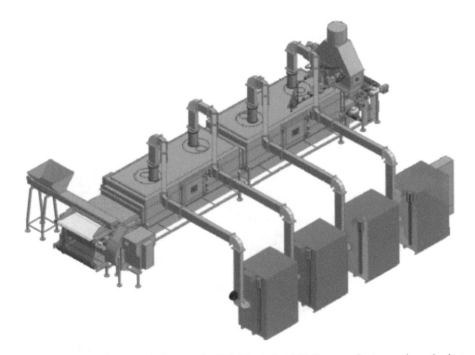

Figure 22. Thermex Thermatron's Conveyorized Modular Industrial Microwave Systems, schematic view.

environment, called dry distillation process. As no oxygen is added to the process, the produced gas will be a concentrated fuel gas with high calorific value. The feedstock is brought into the pyrolysis reactor through air locks purged with inert gas to prevent oxygen to enter the reactor. It is then heated by means of microwaves to a temperature where the bonds between the solids and the volatiles in the material are broken. The volatile fraction consists of a vapor that is separated into gases and fluids by condensation. The process economics and payback periods are shown in Table 7.

94 *Microwave-Mediated Biofuel Production*

Figure 23. Microwave-assisted 915 MHz installation for extraction and synthesis with Batch & Continuous flow reactors by SAIREM.

Figure 24. Continuous flow microwave pyrolysis reactor developed by Scandinavian Biofuels Company.

Table 7. Examples of microwave assisted pyrolysis systems with different feedstock and productivities and economic analysis.

Feedstock	Rice straw	Bagasse	Coconut	Plastic	Car tires
Capacity/yr	27,778 tons	28,409 tons	25,000 tons	25,000 tons	25,000 tons
Water content	20%	20%	15%	5%	1%
Net capacity/yr	25,000 tons	25,000 tons	25,000 tons	25,000 tons	25,000 tons
Output Oil	13,250 tons	16,250 tons	17,500 tons	12,500 tons	5,600 tons
Output Gas	3,500 tons	3,250 tons	3,250 tons	10,000 tons	3,000 tons
Output Carbon	8,250 tons	5,500 tons	4,250 tons	1,240 tons	11,000 tons
Electricity output	32.4 GWh	39 GWh	38.5 GWh	103.4 GWh	44.5 GWh
Electricity use	3.79 GWh	3.79 GWh	3.79 GWh	3.79 GWh	4.57 GWh
Electricity sales	28.66 GWh	35.17 GWh	34.74 GWh	99.65 GWh	39.89 GWh
Payback/yrs	2.9 years	2.6 years	2.1 years	1.4 years	2.2 years

Conclusions

Microwave reactor design and technology and their applications have been interesting to many researchers around the world. A great deal of work and interesting findings have been reported. However, there are many other immediate concerns that need to be overcome in the areas of process configurations, microwave field application and cavity design and large-scale reactor development and demonstration. The near-term future of microwave technology lies in further technical improvements in ease of use, throughput, and materials advancement. Combinations of improvements will allow significant advances in these areas. Enhanced flexibility improvements will enable the treatment of different sample matrices within the same run or batch to be independent of each other in terms of reagents and methodologies, allowing the operator freedom from traditional limitations of matching sample sizes and types within a batch. Thus, a customized program providing the optimum analytical solution for analysis independent of all other samples can essentially treat each sample separately. Improvements in computing power are also allowing finer control of the processes, allowing the instrument to follow the set measurement processes required for application areas such as speciation studies. Vessel technology and improvements in materials are allowing smaller, stronger, and more effective vessels to be developed, which include or incorporate features to control processes that are currently separate and discrete from the dissolution. Automated process control is also continuing to develop and evolve to require minimal operator involvement in the analytical process. Consistency and control are becoming the premise of a new generation of equipment.

References

Abramovitch, R.A., ChangQing, L., Hicks, E. and Sinard, J. 2003. *In situ* remediation of soils contaminated with toxic metal ions using microwave energy. *Chemosphere*. 53(9): pp.1077–1085.

Appleton, T.J., Colder, R.I., Kingman, S.W., Lowndes, I.S. and Read, A.G. 2005. Microwave technology for energy-efficient processing of waste. *Applied Energy*. 81(1): pp.85–113.

Asmussen, J., Lin, H.H., Manring, B. and Fritz, R. 1987. Single mode or controlled multimode microwave cavity applicators for precision materials processing. *Review of Scientific Instruments*. 58(8): pp.1477–1486.

Balasubramanian, S., Allen, J.D., Kanitkar, A. and Boldor, D. 2011. Oil extraction from Scenedesmus obliquus using a continuous microwave system–design, optimization, and quality characterization. *Bioresource Technology*. 102(3): pp.3396–3403.

Barclay, D. 2004. Microwave digestion moves into the 21st century. pp.28–32.

Baxendale, I.R., Hayward, J.J. and Ley, S.V. 2007. Microwave reactions under continuous flow conditions. *Combinatorial Chemistry and High Throughput Screening*. 10(10): pp.802–836.

Bhattacharya, M. and Basak, T. 2006. On the analysis of microwave power and heating characteristics for food processing: Asymptotes and resonances. *Food Research International*. 39(10): pp.1046–1057.

Bogdal, D. 2005. Microwave-assisted organic synthesis: one hundred reaction procedures (Vol. 25). Elsevier.

Bohlmann, J.T., Lorth, C.M., Drews, A. and Buchholz, R. 1999. Microwave high-pressure thermochemical conversion of sewage sludge as an alternative to incineration. *Chemical Engineering and Technology*. 22(5): pp.404–409.

Boldor, D., Balasubramanian, S., Purohit, S. and Rusch, K.A. 2008. Design and implementation of a continuous microwave heating system for ballast water treatment. *Environmental Science and Technology*. 42(11): pp.4121–4127.

Bowman, M.D., Holcomb, J.L., Kormos, C.M., Leadbeater, N.E. and Williams, V.A. 2007. Approaches for scale-up of microwave-promoted reactions. *Organic Process Research and Development*. 12(1): pp.41–57.

Bradshaw, S.M., Van Wyk, E.J. and De Swardt, J.B. 1998. Microwave heating principles and the application to the regeneration of granular activated carbon. *Journal of the South African Institute of Mining and Metallurgy (South Africa)*. 98(4): pp.201–210.

Braşoveanu, M. and Nemţanu, M.R. 2014. Behaviour of starch exposed to microwave radiation treatment. *Starch-Stärke*. 66(1-2): pp.3–14.

Chemat-Djenni, Z., Hamada, B. and Chemat, F. 2007. Atmospheric pressure microwave assisted heterogeneous catalytic reactions. *Molecules*. 12(7): pp.1399–1409.

Choedkiatsakul, I., Ngaosuwan, K., Assabumrungrat, S., Mantegna, S. and Cravotto, G. 2015. Biodiesel production in a novel continuous flow microwave reactor. *Renewable Energy*. 83: pp.25–29.

Církva, V. and Relich, S. 2011. Microwave photochemistry. Applications in organic synthesis. *Mini-Reviews in Organic Chemistry*. 8(3): pp.282–293.

Cleophax, J., Liagre, M., Loupy, A. and Petit, A. 2000. Application of focused microwaves to the scale-up of solvent-free organic reactions. *Organic Process Research and Development*. 4(6): pp.498–504.

Collin, R.E. 1966. Foundation of Microwave Engineering, McGraw-Hill Book Co. New York.

Coronel, P., Simunovic, J., Sandeep, K.P., Cartwright, G.D. and Kumar, P. 2008. Sterilization solutions for aseptic processing using a continuous flow microwave system. *Journal of Food Engineering*. 85(4): pp.528–536.

Cravotto, G. and Cintas, P. 2007. The combined use of microwaves and ultrasound: improved tools in process chemistry and organic synthesis. *Chemistry–A European Journal*. 13(7): pp.1902–1909.

Cravotto, G., Boffa, L., Mantegna, S., Perego, P., Avogadro, M. and Cintas, P. 2008. Improved extraction of vegetable oils under high-intensity ultrasound and/or microwaves. *Ultrasonics sonochemistry*. 15(5): pp.898–902.

Dall'Oglio, E.L., de Sousa Jr, P.T., Campos, D.C., de Vasconcelos, L.G., da Silva, A.C., Ribeiro, F., Rodrigues, V. and Kuhnen, C.A. 2015. Measurement of dielectric properties and microwave-assisted homogeneous acid-catalyzed transesterification in a monomode reactor. *The Journal of Physical Chemistry A*. 119(34): pp.8971–8980.

Dall'Oglio, E.L., Sousa Jr, P.T.D., Oliveira, P.T.D.J., Vasconcelos, L.G.D., Parizotto, C.A. and Kuhnen, C.A. 2014. Use of heterogeneous catalysts in methylic biodiesel production induced by microwave irradiation. *Química Nova*. 37(3): pp.411–417.

Dé, A., Ahmad, I., Whitney, E.D. and Clark, D.E. 1990. Effect of Green Microstructure and Processing Variables on the Microwave Sintering of Alumina. *MRS Proceedings* (Vol. 189, p.283). Cambridge University Press.

Demeuse, M.T. and Johnson, A.C. 1994. Variable Frequency Microwave Processing of Thermoset Polymer Matrix Composites. *MRS Proceedings* (Vol. 347, p.723). Cambridge University Press.

Diaz-Ortiz, A. and Moreno, A. 2005. Activation of organic reactions by microwaves. *Advances in Organic Synthesis*. 1(1): pp.119–171.

Díaz-Ortiz, A., Moreno, A., Sanchez-Migallon, A., Prieto, P., Carrillo, J.R., Vazquez, E., Gomez, M. and Herrero, M.A. 2007. Microwave-assisted reactions in heterocyclic compounds with applications in medicinal and supramolecular chemistry. *Combinatorial Chemistry & High Throughput Screening*. 10(10): pp.877–902.

Díaz-Ortiz, A., de la Hoz, A., Alcázar, J., R Carrillo, J., A Herrero, M., Fontana, A., de M Munoz, J., Prieto, P. and de Cózar, A., 2011. Influence of polarity on the scalability and reproducibility of solvent-free microwave-assisted reactions. *Combinatorial Chemistry & High Throughput Screening*. 14(2): pp.109–116.

Domí, A., Menendez, J.A., Inguanzo, M., Bernad, P.L. and Pis, J.J. 2003. Gas chromatographic–mass spectrometric study of the oil fractions produced by microwave-assisted pyrolysis of different sewage sludges. *Journal of chromatography A*. 1012(2): pp.193–206.

Domínguez, A., Menéndez, J.A., Inguanzo, M. and Pis, J.J. 2005. Investigations into the characteristics of oils produced from microwave pyrolysis of sewage sludge. *Fuel Processing Technology*. 86(9): pp.1007–1020.

Driggers, R.G. (ed.). 2003. Encyclopedia of Optical Engineering: Las-Pho. 2: pp.1025–2048. CRC press.

Drozd, J.M. and Joines, W.T., Industrial Microwave Systems, Inc., 2001. Electromagnetic exposure chamber with a focal region. *U.S. Patent*. 6,265,702.

Franssen, I.S., Irimia, D., Stefanidis, G.D. and Stankiewicz, A.I. 2014. Practical challenges in the energy-based control of molecular transformations in chemical reactors. *AIChE Journal*. 60(10): pp.3392–3405.

Ferguson, J.D. 2003. Focused™ microwave instrumentation from CEM Corporation. *Molecular Diversity*. 7(2): pp.281–286.

Ghaffariyan, S.R. and Methven, J.M. 1990. The Design of a TM 10 Resonant Cavity Microwave Applicator as a Preheating and Crosslinking Die for Pultruded Composites. *MRS Proceedings* (Vol. 189, p.135). Cambridge University Press.

Glasnov, T.N. and Kappe, C.O. 2007. Microwave-assisted synthesis under continuous-flow conditions. *Macromolecular Rapid Communications*. 28(4): pp.395–410.

Groisman, Y. and Gedanken, A. 2008. Continuous flow, circulating microwave system and its application in nanoparticle fabrication and biodiesel synthesis. *The Journal of Physical Chemistry C*. 112(24): pp.8802–8808.

Hill, J.M. and Marchant, T.R. 1996. Modelling microwave heating. *Applied Mathematical Modelling*. 20(1): pp.3–15.

Janney, M.A., Calhoun, C.L. and Kimrey, H.D. 1992. Microwave sintering of solid oxide fuel cell materials: I, Zirconia-8 mol% Yttria. *Journal of the American Ceramic Society*. 75(2): pp.341–346.

Johnson, A.C., Lauf, R.J. and Surrett, A.D. 1994. Effect of Bandwidth on Uniformity of Energy Distribution in a Multimode Cavity. *MRS Proceedings* (Vol. 347, p. 453). Cambridge University Press.
Kappe, C.O. 2004. Controlled microwave heating in modern organic synthesis. *Angewandte Chemie International Edition.* 43(46): pp.6250–6284.
Kappe, C.O., Dallinger, D. and Murphree, S.S. 2009. Practical Microwave Synthesis for Organic Chemists: Strategies, Instruments, and Protocols Wiley.
Kappe, C.O., Stadler, A. and Dallinger, D. 2012. Microwaves in Organic and Medicinal Chemistry. John Wiley & Sons.
Kashyap, S.C. and Wyslouzil, W. 1977. Methods for improving Heating Uniformity of Microwave Owens. *Journal of Microwave Power.* 12(3): pp.224–230.
Kimrey, H.D. and Janney, M.A. 1988. Design Principles for High-Frequency Microwave Cavities. *MRS Proceedings* (Vol. 124, p.367). Cambridge University Press.
Kitchen, H.J., Vallance, S.R., Kennedy, J.L., Tapia-Ruiz, N., Carassiti, L., Harrison, A. and Gregory, D.H. 2013. Modern microwave methods in solid-state inorganic materials chemistry: from fundamentals to manufacturing. *Chemical Reviews.* 114(2): pp.1170–1206.
Komorowska-Durka, M., Barmen't Loo, M., Sturm, G.S., Radoiu, M., Oudshoorn, M., Van Gerven, T., Stankiewicz, A.I. and Stefanidis, G.D. 2013. Novel microwave reactor equipment using internal transmission line (INTLI) for efficient liquid phase chemistries: A study-case of polyester preparation. *Chemical Engineering and Processing: Process Intensification.* 69: pp.83–89.
Kumar, P., Coronel, P., Truong, V.D., Simunovic, J., Swartzel, K.R., Sandeep, K.P. and Cartwright, G. 2008. Overcoming issues associated with the scale-up of a continuous flow microwave system for aseptic processing of vegetable purees. *Food Research International.* 41(5): 454–461.
Lagha, A., Chemat, S., Bartels, P.V. and Chemat, F. 1999. Microwave-ultrasound combined reactor suitable for atmospheric sample preparation procedure of biological and chemical products. *Analusis.* 27(5): pp.452–457.
Leadbeater, N.E. 2010. *Microwave Heating as a Tool for Sustainable Chemistry.* CRC Press.
Lee, B.K., Ellenbecker, M.J. and Moure-Ersaso, R. 2004. Alternatives for treatment and disposal cost reduction of regulated medical wastes. *Waste Management.* 24(2): pp.143–151.
Lehmann, H. and LaVecchia, L. 2005. Evaluation of microwave reactors for prep-scale synthesis in a kilolab. *Journal of the Association for Laboratory Automation.* 10(6): pp.412–417.
Leonelli, C. and Mason, T.J. 2010. Microwave and ultrasonic processing: now a realistic option for industry. *Chemical Engineering and Processing: Process Intensification.* 49(9): pp.885–900.
Letellier, M., Budzinski, H., Garrigues, P. and Wise, S. 1997. Focused microwave-assisted extraction of polycyclic aromatic hydrocarbons in open cell from reference materials (sediment, soil, air particulates). *Spectroscopy-Ottawa-.* 13: pp.71–80.
Lewis, D.A., Summers, J.D., Ward, T.C. and McGrath, J.E. 1992. Accelerated imidization reactions using microwave radiation. *Journal of Polymer Science Part A: Polymer Chemistry.* 30(8): pp.1647–1653.
Li, H., Pordesimo, L.O., Weiss, J. and Wilhelm, L.R. 2004. Microwave and ultrasound assisted extraction of soybean oil. *Transactions of the ASAE.* 47(4): pp.1187–1194.
Li, W., Peng, J., Zhang, L., Yang, K., Xia, H., Zhang, S. and Guo, S.H. 2009. Preparation of activated carbon from coconut shell chars in pilot-scale microwave heating equipment at 60 kW. *Waste Management.* 29(2): pp.756–760.
Lin, Q.H., Cheng, H. and Chen, G.Y. 2012. Preparation and characterization of carbonaceous adsorbents from sewage sludge using a pilot-scale microwave heating equipment. *Journal of Analytical and Applied Pyrolysis.* 93: pp.113–119.
Liu, S.Y., Wang, Y.F., McDonald, T. and Taylor, S.E. 2008. Efficient production of biodiesel using radio frequency heating. *Energy and Fuels.* 22(3): pp.2116–2120.
Matthes, R., Bernhardt, J.H. and Mac Kinlay, A.F. (eds.). 1999. Guidelines on limiting exposure to non-ionizing radiation: A reference book based on the guidelines on limiting exposure to non-ionizing radiation and statements on special applications. International Commission on Non-Ionizing Radiation Protection.
Menéndez, J.A., Arenillas, A., Fidalgo, B., Fernández, Y., Zubizarreta, L., Calvo, E.G. and Bermúdez, J.M. 2010. Microwave heating processes involving carbon materials. *Fuel Processing Technology.* 91(1): pp.1–8.
Meredith, R.J. 1998. Engineers' handbook of industrial microwave heating (No. 25). IET.
Metaxas, A.A. and Meredith, R.J. 1983. Industrial microwave heating (No. 4). IET.
Metaxas, A.C. 1993. Applications for industrial microwave processing. *Ceram. Trans.* 36: pp.549–562.
Morschhäuser, R., Krull, M., Kayser, C., Boberski, C., Bierbaum, R., Püschner, P.A., Glasnov, T.N. and Kappe, C.O., 2012. Microwave-assisted continuous flow synthesis on industrial scale.pp. 281-290.
Moseley, J.D., Lenden, P., Lockwood, M., Ruda, K., Sherlock, J.P., Thomson, A.D. and Gilday, J.P. 2007. A comparison of commercial microwave reactors for scale-up within process chemistry. *Organic Process Research and Development.* 12(1): pp.30–40.
Moseley, J.D. and Woodman, E.K. 2008. Scaling-out pharmaceutical reactions in an automated stop-flow microwave reactor. *Organic Process Research & Development.* 12(5): pp.967–981.
Nüchter, M., Ondruschka, B., Bonrath, W. and Gum, A. 2004. Microwave assisted synthesis–a critical technology overview. *Green Chemistry.* 6(3): pp.128–141.
Özçimen, D. and Yücel, S. 2011. *Novel Methods in Biodiesel Production.* INTECH Open Access Publisher.
Palaith, D., Silberglitt, R., Wu, C.C., Kleiner, R. and Libeld, E.L. 1988. Microwave Joining of Ceramics. *MRS Proceedings* (Vol. 124, p. 255). Cambridge University Press.

Qiu, Y. and Hawley, M.C. 1998. Automated variable frequency microwave processing of graphite/epoxy composite in a single mode cavity. *In*: Thirteenth Technical Conference of the American Society for Composites (p.1998).

Ragaini, V., Pirola, C., Borrelli, S., Ferrari, C. and Longo, I. 2012. Simultaneous ultrasound and microwave new reactor: detailed description and energetic considerations. *Ultrasonicssonochemistry*. 19(4): pp.872–876.

Ramakrishna, D., Travis, S. and Hawley, M.C. 1992. Microwave Processing of Glass Fiber/Vinyl Ester-Vinyl Toluene Composites. *MRS Proceedings* (Vol. 269, p.431). Cambridge University Press.

Rattanadecho, P. 2006. The simulation of microwave heating of wood using a rectangular wave guide: Influence of frequency and sample size. *Chemical Engineering Science*. 61(14): pp.4798–4811.

Regier, M. and Schubert, H. 2001. Microwave processing. Thermal technologies in food processing (ed. P. Richardson). CRC Press, New York. pp.178–207.

Reimbert, C.G., Minzoni, A.A. and Smyth, N.F. 1996. Effect of radiation losses on hotspot formation and propagation in microwave heating. *IMA Journal of Applied Mathematics*. 57(2): pp.165–179.

Risman, P.O., Ohlsson, T. and Wass, B. 1987. Principles and models of power density distribution in microwave oven loads. *J. Microwave Power and Electromagnetic Energy*. 22: pp.193–198.

Rodríguez, A.M., Prieto, P., de la Hoz, A., Díaz-Ortiz, Á., Martín, D.R. and García, J.I. 2015. Influence of Polarity and Activation Energy in Microwave–Assisted Organic Synthesis (MAOS). *ChemistryOpen*. 4(3): pp.308–317.

Saifuddin, N. and Chua, K.H. 2004. Production of ethyl ester (biodiesel) from used frying oils: Optimization transesterification process using microwave irradiation. *Malaysian Journal of Chemistry*. pp.77–82.

Salvi, D., Boldor, D., Ortego, J., Aita, G.M. and Sabliov, C.M. 2010. Numerical modeling of continuous flow microwave heating: a critical comparison of COMSOL and ANSYS. *Journal of Microwave Power and Electromagnetic Energy*. 44(4): pp.187–197.

Sihvola, A.H. 1999. *Electromagnetic Mixing Formulas and Applications* (No. 47). London, United Kingdom: Institution of Electrical Engineers.

Stefanidis, G.D., Munoz, A.N., Sturm, G.S. and Stankiewicz, A. 2014. A helicopter view of microwave application to chemical processes: reactions, separations, and equipment concepts. *Reviews in Chemical Engineering*. 30(3): pp.233–259.

Stein, D.F. 1994. Microwave Processing of Materials, National Materials Advisory Board.

Stein, D.F., Edgar, R.H., Iskander, M.F., Johnson, D.L., Johnson, S.M., Lob, C.G., Shaw, J.M., Sutton, W.H., Tien, P.K. and Munns, T.E. 1994. Microwave Processing-An Emerging Industrial Technology? *In*: Materials Research Society Symposium Proceedings (Vol. 347, pp. 3–3). Materials Research Society.

Sturm, G.S.J. 2013. *Microwave Field Applicator Design in Small-Scale Chemical Processing*. TU Delft, Delft University of Technology.

Surret, A.D., Lauf, R.J., Paulauskas, F.L. and Johnson, A.C. 1994. Polymer Curing Using Variable Frequency Microwave Processing. *MRS Proceedings* (Vol. 347, p.691). Cambridge University Press.

Terigar, B.G. 2009. *Advanced Microwave Technology for Biodiesel Feedstock Processing* (Doctoral dissertation, Louisiana State University).

Terigar, B.G., Balasubramanian, S., Sabliov, C.M., Lima, M. and Boldor, D. 2011. Soybean and rice bran oil extraction in a continuous microwave system: from laboratory-to pilot-scale. *Journal of Food Engineering*. 104(2): pp.208–217.

Thostenson, E.T. and Chou, T.W. 1999. Microwave processing: fundamentals and applications. Composites Part A: Applied Science and Manufacturing. 30(9): pp.1055–1071.

Tian, Y., Zuo, W., Ren, Z. and Chen, D. 2011. Estimation of a novel method to produce bio-oil from sewage sludge by microwave pyrolysis with the consideration of efficiency and safety. *Bioresource Technology*. 102(2): pp.2053–2061.

Tinga, W.R., Xu, J.D. and Vermeulen, F.E. 1995. *Open Coaxial Microwave Spot Joining Applicator* (No. CONF-9505249--). American Ceramic Society, Westerville, OH (United States).

Toukoniitty, B., Mikkola, J.P., Murzin, D.Y. and Salmi, T. 2005. Utilization of electromagnetic and acoustic irradiation in enhancing heterogeneous catalytic reactions. *Applied Catalysis A: General*. 279(1): pp.1–22.

Tran, V.N. 1992. *An Applicator Design for Processing Large Quantities of Dielectric Material*. Deakin Univ Victoria (Australia).

Vogt, G.J. and Unruh, W.P. 1992. Processing Aerosols and Filaments in a TM 010 Microwave Cavity at 2.45 GHz. *MRS Proceedings* (Vol. 269, p.245). Cambridge University Press.

Vogt, G.J., Regan, A.H., Rohlev, A.S. and Curtin, M.T. 1995. *Use of a Variable Frequency Source with a Single-Mode Cavity to Process Ceramic Filaments* (No. LA-UR--95-1440; CONF-950401--2). Los Alamos National Lab., NM (United States).

Waheed ul Hasan, S. and Ani, F.N. 2014. Review of limiting issues in industrialization and scale-up of microwave-assisted activated carbon production. *Industrial & Engineering Chemistry Research*. 53(31): pp.12185–12191.

Wei, J., Hawley, M.C., Delong, J.D. and Demeuse, M. 1993. Comparison of microwave and thermal cure of epoxy resins. *Polymer Engineering and Science*. 33(17): pp.1132–1140.

Wei, J., Delgado, R., Hawley, M.C. and Demeuse, M.T. 1994. Dielectric Analysis of Semi-Crystalline Poly (ethylene terephthalate). *MRS Proceedings* (Vol. 347, p.735). Cambridge University Press.

Wu, X., Thomas Jr, J.R. and Davis, W.A. 2002. Control of thermal runaway in microwave resonant cavities. *Journal of Applied Physics*. 92(6): pp.3374–3380.

Zaini, M.A.A. and Kamaruddin, M.J. 2013. Critical issues in microwave-assisted activated carbon preparation. *Journal of Analytical and Applied Pyrolysis*. 101: pp.238–241.

Zuo, W., Tian, Y. and Ren, N. 2011. The important role of microwave receptors in bio-fuel production by microwave-induced pyrolysis of sewage sludge. *Waste Management*. 31(6): pp.1321–1326.

4

Microwave Mediated Biodiesel Production

Introduction

Energy, economic and environmental security related concerns at local, regional, national and global levels have invigorated the renewable energy research in recent years (Gude et al. 2013). Present petroleum consumption is 10^5 times faster than what nature can create (Satyanarayana et al. 2011) and at this rate of consumption, it is estimated that the world's fossil fuel reserves will be diminished by 2050 (Demirbas 2009). Apart from this, the fuel consumption is expected to rise by 60% in the next 25 years (Rittmann 2008). In an effort to achieve independence from the fossil fuel sources and imports from oil-rich countries and maintain environmental sustainability, many countries have committed to increase renewable energy production and/or reduce greenhouse gas emission at national and international levels (Fabbri et al. 2007). Policy amendments and changes in energy management strategies have been considered as well.

Renewable energy sources such as solar thermal and photovoltaic collectors, wind and geothermal sources have limitations related to location, availability, and intensity. These energy sources are not available in liquid form to serve as fuels. Because most of the transportation and industrial sectors need liquid fuels to drive the machinery and engines, more emphasis has been devoted to alternative liquid fuel sources such as biodiesel. Biodiesel is composed of methyl or ethyl esters produced from vegetable oil or animal oil and has fuel properties similar to diesel fuel which renders its useful as a biofuel. Biodiesel offers many benefits (Gude et al. 2013):

a) Serves as an alternative to petroleum-derived fuel, minimizing dependence on crude oil foreign imports;
b) Provides favorable energy returns on energy invested, i.e., high net energy balance;
c) Reduces greenhouse emissions;
d) Lowers harmful gaseous emissions such as nitric oxide; and
e) Is a biodegradable and nontoxic fuel, with minimal effect on reservoirs, lakes, marine life, and other environmentally sensitive areas (Georgoginanni et al. 2007, Yang et al. 2011, Hoekman 2009).

Local biodiesel production can address challenges related to energy independence, economic prosperity, and environmental sustainability in any nation. There are two major challenges that inhibit biodiesel production—high feedstock costs and conversion process of oils to biodiesel. Low cost feedstock and recycling of waste cooking oils and animal fats can be considered to minimize the feedstock costs. Process improvements and optimization can help reduce the biodiesel conversion process capital and operational costs. Biodiesel production involves two main steps: (1) feedstock preparation and extraction of oils, and (2) conversion (esterification and or transesterification) of oils (free fatty acids, FFAs) to biodiesel (fatty acid alkyl esters, FAAE). Common methods employed to demonstrate these two steps simultaneously or in series include conventional heating, high pressure and temperature reactions such as thermal liquefaction and pyrolysis. These methods are employed based on the feedstock type and quality (Haas et al. 2006). These methods are not energy-efficient and are expensive and offer scope for further improvements. Several process modifications and improvements were reported both at laboratory research and industrial levels in recent years (Harvey et al. 2003, Haas et al. 2006, Cintas et al. 2010). These include application of radio frequency and ultrasound (Fabbri et al. 2007, Georgogianni et al. 2007). Biodiesel production using ultrasound is an attractive option because it improves extraction and transesterification processes. However, this method may require longer reaction times and larger volumes of solvents possibly with excess energy consumption when compared with microwave enhanced process (Cintas et al. 2010).

Microwaves have been studied extensively due to their ability to complete chemical reactions in very short times. Few advantages with microwave processing can be listed as: rapid heating and cooling; cost savings due to energy, time and work space savings; precise and controlled processing; selective heating; volumetric and uniform heating; reduced processing time; improved quality and properties; and effects not achievable by conventional means of heating (Giguere et al. 1986, Clark and Sutton 1996, Varma 1999, Ku et al. 2002, de la Hoz et al. 2005, Roberts et al. 2005, Caddick and Fitzmaurice 2009). Microwaves have been used by many researchers around the world in many organic and inorganic syntheses at exploratory levels as discussed in Chapter 1. Recently, many industries have successfully implemented microwave based processes. Examples include ceramic/ceramic matrix composite sintering and powder processing, polymers and polymer-matrix composites processing, microwave plasma processing of materials, and minerals processing (Clark and Sutton 1996). Microwaves have the ability to induce reactions even in solvent-free conditions offering sustainable solutions to many environmental problems related to hazardous and toxic contaminants (Varma 1999). Due to these advantages, microwaves provide for tremendous opportunities to improve biodiesel conversion processes from different feedstocks and oils. This chapter will provide the basics of microwave energy applications specific to biodiesel preparation and processing, and explanation of microwave effect on the chemical reactions involved in extraction and transesterification, update on process utilization and improvements, and information related to different process configurations and reactor designs available for biodiesel production.

Biodiesel Production Technologies

Commercial biodiesel production processes are based on either conventional or supercritical heating methods. Commonly used methods are: (i) Pyrolysis; (ii) Micro-emulsions; (iii) Dilution; and (iv) Transesterification of oils to esters (Akoh et al. 2007, Maher and Bressler 2007, Sharma et al. 2008). Among these methods, transesterification is the simplest and most economical route to produce biodiesel, with physical characteristics similar to fossil diesel and little or no deposit formation when used in diesel engines. Transesterification of oils from any feedstock is to simply reduce the viscosity of the oils derived from them. Transesterification is a process in which an alcohol (methanol or ethanol) in the presence of a catalyst (acid or alkali or enzyme) is used to chemically break the molecule of the vegetable oils or animal fats into methyl or ethyl esters of the renewable fuel. The overall transesterification process is a sequence of three consecutive and reversible reactions, in which di- and mono-glycerides are formed as intermediates, yielding one ester molecule in each step as shown in Scheme 1.

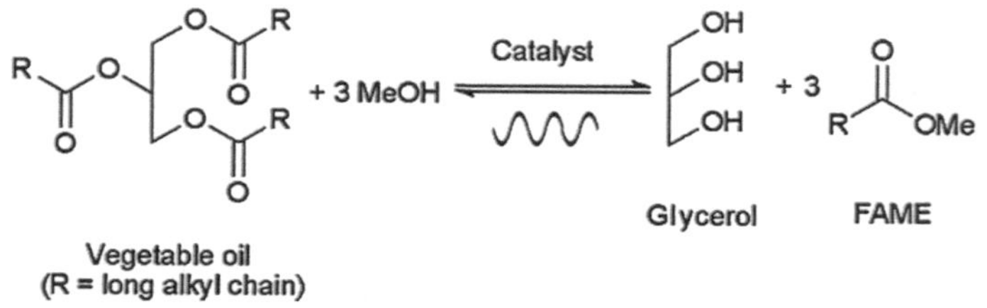

Scheme 1. Transesterification reaction from vegetable oils (triglycerides) to fatty acid methyl esters.

The stoichiometric reaction requires 1 mole of a triglyceride and 3 moles of alcohol. However, excess amount of alcohol is used to increase the yields of the alkyl esters by shifting the equilibrium towards the formation of esters and to allow its phase separation from the glycerol formed as a by-product. The product of transesterification process is known as "biodiesel".

While transesterification of oils to produce biodiesel is a well-established method, process conversion and energy utilization inefficiencies increase the biodiesel product costs. These are mainly associated with the heating method employed in the process. A general process diagram for biodiesel production via transesterification reaction is shown in Figure 1. The process starts with oils extraction from feedstock and mixed with methanol and catalyst prior to entering the transesterification reactor. If the feedstock has a free fatty acid content of greater than 3%, a two-step process including esterification and transesterification should be considered. Esterification is usually performed with an acid catalyst while the transesterification reaction can be performed using an acid or a base catalyst. When the FFA content is less than 3%, it is possible to directly convert the oils (triglycerides) into biodiesel (fatty acid alkyl esters).

Transesterification reaction can be performed by the following methods: (1) conventional heating with acid, base catalysts and co-solvents (Gude et al. 2013, Srivastava and Prasad 2000, Kaieda et al. 2001, Machek and Skopal 2001, Zhang et al. 2003a,b, Suppes et al. 2004, Meher et al. 2006, Patil and Deng 2009, Patil et al. 2010); (2) sub- and super-critical methanol conditions with co-solvents and without a catalyst (King et al. 1999, Demirbas 2002, 2003, 2005, Kusdiana and Saka 2001, 2004a, b); (3) enzymatic method using lipases (Hernandez-Martin and Otero 2008, Hsu et al. 2004, Deng et al. 2005, Dossat et al. 1999, Kumari et al. 2009, Roy and Gupta 2003); and (4) microwave irradiation with acid, base and heterogeneous catalysts (Azcan and Danisman 2007, Refaat et al. 2008a, b, Melo-Junior et al. 2009). Among these methods, conventional heating method requires longer reaction times with higher energy inputs and losses (Refaat et al. 2008a). Super and sub-critical methanol process operates in expensive reactors at high temperatures and pressures resulting in higher energy inputs and higher production costs (Demirbas 2002, Melo-Junior et al. 2009, Yin et al. 2008). The enzymatic method, though operates at much lower temperatures, requires much longer reaction times (Akoh et al. 2007). Microwave-assisted transesterification, on the other hand, is energy-efficient and quick process to produce biodiesel from different feedstocks (Refaat et al. 2008b).

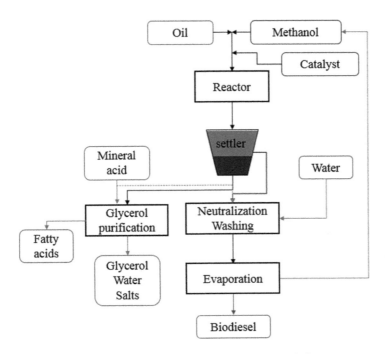

Figure 1. Generic schematic of a typical homogeneous catalytic process.

Microwaves in Biodiesel Production

Conventional methods for biodiesel production include pyrolysis, thermochemical liquefaction, supercritical reactors, oil and sand baths, and jacket type heating. In recent years, many researchers have tested application of microwaves in biodiesel production and optimization studies with various feedstocks. Microwave energy, a non-conventional heating method, can be utilized in biodiesel production in two main stages: (1) pretreatment of feedstock or oil extraction and (2) chemical esterification or transesterification reaction (Figure 2). It can be beneficial to combine the above two steps to perform a single-step extractive transesterification reaction as discussed in later sections. Biodiesel production involves mixing of appropriate ratios of oil, methanol (solvent) and catalysts. The mixture is then processed through a microwave reactor followed by separation of products to yield biodiesel and glycerin.

A comparison between the conventional, supercritical and microwave heating methods for biodiesel production is shown in Table 1. It can be noted that there are clear differences in operating parameters such as temperatures, reaction time and pressures. Supercritical heating requires shorted reaction time compared to conventional method while microwave heating produces quickest results. This could be an important advantage in process design, footprint and scale up. Supercritical and microwave heating methods provide for the possibility of conducting non-catalytic and solvent-free reactions which cannot be achieved by conventional heating.

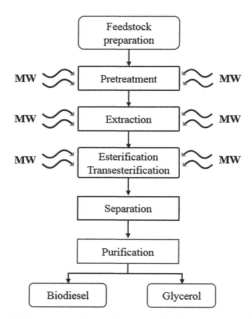

Figure 2. Microwave-enhanced transesterification of oils to biodiesel.

Microwaves in Biodiesel Chemistry

Biodiesel is composed of methyl or ethyl esters produced from vegetable oil or animal fats and has fuel properties similar to diesel fuel which renders its useful as a biofuel. Figure 3 shows the schematic of conversion of triacylglyderides and fatty acids to hydrocarbons or fatty acid methyl esters (FAME), i.e., biodiesel. Triacylglycerides can be hydrogenated at high temperatures to produce hydrocarbons or converted to biodiesel. In a similar manner, fatty acids can be transformed into biodiesel through esterification and/or transesterification reactions or hydrogenated to produce hydrocarbons.

Table 1. Comparison between three kinds of heating for biodiesel production (Gude et al. 2013).

Characteristic/parameter	Conventional heating	Supercritical heating	Microwave heating
Reaction time	Long (1–2 hr)	Short (< 1 hr)	Very short (0.05–0.1 hr)
Reaction temperature	40–100°C	250–400°C	40–100°C
Reaction pressure	Atmospheric	High pressure 35–60 MPa	Atmospheric*
Catalyst required	Yes	No (may be)	Yes/No
Heat losses	High	Moderate	Low
Form of energy	Electrical energy converted to thermal energy	Electrical energy converted to thermal energy	Electrical energy applied through microwaves
Process efficiency	Low	Moderate	High
Catalyst removal	Yes	No	Yes
Soap removal	Yes	No	Yes
Advantages	Simple operation, use of low energy source	Short reaction time (compared to conventional), easy product separation	Short reaction time, cleaner products, and energy efficient
Limitations	High energy requirements, saponified products	High capital costs, pressure vessel safety	May not be efficient with feedstock containing solids

*reactions at high pressure and temperatures without catalyst are possible.

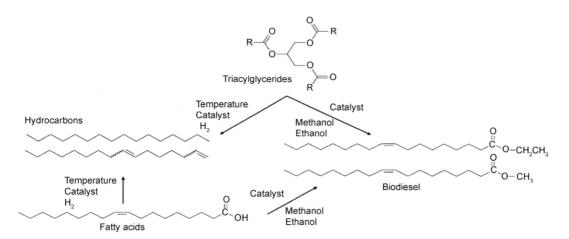

Figure 3. Biofuels (hydrocarbons and biodiesel) obtained from triacylglycerides and fatty acids (modified after Brandao et al. 2009).

Microwave mediated lipid extraction

Lipid extraction is an important step in biodiesel production. Earlier applications of microwave radiation were limited to the digestion of samples for measuring trace metals (Huffer et al. 1998) and extraction of organic contaminants (Marcato and Vianello 2000). The feasibility of extracting lipids using microwave irradiation was first reported in the mid-1980s (Ganzler et al. 1986). A microwave extraction technique for isolating lipids and pesticides from seeds, foods, feeds, and soil was developed which was more effective than the conventional procedures. This allowed further development of rapid, safe, and economical microwave based methods for extracting lipids. Similarly, use of microwave remains the most simple and most effective method among the other tested methods for microalgal lipid extraction (Lee et al. 2010).

A dielectric or polar material introduced in a rapidly oscillating electric field, such as that produced by microwaves, will generate heat because of the frictional forces arising from inter- and intra-molecular movements (Amarni and Kadi 2010). Intracellular heating results in the formation of water vapor, which disrupts the cells from within. This in turn leads to the electroporation effect, which further opens up the cell membrane, thereby rendering efficient extraction of intracellular metabolites (Rosenberg and Bogl 1987). Thus, rapid generation of heat and pressure within the biological system forces out compounds from the cell matrix, resulting in the production of good-quality extracts with better target compound recovery (Hemwimon et al. 2007). Šoštarič et al. (2012) suggested that microwave-pretreated microalgae have higher bio-oil yields because of the presence of several micro-cracks in the cell wall. Microwaves can also be used to extract and transesterify the oils into biodiesel. Microwave extraction is the preferred method at present because of the economics involved in the above process; it is expected to be attractive due to short reaction time, low-operating costs, and efficient extraction of algal oils. It was also reported that the recovery of biodiesel from the reaction mixture in a microwave-assisted process is approximately 15–20 min, which is far quicker when compared to the 6-h period in the conventional heating method (Refaat et al. 2008b). However, the disadvantage with the microwave-assisted process is the maintenance cost involved, particularly on a commercial scale (Ranjith kumar et al. 2015).

Selection of a suitable solvent for extraction is critical to achieving energy efficiency and cost objectives while reducing environmental burdens. Solvents should be environmentally benign, safe and non-toxic. It is unclear what it constitutes to be an environmentally friendly solvent because the organic solvents commonly used in lipid extraction such as hexane are well known for their toxic effects on human health and ecosystem. The volatile nature of these solvents at lower temperature further reduces their contact with analytes to be extracted and excessive quantities have to be used to ensure contact between the solvent and solute for efficient extraction (Iqbal and Theegala 2013). An ideal or a green solvent should possess the following characteristics or properties as defined by Gu and Jérôme (2013):

Availability—a green solvent should be available in large quantities and its production capacity should not vary with seasonal fluctuations in order to ensure a constant availability of the solvent in the market.

Price—green solvents have to be competitive in terms of price to ensure sustainability of the chemical processes.

Recyclability—it is desirable to recycle the solvent in any chemical process, but the recovery process should follow environmental-friendly procedures.

Grade—technical grade solvents are preferred in order to avoid energy-consuming purification processes required to obtain highly pure solvents.

Synthesis—green solvents should be prepared through an energy-saving process and the synthetic reactions should have high atom-economy, i.e., less waste generation.

Toxicity—green solvents have to exhibit negligible toxicity in order to reduce all risks when manipulated by humans or released in nature when used for personal and home care, paints, etc.

Biodegradability—green solvents should be biodegradable and should not produce toxic metabolites.

Performance—to be eligible, a green solvent should exhibit similar and even superior performance (viscosity, polarity, density, etc.) compared to currently employed solvents.

Stability—for use in a chemical process, a green solvent has to be thermally and (electro- and photo-) chemically stable.

Flammability—for safety reasons during manipulation, a green solvent should not be flammable.

Storage—a green solvent should be easy to store and should fulfill all legislations to be safely transported either by road, train, boat or plane.

Renewability—the use of renewable raw materials for the production of green solvents should be favored with respect to the carbon footprint.

Extraction processes are faced with stringent regulatory requirements and environmental regulations and cost and safety objectives (Chemat et al. 2015). These processes are expected to meet the production demands and fulfill regulations while increasing extraction efficiency (yield and selectivity toward compounds of interest), reducing or eliminating petrochemical solvents, together with moderate energy consumption. Adopting green extraction concepts is beneficial from several perspectives. These concepts should be based on green chemistry and engineering principles established earlier. As per Chemat et al. (2012) *"Green Extraction is based on the discovery and design of extraction processes which will reduce energy consumption, will allow use of alternative solvents and renewable natural products, and will ensure a safe and high quality extract/product."* They also suggested a list of principles for green extraction (Chemat et al. 2012) which are given below:

Principle 1: Innovation by selection of varieties and use of renewable plant resources.

Principle 2: Use of alternative solvents and principally water or agro-solvents.

Principle 3: Reduce energy consumption by energy recovery and using innovative technologies.

Principle 4: Production of co-products instead of waste to include the bio- and agro-refining industry.

Principle 5: Reduce unit operations and favor safe, robust and controlled processes.

Principle 6: Aim for a non-denatured and biodegradable extract without contaminants.

These principles were identified and described not as rules but more as innovative examples to follow, discovered by scientists and successfully applied by industry. The listing of the six principles of green extraction of natural products should be viewed for industry and scientists as a guideline to establish an innovative and green label, charter, and standard, and as a reflection to innovate not only in process but in all aspects of solid–liquid extraction.

An ideal solvent should be cheap, have high polarity, high selectivity, low toxicity and less safety concerns. Often the extraction technique will determine the ideal solvent for the intended purpose. A combination of solvent and extraction techniques are shown in Table 2 below (Chemat et al. 2012). Microwaves are preferable with solvents that have high polarity and even with water as solvent. Microwaves

Table 2. A comparison of solvent-extraction techniques.

Solvent	Extraction technique (Application)	Solvent Power Polar	Solvent Power Weakly Polar	Solvent Power Non-Polar	Health & Safety	Cost	Environmental Impact
Solvent-free	*Microwave Hydrodiffusion and Gravity (antioxidants, essential oils)*	+++	+		+++	+	+++
	Pulse Electric Field (antioxidants, pigments)	+++	+		+++	+	+++
Water	Steam distillation (essential oils)	++	+		+	++	+
	Microwave-assisted distillation (essential oils)	+++	+++	+	+	+	++
CO_2	Extraction by sub-critical water (Aromas)	+	++		+	+	+
	Supercritical fluid extraction (decaffeination of tea and coffee)	−	+	+++	+	+	+
Ionic liquids	Ammonium salts (Artemisinin)	−	+	+++	−	−	++
Agrosolvents	Ethanol (pigments and antioxidants)	+	+	−	−	++	+
	Glycerol (polyphenols)	+	+	−	−	+	+
	Terpenes such as *d*-limonene (fats and oils)	−	−	++	−	+	+
Petrochemical solvents	*n*-Hexane (fats and oils)	−	+	+++	− −	++	− −

also allow for solvent-free reactions over dry media due to their unique interaction capability with catalysts and supporting media. The reaction safety is comparable or better than other conventional options. Ionic liquids are a good combination for microwave mediated extraction. Some techniques are more favorable for extraction of specific compounds while the others are more preferred for oil extraction. It is important to note the advantages and drawbacks of these techniques in developing appropriate biorefinery schemes.

Esterification and Transesterification Reactions

Biodiesel is also made through esterification of free fatty acids (FFAs) derived from vegetable oils, wherein the water by-product is separated from the FAME (Scheme 2). For fatty acids (usually from vegetable sources), the acid-catalyzed esterification reaction is a preferred method. The base-catalyzed esterification to methyl and ethyl esters is not a suitable mechanism because the catalyst is deactivated by reaction with the carboxylic acid group to form a carboxylic acid salt, often referred to as soap. Direct acid or base catalyzed transesterification reaction can be conducted for feedstock containing negligible amount of free fatty acids (< 1%) as shown in Scheme 3.

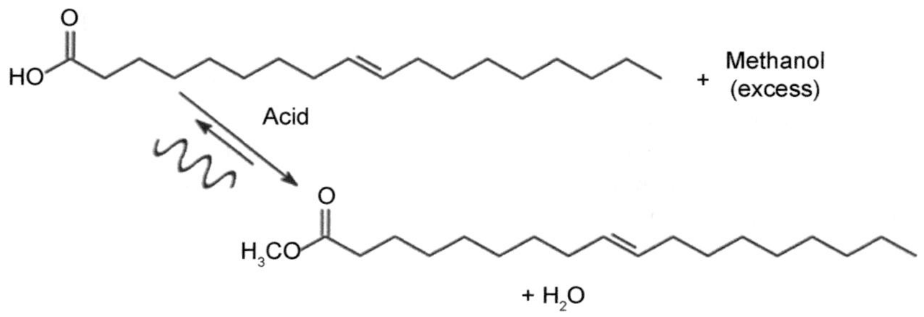

Scheme 2. Acid-catalyzed esterification of fatty acids to FAMEs. The trans-isomers are shown in the diagram for convenience, but in reality they are a mixture of cis- and trans-isomers. It should be noted that most naturally occurring fatty acids are cis-isomers (Earle et al. 2009).

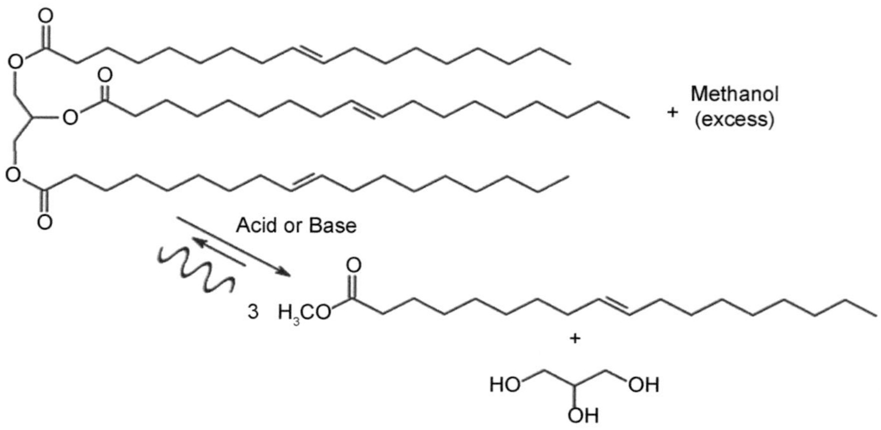

Scheme 3. Acid- or base-catalyzed transesterification of triglycerides, and the esterification of fatty acids to FAMEs. The trans-isomers are shown in the diagram for convenience, but in reality they are a mixture of cis- and transisomers. It should be noted that most naturally occurring fatty acids are cis-isomers and double bonds can occur in various positions in the chains. The methyl ester of cis-octadec-9-enoic acid is normally referred to as methyl oleate (Earle et al. 2009).

Reaction Mechanism

The base catalysis mechanism for triglycerides using alkali hydroxide as catalyst is shown in Figure 4 (Schuchardt et al. 1998, Lotero et al. 2005). The mechanism can be assumed to follow the commonly reported route under conventional transesterification since microwaves do not have the capability to cause chemical bond alteration at molecular levels. The microwaves may cause electron excitation but do not possess the energy required to break the chemical bonds. However, localized superheating created by microwaves (due to dipole moment and ionic conduction) will allow for expedited reactions. Therefore, reactions enhanced by microwaves will be similar to the route followed by the conventional transesterification method except that the reaction speed increases tremendously due to the aforementioned special effects.

Base-catalyzed transesterification mechanism follows essentially four important steps (Figure 4); (1) A catalytic reaction with alcohol, produces an alkoxide. The nucleophilic attack of the alkoxide to the carbonyl group of the triglyceride generates a tetrahedral intermediate compound (2) from which the alkyl ester is formed and the corresponding anion of triglyceride (3). Finally, the catalyst is deionized to regenerate the active compound (4), which allows for it to react with a new molecule of alcohol, beginning a new catalytic cycle. In the notation used, B is the base catalyst, R_1, R_2 and R_3 are the carbonyl groups of fatty acids and R is the functional group of alcohol (Schuchardt et al. 1998, Lotero et al. 2005). Currently, methanol is a commonly used reagent for transesterification, but ethanol is attractive for long term sustainability since

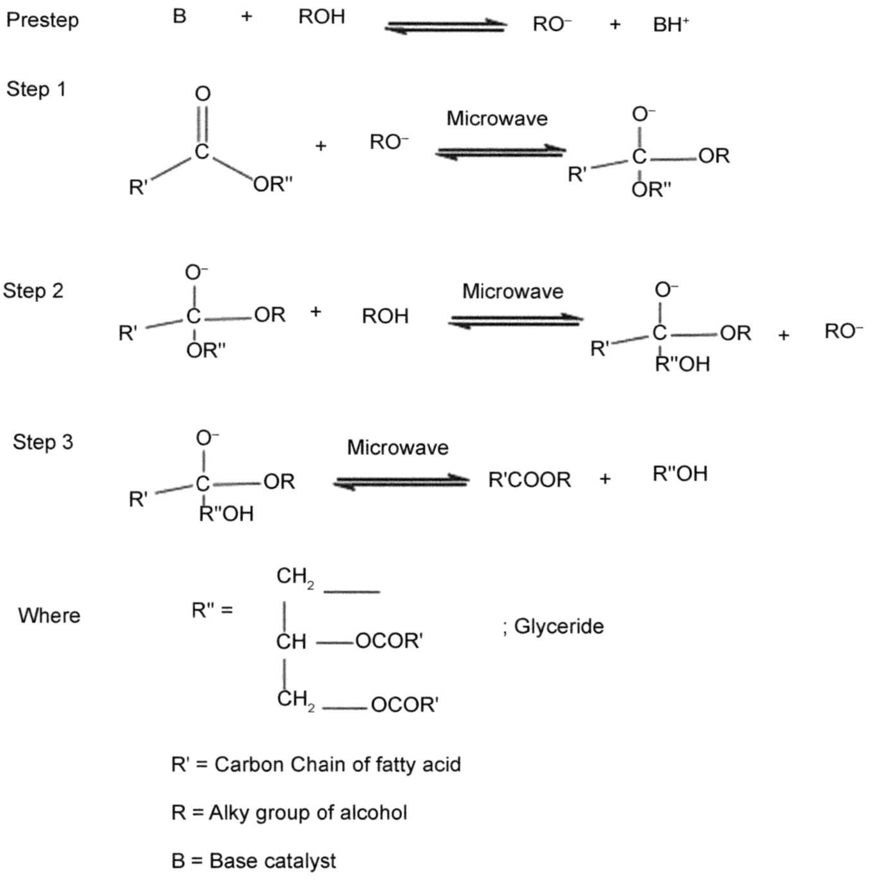

Prestep: B + ROH ⇌ RO⁻ + BH⁺

Step 1: R'COOR'' + RO⁻ ⇌ (Microwave) tetrahedral intermediate

Step 2: intermediate + ROH ⇌ (Microwave) intermediate + RO⁻

Step 3: intermediate ⇌ (Microwave) R'COOR + R''OH

Where R'' = Glyceride (CH₂—, CH—OCOR', CH₂—OCOR')

R' = Carbon Chain of fatty acid

R = Alky group of alcohol

B = Base catalyst

Figure 4. Base catalyzed microwave transesterification reaction (Patil et al. 2011a).

it can be derived from renewable sources, and is less toxic than methanol (Demirbas 2009a). Fatty acid ethyl esters (FAEE) typically demonstrate slightly higher cetane numbers, improved low temperature operability, and greater oxidative stability when compared to fatty acid methyl esters (FAME) (Demirbas 2009, Knothe 2005). In addition, the majority of the world methanol production is based on natural gas. This is one of the reasons behind the dependence of methanol costs on the price of crude oil. The higher toxicity of methanol relative to ethanol also raises additional concerns regarding its transportation and storage (Zanin et al. 2013).

Acid-catalyzed reaction mechanism

Acid catalyzed transesterification involves the protonation of the carbonyl group of the ester leading to the carbocation II which, after a nucleophilic attack of the alcohol, produces the tetrahedral intermediate III, which eliminates alcohol to form the new ester IV, and to regenerate the catalyst H⁺ (Figure 5). According to this mechanism, carboxylic acids can be formed by reaction of the carbocation II with water present in the reaction mixture. This suggests that an acid-catalyzed transesterification should be carried out in the absence of water, in order to avoid the competitive formation of carboxylic acids which reduce the yields of alkyl esters.

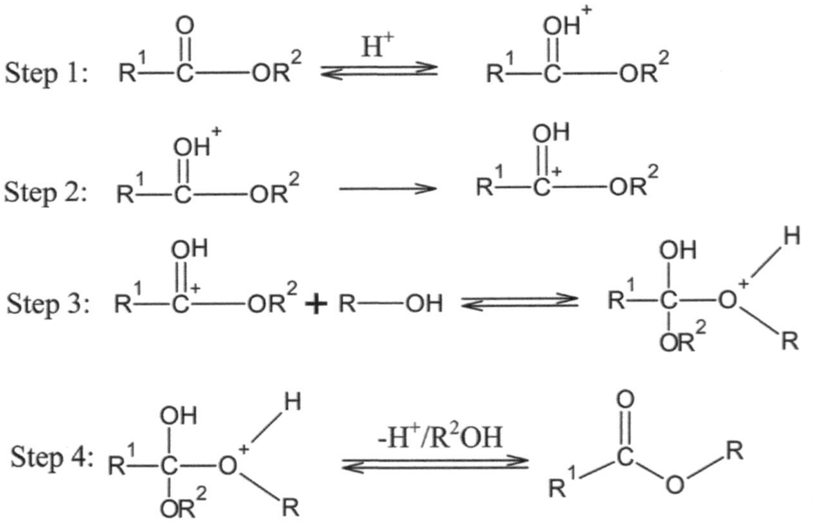

Figure 5. Acid catalyzed microwave transesterification reaction.

Microwave Mediated Transesterification Reaction

The schematic diagram of transesterification using methanol is shown in Figure 6. The interaction of microwaves with fatty acids or triacylglycerides and methanol are depicted. R^1, R^2 and R^3, in this equation represents alkenyl and alkyl groups. They can change into the number of carbon atoms and their double bonds (Sajjadi et al. 2014). There are several C–O and C–C single bonds near the three-ester bonds in the triglyceride molecule. Different types of conformational isomers with different momentary dipole can be produced by rotation of these bonds. Therefore, different directions of three oxygen atoms of C=O bonds can be produced. Under the microwave irradiation, rotation around C–O or C–C bonds occurs in the case of polar molecules. Thus the polarization of C=O bonds, which have higher electronegativity within the molecule and are active under microwave irradiation, occurs in the molecule. Then C=O bonds try to be scattered in an electromagnetic field as shown in Figure 6. This reduces the dipole moment of a triglyceride

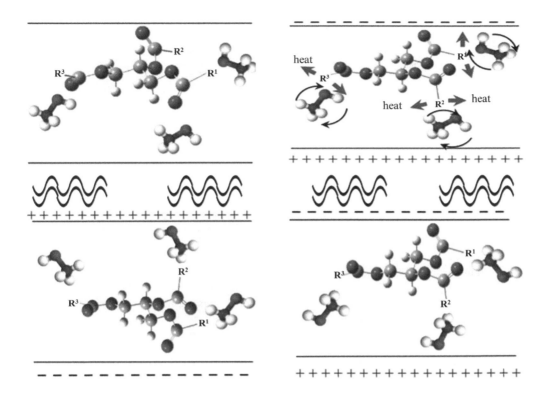

Figure 6. Microwave interaction with fatty acids or triacylglycerides during esterification-transesterification reactions (adapted from Sajjadi et al. 2014).

molecule under microwave irradiation. About 26.9 kcal/mol energy is required to enhance the momentary dipole to the higher level, and after stopping the irradiation, the lower molecular structure with the lower dipole moment is retained momentarily. Polar groups under this situation (microwave field with the frequency of 2.45 GHz) rotate frequently. Since the electronegativity of the oxygen is much higher than that of an alkyl group, the R groups do not have any relative effect on the different dipole moment of the molecule (Asakuma et al. 2011).

Microwaves Under Supercritical Conditions

Feedstock containing high percentage of water content cannot be processed efficiently under conventional process conditions as water content often interacts strongly with the catalysts or other solvents in the reaction. Instead it can be turned into a beneficial medium for these reactions. Water acts as an organic solvent under supercritical conditions. Supercritical conditions (under high pressures and temperatures) can be applied to conduct direct extraction and transesterification of lipids from biomass. For example, algal biomass can be processed under supercritical methanol or ethanol conditions to extract the lipids and transesterify simultaneously without the requirement for a catalyst. Thus supercritical conditions allow for non-catalytic extraction and transesterification of lipids from various biomass feedstock. Microwaves can be used to create supercritical conditions.

The reaction mechanism for transesterification of vegetable oil under supercritical methanol or ethanol can be described by the hypothesis proposed by Kusdiana and Saka (2004) which is shown in Scheme 4. This hypothesis was based on the mechanism developed by Krammer and Vogel (2000) for the hydrolysis of esters in sub/supercritical water. In addition, Kusdiana and Saka (2004a) suggested that methanol and ethanol behave as acid catalysts in the supercritical synthesis. In a catalyst-free transesterification with

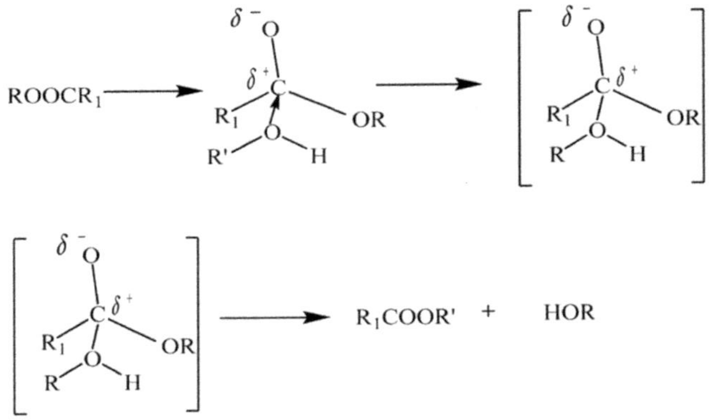

Scheme 4. Microwave-mediated supercritical alcohol transesterification mechanism for triglycerides (R0 is a diglyceride group and R1 is a fatty acid chain) (Patil et al. 2013). R'—alkyl group, ROH—diglyceride.

supercritical methanol, it is assumed that the hydroxyl group of the alcohol molecule directly attacks the carbonyl group of the triglyceride. Liquid methanol is a polar solvent that has hydrogen bonds between the OH oxygen and the OH hydrogen to form the methanol clusters. In the supercritical state, depending on pressure and temperature, the intermolecular hydrogen bonding would be significantly decreased, thus the polarity of methanol decreases in the supercritical state, which would allow methanol to act as a free monomer; therefore, supercritical methanol has a hydrophobic nature with a lower dielectric constant. As a result, nonpolar triglycerides can be well solvated with supercritical methanol to form a single phase of vegetable oil/methanol mixture. Saka and Kusdiana (2001) reported that this phenomenon promotes the transesterification of vegetable oils and is completed in supercritical methanol/ethanol via a methoxide transfer, whereby the fatty acid methyl ester and diglyceride are formed. In a similar way, diglyceride is transesterified to form methyl ester and monoglyceride, which is converted further to methyl ester and glycerol in the last step.

Thermodynamic Justification

The advantage of microwave assisted reactions clearly reflects in short reaction times by rapid heating and cooling. Perhaps, the motivation for microwave reactions was derived from the desire to reduce reaction times and produce cleaner reaction products. A very high increase in (5–1000 times) reaction rates was reported by early researchers (Demirbas 2002, Gedye et al. 1988, Wiesbrock et al. 2004, Sinnwell and Ritter 2007). It is also possible to observe different product composition under microwave and conventional heating. Probable explanation for this phenomenon is that microwave heating significantly increases the reaction temperature and it is possible that the reaction temperature (due to dielectric heating) could exceed the ignition temperature for an additional reaction, which is not possible at the lower temperatures achieved by conventional heating. Many theories attempt to elaborate on the special microwave effects of heating. Since reactions involve thermodynamics of materials, fundamental thermodynamic equation for reactions (the Arrhenius equation) can be taken as a basis to explain the special microwave heating effect (Lidström et al. 2001):

$$K = A \times e^{\Delta G/RT} \qquad [1]$$

From the above equation, it can be noted that there are only two possible ways to increase the rate of reaction. First, by increasing the pre-exponential factor "A" which is the molecular mobility that depends on frequency of the vibrations of molecules at the reaction interface (Perreux and Loupy 2001). This relates to the microwave effects of dipolar polarization and ionic conduction mechanisms explained earlier. The pre-exponential factor "A" is expressed as:

$$A = \pi_\gamma \lambda^2 \Gamma \qquad [2]$$

where γ = number of neighbor jump sites, λ = jump distance, and Γ = jump frequency (Binner et al. 1995).

The other way is to decrease the activation energy, ΔG, which is given in terms of enthalpy and entropy ($\Delta G = \Delta H - T\Delta S$). In microwave assisted reactions, entropy generation is higher due to quick and random dipolar movement and molecular level microwave interactions which increases the value of second term in the equation. The expedited superheating can also contribute to reduction in activation energy as shown in Figure 7 (Perreux and Loupy 2001). Kappe mentioned that non-thermal effects essentially result from a direct interaction of the electric field with specific molecules in the reaction medium. It has been argued that the presence of an electric field leads to orientation effects of dipolar molecules and hence changes the pre-exponential factor "A" or the activation energy (entropy term) in the Arrhenius equation. A similar effect should be observed for polar reaction mechanisms, where the polarity is increased from the ground state to the transition state, thus resulting in an enhancement of reactivity by lowering the activation energy (Kappe 2004, 2008).

Microwave effects result from material-electromagnetic wave interactions and due to the dipolar polarization phenomenon, the greater the polarity of a molecule (such as the solvent), the more pronounced will be the microwave effect when thermal effects are considered (Kappe 2008). In terms of reactivity and kinetics, the specific effect has therefore to be considered according to the reaction mechanism and particularly with regard to how the polarity of the system is altered during the progress of the reaction. When polairty is increased during the reaction from the ground state towards the transition state, specific microwave effects can be expected for the polar mechanism. The outcome is essentially dependent on the medium and the reaction mechanism. If stabilization of the transition state (TS) is more effective than that of the ground state (GS), this results in an enhancement of reactivity by a decrease in the activation energy as shown in Figure 8 (Demirbas 2005). Alteration of esterification kinetics under microwave irradation was reported by Jermolovicius et al. (Jermolovicius et al. 2006).

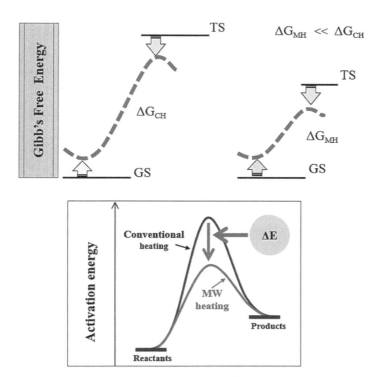

Figure 7. Gibb's free energy differences in conventional and microwave heating methods (GS – ground state; TS – transition state; CH – conventional heating; and MH – microwave heating).

Methanol is the most commonly used solvent in transesterication reaction, partly due to its low cost and high availability. Methanol has high microwave absorption capacity because it is an organic solvent with high polarity. It can therefore be understood that oil-methanol-catalyst involved transesterification reaction can be enhanced by microwave interactions through dipolar polarization and ionic conduction. In biodiesel reactions with water containing feedstock, microwave assisted supercritical reactions can turn the water content into an organic solvent because water molecules possess a moderate dipole moment but higher than triglycerides. A dipole is sensitive to external electrical fields and will attempt to align itself with the field by rotation to generate local superheating.

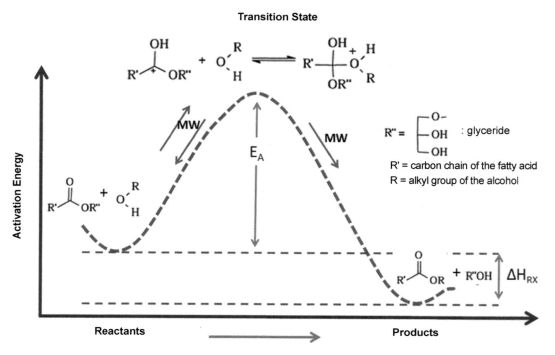

Figure 8. Activation energy reduction in microwave heating method (acid-catalyzed transesterification reaction).

Microwave Mediated Biodiesel Production

Various types of feedstock were utilized in microwave enhanced biodiesel production. In general, the feedstock can be categorized into (1) edible oils, (2) non-edible oils, and (3) oils from algae and other microbial feedstock such as wastewater sludge and cellulose based renewable feedstock. Economics of biodiesel production mainly depends on the feedstock costs. The cost of feedstock can range from 60–80% of the biodiesel cost.

Edible oils (first generation)

Rudolph diesel first tested the engine by using peanut oil and discovered that vegetable oils are excellent alternatives for engine fuels after further treatment and upgrading. Edible oils commonly used as feedstock for the biodiesel production have been soybean, canola, corn, coconut, palm tree, rapeseed, rice bran, sunflower, safflower, *Camelina* and cottonseed oils to just name a few. Among these, soybean oil is the dominant feedstock due to its versatility in geographic locations while palm tree can produce highest quantity of oils per cultivated area (Sim et al. 1988, Patil et al. 2009). Rapeseed and sunflower oils are predominant in the European Union. Although use of vegetable oils to prepare biodiesel was well received in early stages, soon it turned out be a food versus fuel issue. This conflict arose due to increase in vegetable oil demand and price and supply-demand imbalances all over the world. Table 3 shows a long list of edible oils evaluated for biodiesel production (Demirbas 2009b).

Table 3. Oil feedstock type and sources for biodiesel production (Demirbas 2009c, Gude et al. 2013).

Type	Source of oil
Major oils	Coconut (copra), corn (maize), cottonseed, canola (a variety of rapeseed), olive, peanut (groundnut), safflower, sesame, soybean, and sunflower
Nut oils	Almond, cashew, hazelnut, macadamia, pecan, pistachio and walnut
Other edible oils	Amaranth, apricot, argan, artichoke, avocado, babassu, bay laurel, beechnut, ben, Borneo tallow nut, carob pod (algaroba), cohune, coriander seed, false flax, grapeseed, hemp, kapok seed, lallemantia, lemon seed, macauba fruit (Acrocomia sclerocarpa), meadowfoam seed, mustard, okra seed (hibiscus seed), perilla seed, pequi (Caryocar brasiliensis seed), pine nut, poppy seed, prune kernel, quinoa, ramtil (Guizotia abyssinica seed or Niger pea), rice bran, tallow, tea (camellia), thistle (Silybum marianum seed), and wheat germ
Inedible oils	Microalgae, babassu tree, copaiba, honge, jatropha or ratanjyote, jojoba, karanja or honge, mahua, milk bush, nagchampa, neem, petroleum nut, rubber seed tree, silk cotton tree, and tall oil Almond, Abutilon muticum, Andiroba, Babassu, Brassica carinata, B. napus, Camelina, Cumaru, Jatropha Curcas, Pongamia glabra, Cynara cardunculus, Jatropha nana, Laurel Lesquerella fendleri, Mahua Piqui, Palm, Karang, Tobacco seed, Rubber plant, Rice bran, Sesame, Salmon oil
	Animal fats Lard, Tallow, Poultry fat, Fish oil
Other oils	Castor, radish, and tung
Microbial oils	Bacteria, Cyanobacteria, Fungi, Macroalgae, Tarpenes, Microalgae (examples—*Chlorella Vulgaris*, *Nannochloropsis*, and *Scenedesmus* sp.), sludge microorganisms

Non-edible oils (second generation)

Among possible alternative biodiesel feedstocks are oils of non-edible crops such as jatropha, castor, neem, karanja, rubber seed, used frying oils (waste cooking oil), animal fats, beef and sheep tallow (Schenk et al. 2008). pongamia pinnata, maize, yellow grease, poultry fat, castor, and Chinese tallow tree. While these feedstock do not conflict with food interest, they conflict with other commercial products such as cosmetics and industrial products. These feedstock can reduce the biodiesel costs by upto 80%.

Microalgae and other microbial feedstock (third generation)

Third generation biodiesel feedstock are those that do not conflict with any food, feed or cosmetic related human consumption interests. Macro and microalgae, cyanobacteria, wastewater treatment plant activated sludge, switch grass and other microbial communities belong to this type. Among these, microalgae seem to be a superior feedstock and offer following advantages:

- Microalgae can utilize non-arable land;
- Oil content in algae is orders of magnitude higher than from other feedstocks such as corn, sugar cane, jatropha, etc.;
- Microalgae need CO_2 to photosynthesize and can be used to sequester CO_2 from industrial sources of flue and flaring gas;
- Microalgae-based fuels are carbon-neutral or even more carbon-capturing than releasing;
- Microalgae can be used to remediate high-nutrient water sources such as sewage treatment plant and agricultural runoff;
- End-products include biodiesel and/or other higher value feed (protein), pharmaceutical, and health-related products.
- Different species of microalgae can be grown in polluted, saline, brackish, and freshwater;

- Co-location of algal ponds with industrial production plants for potential recycling of CO_2 and impaired waters.

Algal biofuels are thus renewable, sustainable, and environmentally-benign (Sheehan et al. 1998, Pienkos and Darzins 2009, Gressel 2008, Revellame et al. 2010).

Oil Extraction from Biomass Feedstock

Extraction of lipids from biofuel feedstock is the first step in the process. Several techniques based on mechanical and non-mechanical principles have been developed over the years (Figure 9). Mechanical methods are based on agitation, and intense mixing such as in homogenization of reaction mixtures. Mechanical methods are also similar in nature to physical methods. High pressure and high temperatures are used for extracting the lipids from biofuel feedstock. Chemical methods are based on addition of acids or bases, organic solvents and other ionic liquid solvents specifically designed for easy separation and reuse. Enzymatic methods appear to be environmentally friendly alternatives but they are not cost-effective and slow in extraction.

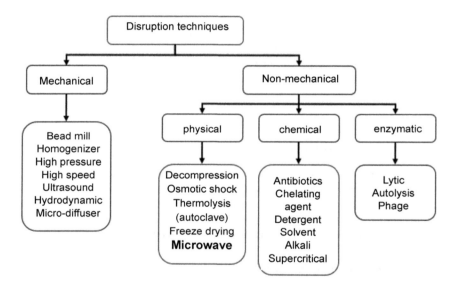

Figure 9. Biofuel feedstock disruption techniques for oil extraction.

Microwave Mediated Oil Extraction

Microwaves can be used either as a thermal pretreatment or process enhancement technique for extraction of oils and lipids from biodiesel feedstock (Giese 1992). Microwave extraction is more efficient than other conventional extraction methods in many ways. Microwaves allow for rapid and selective extraction of organic compounds with low solvent and energy consumptions (Paré et al. 1994, Letellier and Budzinski 1999). In conventional extraction, the extractability of different components depends mainly on solubility of the compound in the solvent, mass transfer kinetics of the product and matrix interactions (Spigno and De Faveri 2009), whereas under microwave mediated extraction conditions, localized superheating rate plays an important role in extraction efficiency. This heating rate is influenced by factors such as microwave power level, frequency, initial temperature and design of microwave applicator. Microwaves have been successfully applied for the extraction of natural compounds from foodstuffs like flavonoids and polyphenol compounds from tea (Pan et al. 2003) and grape seeds (Hong et al. 2001), constituents from herbals (Guo et al. 2001), pigments from paprika (Kiss et al. 2000), antioxidants from rice bran (Zigoneanu et al. 2008), isoflavones from soybeans (Rostagno et al. 2007, Terigar et al. 2010a) and also for trace analysis of organic compounds in solid and liquid samples (Eskilsson and Björklund 2000, Bhattacharya and Basak 2006, Hemwimon et al. 2007). Microwaves may also allow for solvent free extraction of essential oils from plant materials (Lucchesi et al. 2004).

Selection of solvent is another important consideration in microwave extraction. Microwaves are effective on materials that have high dielectric properties, an intrinsic property of the material that requires empirical measurement but is mostly influenced by the moisture liquid/solid mixture content and spatial distribution of the water and other polar/ionic compounds in the matrix. The dielectric properties of materials are defined in terms of their relative complex permittivity. For a solvent/matrix to heat up rapidly under the microwave radiation, it has to have a high dielectric constant, associated with the potential for electrical energy storage in the material, and a high dielectric loss which is related to the electrical energy dissipation in the material (Nelson 1994). The heating of a dielectric material in the presence of an electromagnetic field is based on intermolecular friction that arises due to ionic conduction and dipolar rotation (Nüchter et al. 2004). n-hexane is widely used as solvent for extraction with other commonly used solvents such as isopropanol, methanol, ethanol, acetone and water (Pan et al. 2003, Duvernay et al. 2005, Virot et al. 2007, Spigno and De Faveri 2009).

Extraction of lipids and oils from plant leaves and seeds depends on the microwave penetration ability. Disruptions of the oilseed cells take place when temperature of water molecule inside the cells reaches the boiling point leading to high pressure gradients and rupture of cell walls, causing migration of selected compounds from sample matrix into the extraction solvent. This particularity makes the technology appealing for biodiesel, as biodiesel is produced from vegetable oil. The microwave thermal effects (localized microscopic super heating) naturally match the requirements for the disruption process of tissues and could be used to induce rupture of cells for efficient extraction of oils and other components from plants. The above mechanism of extraction applies to algal cells as well. In a recent study, lipid extraction from microalgae was tested by various methods including autoclaving, bead-beading, microwaves, sonication, and a 10% NaCl solution. Microwave based extraction proved to be the most simple, easy and effective method for disruptive extraction of lipids from *Botryococcus* sp., *Chlorella vulgaris*, and *Scenedesmus* sp. (Lee et al. 2010).

Extraction by microwaves can be fast and simple. Kanitkar conducted microwave assisted extraction of oils from soybeans, rice bran and Chinese tallow tree seeds. About 95% of recoverable oils were extracted from these seeds by microwave extraction process in just 20 minutes which would otherwise have taken hours of processing using other solvent and mechanical extraction methods. It was observed that the enhanced extraction was due to the specific interaction of the microwave field with the solvent-feedstock matrix, where higher temperature and pressure gradients develop at the microscopic level, leading to enhanced mass transfer coefficients (Kanitkar 2010).

Extraction kinetics can be explained using the Arrhenius equation. An explanation provided by Cooney et al. (2009) is as follows. Solvent extraction of bio-oils from biomass is a process whereby the target analyte is transferred from one phase (e.g., a solid phase in the case of dried biomass and an aqueous liquid phase in the case of wet biomass) to a second immiscible phase (e.g., an alcohol such as methanol or an alkyl halide such as chloroform). In other words, the analyte (i.e., lipid) molecule must dissolve into the solvent and form a solution. The solubility of the analyte in the solvent is governed by the Gibbs free energy of the dissolution process, which is directly related to the equilibrium constant governing the concentration of the analyte in either phase.

$$\Delta G = -RT \ln \frac{[analyte]^{solvent\ phase} \times [solvent]^{solvent\ phase}}{[analyte]^{analyte\ phase} \times [solvent]^{analyte\ phase}} = \Delta H - T\Delta S \quad [3]$$

As more of the analyte dissolves into the solvent phase, the natural logarithm of the quotient becomes positive and the Gibbs free energy for this reaction becomes negative, indicating that the reaction has proceeded more favorably in the direction of the analyte dissolving into the solvent. As the analyte fully dissolves into the solvent phase, the quotient approaches infinity and the equilibrium lies totally to the right, and the target analyte (i.e., lipid) is considered fully extracted into the solvent phase.

The solubility of the target analyte in various solvents is governed by two independent parameters (which may, or may not, work together): the enthalpy of mixing (ΔH) and the entropy of mixing (ΔS) (Cooney et al. 2009). The solubilization of the analyte in the solvent is therefore favored when the dissolution process gives off energy (i.e., ΔH) and/or when the dissolution process increases entropy (ΔS). Since these two properties are interdependent, a favorable change in one may (or may not) offset an unfavorable change in the other. How the analyte molecule chemically interacts with the selected solvent will dictate whether the change in enthalpy is positive or negative, whether the change in entropy is positive or negative, and whether their combined sum yields a favorable Gibbs free energy of dissolution. The overall sum of these two terms is defined by the total relative contribution of all intermolecular forces that occur between the analyte and solvent molecules: Electrostatic, London forces, hydrogen bonds, and hydrophobic bonding. Consequently, the development of any solvent based extraction process must comprise a choice of solvent (or co-solvent mixture) that yields a set of chemical interactions between the analyte and solvent molecules that is more favorable than the chemical interactions between (i) the solvent molecules themselves (i.e., self-association), and (ii) the analyte with the matrix it was already associated with. As a general rule analytes that strongly self-associate dissolve best in strongly associated solvents, while analytes that weakly associate dissolve best in weakly associated solvents. In other words, polar solutes will dissolve in similar polar solvents and non-polar solutes will dissolve better in similar non-polar solvents.

An improved process of Soxhlet extraction assisted by microwave, called microwave-integrated Soxhlet (MIS) was evaluated for the extraction of oils and fats from different food matrixes such as oleaginous seeds, meat and bakery products. Results have shown that MIS parameters do not affect the composition of the extracts. For the generalization of the study with several food matrices, MIS extraction results obtained were then compared to conventional Soxhlet extraction in terms of crude extract and fatty acid composition and shown that the oils extracted by MIS were quantitatively and qualitatively similar to those obtained by conventional Soxhlet extraction. MIS labstation can be considered as a new and general alternative for the extraction of lipids by using microwave energy (Demirbas 2007).

Microwave Enhanced Extraction of Algal Lipids

The intracellular location of the lipids requires algal cell wall breakage for oil recovery. Extraction of intracellular lipids from intact microalgae is difficult as the lipids are bound within cell membranes and cell disruption is required to maximize lipid recovery. For this reason, conventional oil extraction methods such as physical straining, mechanical pressing, and solvent extraction are not suitable for microalgae oil extraction. Mechanical pressing involves high specific energy consumption due to high mechanical strength of algal cell walls which may exceed the extractable energy in microalgae lipids. Supercritical fluid extraction using dry CO_2 is another well-known technique. This technique requires high temperatures and high pressures but relatively lower chemical solvents and reduced reaction times. But high capital costs and process safety are the major concerns with this technique.

Novel techniques with ability to enhance both the diffusive (diffusion of solvent into the cells) and disruptive (disrupt the cell walls to force-release the lipids) mechanisms are essential for efficient algal oil extraction. The key to algal lipid extraction is an effective solvent which can firstly penetrate the solid matrix enclosing the lipid, secondly physically contact the lipid and thirdly solvate the lipid (Cooney et al. 2009). Since the microalgae are protected by a cell wall that limits the solvents' access to the lipid, microwaves may play an important role to enhance both disruptive and diffusive mechanisms (Figure 10). In general, two basic mechanisms by which extraction of a lipid can possibly occur are (1) diffusion of lipids across the cell wall, if the algal biomass is suspended in the solvent with higher selectivity and solubility (or large partition coefficient) for lipids and (2) disruption of the cell wall with release of cell contents in the solvent. The relative contribution of each of these mechanisms depends on the extraction technique. It could be easily perceived that diffusive mechanism will have less efficiency (in terms of long extraction time and smaller yield of lipid) due to the slow diffusion of lipid molecules

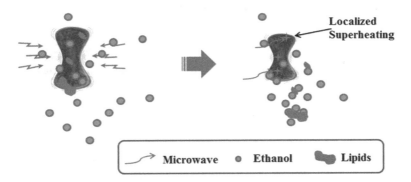

Figure 10. Extractive-transesterification (*in situ*) of algal lipids.

across the cell wall. On the other hand, a disruptive mechanism is likely to cause faster extraction of lipids with high yields, as it involves the direct release of the lipid droplets in cytoplasm into the bulk liquid with the rupture of the cell wall (Ranjan et al. 2010). Diffusive mechanism is more predominant in extraction methods such as solvent extraction, soxhlet extraction and others. Disruptive mechanism refers to mechanical breakdown of the cell as in mechanical pressing and supercritical high pressure and high temperature treatment. However, it has been reported that mechanical pressing is an inefficient method of extraction for algal biomass due to their rigid wall structure. It is advantageous to perform extraction and transesterification reactions simultaneously. Biodiesel production from microalgae requires extraction of oils and lipids from the cellular mass prior to their transesterification. Microwaves can be used as efficient medium to perform these two tasks simultaneously.

The effect of microwave irradiation on the extraction of algal lipids has been reported in many studies. Table 4 summarizes the results reported from various studies using different algal species.

Table 4. Microwave enhanced extraction of algal lipids (Gunerken et al. 2015).

Microalgae	Conditions	Scale	Outcome	Reference
Chlorella sp. Total lipid	2450 MHz, 100°C, 5 min, 0.5% DCW	100 ml	2.6 fold more extraction than untreated cells	Prabakaran and Ravindran (2011)
Nostoc sp.			2.2 fold more extraction than untreated cells	Prabakaran and Ravindran (2011)
Tolypothrix sp.			5.3 fold more extraction than untreated cells	Prabakaran and Ravindran (2011)
Chlorella vulgaris	2450 MHz, 100°C, 5 min	Laboratory (N250 ml), Volume not given	3.9 fold more extraction than untreated cells	Zheng et al. (2011)
Synechocystis PCC 6803	1.4 kW, 57°C, 1 min	Analytical, Volume not given	1.1 fold more extraction than untreated cells	Sheng et al. (2012)
	1.4 kW, 26°C, 30 s treatment 30 s pause		1.1 fold more extraction than untreated cells	Sheng et al. (2012)
Scenedesmus obliquus	1.2 kW, 2450 MHz, 7.6% DCW, 95°C, 30 min	Laboratory (N50 ml), Volume not given	77% of recoverable oil (1.64 fold of only heating method) extracted	Balasubramanian et al. (2011)
Botryococcus sp.	100°C, 2450 MHz, 5 min, 0.5% DCW	100 ml	28.6% (w/w) Lipid extraction	Lee et al. (2010)
Chlorella vulgaris sp.			10% (w/w) Lipid extraction	Lee et al. (2010)
Scenedesmus sp.			10.4% (w/w) Lipid extraction	Lee et al. (2010)

Microwave Mediated Transesterification

The chemical conversion of the oil to its corresponding fatty ester (biodiesel) is called transesterification. Transesterification is the process of using a monohydric alcohol in the presence of an alkali catalyst, such as sodium hydroxide (NaOH) or potassium hydroxide (KOH), to break chemically the molecule of the raw renewable oil into methyl or ethyl esters of the renewable oil with glycerol as a byproduct (Demirbas 2007). Microwave effect on the transesterification reaction can be two-fold: (1) enhancement of reaction by a thermal effect, and (2) evaporation of methanol due to the strong microwave interaction of the material (Loupy et al. 1993, Yuan et al. 2008). The microwave interaction with the reaction compounds (triglycerides and methanol) results in large reduction of activation energy due to increased dipolar polarization phenomenon (Perreux and Loupy 2001). This is achieved due to molecular level interaction of the microwaves in the reaction mixture resulting in dipolar rotation and ionic conduction (Lidström et al. 2001, Boldor et al. 2010). The amount, by which the activation energy is reduced, is essentially dependent on the medium and reaction mechanism (Perreux and Loupy 2001). Methanol is a strong microwave absorption material and in general, the presence of an -OH group attached to a large molecule behaves as though it were anchored to an immobile raft and the more localized rotations result in localized superheating which assists the reaction to complete faster. For this reason, methanol is preferred over ethanol for microwave-assisted transesterification process. The yield of biodiesel is affected by some critical parameters in the transesterification reaction. In order to obtain the maximum conversion yield, these variables should be optimized. The most important variables in the transesterification reaction are listed below.

- FFA of the oil
- Moisture and water content in the oil
- Reaction temperature
- Reaction time
- Molar ratio of alcohol to oil and type/chemical structure of alcohol
- Catalyst type and concentration
- Mixing intensity (stirring)
- Use of organic co-solvents

Effect of water and free fatty acid (FFA) contents

An acid value less than 1% and anhydrous chemicals are highly desirable for transesterification reaction. If the acid value is greater than 1, higher amounts of NaOH or KOH will be required to neutralize the free fatty acids. Figure 11 shows the plots for yields of methyl esters as a function of free fatty acid content in biodiesel production (Kusdiana and Saka 2004). It can be noted that methyl ester yields generally decrease with increasing free fatty acid (%) content except for supercritical methanol conditions which are more tolerant to the presence of FFAs.

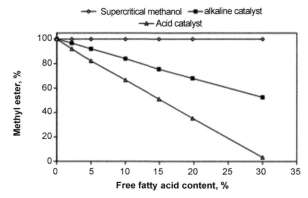

Figure 11. Methyl ester yields as a function of free fatty acids content in transesterification of triglycerides (Kusdiana and Saka 2004a).

Water can cause soap formation and frothing. The resulting soaps can increase the viscosity, formation of gels and foams making glycerol separation difficult. Water content is an important factor in the conventional catalytic transesterification of vegetable oil. In the conventional transesterification of fats and vegetable oils for biodiesel production, free fatty acids and water always produce negative effects since the presence of free fatty acids and water causes soap formation, consumes catalyst, and reduces catalyst effectiveness. Figure 12 shows the plots for yields of methyl esters as a function of water content in the transesterification of triglycerides for acid-catalyzed, alkaline-catalyzed, and supercritical methanol processes.

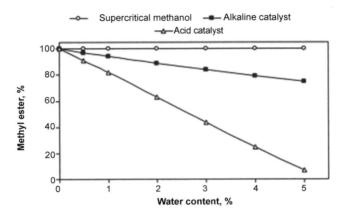

Figure 12. Methyl ester yields as a function of water content in transesterification of triglycerides for acid-catalyzed, alkaline-catalyzed, and supercritical methanol processes (Kusdiana and Saka 2004a).

Catalyst and alcohol-oil ratio

Potassium hydroxide (KOH) and sodium hydroxide (NaOH) flakes are the most commonly used alkaline catalysts in the biodiesel industry because they are inexpensive, easy to handle in transportation and storage, and preferred by small producers. Alkyl oxide solutions of sodium methoxide or potassium methoxide in methanol, which are now commercially available, are the preferred catalysts for large continuous-flow production processes. However, both NaOH and KOH catalysts make separation and purification a difficult process due to their high solubility in the both biodiesel and glycerin (Gude et al. 2013). Biodiesel with the best properties was obtained using sodium hydroxide as catalyst in many studies. On the other hand, many other studies achieved best results using potassium hydroxide (Refaat et al. 2008a). Refaat used 500 mL reactor at a reaction temperature of 65°C with a microwave power of 500 W controlled by a microprocessor. A methanol/oil molar ratio of 6:1 was employed, and potassium hydroxide (1%) was used as a catalyst. Barium hydroxide was also used as a homogeneous catalyst. The range of homogeneous catalysts applied was between 0.1 and 5% (Table 5). Slightly higher concentrations of KOH are required compared to NaOH catalyst due to its higher molecular weight. For feedstock containing high free fatty acid content such as animal fats and used vegetable oils, KOH proved to be a better performer (Brunschwig et al. 2012). Transesterification reaction depends on the type of oil and catalyst applied and the effects of catalysts vary with types of oils.

Homogeneous catalysts are advantageous in terms of fast reaction rates. However, the reaction products require longer separation and purification times. Use of heterogeneous catalysts can be advantageous in microwave mediated transesterification reactions since the catalyst can provide locations for hotspots for rapid heating. In addition, they are recyclable and reusable with acceptable performance. Patil et al. (2009b) employed heterogeneous catalysts such as BaO, CaO, MgO, and SrO for transesterification of *Camelina Sativa* oil into biodiesel. They reported the kinetic rate constants for different catalysts. Two orders of magnitude of difference in the kinetic rate constants between the conventional heating method

Table 5. Summary of microwave-enhanced biodiesel production studies (Gude et al. 2013).

Oil Feedstock	Reaction Time (min)	Reaction Temp. (°C)	Catalyst	Oil to alcohol ratio	FAEE/FAME Conv. (%)	Equipment	Mode	Reference
Vegetable/Edible Oils								
Coconut Oil*	0.5	79.5	NaOH	1:9 Ethanol	100	Domestic Microwave	Continuous	Lertsathapornsuk et al. 2003, 2005
	0.75	82.2	NaOH	1:9 Ethanol	100		Continuous	
	1	83.4	NaOH	1:9 Ethano	100		Continuous	
Rice Bran Oil*	0.5	77.8	NaOH	1:9 Ethanol	93.5		Continuous	
	0.75	80.4	NaOH	1:9 Ethanol	93.2		Continuous	
	1	83.4	NaOH	1:9 Ethanol	93.1		Continuous	
	0.75	81.2	NaOH	1:9 Ethanol	83.9		Continuous	
	1	84.1	NaOH	1:9 Ethanol	90.6		Continuous	
Vegetable oil (Triolein)	3.5	50	NaOH	1:6 Methanol	98		Batch	Leadbeater et al. 2008
Soybean oil	10	65	Ba(OH)$_2$, H$_2$O	1:9 Methanol	97.8	Milestone Ethos 1600, 1000 W	Batch	Mazzocchia et al. 2004
Rapeseed oil	10	103	Ba(OH)$_2$, H$_2$O (1.5%)	1:9 Methanol	99		Batch	Mazzocchia 2006
Rapeseed oil	15	60	Ba(OH)$_2$, H$_2$O (1.5%)	1:9 Methanol	98		Batch	Mazzocchia 2006
Rapeseed oil	0.5	65	NaOH (0.1%)	1:30 Methanol	89		Batch	Lertsathapornsuk et al. 2003
	3	65	NaOH (0.1%)	1:30 Methanol	92		Batch	
	5	65	NaOH (0.1%)	1:30 Methanol	94		Batch	
	16	65	NaOH (0.1%)	1:30 Methanol	99		Batch	
Sunflower	16	65	NaOH (1.0%)	1:30 Methanol	99		Batch	
Soybean oil	20	60	NaOH (0.15%)	1:5 Methanol	98		Batch	Kanitkar 2010
Soybean oil	20	65	NaOH (0.15%)	1:9 Ethanol	98		Batch	Kanitkar 2010
Rice Bran oil	20	60	NaOH (0.15%)	1:5 Methanol	98		Batch	Kanitkar 2010
Rice Bran oil	20	60	NaOH (0.15%)	1:9 Ethanol	97		Batch	Kanitkar 2010
Rice Bran oil	10	50	NaOH (0.6%)	1:5 Ethanol	99		Continuous	Terigar et al. 2011

Sample	Reaction Time (min)	Reaction Temp. (°C)	Catalyst	Oil to alcohol ratio	FAEE/ FAME Conv. (%)	Equipment	Mode	Reference
Rice Bran oil	10	73	NaOH (0.6%)	1:5 Ethanol	99		Continuous	Terigar 2010b
Soybean oil	10	50	NaOH (0.6%)	1:5 Ethanol	98		Continuous	Terigar 2011
Soybean oil	10	73	NaOH (0.6%)	1:5 Ethanol	99		Continuous	Terigar 2010c
Vegetable oil	2	50	KOH (0.6%)	1:6 Methanol	98	CEM Mars	Continuous	Terigar 2010c
Cottonseed	7	333	KOH (1.5%)	1:6 M	92.4	21% of 1200 W		Azcan and Danisman 2007
Safflower seed oil	6	333	NaOH (1%)	1:10 M	98.4	300 W		Duz et al. 2011
Rapeseed & soybean	1	333	NaOH (1.3%)	1:18 M	97	300 W		Hernando et al. 2007
Soybean	1	333	NaOH (1.3%)	1:27 M	95	300 W		Hernando et al. 2007
Corn	20	150 C	Diphenylammonium salts:DPAMs (Mesylate) (10 molar) DPABs (Benzenesulfonate) (10 molar)	2 goil: 5g methanol	100 96			Majewski et al. 2009
Soybean	20	150	DPATs (Tosylate) (10 molar) DPAMs (10 molar) DPABs (10 molar)	2 goil: 5g methanol	100 92 97			Majewski et al. 2009
Soybean	60	338	Nan CaO (heterogeneous)	1:7	96.6	-		Hsiao et al. 2011
Soybean	20	60	Sulfated zirconia (5%)	1:20	90			Kim et al. 2011
Oleic acid	20	60	Sulfated zirconia (5%)	1:20	90			Kim et al. 2011
Canola	5	100	ZnO/La$_2$O$_2$CO$_3$(1%)	1:1 (W/W)	95	-		Jin et al. 2011
Camelina	-	-	BaO (1.5%), SrO (2%)	1:9	94.8	800 W		Patil et al. 2009b
Soybean	2	333	NaOH (1%)	1:6	97.7	900 W		Hsiao et al. 2010
Sunflower	45	-	H2SO4 (0.05%)		96.2	400		Han et al. 2008
Sunflower	25	-	TiO$_2$/SO$_4$ (0.02%)	1:12	94.3	300		Kong et al. 2009
Vegetable oil	2	50	KOH (0.6%)	1:6 Methanol	98	CEM Mars	Continuous	Barnard et al. 2007

Table 5. contd....

Table 5. contd.

Sample	Reaction Time (min)	Reaction Temp. (°C)	Catalyst	Oil to alcohol ratio	FAEE/FAME Conv. (%)	Equipment	Mode	Reference
Waste vegetable oil (domestic)	1	65	KOH	1:9 Methanol	96	Start S Milestone	Batch	Refaat et al. 2008a
Waste vegetable oil (restaurant)	1	65	KOH	1:9 Methanol	94.5		Batch	
Kerosene Used Palm Oil mixture*	0.5		NaOH	1:9 Ethanol	70.9			Lertsathapornsuk 2003
	0.75	76.5	NaOH	1:9 Ethanol	91.5	Domestic Microwave	Continuous	
	1	80.2	NaOH	1:9 Ethanol	91.6		Continuous	
Used vegetable oil	0.15		NaOH (1%)	1:9 Ethanol	100			Lertsathapornsuk 2005
Used Palm Oil*	0.5	77.5	NaOH	1:9 Ethanol	82.5		Continuous	
Unknown	1	60	NaOH	2.3:1.27 (g:mL) Methanol	97	CEM Explorer	Batch	Moseley and Woodman 2009
Waste vegetable oil	6	50	KOH	1:6 Methanol	96	CEM Explorer	Batch	Barnard et al. 2007
Waste frying oil	5	64	NaOH (1%)	1:6 m	93.36	600 W		Rahmanlar et al. 2101
Macauba	15	30	Novozyme 435 (2.5%)	1:9 E	45.2			
	5	40	Lipozyme IM (5%)	1:9 E	35.8			
Waste frying oil	0.5	-	NaOH (3%)	1:12 E	97	800 W		Nogueira et al. 2010
Rapeseed	5	323	KOH (1%)	1:6 M	93.7	67% of 1200 W		Azcan and Danisman 2008
Rapeseed	3	313	NaOH (1%)	1:6	92.7	67% of 1200 W		Azcan and Danisman 2008
Karanja	150s	-	KOH	33.4% (W/W)	89.9	180 W		Kamath et al. 2011
Jatropha	2	65	KOH (1.5%)	1:7.5	97.4	-		El Sherbiny et al. 2010

Table 5. contd....

Microwave Mediated Biodiesel Production

Oil Feedstock	Reaction Time (min)	Reaction Temp. (°C)	Catalyst	Oil to alcohol ratio	FAEE/FAME Conv. (%)	Equipment	Mode	Reference
Palm oil	5	70	KOH (1.5%)	1:8.5 E	98	70 W		Suppalakpanya et al. 2010
Yello horn	10	60	Heteropolyacid (HPAs) (1%)	1:12	96.2	500 W		Zhang et al. 2010
Castorbean	5	-	Al$_2$O$_3$/50%KOH (1%)	1:6	95	40		Perin et al. 2008
Castorbean	30	-	SiO$_2$/50%H$_2$SO$_4$ (1%)	1:6	95	40		Perin et al. 2008
Castorbean	25	-	SiO$_2$/30%H$_2$SO$_4$ (1%)	1:6	95	220		Perin et al. 2008
Castor	60	338	H$_2$SO$_4$	1:12	94	200		Yuan et al. 2008
Triolin	1	323	KOH (5%)	1:6	98	25		Leadbeater and Stencel 2006
	1	323	NaOH (5%)	1:6	98	25		Leadbeater and Stencel 2006
Frying oil	4	60	NaOH (0.5%)	1:6	87	50% of 750 W		Duz et al. 2011
Rapeseed	4 hr	310	-	1:2.5	91	-		Geuens et al. 2007
Safflower	16	60	NaOH (1%)	1:10	98.4	300 W		Duz et al. 2011
Maize	-	-	NaOH (1.5%)	1:10	98	-		Ozturk et al. 2010
Jatropha	7	328	NaOH (4%)	1:30	86.3	-		Yaakob et al. 2008

* domestic microwave

and microwave heating methods was reported in their study. Sol gel type catalysts were also developed and tested by the same group of researchers. Heterogeneous catalysts also reportedly provide for cleaner products and easier separation of the end products. Variety of heterogeneous catalysts were tested. Few examples include: diphenylammonium salts—DPAMs (mesylates), DPABs (benzenesulfonate), DPATs (tosylate), sulfated zirconia, $ZnO/La_2O_2CO_3$, TiO_2/SO_4, heteropolyacids, aluminum oxides with sulfuric acid. Whether reactions involve homogeneous or heterogeneous catalysts, when the reaction is carried out under microwaves, transesterification is efficiently activated, with short reaction times, and as a result, a drastic reduction in the quantity of by-products and a short separation time is obtained (> 90% reduction in separation time), and all with a reduced energy consumption. The rate acceleration in solid-state catalytic reactions, on exposure to microwave radiation, is attributed to high temperatures on the surface of the catalyst. The increase in the local surface temperature of the catalyst results in enhancement of the catalytic action, leading to an enhanced rate of reaction. It has been observed that when the catalyst is introduced in a solid granular form, the yield and rate of the heterogeneous oxidation, esterification and hydrolysis reactions increases with microwave heating, compared to conventional heating under the same conditions (Zabeti et al. 2009). Solid base catalysts are more efficient than solid-acid catalysts. The advantage with the solid catalysts is that they are not sensitive to the presence of water in the reactants. Breccia et al. reported on the use of a domestic microwave apparatus for the synthesis of biodiesel by reaction between methanol and commercial seed oils (Breccia et al. 1999). In this work, they found that the reaction was complete in less than 2 min under microwave irradiation. Activities of several catalysts such as sodium methylate, sodium hydroxide, sodium carbonate, sulfuric acid, benzensulfonic acid and boron carbide were also briefly discussed in their study.

The transesterification reaction is governed by the amount and type of alcohol participating in the reaction. Considering the type of the alcohol, the use of methanol is advantageous as it allows the simultaneous separation of glycerol. The same reaction using ethanol is more complicated as it requires a water-free alcohol, as well as an oil with a low water content, in order to obtain glycerol separation (Schuchardt et al. 1998). Methanol is the most commonly used reactant both in conventional and microwave assisted transesterification reactions. Ethanol is more sensitive to the presence of moisture content in the oil causing soap formation and has less dielectric constant compared to methanol. Ethanolysis proceeds at a slower rate than methanolysis because of the higher reactivity of the methoxide anion in comparison to ethoxide. As the length of the carbon chain of the alkoxide anion increases, a corresponding decrease in nucleophilicity occurs, resulting in a reduction in the reactivity of ethoxide in comparison to methoxide. An example of this phenomenon is the transesterification (at 25°C) of canola oil with a 1:1 mixture of ethanol and methanol (to provide an overall molar ratio of alcohol to oil of 6:1) that results in 50% more methyl than ethyl esters (Kulkarni et al. 2007, Moser 2009). Therefore, for microwave assisted reactions, it is more favorable to use methanol as a solvent. On the other hand, ethanol has environmental acceptance due to its environmental friendly production from biomass. Since the transesterification reaction is an equilibrium reaction, excess amounts of alcohols need to be added to drive the reaction to completion within reasonable time. Alcohol:oil ratios of wide ranges (30:1) have been tested by many researchers with most common ratio being 9:1 (Gude et al. 2013).

Effect of reaction temperature

Higher reaction temperatures below the boiling points, in general, favor the transesterification reaction and the biodiesel yields. Reaction temperatures above the alcohol boiling point may result in soap formation due to catalyst effect. Under supercritical conditions, increasing the reaction temperature can have a favorable influence on the biodiesel yields (Demirbas 2009a). In the alkali (NaOH or KOH) transesterification reaction, the temperature maintained by the researchers during different steps range between 45 and 65°C. A study (Leung and Guo 2006) showed that temperature higher than 50°C had a negative impact on the product yield for pure oil, but had a positive effect for waste oil with higher viscosities.

Effect of reaction time

Camelina Sativa oil was evaluated as a biodiesel feedstock (Patil et al. 2009a,b). These studies included different methods of heating such as conventional, supercritical and microwave methods. Among which the microwave method proved to be superior due to inherent advantages of shorter reaction time and lower energy requirements. Microwave assisted reactions not only reduce the reaction time and increase the biodiesel yield but also reduce the product separation time significantly (Refaat et al. 2008b). It was reported that the product separation in conventional heating method required 480 minutes which was around 30 minutes in microwave assisted heating method. Microwave irradiation resulted in reduction of the reaction time by about 97% and the separation time by about 94%. Saifuddin and Chua (2004) reported that the separation time was between 45–60 min for ethyl esters.

Continuous preparation of fatty acid ethyl esters (FAEE) from coconut, rice bran and used frying (palm) oils in a modified conventional microwave oven (800 Watts) was reported by Lertsathapornsuk et al. (2003). In a continuously mixed batch reactor system, rapid reaction rate and higher conversion yield of FAEE in the presence of alkali catalyst of three vegetable oils was observed with excess amounts of alcohol. The reaction time was reduced to 30–60 seconds which was 30–60 times higher when compared with conventional and super critical methods. Refaat and Sheltawy (2008b) reported that microwave irradiation also allows for use of high free fatty acid (FFA) containing feedstocks, including animal fats and used cooking oils, in existing transesterification processes by promoting the removal of the fatty acid. Radio frequency microwave energy further improves product recovery in the separation of the biodiesel product from alcohol and glycerin in the reaction mixture (Leadbeater et al. 2008). Mazzocchia et al. (2004) reported that microwave irradiation method prevented product degradation, when barium hydroxide was employed as a catalyst. The separation of the reaction products was quick and increased with $Ba(OH)_2$ H_2O when anhydrous and barium hydroxide is employed (Mazzocchia et al. 2004). The total microwave irradiation power on the non-catalytic reaction indicated conversion up to 60% in 60 min of reaction in the esterification of oleic acid (C_{18}). The effects of alcohol type (methanol or ethanol), temperature (150–225°C) and molar ratio of alcohol/fatty acid (3.5–20) on the ester yield were studied in detail (Melo-Junior et al. 2009). To enhance the synthesis process for biodiesel from castor oil (fatty acid methyl ester, FAME), microwave absorption solid acid catalysts (H_2SO_4/C) were used for transesterification under microwave radiation (Yuan et al. 2008). A maximum yield of 94% was obtained using 12:1 (MeOH to Oil), 5 wt% catalyst, and 55 wt% H_2SO_4 loading amounts of catalyst at 338 K under microwave radiation after 60 min.

Zhang and co-workers studied biodiesel production from yellow horn (*Xanthoceras sorbifolia*) oil with a heteropolyacid (HPA) catalyst namely $Cs_{2.5}H_{0.5}PW_{12}O_{40}$. A conversion yield higher than 96% was achieved by using a lower catalyst amount (1% w/w of oil) and a lower molar ratio of methanol/oil (12:1) in a relatively shorter reaction time (10 min) at 60°C (Zhang et al. 2010). The transesterification of high FFA jatropha curcas oil was carried out using microwave irradiation with homogenous catalyst. Biodiesel with 99% conversion can be achieved at 7 minutes reaction time (Yaakob et al. 2008). It was studied that rapeseed oil can be converted to fatty acid butyl esters by means of microwave irradiation without using a catalyst or supercritical conditions of the alcohol (Geuens et al. 2007). The microwave assisted solvent extraction was studied effectively for Tallow tree. The major advantage of this implemented process was the reduced time of extraction required to obtain total recoverable lipids, with corresponding reduction in energy consumption costs per unit of lipid extracted (Boldor et al. 2010).

Moseley and Woodman compared the energy efficiency of microwave- and conventionally heated reactors at meso scale for organic reactions. They reported that at meso scale, microwave heating was more energy-efficient than conventional heating (Moseley and Woodman 2009). Barnard et al. developed a continuous-flow approach for the preparation of biodiesel using microwave heating. The methodology used for this process allows for the reaction to be run under atmospheric conditions and performed at flow rates of up to 7.2 L/min using a 4 L reaction vessel. This study assessed a range of different processing techniques for the scale-up of microwave-promoted reactions, taking them from the milligram to at least the multigram level for batch and continuous flow processing (Barnard et al. 2007, Bowman et al. 2007). Microwave assisted extraction and transesterification was performed using various types of feedstock ranging from edible oils to non-edible and waste frying oils. The experimental studies are summarized in Table 2 (Gude et al. 2013).

Algal Biodiesel Production

Algal biodiesel production essentially involves five steps: (1) cultivation; (2) harvesting; (3) extraction of lipids; (4) transesterification; and (5) separation of products. Microwaves can be utilized in processing stage of the process, i.e., for extraction and transesterification of oils. High lipid yielding microalgae are cultivated and grown either in open or closed raceway ponds or in photobioreactors. Photobioreactors are designed to maximize the lipid yield and to minimize contamination and to improve the efficiency of the process. Algae are harvested by coagulation, flocculation, sedimentation and filtration methods followed by extraction and transesterification steps. The algal culture is usually concentrated to 15–20% by volume from its original concentration of 0.02–5% concentration in the cultivation ponds. Once concentrated microalgae biomass is obtained, it is subjected to drying (if required), oil extraction, transesterification and separation and purification processes.

Direct or "in situ" transesterification

Microalgae biodiesel is produced from extracted lipids via a traditional extraction–conversion approach (Krohn et al. 2011, Miao and Wu 2006, Umdu et al. 2009). However, this conventional route heavily relies on organic solvent extraction efficiency, which is a major drawback in several recent reports due to incomplete extraction of oils (Johnson et al. 2009, McNichol et al. 2012). The "*in situ* transesterification" of algal lipids may overcome this limitation. In this method, algal lipids are simultaneously extracted and converted to biodiesel (Fatty Acid Alkyl Esters, FAAEs). Since the *in situ* approach integrates the extraction and conversion in one step, it eliminates the need to first isolate and refine the lipid before converting it into biodiesel which could lead to a reduction in product cost. Moreover, besides serving as a reactant in the *in situ* process, the alcohol may weaken the cellular and lipid body membranes to facilitate the FAME conversion (Haas and Wagner 2011). Recently, the *in situ* process has been applied to prepare biodiesel from various microalgal species (biomass) with H_2SO_4 (Ehimen et al. 2010, Wahlen et al. 2011), KOH (Patil et al. 2011), and SrO (Koberg et al. 2011, Rawat et al. 2013) as catalyst using conventional, microwave and supercritical methods (Dong et al. 2013, Cui and Liang 2014).

Supercritical conditions can be applied in direct extractive-transesterification of vegetable oils and algal oils. Water at supercritical conditions can act as organic solvent, thus eliminating the need for solvent use. Many studies have focused on this method to extract and transesterify bio-oils from different feedstock. The process operates at high temperatures and high pressures close to sub and supercritical conditions of water or solvent. In these studies, it was observed that higher temperatures favored extraction and transesterification process, however, at certain temperatures decomposition of biomass was inevitable (Banapurmath et al. 2008, Demirbas 2008, Hawash et al. 2009, Madras et al. 2004, Rathore et al. 2007). Apart from it, safety of pressurized vessels is another concern. Advantages of this process are high quality extracts and end products which require easy separation (Cooney et al. 2009).

Direct transesterification of freeze-dried microalgae in various solvents and using various catalysts was conducted by Cooney and co-workers under various experimental conditions. A 100% conversion of lipids (triglycerides) to FAMEs was observed. They also executed this reaction in a novel ionic liquid based co-solvent that replaces the organic (i.e., chloroform) of the Bligh and Dyer co-solvent system with a hydrophilic ionic liquid (e.g., 1-ethyl-3-methylimidazolium methyl sulfate). It is proposed that the methanol facilitates the permeabilization of the cell wall and intracellular extraction of the lipids, while the ionic liquid facilitates the auto partitioning of the lipids to a separate immiscible phase (Cooney et al. 2009, 2011, Young et al. 2010). Johnson and Wen have attempted extraction and transesterification of oils from *Schizochytrium limacinum*, heterotrophic microalga. They conducted their experiments by two methods: (1) oil extraction followed by transesterification (a two-stage method) or direction transesterification of algal biomass (a one-stage method). When freeze-dried biomass was used as feedstock, the two-stage method resulted in 57% of crude biodiesel yield (based on algal biomass) with a fatty acid methyl ester (FAME) content of 66.4%. The one-stage method (with chloroform, hexane, or petroleum ether used in transesterification) led to a high yield of crude biodiesel, whereas only chloroform-based transesterification led to a high FAME content. When wet biomass was used as feedstock, the one-stage method resulted in a much-lower biodiesel yield. The biodiesel prepared via the direct transesterification of dry biomass has met the ASTM standards. Different schemes using different solvents for one stage and two stage methods were also presented (Johnson and Wen 2009).

Aresta et al. (2005) conducted thermochemical liquefaction using wet algal biomass and supercritical CO_2 extraction using dry algal biomass for direct transesterification of bio-oils. Both of the processes seem to be energy intensive by the reaction conditions they reported (thermochemical liquefaction conditions: 250–395°C for 1 h and supercritical CO_2 extraction conditions: 50°C, 2.60 MPa for 7 h). The two technologies resulted in different extraction capacities; the extraction with sc-CO_2 allows to obtain a higher amount of long chain FA, while the liquefaction gives a higher amount of oily material. Also, the isolated yield of poly-unsaturated species (18:2, 20:4, 20:5) is higher with the sc-CO_2 extraction compared to thermochemical liquefaction. Thermochemical liquefaction requires temperature around 350 and 395°C in order to have the optimal amount of extracted oil. However, as explained earlier, its composition depends on the working temperature and the content of long chain FA is higher at lower temperature as decomposition may occur at higher temperatures. Between these two technologies, the thermochemical liquefaction seems to be more efficient than the extraction with sc-CO_2 from the quantitative results but decomposition of the fatty acids may occur under those operating conditions.

Simultaneous extraction and transesterification (*in situ* transesterification) of the wet algal biomass in supercritical methanol conditions was conducted by Patil et al. (2011b). In a microwave-assisted extraction and transesterification process, as it has been demonstrated in many organic and biodiesel synthesis studies, it is anticipated that the reaction can be conducted at atmospheric pressures and temperatures merely close to the boiling point of methanol with much shorter reaction time. The same group also performed direct extractive-transesterification of dry algal biomass and optimized process parameters using microwave heat source. Response surface methodology (RSM) was used as an optimization technique to analyze the influence of the process variables (dry microalgae to methanol (wt/vol) ratio, catalyst concentration, and reaction time) on the fatty acid methyl ester conversion. From experimental results and RSM analysis, they reported the optimal conditions as: dry microalgae to methanol (wt/vol) ratio of around 1:12, catalyst concentration about 2 wt.%, and reaction time of 4 min. The algal biodiesel samples were analyzed by GC–MS and thin layer chromatography (TLC) methods. Transmission electron microscopy (TEM) images of the algal biomass samples before and after the extraction/transesterification reaction were also presented which are shown in Figure 13.

The response contours for the effect of different process parameters, namely microalgae to methanol (wt./vol.) ratio, catalyst concentration expressed as wt.% of dry microalgae, and reaction time (min) on the fatty acid methyl ester (FAME) contents are shown in Figure 14. The effect of methanol on the simultaneous extraction and transesterification reaction is significant with increasing dry microalgae to methanol ratios up to 1:12 (wt./vol.).

In this reaction, methanol acts both as a solvent for extraction of the algal oils/lipids as well as the reactant for transesterification of esters. Methanol is a good microwave radiation absorption material (loss factor, $tan\delta$ = 0.659 at 2.45 GHz) which absorbs most of the microwave effect in its entire spectrum to produce localized superheating in the reactants and assists the reaction to complete faster. However, higher volumes of methanol may also result in excess loss of the solvent or aggravated rates of solvent recovery. In addition, excessive methanol amounts may reduce the concentration of the catalyst in the reactant mixture and retard the transesterification reaction (Zhang et al. 2010). From the counter plot for catalyst concentration effect on FAME yields (Figure 17), catalyst concentrations up to 2% (wt.%) shows a positive effect on the transesterification reaction. As this is a two-phase reaction mixture, the oil/lipid concentration in the methanol phase is low at the start of the reaction, leading to mass transfer limitations. As the reaction continues, the concentration of oil/lipid in the methanol phase increases, leading to higher transesterification rates with increased catalyst concentrations (Boocock et al. 1998). However, higher

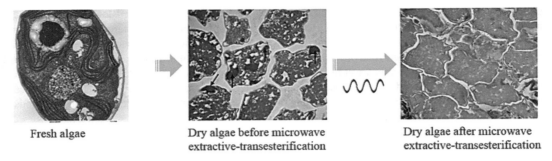

Figure 13. By transmission electron microscopy, areas of thin sections from dry microalgae sample (a) and microwave processed sample (b). Many cell profiles in (a) include electron-dense inclusions representing lipid (arrows), but after microwave processing, cell profiles do not contain electron-dense inclusions. Scale bar is one micrometer (Gude et al. 2013).

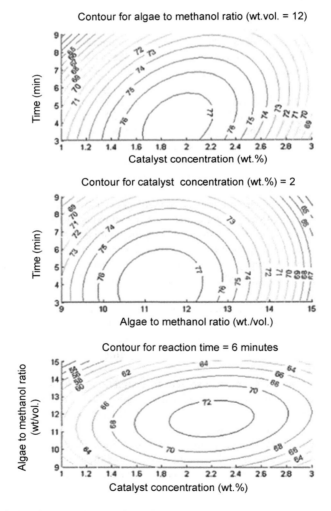

Figure 14. The effect of microalgae to methanol ratio (wt./vol), catalyst concentration (wt.%), and reaction time (min) on the fatty acid methyl ester (FAME) content using RSM (Patil et al. 2011b).

concentrations of catalyst above 2% (wt.%) did not show any positive effect on the biodiesel conversion. This may be due to the interaction of the other compounds resulting in byproducts. Other disadvantages of high basic catalyst concentrations, in general, are their corrosive nature and tendency to form soap which hinders the transesterification reaction (Leadbeater et al. 2008). The reaction time has a significant effect on the FAME content. Generally, extended reaction times provide for enhanced exposure of microwaves to the reaction mixture which result in better yields of extraction and biodiesel conversion. Lower reaction times do not provide sufficient interaction of the reactant mixture to penetrate and dissolve the oils into the reaction mixture. Koberg et al. (2011) demonstrated the direct production (extraction and transesterifcation) of biodiesel from *Nannochloropsis*. Microwave irradiation and ultrasonication were used separately and compared to identify the most effective biodiesel production method. They concluded that the microwave method was the most simple and efficient method for the one-stage direct transesterification of the microalgae biomass.

Effect of co-solvents

To be successful, any extracting solvent must be able to (1) penetrate through the matrix enclosing the lipid material, (2) physically contact the lipid material, and (3) solvate the lipid. As such the development of any extraction process must also account for the fact that the tissue structure and cell walls may present formidable barriers to solvent access. This generally requires that the native structure of the biomass must be disrupted prior to extraction. It is important to select a solvent with high extracting power and strong interaction with the microwaves and the analyte (oils). Polar molecules and ionic solutions (typically acids) strongly absorb microwave energy because of the permanent dipole moment. On the other hand, when exposed to microwaves, non-polar solvents such as hexane will not heat up but they will contribute to mass transfer of analytes. Solvents that are transparent to microwaves do not heat up under irradiation. Hexane is an example of microwave-transparent solvent whereas ethanol is an excellent microwave-absorbing solvent. Therefore a combination of these two (polar and non-polar solvents) can be used in microwave assisted extraction if mass/heat transfer controlled reactions need to be achieved (Martinez-Guerra et al. 2014a). The main concern when using microwave irradiation as a heat source is the rapid heating of solvent mixtures which may cause product degradation and byproduct formation.

Martinez-Guerra et al. (2014b) investigated the effect of solvent n-hexane along with ethanol on extractive-transesterification of algal lipids (*Chlorella* sp.). Two different single-step extractive-transesterification methods (1) with ethanol as solvent/reactant and sodium hydroxide catalyst; and (2) with ethanol as reactant and hexane as solvent (sodium hydroxide catalyst) were studied. Biodiesel (fattyacid-ethyl-esters, FAEE) yields from these two methods were compared with the conventional Bligh and Dyer (BD) method which followed a two-step extraction-transesterification process. The maximum lipid yields for MW, MW with hexane and BD methods were 20.1%, 20.1%, and 13.9%, respectively; while the FAEE conversion of the algal lipids were 96.2%, 94.3%, and 78.1%, respectively. The microalgae-biomass:ethanol molar ratio of 1:250–500 and 2.0–2.5% catalyst with reaction times around 6 min were determined as optimum conditions for both methods.

Temperature profiles (Figure 15a) at different microwave power levels between 10% and 100% at 10% intervals show that the boiling temperatures of the bulk medium are not required for the enhanced extractive-transesterification by MW method. This is because the microwaves produce localized superheating at molecular levels which provide for enhanced heat and mass transfer in the extractive-transesterification reaction. Figure 15b shows the temperature profiles for this method where different volumetric ratios (mL) of ethanol and hexane were added. As it can be seen, the reaction temperatures are in the range of 60–70°C which is lower than the MW method with ethanol alone (70–80°C). This is due to the non-polarity of n-hexane which does not allow n-hexane to heat up under the microwave effect. As a result, the overall temperature of the reaction mixture is maintained low during the reaction. This suggests that addition of n-hexane as co-solvent will reduce the reaction condition severities under microwave effect.

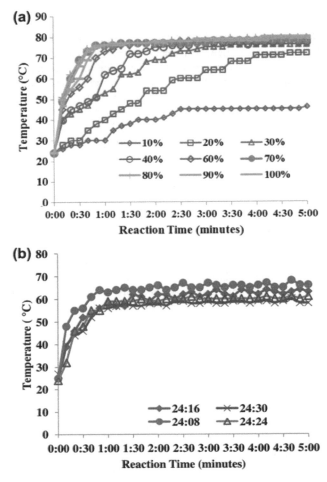

Figure 15. (a) Temperature profiles for MW power effect (process conditions: 4 g dry microalgae; 2% catalyst and 5 min reaction time); and (b) temperature profiles for ethanol and hexane mixtures (vol:vol in mL) (process conditions: 4 g dry microalgae; 2% catalyst and 5 min reaction time).

Microwave Mediated Supercritical Conditions

The basic idea of supercritical treatment is a relationship between pressure and temperature upon thermo-physical properties of the solvent such as dielectric constant, viscosity, specific weight, and polarity. The transesterification of triglycerides by supercritical methanol (SCM), ethanol, propanol and butanol, has proved to be the most promising process. Table 6 shows critical temperatures and critical pressures of various alcohols. A non-catalytic BD production route with supercritical methanol has been developed that allows a simple process and high yield because of simultaneous transesterification of triglycerides and methyl esterification of fatty acids (Demirbas 2005).

Microwave reactors that operate at supercritical conditions (high temperatures and pressures) are available for laboratory as well as commercial uses. In a microwave transparent pressurized vessel, batch heating of reactants can be extremely rapid. Usually these applicators involve pressures from 2–3 to 8–10 MPa and temperatures from 150 to 250°C. They are used for hydrothermal and solvothermal preparation, staining, sterilization and other applications. The combination of microwave and vapor pressure has been already applied successfully to polymers-composite reticulation in large (7500 liters) multi-mode cavities (Geuens et al. 2008, Leonelli and Mason 2010).

Table 6. Critical temperatures and critical pressures of various alcohols.

Alcohol	Critical temperature (K)	Critical pressure (MPa)
Methanol	512.2	8.1
Ethanol	516.2	6.4
1-Propanol	537.2	5.1
1-Butanol	560.2	4.9

C18 fatty acids (Oleic acid) esterification at supercritical conditions (elevated pressures and temperatures) without catalyst was performed recently. The operating temperature was 150–225°C and the pressure was 20 bar. Methanol and ethanol were used as solvents in this non-catalytic reaction. A microwave batch reactor (synthos 3000, Anton-Paar) equipped with two magnetrons, 1400 W of continuous microwave power at 2.45 GHz, a rotor system in which 8 quartz vessels with 80 mL of capacity can be inserted at one time, and a magnetic stirrer for agitation of the sample in each vessel (up to 600 rpm) was used. The equipment is projected to operate up to 300°C and 80 bar. The non-catalytic esterification of the oleic acid resulted in 60% conversion in 60 minutes which is similar to conventional heating (Melo-Junior et al. 2009). A microwave based high pressure thermo-chemical conversion of sewage sludge as an alternative to incineration was performed by Bohlmann (Bohlmann 1999). A maximum oil yield of 30.7% with a heating value of 36.4 MJ/kg was achieved in this study. Microwave-assisted catalyst-free transesterification of triglycerides with 1-butanol under supercritical conditions was conducted by Geuens et al. (2008). Microwave based pyrolysis of sewage sludge to recover bio-oils was also studied by many researchers in last few years (Domí et al. 2003, Domínguez et al. 2005, Tian et al. 2011, Zuo et al. 2011).

A catalyst-free conversion of rapeseed oil into fatty acid butyl esters under microwave irradiation was demonstrated (Geuens et al. 2007). High conversions were reached when the transesterification of triglycerides with 1-butanol was performed under near-critical or supercritical conditions. Since no catalyst is used, the byproduct (glycerol) is free of salts and can be used without the need of purification. The excess butanol can easily be recycled, thus eliminating waste. Best results were obtained employing supercritical conditions (310°C, ca. 80 bar) using a microwave autoclave. 1-butanol has low microwave absorptivity at high temperatures under supercritical conditions, passive heating elements made from silicon carbide were used to facilitate microwave heating of the reaction mixture. Microwave energy absorbers such as silicon carbide (SiC) are commonly used as passive heating elements to increase the rate of temperature rise (Patil et al. 2013).

An *in situ* transesterification approach was demonstrated for converting lipid-rich wet microalgae (*Nannochloropsis salina*) into fatty acid ethyl esters (FAEE) under microwave-mediated supercritical ethanol conditions. The possible reaction mechanism under supercritical conditions are shown in Figure 16.

Typical pressure, temperature and power supply profiles are shown in Figure 17. Direct extractive-transesterification of wet algal biomass may still preserve the nutrients and other valuable components in the cells.

Based on the experimental analysis for wet microalgae biodiesel, the optimal conditions for maximum yield of 30.9% are reported as: wet microalgae/ethanol (wt./vol.) ratio of around 1:9, reaction time of about 25 min, respectively, at controlled power dissipation levels. The temperature noted at these optimum reaction conditions was around 260°C. It was observed that the nutrient values of both fresh and the lipid extracted wet microalgae (Figure 18) are high enough to be considered as potential animal feed. More in-depth studies are needed to verify these preliminary results. Some of the high value bio-products that can be extracted and analyzed by GC–MS from *N. salina* microalgae are phycobiliproteins carotenoids (e.g., indole, oxalic acid, naphthalene); polyunsaturated fatty acids (e.g., EPA, DHA, arachidonic acid); and vitamins (e.g., ascorbic acid, alpha-tocopherol). Other valuable compounds found in microalgae are vaccenic acid and stigmastan-3,5-diene. Omega-3 fatty acids, eicosapentanoic acid (EPA) and decosahexaenoic acid (DHA) can be purified to provide a high-value food supplement and can play a vital role in medicine (Patil et al. 2011b).

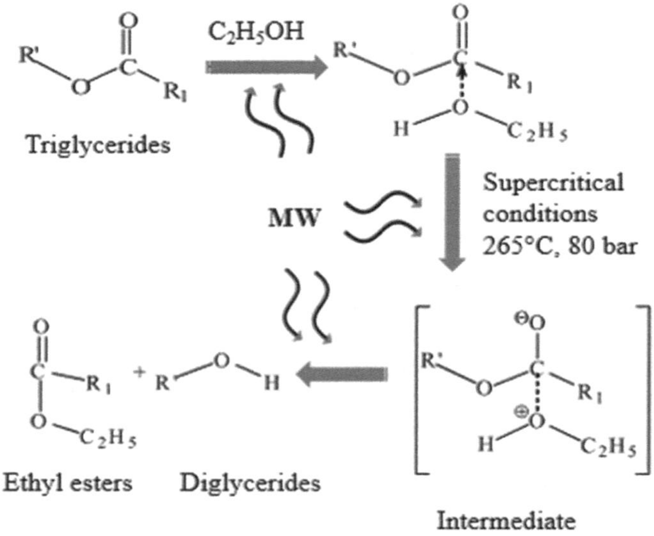

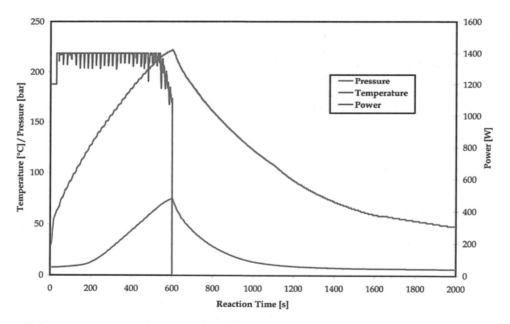

Figure 16. Schematic of a microwave-supercritical transesterification reaction scheme.

Figure 17. Pressure, temperature and power supply profiles for a microwave based supercritical reaction (Patil et al. 2012).

Other Uses of Fatty Acid Methyl Esters

Fatty acid methyl esters can be transformed into a variety of useful chemicals, and raw materials for further synthesis, as shown in Figure 19 (Schuchardt et al. 1998). The alkanolamides, whose production consumes the major part of the methyl esters produced in the world, have a direct application as non-ionic surfactants, emulsifying, thickening and plastifying agents, etc. The fatty alcohols are applied as pharmaceutical and cosmetics additives (C16-C18), as well as lubricants and plastifying agents (C6-C12), depending on the

Microwave Mediated Biodiesel Production 133

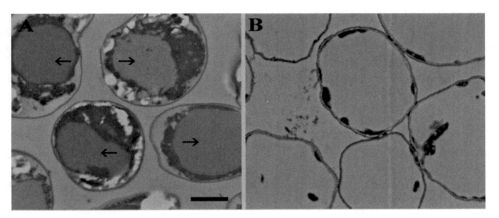

Figure 18. Fresh and lipid extracted wet microalgae after microwave-supercritical conditions.

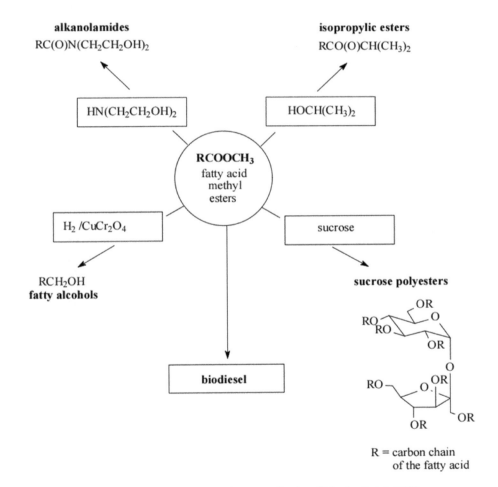

Figure 19. Some fatty acid methyl esters applications (Schuchardt et al. 1998).

length of their carbon chain. The isopropyl esters are also applied as plastifying agents and emollients. However, they cannot be produced in a convenient way by esterification of fatty acids, because an azeotrope formed by water and isopropanol avoids the recycling of the alcohol. Fatty acid methyl esters are further used in the manufacture of carbohydrate fatty acid esters (sucrose polyesters), which can be applied as non-ionic surfactants or edible non-calorific oils, and can be used as an alternative fuel substitute for diesel engines which is called biodiesel.

Glycerol Byproduct

Biodiesel production by transesterification reaction involves glycerol formation as a byproduct. Glycerol has commercial value and it is used as a raw material in many chemical manufacturing industries. It dissolves many organic and inorganic compounds including transition metal complexes. It is immiscible with organic solvents such as hydrocarbons, ethers and esters rendering easy separation from the reaction products and further use in other reactions. Glycerol has high viscosity (1200 cP at 20°C) which limits its applications as a solvent. Its low solubility of highly hydrophobic compounds and gases limits its possible applications. The drawback of viscosity is usually overcome by heating above 60°C or by using co-solvents, but also by using high-intensity ultrasound or microwaves activation in a standalone or combined manner. Glycerol is a well-suited solvent for these kinds of activation, since it allows excellent acoustic cavitation even at high temperatures, and its dielectric properties (dipole moment = 2.67 D; dielectric permittivity = 41.01 at room temperature) result in a quick and efficient heating, favoring both the mass transfer (by decreasing the viscosity of the medium) and the reaction kinetics. Moreover, its high boiling point allows carrying out reactions at high temperatures without losing solvent by evaporation.

Glycerol has important applications in cosmetics, toothpastes, pharmaceuticals, food, lacquers, plastics, alkyl resins, tobacco, explosives, cellulose processing among others. Glycerol is used as a reaction medium in organic synthesis, both with and without a catalyst. Glycerol can be used as a single solvent, as co-solvent, being part of a deep eutectic mixture, or playing the double role of solvent and reagent. Esters derived from glycerol can be used as solvents. The synthesis of glycerol esters can be carried out either by esterification reaction with carboxylic acids, usually catalyzed by acids, or by transesterification reactions catalyzed by bases as shown in Figure 20.

Glycerol-derived solvents can be classified into two very different groups. The first group called "classical" glycerol derivatives has been traditionally used as solvents, such as esters (like acetins), carbonates (glycerol carbonate), acetals (glycerol formal) and ketals (solketal). The second group, organic solvents that are usually prepared from other sources such as propylene glycol, 1,3-propanediol, ethyl lactate or butanol are synthesized from glycerol. These have the potential to become competitive as the price of glycerol decreases. Glycerol derivatives such as acetal, ketal (acetalization of glycerol with aldehydes and ketones gives respective 1,3-dioxalane and 1,3-dioxane, commercial important compounds) ether, and ester find applications in the cosmetics, plastic, pharmaceuticals, detergents, fuel additives and fine chemical industries (Pawar et al. 2014). Several catalytic systems feasible with or without solvents have been developed to perform the acetalization of glycerol ranging from harsh mineral acids like H_2SO_4 and mild organic acids like p-toluenesulfonic acid (PTSA) to solid acid catalysts such as alluminosilicates, resins, zeolites and others. In addition to prolonged reaction time and complicated synthesis protocols, these systems suffer either stringent environmental regulations or operational disadvantages like moisture-sensitivity of solid acid catalysts. This type of difficult reactions can be accelerated in simplified microwave enhanced reaction environment. Microwaves can provide a quick and easy alternative to carry out these reactions as shown Figure 21. Other beneficial products that can be derived from glycerol and corresponding processes or reaction pathways are shown in Table 7.

The microbial conversion of glycerol to various compounds are investigated by different authors with particular focus on the production of 1,3 propanediol, which can be applied as a basic ingredient of polyester (Günzel et al. 1991, Beibl et al. 1992). Moreover, biological production of hydrogen and ethanol from glycerol is also attractive because hydrogen is expected to be future of the clean energy source and ethanol can be used as a raw material and a supplement to gasoline (Ito et al. 2004). The fermentation of glycerol to 1, 3-propanediol has been studied using a number of microorganisms such as *Klebsiella*

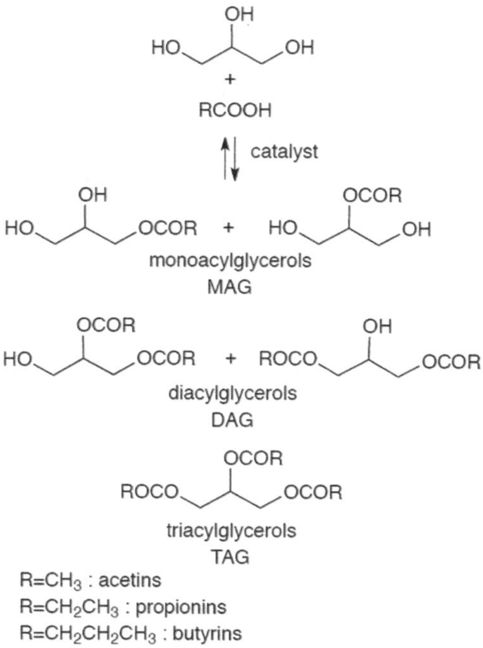

Figure 20. Esterification of glycerol.

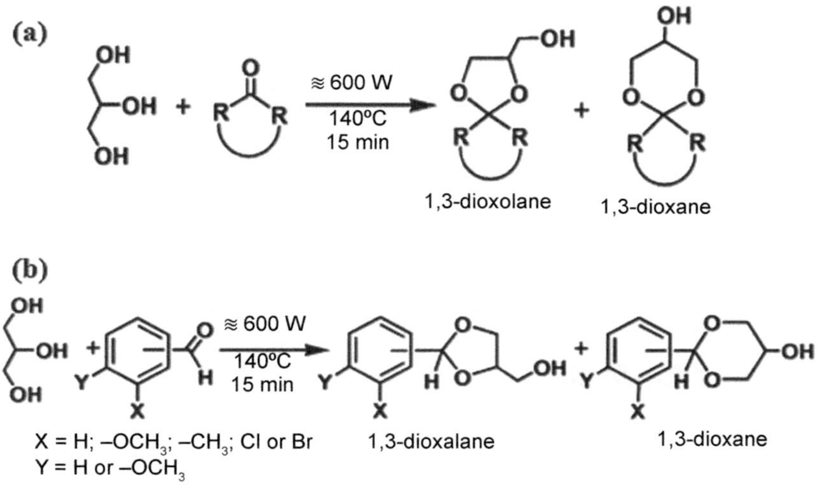

Figure 21. Production of acetals and ketals from glycerol under microwave irradiation.

pneumoniae (Solomon et al. 1994), *Citrobacter freundii* (Boenigk et al. 1993), *Clostridium butyricum* (Forsberg 1987), *Enterobacter agglomerans* (Barbirato et al. 1995), *Enterobacter aerogenes* HU-101 (Ito et al. 2004) and so on.

Market value of glycerol produced as a by-product is an important factor in biodiesel economics. Glycerol market is limited; any major increase in biodiesel capacity would undoubtedly lead to glycerol prices to decline, thereby affecting the overall economics of biodiesel. Notwithstanding the potential of microbial conversion of glycerol into biohydrogen or bioethanol, there are some problems that need to be solved before this technology can have practical applications. For example it is necessary to increase glycerol

Table 7. Valuable products from glycerol (Quispe et al. 2013).

Products	Process	Uses
Hydrogen	Steam reforming, partial oxidation, auto thermal reforming, aqueous-phase reforming and supercritical water, photofermentation using a photosynthetic bacterium	New fuel and energy carrier that could be used in the transport sector, power generation, chemical industry, photovoltaic cells
Fuel additive	Reactive of glycerol with acetic through acetylation or esterification process. Reaction of glycerol with ether substrate through etherification process. Reaction of glycerol with acetone and acid anhydride through acetalation process. Glycerol fermentation by Clostridium pateuriunum	These products may have suitable properties for use as solvents or additives in gasoline/petroleum engines without changes in design. Uses as brake fluids, as perfume based, as paint thinner and hydraulic fluid
Methanol	Via synthesis gas	Chemical feedstock, medical and industrial application
Ethanol	Bioconversion of raw glycerol (glycerol fermentation by *E. coli*	Used as fuel in space, industrial and transportation sectors. Largely use in alcoholic beverages, medical applications and chemical feedstock
Chemical industry products	Glycerol fermentation by klepsiella Pneumoniae. Glycerol selective dehydroxylation. Glycerol hydrogenolysis. Glycerol with CO_2 (glycerol carbonate), Glycerol with heteropolyacid (DCP), glycerol with hydrochloric acid catalyzed by acetic acid as acid catalyst	• Polymer industry (use as monomer in the synthesis of several polyester and polymers, unsaturated polyester) • Plastic industry (polyglycerol methacrylates) • Textile industry (as a substitute for petroleum-based polypropelene, sizing and softening to yarn and fabric) • Explosives industry (nitroglycerin)—Antifreeze liquid • Additive for liquid detergent
Pharmaceutical products	Glycerol oxidation for produce Dihydroxyacetone and crude glycerol with microalgal culture (DHA)	Used as a tanning agent in cosmetics industries, additive in drugs, health supplements and nutrient
Biogas Fuel	Co-digested in anaerobic digesters, syngas production	Fuel

concentration used in the production of H_2 and ethanol because an excessive dilution of biodiesel wastes using the medium increases the cost for the recovery of ethanol and waste water treatment [Ito et al. 2004].

Biodiesel Purification Methods

Products obtained from transesterification reaction (FAMEs and glycerol) need separation and purification prior to their use. All the chemicals used in the process have the potential to show significant effects in various end uses. Table 8 shows a list of chemicals and their potential effects on the end uses of biodiesel.

There is a possibility for the catalysts used in the process to be trapped and carried into the biodiesel end uses. Homogenous catalysts are usually separated through washing the biodiesel product to remove the solvent and glycerol. They are usually soluble in water and easily removed. Heterogeneous catalysts such as metal oxide derivatives may also dissolve in the biodiesel products and cause concerns in their uses. Singh and Fernando (2008) reported the metal content in the biodiesel and glycerol fractions of the transesterification reaction products for various metal oxide catalysts as shown in Table 9. FAME fraction and glycerol fraction of each of the samples were subjected to elemental analysis via flame atomic absorption analysis (FLAA) after all samples were centrifuged at 400 rpm for 20 min. It is apparent that thallium oxide had the largest residual elements in both biodiesel and glycerol samples, whereas zinc oxide had the minimum. Pb_3O_4, despite having the lowest surface area and comparatively low leaching tendency, rendered one of the best FAME yields.

Many purification methods are based on the distillation of the glycerol phase to strip alcohol contaminants from glycerol or using mechanically-agitated thin film processor to continuously evaporate and distill glycerol and other heat sensitive solids containing products. Glycerol can also be recovered by heating a glycerol containing effluent stream having low molecular weight alcohol, water and fatty acid esters. The reaction mixture is sparged with nitrogen to help remove water and low molecular weight alcohol. The crude glycerol phase can be purified by ion exchange on the strong acid resin Amberlite–252 and it has been suggested that the macroporous Amberlite could be useful for removal of sodium ions from glycerol/water solutions with a high salt concentration. Another technology of purification of glycerol with high salt content is by ion-exchange Ambersep BD50. Use of ionic liquids has also been explored as a solvent for extraction. A Lewis basic mixture of quaternary ammonium salts—a eutectic based ionic liquid with glycerol was used to extract excess glycerol from biodiesel formed from the reaction of triglycerides with ethanol in the presence of KOH. The effect of the cation on the partition coefficient of glycerol was determined, together with the time taken for the systems to reach equilibrium.

Microwave Based Biodiesel Composition and Properties

Microalgae are mainly composed of carbohydrates, proteins, nucleic acids, and lipids, where carbohydrates and lipids are responsible for the energy storage. The lipid content depends on the growth and the nutrient

Table 8. Negative effects of contaminants on biodiesel and engines (Atadashi et al. 2011).

Contaminants	Negative effect
Methanol	Deterioration of natural rubber seals and gaskets, lower flash points (problems in storage, transport, and utilization, etc.), Lower viscosity and density values, Corrosion of pieces of Aluminum (Al) and Zinc (Zn)
Water	Reduces heat of combustion, corrosion of system components (such as fuel tubes and injector pumps) failure of fuel pump, hydrolysis (FFAs formation), formation of ice crystals resulting to gelling of residual fuel, Bacteriological growth causing blockage of filters, and Pitting in the pistons
Catalyst/soap	Damage injectors, pose corrosion problems in engines, plugging of filters and weakening of engines
Free fatty acids (FFAs)	Less oxidation stability, corrosion of vital engine components
Glycerides	Crystallization, turbidity, higher viscosities, and deposits formation at pistons, valves and injection Nozzles
Glycerol	Decantation, storage problem, fuel tank bottom deposits Injector fouling, settling problems, higher aldehydes and acrolein emissions, and severity of engine durability problems

Table 9. Leaching of Metals from their respective metal oxide in biodiesel and glycerol samples (Singh and Fernando 2008).

Catalyst	Type	Leaching in glycerol (mg/kg of glycerol)	Leaching in biodiesel (mg/kg of biodiesel)
MgO	basic	460	8200
CaO	basic	1500	6800
ZnO	amphoteric	45	110
PbO	amphoteric	2100	13000
PbO_2	amphoteric	4400	710
Pb_3O_4	basic	8100	760
Tl_2O_3	basic	35000	19000

conditions of microalgae. Various techniques are employed for obtaining elemental or molecular information of an algal strain. For instance, nuclear magnetic resonance (NMR) spectroscopy was applied to the analysis of plants, fungi and microalgae by Martin (1985). A non-invasive *in vivo* measurement employing NMR spectroscopy revealed details of the nitrogen and carbon metabolism in real time (Schneider 1997). NMR spectroscopy was utilized to give characteristic fingerprints of the lipid extractions from algal samples, while marine algal strains and samples from the Lagoon of Venice were compared (Pollesello et al. 1992). Danielewicz et al. (2013) studied the intact triacylglycerol composition of four microalgae species using MALDI-TOF-MS (matrix-assisted laser desorption and ionization time-of-flight mass spectrometry) and 1H-NMR spectroscopy. This technique was employed in other studies for comparison of various algal strains. The fast estimation of the microalgae lipid content is possible by employing Raman spectroscopy. Raman spectra—in the sense of a fingerprint—give information about the saturated and unsaturated fatty acids in the lipid body. Samek et al. (2010) showed that it is feasible to calculate the iodine value (IV) from Raman spectra. IV quantifies the degree of unsaturation and is mainly used in the biodiesel industry (Schober and Mittelbach 2007) as shown in Figure 22. Moreover, the analysis of fatty acid composition in microalgae by gas chromatography—mass spectrometry (GC-MS) is also possible; however, it is a time-consuming technique (Gouveia et al. 2009).

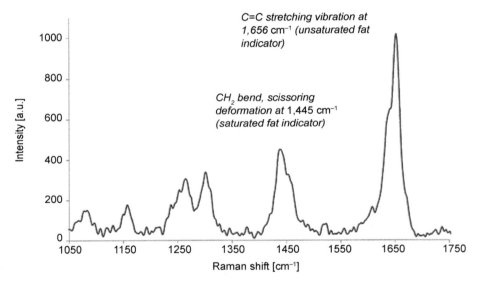

Figure 22. Raman spectrum of a lipid body within the microalgae (*Trachydiscus minutus*). From the ratio of intensities I1656/I1445, the iodine value (IV) can be estimated. Here algal lipid content is close to IV ~ 230 (Porizka et al. 2014).

Thin Layer Chromatography

Thin layer chromatography (TLC) is a useful technique for the separation and identification of compounds in mixtures. TLC is used routinely to follow the progress of reactions by monitoring the consumption of starting materials and the appearance of products. Commercial applications of TLC include the analysis of urine for evidence of "doping", the analysis of drugs to establish purity or identity of the components, and analysis of foods to determine the presence of contaminants such as pesticides. TLC is commonly used for qualitative analysis. It is fast and effective, and also can be used to verify the conversion of the oil during the reaction. TLC can be used as a quick and easy test to detect the reaction products prior to testing and quantification by GC. Some solvent systems and detecting reagents commonly used for evaluation of starting material conversion during the reaction are shown in Table 10. As shown in Figure 23a, various reaction products can be identified while the reaction is taking place. Figure 23b shows a comparison of the FTIR results for camelina and microalgae oil derived biodiesel which are compared against the conventional petrodiesel for FAME functional group.

Gas Chromatography (GC) and High-Performance Liquid Chromatography (HPLC) are the most commonly employed methods for evaluating biodiesel composition (Pinto et al. 2005). The major advantages of HPLC over GC are lower temperatures during analysis, which reduces the risk of isomerization of double bonds, and the possibility of collecting fractions for further investigation (Li et al. 2001). The speed of analysis, selectivity and sensitivity are important parameters in HPLC that may be improved with derivatization. The elution order of FAs, in both methods is based on chain length and degree of unsaturation. FA retention times increase with carbon number (Sanches-Silva et al. 2004). Sample preparation with HPLC is longer than the GC method. Nevertheless this disadvantage is largely compensated by run time. H-NMR spectroscopy is a method used to quantify the conversion of vegetable oils in methyl esters by transesterification reaction. The relevant signals chosen for integration are those of methoxy groups in the methyl esters at 3.7 ppm (singlet) and of the α-carbonyl methylene groups present in all fatty ester derivatives at 2.3 ppm. The latter appears as a triplet, so accurate measurements require good separation of this triplet from the multiplet at 2.1 ppm, which is related to allylic protons. Figure 24 shows the GC chromatogram for microalgal biodiesel with a mass spectrum of methyl esters, while Figure 25 shows the high-value chemicals that can be derived from microalgae using GC-MS.

Table 10. Solvent systems and detecting reagents parameters by TLC (Pinto et al. 2005).

Solvent system (v/v/v)	Detecting reagents	Reference
Hexane:ethyl acetate:acetic acid (90:10:1)	iodine vapour	Shah et al. 2004
Isohexane:diethyl ether:acetic acid (80:20:1)	ultraviolet light	Gandhi and Mukherjee 2000
Chloroform:petroleum ether (1:3)	-	Stavarache et al. 2005
Petroleum ether:diethyl ether:acetic acid (85:15:1)	iodine vapour	Tomasevic and Siler-Marinkovic 2003
Hexane:ethyl acetate:acetic acid (90:10:1)	sulphuric acid:methanol 1:1	Samukawa et al. 2000

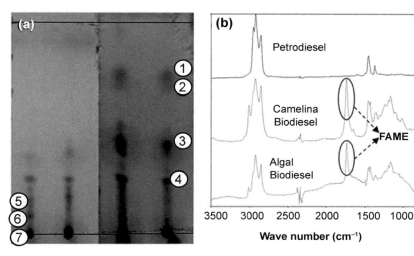

Key: 1. FAMEs; 2. Oil (Triglycerides); 3. Fatty acids; 4. Ascorbic acid, phytols; 5. Di-glycerides; 6. Mono-glycerides; 7. Biodiesel

Figure 23. TLC chromatograms of algal biodiesel samples (left) initial development of the 18 TLC plate; and (right) Silica-gel plate showing the position of methyl esters, 19 triglycerides and fatty acids found in *Nannochloropsis algal* sp.; (b) FTIR results of algal biodiesel, petro-diesel, and *Camelina Sativa* biodiesel.

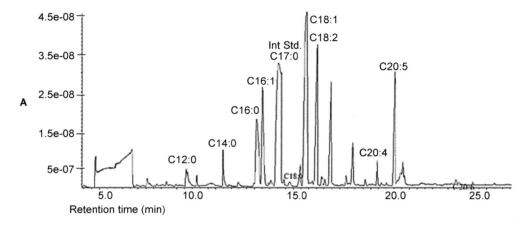

Figure 24. GC Chromatogram obtained for algal biodiesel; mass spectrums of PUFA methyl esters.

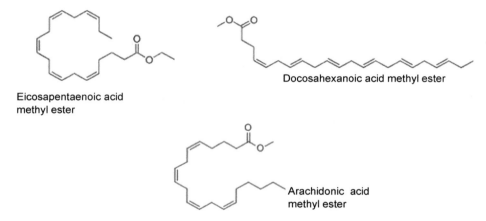

Figure 25. Valuable chemicals from the algal lipids, mass spectrums of PUFA methyl esters: Eicosapentaenoic acid (EPA) methyl ester [C20:5]; Docosahexanoic acid (DHA) methyl ester [C20:6]; and Arachidonic acid methyl ester 14 [C20:4].

Microwave Based Biodiesel Composition and Properties

The use of 100% pure vegetable or animal fats to power diesel engines is not recommended due to several drawbacks such as high fuel viscosity, low power output, thickening or gelling of the lubricating oil, oxidative stability, and low volatility resulting in carbon deposits by incomplete combustion. When biodiesel is used in its 100% purity, it is referred to as B100 or "neat" fuel. Blended biodiesel means pure biodiesel is blended with petrodiesel. Biodiesel blends are referred to as BXX. The XX indicates the amount of biodiesel in the blend (i.e., a B80 blend is 80% biodiesel and 20% petrodiesel). Commercially, these blends are named as B5, B20 or B100 to represent the volume percentage of biodiesel component in the blend with petro diesel as 5, 20 and 100 vol.%, respectively. Biodiesel obtained by microwave heating process compares well with that obtained by other conventional methods of production. A summary of biodiesel properties obtained by microwave processing are shown in Table 11 (Gude et al. 2013).

Biodiesel cetane number depends on the feedstock used for its production. The longer the fatty acid carbon chains and the more saturated the molecules, the higher will be the cetane number (Bajpai and Tyagi 2006, Dermibas 2005, Knothe et al. 1998). Figure 26 shows the correlation between the cetane number and the degree of unsaturation. It is clear that the cetane number fits linearly with the degree of unsaturation. The cetane number for biodiesel should be a minimum of 51 (UNE-EN 14214). According to this, biodiesels of soybean, sunflower and grape seed oil were out of specification. Therefore, a higher degree of unsaturation than 137 make oils did not meet the European Standard for the cetane number. Low cetane numbers have been associated with more highly unsaturated components such as the esters of linoleic (C18:2) and linolenic (C18:3) acids (Dunn and Knothe et al. 2003). Polyunsaturated fuels that contain high levels of C18:2 and C18:3 fatty acids include soybean, sunflower and grape seed oils (Ramos et al. 2009).

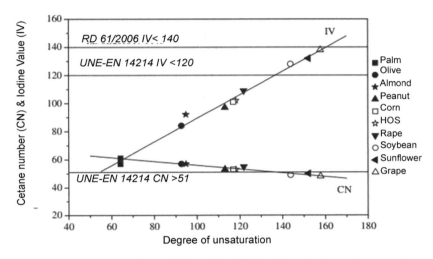

Figure 26. Cetane number and iodine value of biodiesels vs. the degree of unsaturation.

Table 11. Microwave processed biodiesel properties.

Property	Units	ASTM D6751 Std.	EN 14214 Std.	Soybean	Rapeseed	Yellow horn	Pongamia pinnata	Karanja	Waste frying palm oil	Algal* biodiesel
Specific gravity	-	-	0.86–0.9	0.877	0.86	0.882	0.9	0.89	0.87	0.86
Viscosity	cSt @40°C	1.9–6	3.5–5	4.22	4.5	4.4	5.38	4.3	6.3	5.2
Pour point									10	−12
Sulfur content		0.05% w, max	15 max			2	15		0.008	
Carbon residue			0.5 max			0.03			0.05	
Flash point	°C	> 93	> 101	173	136	165	195	145	130	115
Cetane index		> 47	> 51	50.9				56.3	62	
Oxidation stability		> 3	> 6					7.5		
Copper strip Corrosion index		No 3 max	Class 1					2	1a	
Iodine value	g I$_2$/100 g	-	< 120	115.3				83		
Heating value	MJ/k g	-	32.5–36.1		38.8	35.8	39.24		39.9	41
Saponification value	mg KOH/g	-	-	181.3				195		
Acid value	Mg KOH/g	< 0.5	< 0.5	0.14			0.276	0.405		0.374
Ester content	% (w/w)		Min 96.5	99.7	99.4	99.4				

* hot water bath as heat source

Energy scenario of biodiesel production

A viable alternative fuel as a substitute to fossil fuel (ex: biodiesel) will not only provide comparable or superior environmental performance but also will result in an energy gain in the overall process (Hill et al. 2006, Scott et al. 2010). For instance, among current food-based biofuels, biodiesel provides 93% more usable energy than the fossil energy needed for its production, reduces greenhouse gas emissions by 41% compared with diesel, reduces several major air pollutants, and has minimal impact on human and environmental health through N, P, and pesticide release. Sustainability of biodiesel production can be evaluated by a new concept called "Net energy balance" ratio. Net energy balance simply means the ratio of energy derived from the renewable feedstock (energy-out) to the energy invested (energy-in) in the process. The following expression can be used to represent the net energy balance (NEB) ratio (Lardon et al. 2009). The overall savings in energy and greenhouse gas emissions over the lifecycle of the biofuel may be less than anticipated; for example for biodiesel from oilseed rape and soya, the input of energy required over the life-cycle is ~50% of the energy contained in the fuel (Scott 2010).

$$Net\ Energy\ Balance\ (NEB) = \frac{E_{out}}{E_{in}} = \frac{Energy\ produced\ by\ biomass}{Energy\ invested\ in\ the\ process} \quad [4]$$

Since the energy invested in the biodiesel production (energy required for farming, harvesting, processing, transport, etc.) is derived from non-renewable energy sources such as fossil fuels, the net energy balance can also be written as follows (Schenk et al. 2008):

$$Fossil\ energy\ ratio\ (FER) = \frac{E_{RE}}{E_{FE}} = \frac{Renewable\ fuel\ energy\ output}{Fossil\ energy\ input} \quad [5]$$

Sheehan reported that the fossil energy ratio of biodiesel is equal to 3.2. In other words, biodiesel yields 3.2 units of energy for every unit of fossil energy consumed over its life cycle. In comparison, it was found that petroleum diesel's life cycle yielded only about 0.84 units of energy per unit of fossil energy consumed. Few other studies have reported similar results (Pimentel and Patzek 2005, Janulis 2004). It may be more appealing and sustainably acceptable alternative if renewable energy sources can be utilized to produce biodiesel. Fossil fuel energy input can be replaced by other renewable sources such as solar thermal, photovoltaic, geothermal and wind energy. The substitution can be in part or as a whole, wherever applicable.

Microwave energy efficiency and requirements

Energy generation efficiency of microwaves from electrical energy is in the range of 50–65%. This means 35–50% of electrical energy is not converted into microwave energy. Again, in chemical reactions, it is an assumption that all of the microwave energy has been absorbed by the materials participating in reaction. Although microwaves have shown to increase reaction rates by 1000 times in particular chemical synthesis, the downside of it is that the energy generation process is not competitive with conventional steam based production plants with energy conversion efficiencies in the range of 65%–90% (Electricity to steam conversion –90%; fossil fuel to steam –65%) (Moseley and Woodman 2009).

The energy efficiency of a microwave assisted reaction can be calculated using the following equations. Equation 4 represents the heat energy supplied by the microwaves which is given in terms of the power dissipation and the time of exposure. The power dissipation level of the microwave device is usually reported by the manufacturer. Equation 5 quantifies the thermal effect caused by the microwave radiation in the sample volume (i = reactant; ex: oil, catalyst, and solvent) which is simply the product of the mass of the sample multiplied by the specific heat of the material and the temperature gain during the reaction. Energy efficiency of the microwave energy is the ratio of the observed resultant temperature effect to the total energy supplied to the sample as in Eq. 6.

$$Q_{mw} = P_{mw} \times t \quad [6]$$

$$Q_{th} = \sum m_i c_{pi} \Delta T \quad [7]$$

$$\eta = \frac{Q_{th}}{Q_{mw}} \qquad [8]$$

The energy efficiency of the microwave assisted reactions depends on several factors such as the sample volume, nature of the medium (solvents), dissipation level of the microwave device and the penetration depth of the microwaves required in the reaction sample volume. Poor efficiencies can be observed when a high power microwave device is used for a very small sample volume. It is very important to consider the effective level of power dissipation in microwave assisted chemical synthesis to eliminate the energy losses to the surroundings. Patil et al. (2010) found that transesterification of the *Camelina Sativa* oil was even successful at reduced microwave power levels using domestic microwave unit. This observation suggests that effective utilization of microwave power can lead to process energy savings.

Energy calculations for microwave based process need to consider the actual microwave power applied into the process. Leadbeater (2008) conducted batch and continuous flow microwave experiments using 4.6 L batch vessel and flow rates 2 L/min, 7.2 L/min. Energy consumption rates reported from this study are comparable to energy consumption by conventional method. Process energy requirements were calculated based on both actual power consumed and actual microwave power delivered (65% of the power setting) by the system. Overall conversion (oil to FAMEs) rates of 97.9 and 98.9% were reported for these tests. For instance, considering preliminary analysis for 2 L continuous flow conditions, the initial assumption was that the microwave unit would operate at an average of 66% of maximum power (1100 W microwave input; power consumed 2600 W) as observed when the reaction was performed. On the basis of this, energy consumption would be 60.3 kJ/L of biodiesel prepared. If the microwave was operating at full power (1600 W; power consumed 2600 W), energy consumption would be 92.3 kJ/L of biodiesel prepared. For a batch process, calculations were based on the process to heat a 4.6 L reaction mixture to the target temperature of 50°C which takes 3.5 min using a microwave power of 1300 W. With a hold time of 1 min at 50°C, a total reaction time of 4.5 min is given. Assuming that the microwave power remains constant at 1300 W throughout the process, the energy consumption would be 90.1 kJ/L of biodiesel prepared. In reality, the power drops once the target temperature is reached. Thus, this is an overestimation of energy consumption (Loupy et al. 1993).

While few other studies attempted to report the energy efficiency and requirements for the microwave based biodiesel production, they are based on some rudimentary assumptions and calculations (Terigar 2009). Some energy requirements are based on milliliter volumes without a measure of scale in laboratory studies. This is one of the most serious drawbacks for the microwave based biodiesel process. A pilot scale demonstration study at a biodiesel production capacity of 1 ton/d may provide an estimate of actual energy requirements of the process. Results compiled from recent studies are shown in Table 12. Chand et al. conducted biodiesel conversion process using ultrasonication method. They estimated an energy consumption of 91–100 kJ/L for the transesterification process with total energy requirements around 137.5 kJ/L. Their estimates at a large scale level are comparable to the conventional method (Chand et al. 2010).

Energy Return on Investment (EROI) from Microwave Process

Energy return on investment—Case study 1

Motasemi and Ani presented an energy return analysis for microwave biodiesel production (Motasemi and Ani 2012). The energy payback of the system was calculated using electrical energy requirements for production of 1 kg biodiesel under microwave irradiation and the potential electricity production of the same quantity of produced biodiesel. An average microwave exit power of 300 W, for at least 9 min with mixing intensity of 300 rpm produced 81% yield of biodiesel which was used as a basis for calculations. An energy analysis on the system revealed that an average of 2.13 kWh electrical energy can be produced from one kg biodiesel, while production of one kg biodiesel needs only 0.4681 kWh electricity. Therefore, microwave-assisted production of biodiesel can save the energy significantly. This system has the potential

144 Microwave-Mediated Biofuel Production

Table 12. Energy requirements for microwave-enhanced biodiesel production.

Type of heating	Conditions	Energy consumption (kJ/L)	Reference
Conventional	Continuous	94.3	Chand et al. 2010
Microwave	Continuous, 7.2 L/min	26	Barnard et al. 2007
Microwave	Continuous, 2 L/min (a power consumption of 1700 W and a microwave input of 1045 W)	60.3	
	Continuous, 2 L/min (a power consumption of 2600 W and a microwave input of 1600 W	92.3	
Microwave	Batch, 4.6 L (a power consumption of 1300 W, a microwave input of 800 W, a time to reach 50°C of 3.5 min, and a hold time at 50°C of 1 min)	90.1	
Microwave	Supercritical, 10 ethanol/Oleic acid molar ratio, 150°C, 3.6 min	265	Melo-Junior et al. 2009
	Supercritical, 10 ethanol/Oleic acid molar ratio, 200°C, 5.7 min	762	
	Supercritical, 10 methanol/Oleic acid molar ratio, 150°C, 3.7 min	251	
	Supercritical, 20 methanol/Oleic acid molar ratio, 200°C, 3.7 min	609	
	Supercritical, 10 methanol/Oleic acid molar ratio, 200°C, 5.5 min	753	
	Supercritical, 5 methanol/Oleic acid molar ratio, 200°C, 5.1 min	804	
Microwave	Soybean methyl ester, 80°C, 20 min, 98.64% conversion	180.42 (kJ/kg)	Kanitkar 2010
	Soybean ethyl ester, 80°C, 20 min, 98.32% conversion	181.01 (kJ/kg)	
	Rice bran methyl ester, 80°C, 20 min, 98.82% conversion	153.26 (kJ/kg)	
	Rice bran ethyl ester, 80°C, 20 min, 97.78% conversion	191.14 (kJ/kg)	
Ultrasound		137.5	Chand et al. 2008
		185*	

*industry reported data

for production of 1.67 kWh additional electrical energy per kg biodiesel, which proves the techno-economic feasibility of system.

Energy return on investment—Case study 2

Energy return on investment analysis of an *in situ* extractive transesterification (acid catalyzed) of *Rhodotorula glutinis* microalgae lipids was reported (Chuck et al. 2014). The microwave energy demands and corresponding biodiesel yields profiles are shown in Figure 27. Microwave energy did not alter the FAME profile and over 99% of the lipid was esterified when using 25 wt% H_2SO_4 over 20 min at 120°C. Similar yields were achieved over 30 s at higher catalyst loads. Equivalent amounts of FAME were recovered in 30 s using the microwave method when compared with a 4 h Soxhlet extraction, run with the same solvent system. The energy requirements for the microwave were less than 20% of the energy content of the biodiesel produced. Increasing the temperature did not change the energy return on investment substantially; however, longer reaction times used an equivalent amount of energy to the total energy content of the biodiesel.

Higher temperatures and longer reaction times increase the energy requirements at 120°C over 20 minutes. In this case, energy expenditure was higher than the energy that can be extracted from the lipids. However, shorter reaction times resulted in a large increase in the EROI. For example, extraction at 80°C over 30 seconds, produced lipid with 6x the amount of energy than was consumed in its extraction.

The effect of the lipid content on the EROI was also extrapolated. It was reported that higher lipid contents changed the EROI substantially, though this effect has arguably less impact on the energy invested than the time of reaction. While it seems logical to aim for an oil content of 60–70% of the cell, Ratledge and Wynn (2002) reasoned that due to the increased level of nutrients and decreased production of biomass, the optimal level of lipid production for heterotrophic organisms is roughly 40% dry weight.

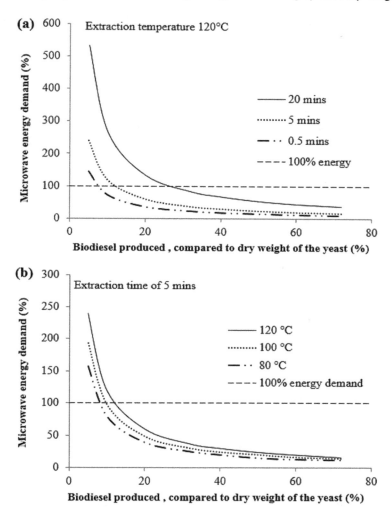

Figure 27. Energy Return On Investment (EROI) of the microwave step, using (a) an extrapolation based on a variable lipid content at 120°C, and (b) an extrapolation based on a variable lipid content over 5 minutes.

Real Time Monitoring of Biodiesel Quality

The biodiesel industry is developing rapidly in recent years. It is very important to determine the FAME content in reaction mixtures for industrial purposes, such as quality control and monitoring the conversion of oil to FAME, and for academic and research development, such as reaction kinetics studies. However, these transesterification systems contain triglyceride (TG) and alcohol reactants, diglyceride (DG) and mono-glyceride (MG) intermediates, and FAME and glycerin products, which makes this task challenging. Various analytical methods, including gas chromatography (GC), gel permeation chromatography, high-pressure liquid chromatography, and nuclear magnetic resonance (NMR), have been used to analyze the FAME products. However, the measurements are time-consuming and involve complex sample preparation, limiting their utility for real-time analyses.

Infrared (IR) spectroscopy (mid-IR and near-IR) is advantageous because of its rapid data acquisition and easy sample preparation, and it is generally employed to cover the shortcomings of the above analytical methods. In principle, the concentrations of reaction components can be determined using Beer's law, provided that their characteristic IR bands are well separated. In transesterification systems, however, reactants, products, and intermediates possess similar functional groups, which results in extensive overlap of their characteristic IR bands. In industrial methods, multivariate calibration models are often used to reduce the calibration error. The multivariate calibration model for a particular biodiesel reaction system requires extensive retesting when it is used for different reactant oils or catalysts, because the IR absorption patterns may vary greatly. To simplify data processing, much effort has been devoted to searching for potential non interfering IR bands that correspond solely to FAME. Figure 28 shows an example of the spectra profiles for a lipase-catalyzed transesterification reaction (Zhang et al. 2013).

Other studies by Natalello and co-workers reported a marker band around 1435 cm^{-1} that was attributed to the CH$_3$ asymmetric deformation, an indication of the formation of FAME (Natalello et al. 2013).

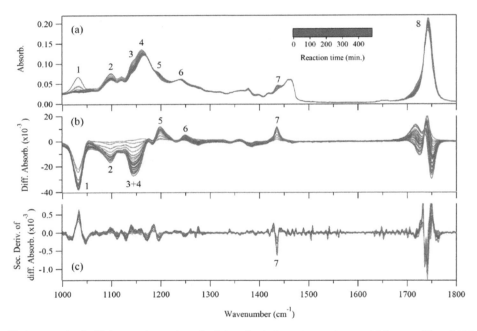

Figure 28. An example of utilizing spectroscopic methods to estimate the process outcomes: (a) Superposition of ATR-FTIR (attenuated total reflection, ATR, Fourier-transform (FT) IR spectroscopy spectra recorded during reaction 4. (b) Superposition of difference spectra obtained by subtracting the spectrum at 0 min. (c) Superposition of the second-order derivatives of the difference spectra in part b.

The peak height of the second derivative of the band gave a highly linear calibration curve for FAME concentration. Similarly, de Souza and Cajaiba da Silva (2012) reported a linear relationship between the variations of the peak height of the C–O stretching band around 1025 cm^{-1} of the methanol substrate and FAME concentration. *In situ* spectroscopic methods can be useful to monitor the esterification and transesterification reactions in biodiesel production. More detailed and dedicated studies are necessary to understand the effect of various intermediate compounds and byproducts in the overlapping of their characteristic infrared bands. It should be noted that the calibration and relationships may vary from process to process depending on the process scheme, feedstock and operating parameters.

Concluding Remarks

Microwave mediated organic and inorganic syntheses are considered green chemistry and preferred methods due to several advantages such as lower energy consumption, substantial reduction in reaction times and solvent requirements, enhanced selectivity, and improved conversions with less byproduct formation. Many reactions that do not occur under classical methods of heating can be carried out with high yields under microwave irradiation. Microwaves have the potential for large scale applications, specifically in biodiesel production due to their ability to interact with a variety of reagents. Laboratory scale results in both batch and continuous conditions are encouraging and few pilot scale studies need to be developed to test their ability and efficiency for large scale adaptability. The reactor design, configurations, flow patterns, reactor safety and operational logistics are yet to be developed.

Understanding the effect of microwaves on biomass extraction and transesterification reactions can be beneficial in the reactor design. Similarly, understanding microwave effect on different catalysts and solvents is crucial to develop safe reactors. Specific areas of challenges that need critical attention prior to large scale development are: controlled heating because biodiesel process is sensitive to temperature variations, efficient transfer of microwave energy into work area with fewer losses to the reactor walls and environment, compatibility of the process with rest of the process pipeline which includes biodiesel product separation and purification. Other important areas are better fundamental understanding and modeling of microwave-material interactions, better preparation of reaction mixtures and compositions tailored specific to microwave processing, better process controls, electronic tuning and automation (smart processing). Finally, availability of low-cost equipment, supporting technologies and other processing support hardware is to be considered.

References

Abbott, A.P., Cullis, P.M., Gibson, M.J., Harris, R.C. and Raven, E. 2007. Extraction of glycerol from biodiesel into a eutectic based ionic liquid. *Green Chemistry*. 9(8): pp.868–872.
Akoh, C.C., Chang, S.W., Lee, G.C. and Shaw, J.F. 2007. Enzymatic approach to biodiesel production. *Journal of Agricultural and Food Chemistry*. 55(22): pp.8995–9005.
Amarni, F. and Kadi, H. 2010. Kinetics study of microwave-assisted solvent extraction of oil from olive cake using hexane: comparison with the conventional extraction. *Innovative Food Science & Emerging Technologies*. 11(2): pp.322–327.
Aresta, M., Dibenedetto, A., Carone, M., Colonna, T. and Fragale, C. 2005. Production of biodiesel from macroalgae by supercritical CO2 extraction and thermochemical liquefaction. *Environmental Chemistry Letters*. 3(3): pp.136–139.
Asakuma, Y., Ogawa, Y., Maeda, K., Fukui, K. and Kuramochi, H., 2011. Effects of microwave irradiation on triglyceride transesterification: experimental and theoretical studies. Biochemical engineering journal, 58, pp.20-24.
Atadashi, I.M., Aroua, M.K., Aziz, A.A. and Sulaiman, N.M.N. 2011. Refining technologies for the purification of crude biodiesel. *Applied Energy*. 88(12): pp.4239–4251.
Azcan, N. and Danisman, A. 2007. Alkali catalyzed transesterification of cottonseed oil by microwave irradiation. *Fuel*. 86(17): pp.2639–2644.
Azcan, N. and Danisman, A. 2008. Microwave assisted transesterification of rapeseed oil. *Fuel*. 87(10): pp.1781–1788.
Bajpai, D. and Tyagi, V.K. 2006. Biodiesel: source, production, composition, properties and its benefits. *Journal of Oleo Science*. 55(10): pp.487–502.
Banapurmath, N.R., Tewari, P.G. and Hosmath, R.S. 2008. Experimental investigations of a four-stroke single cylinder direct injection diesel engine operated on dual fuel mode with producer gas as inducted fuel and Honge oil and its methyl ester (HOME) as injected fuels. *Renewable Energy*. 33(9): pp.2007–2018.
Barbirato, F., Camarasa-Claret, C., Grivet, J.P. and Bories, A. 1995. Glycerol fermentation by a new 1, 3-propanediol-producing microorganism: Enterobacter agglomerans. *Applied Microbiology and Biotechnology*. 43(5): pp.786–793.
Barnard, T.M., Leadbeater, N.E., Boucher, M.B., Stencel, L.M. and Wilhite, B.A. 2007. Continuous-flow preparation of biodiesel using microwave heating. *Energy & Fuels*. 21(3): pp.1777–1781.
Baxendale, I.R., Hayward, J.J. and Ley, S.V. 2007. Microwave reactions under continuous flow conditions. *Combinatorial Chemistry & High Throughput Screening*. 10(10): pp.802–836.
Bender, M. 1999. Economic feasibility review for community-scale farmer cooperatives for biodiesel. *Bioresource Technology*. 70(1): pp.81–87.
Bhattacharya, M. and Basak, T. 2006. On the analysis of microwave power and heating characteristics for food processing: Asymptotes and resonances. *Food Research International*. 39(10): pp.1046–1057.
Biebl, H., Marten, S., Hippe, H. and Deckwer, W.D. 1992. Glycerol conversion to 1, 3-propanediol by newly isolated clostridia. *Applied Microbiology and Biotechnology*. 36(5): pp.592–597.

Binner, J.G.P., Hassine, N.A. and Cross, T.E. 1995. The possible role of the pre-exponential factor in explaining the increased reaction rates observed during the microwave synthesis of titanium carbide. *Journal of Materials Science*. 30(21): pp.5389–5393.

Biodiesel, D. 2009. Green Accounting 2007/2008 Daka Biodiesel amba Daka Biodiesel.

Boenigk, R., Bowien, S. and Gottschalk, G. 1993. Fermentation of glycerol to 1, 3-propanediol in continuous cultures of Citrobacter freundii. *Applied Microbiology and Biotechnology*. 38(4): pp.453–457.

Bogdal, D. 2005. *Microwave-Assisted Organic Synthesis: One Hundred Reaction Procedures* (Vol. 25). Elsevier.

Bohlmann, J.T., Lorth, C.M., Drews, A. and Buchholz, R. 1999. Microwave high-pressure thermochemical conversion of sewage sludge as an alternative to incineration. *Chemical Engineering & Technology*. 22(5): pp.404–409.

Boldor, D., Balasubramanian, S., Purohit, S. and Rusch, K.A. 2008. Design and implementation of a continuous microwave heating system for ballast water treatment. *Environmental Science & Technology*. 42(11), pp.4121–4127.

Boldor, D., Kanitkar, A., Terigar, B.G., Leonardi, C., Lima, M. and Breitenbeck, G.A. 2010. Microwave assisted extraction of biodiesel feedstock from the seeds of invasive Chinese tallow tree. *Environmental Science & Technology*. 44(10): pp.4019–4025.

Bowman, M.D., Holcomb, J.L., Kormos, C.M., Leadbeater, N.E. and Williams, V.A. 2007. Approaches for scale-up of microwave-promoted reactions. *Organic Process Research & Development*. 12(1): pp.41–57.

Breccia, A., Esposito, B., Fratadocchi, G.B. and Fini, A. 1999. Reaction between methanol and commercial seed oils under microwave irradiation. *Journal of Microwave Power and Electromagnetic Energy*. 34(1): pp.2–7.

Brune, D.E., Lundquist, T.J. and Benemann, J.R. 2009. Microalgal biomass for greenhouse gas reductions: potential for replacement of fossil fuels and animal feeds. *Journal of Environmental Engineering*. 135(11): pp.1136–1144.

Brunschwig, C., Moussavou, W. and Blin, J. 2012. Use of bioethanol for biodiesel production. *Progress in Energy and Combustion Science*. 38(2): pp.283–301.

Bunyakiat, K., Makmee, S., Sawangkeaw, R. and Ngamprasertsith, S. 2006. Continuous production of biodiesel via transesterification from vegetable oils in supercritical methanol. *Energy & Fuels*. 20(2): pp.812–817.

Canakci, M. and Van Gerpen, J. 1999. Biodiesel production via acid catalysis. *Transactions of the ASAE*. 42(5): p.1203.

Carmona, M., Valverde, J.L., Pérez, A., Warchol, J. and Rodriguez, J.F. 2009. Purification of glycerol/water solutions from biodiesel synthesis by ion exchange: sodium removal Part I. *Journal of Chemical Technology and Biotechnology*. 84(5): pp.738–744.

Chai, F., Cao, F., Zhai, F., Chen, Y., Wang, X. and Su, Z. 2007. Transesterification of vegetable oil to biodiesel using a heteropolyacid solid catalyst. *Advanced Synthesis & Catalysis*. 349(7): pp.1057–1065.

Chand, P., Reddy, C.V., Verkade, J.G. and Grewell, D. 2008. Enhancing Biodiesel Production from Soybean Oil using Ultrasonics (p. 1). *American Society of Agricultural and Biological Engineers*.

Chemat, F., Vian, M.A. and Cravotto, G. 2012. Green extraction of natural products: Concept and principles. *International Journal of Molecular Sciences*. 13(7): pp.8615–8627.

Chemat, F., Fabiano-Tixier, A.S., Vian, M.A., Allaf, T. and Vorobiev, E. 2015. Solvent-free extraction of food and natural products. *TrAC Trends in Analytical Chemistry*. 71: pp.157–168.

Chen, W., Wang, C., Ying, W., Wang, W., Wu, Y. and Zhang, J. 2008. Continuous production of biodiesel via supercritical methanol transesterification in a tubular reactor. Part 1: Thermophysical and transitive properties of supercritical methanol. *Energy & Fuels*. 23(1): pp.526–532.

Chen, W., Sommerfeld, M. and Hu, Q. 2011. Microwave-assisted Nile red method for *in vivo* quantification of neutral lipids in microalgae. *Bioresource Technology*. 102(1): pp.135–141.

Chisti, Y. 2007. Biodiesel from microalgae. *Biotechnology Advances*. 25(3): pp.294–306.

Cintas, P., Mantegna, S., Gaudino, E.C. and Cravotto, G. 2010. A new pilot flow reactor for high-intensity ultrasound irradiation. Application to the synthesis of biodiesel. *Ultrasonics Sonochemistry*. 17(6): pp.985–989.

Clark, D.E. and Sutton, W.H. 1996. Microwave processing of materials. *Annual Review of Materials Science*. 26(1), pp.299–331.

Cleophax, J., Liagre, M., Loupy, A. and Petit, A. 2000. Application of focused microwaves to the scale-up of solvent-free organic reactions. *Organic Process Research & Development*. 4(6): pp.498–504.

Cooney, M., Young, G. and Nagle, N. 2009. Extraction of bio-oils from microalgae. *Separation & Purification Reviews*. 38(4): pp.291–325.

Cooney, M.J., Young, G. and Pate, R. 2011. Bio-oil from photosynthetic microalgae: case study. *Bioresource Technology*. 102(1): pp.166–177.

Corma, A., Huber, G.W., Sauvanaud, L. and O'Connor, P. 2008. Biomass to chemicals: catalytic conversion of glycerol/water mixtures into acrolein, reaction network. *Journal of Catalysis*. 257(1): pp.163–171.

Cravotto, G. and Cintas, P. 2007. The combined use of microwaves and ultrasound: improved tools in process chemistry and organic synthesis. *Chemistry–A European Journal*. 13(7): pp.1902–1909.

Cravotto, G., Boffa, L., Mantegna, S., Perego, P., Avogadro, M. and Cintas, P. 2008. Improved extraction of vegetable oils under high-intensity ultrasound and/or microwaves. *Ultrasonics sonochemistry*. 15(5): pp.898–902.

Cui, Y. and Liang, Y. 2014. Direct transesterification of wet Cryptococcus curvatus cells to biodiesel through use of microwave irradiation. *Applied Energy*. 119: pp.438–444.

Danielewicz, M.A., Anderson, L.A. and Franz, A.K. 2011. Triacylglycerol profiling of marine microalgae by mass spectrometry. *Journal of Lipid Research*. 52(11): pp.2101–2108.

Demirbaş, A. 2002. Biodiesel from vegetable oils via transesterification in supercritical methanol. *Energy Conversion and Management.* 43(17): pp.2349–2356.

Demirbaş, A. 2003. Biodiesel fuels from vegetable oils via catalytic and non-catalytic supercritical alcohol transesterifications and other methods: a survey. *Energy Conversion and Management.* 44(13): pp.2093–2109.

Demirbas, A. 2005. Biodiesel production from vegetable oils via catalytic and non-catalytic supercritical methanol transesterification methods. *Progress in Energy and Combustion Science.* 31(5): pp.466–487.

Demirbas, A. 2006. Biodiesel production via non-catalytic SCF method and biodiesel fuel characteristics. *Energy Conversion and Management.* 47(15): pp.2271–2282.

Demirbas, A. 2007. Biodiesel from sunflower oil in supercritical methanol with calcium oxide. *Energy Conversion and Management.* 48(3): pp.937–941.

Demirbas, A. 2008. Studies on cottonseed oil biodiesel prepared in non-catalytic SCF conditions. *Bioresource Technology.* 99(5): pp.1125–1130.

Demirbas, A. 2009. Global renewable energy projections. *Energy Sources, Part B.* 4(2): pp.212–224.

Demirbas, A. 2009a. Biodiesel from waste cooking oil via base-catalytic and supercritical methanol transesterification. *Energy Conversion and Management.* 50(4): pp.923–927.

Demirbas, A. 2009b. Production of biodiesel fuels from linseed oil using methanol and ethanol in non-catalytic SCF conditions. *Biomass and Bioenergy.* 33(1): pp.113–118.

Demirbas, A. 2009c. Progress and recent trends in biodiesel fuels. *Energy Conversion and Management.* 50(1): pp.14–34.

Deng, L., Xu, X.B., Haraldsson, G.G., Tan, T.W. and Wang, F. 2005. Enzymatic production of alkyl esters through alcoholysis: A critical evaluation of lipases and alcohols. *J. Am. Oil. Chem. Soc.* 82(5): pp.341–347.

de Souza, A.V.A. and Cajaiba da Silva, J.F. 2012. Biodiesel Synthesis Evaluated by Using Real-Time ATR-FTIR. *Organic Process Research & Development.* 17(1): pp.127–132.

DOE, U. 2010. National algal biofuels technology roadmap (technology roadmap). US Department of Energy, Office of Energy Efficency and Renewable Energy. *Biomass Program.*

Domí, A., Menendez, J.A., Inguanzo, M., Bernad, P.L. and Pis, J.J. 2003. Gas chromatographic–mass spectrometric study of the oil fractions produced by microwave-assisted pyrolysis of different sewage sludges. *Journal of Chromatography A.* 1012(2): pp.193–206.

Domínguez, A., Menéndez, J.A., Inguanzo, M. and Pis, J.J. 2005. Investigations into the characteristics of oils produced from microwave pyrolysis of sewage sludge. *Fuel Processing Technology.* 86(9): pp.1007–1020.

Dong, T., Wang, J., Miao, C., Zheng, Y. and Chen, S. 2013. Two-step in situ biodiesel production from microalgae with high free fatty acid content. *Bioresource Technology.* 136: pp.8–15.

Dossat, V., Combes, D. and Marty, A. 1999. Continuous enzymatic transesterification of high oleic sunflower oil in a packed bed reactor: influence of the glycerol production. *Enzyme and Microbial Technology.* 25(3): pp.194–200.

Du, W., Li, W., Sun, T., Chen, X. and Liu, D. 2008. Perspectives for biotechnological production of biodiesel and impacts. *Applied Microbiology and Biotechnology.* 79(3): pp.331–337.

Dunn, R.O. and Knothe, G. 2003. Oxidative stability of biodiesel in blends with jet fuel by analysis of oil stability index. *Journal of the American Oil Chemists' Society.* 80(10): pp.1047–1048.

Duvernay, W.H., Assad, J.M., Sabliov, C.M., Lima, M. and Xu, Z. 2005. Microwave extraction of antioxidant components from rice bran. *Pharmaceutical Engineering.* 25(4): p.126.

Duz, M.Z., Saydut, A. and Ozturk, G. 2011. Alkali catalyzed transesterification of safflower seed oil assisted by microwave irradiation. *Fuel Processing Technology.* 92(3): pp.308–313.

Earle, M.J., Plechkova, N.V. and Seddon, K.R. 2009. Green synthesis of biodiesel using ionic liquids. *Pure and Applied Chemistry.* 81(11): pp.2045–2057.

Ehimen, E.A., Sun, Z.F. and Carrington, C.G. 2010. Variables affecting the in situ transesterification of microalgae lipids. *Fuel.* 89(3): pp.677–684.

El Sherbiny, S.A., Refaat, A.A. and El Sheltawy, S.T. 2010. Production of biodiesel using the microwave technique. *Journal of Advanced Research.* 1(4): pp.309–314.

Encinar, J.M., González, J.F., Martínez, G., Sánchez, N. and Pardal, A. 2012. Soybean oil transesterification by the use of a microwave flow system. *Fuel.* 95: pp.386–393.

Eskilsson, C.S. and Björklund, E. 2000. Analytical-scale microwave-assisted extraction. *Journal of Chromatography A.* 902(1): pp.227–250.

Fabbri, D., Bevoni, V., Notari, M. and Rivetti, F. 2007. Properties of a potential biofuel obtained from soybean oil by transmethylation with dimethyl carbonate. *Fuel.* 86(5): pp.690–697.

Fjerbaek, L., Christensen, K.V. and Norddahl, B. 2009. A review of the current state of biodiesel production using enzymatic transesterification. *Biotechnology and Bioengineering.* 102(5): pp.1298–1315.

Forsberg, C.W. 1987. Production of 1, 3-propanediol from glycerol by Clostridium acetobutylicum and other Clostridium species. *Applied and Environmental Microbiology.* 53(4): pp.639–643.

Gandhi, N.N. and Mukherjee, K.D. 2000. Specificity of papaya lipase in esterification with respect to the chemical structure of substrates. *Journal of Agricultural and Food Chemistry.* 48(2): pp.566–570.

Ganzler, K., Salgó, A. and Valkó, K. 1986. Microwave extraction: A novel sample preparation method for chromatography. *Journal of Chromatography A.* 371: pp.299–306.

Gedye, R.N., Rank, W. and Westaway, K.C. 1991. The rapid synthesis of organic compounds in microwave ovens. II. *Canadian Journal of Chemistry.* 69(4): pp.706–711.

Gedye, R.N., Smith, F.E. and Westaway, K.C. 1988. The rapid synthesis of organic compounds in microwave ovens. *Canadian Journal of Chemistry.* 66(1): pp.17–26.

Georgogianni, K.G., Kontominas, M.G., Tegou, E., Avlonitis, D. and Gergis, V. 2007. Biodiesel production: reaction and process parameters of alkali-catalyzed transesterification of waste frying oils. *Energy & Fuels.* 21(5): pp.3023–3027.

Geuens, J., Kremsner, J.M., Nebel, B.A., Schober, S., Dommisse, R.A., Mittelbach, M., Tavernier, S., Kappe, C.O. and Maes, B.U. 2007. Microwave-assisted catalyst-free transesterification of triglycerides with 1-butanol under supercritical conditions. *Energy & Fuels.* 22(1): pp.643–645.

Giese, J. 1992. Advances in microwave food processing. *Food Technology.* 46(9): pp.118–123.

Giguere, R.J., Bray, T.L., Duncan, S.M. and Majetich, G. 1986. Application of commercial microwave ovens to organic synthesis. *Tetrahedron Letters.* 27(41): pp.4945–4948.

Glasnov, T.N. and Kappe, C.O. 2007. Microwave-assisted synthesis under continuous-flow conditions. *Macromolecular Rapid Communications.* 28(4): pp.395–410.

Gouveia, L., Marques, A.E., Da Silva, T.L. and Reis, A. 2009. Neochloris oleabundans UTEX# 1185: a suitable renewable lipid source for biofuel production. *Journal of Industrial Microbiology & Biotechnology.* 36(6): pp.821–826.

Gressel, J. 2008. Transgenics are imperative for biofuel crops. *Plant Science.* 174(3): pp.246–263.

Gu, Y. and Jérôme, F. 2013. Bio-based solvents: an emerging generation of fluids for the design of eco-efficient processes in catalysis and organic chemistry. *Chemical Society Reviews.* 42(24): pp.9550–9570.

Gude, V.G., Patil, P., Martinez-Guerra, E., Deng, S. and Nirmalakhandan, N. 2013. Microwave energy potential for biodiesel production. *Sustainable Chemical Processes.* 1(1): p.1.

Günerken, E., D'Hondt, E., Eppink, M.H.M., Garcia-Gonzalez, L., Elst, K. and Wijffels, R.H. 2015. Cell disruption for microalgae biorefineries. *Biotechnology Advances.* 33(2): pp.243–260.

Günzel, B., Yonsel, S. and Deckwer, W.D. 1991. Fermentative production of 1, 3-propanediol from glycerol by Clostridium butyricum up to a scale of 2m3. *Applied Microbiology and Biotechnology.* 36(3): pp.289–294.

Guo, Z., Jin, Q., Fan, G., Duan, Y., Qin, C. and Wen, M. 2001. Microwave-assisted extraction of effective constituents from a Chinese herbal medicine Radix puerariae. *Analytica Chimica Acta.* 436(1): pp.41–47.

Haas, M.J., McAloon, A.J., Yee, W.C. and Foglia, T.A. 2006. A process model to estimate biodiesel production costs. *Bioresource Technology.* 97(4): pp.671–678.

Haas, M.J. and Wagner, K. 2011. Simplifying biodiesel production: the direct or in situ transesterification of algal biomass. *European Journal of Lipid Science and Technology.* 113(10): pp.1219–1229.

Hájek, M. and Skopal, F. 2010. Treatment of glycerol phase formed by biodiesel production. *Bioresource Technology.* 101(9): pp.3242–3245.

Halim, R., Harun, R., Danquah, M.K. and Webley, P.A. 2012. Microalgal cell disruption for biofuel development. *Applied Energy.* 91(1): pp.116–121.

Han, X., Chen, L. and Peng, Q. 2008. Preparation of biodiesel from sunflower oil under microwave irradiation by ionic liquids H2SO4. *Journal of Zhengzhou University (Engineering Science).* 4.

Harding, K.G., Dennis, J.S., Von Blottnitz, H. and Harrison, S.T.L. 2008. A life-cycle comparison between inorganic and biological catalysis for the production of biodiesel. *Journal of Cleaner Production.* 16(13): pp.1368–1378.

Harvey, A.P., Mackley, M.R. and Seliger, T. 2003. Process intensification of biodiesel production using a continuous oscillatory flow reactor. *Journal of Chemical Technology and Biotechnology.* 78(2-3): pp.338–341.

Hawash, S., Kamal, N., Zaher, F., Kenawi, O. and El Diwani, G. 2009. Biodiesel fuel from Jatropha oil via non-catalytic supercritical methanol transesterification. *Fuel.* 88(3): pp.579–582.

Hemwimon, S., Pavasant, P. and Shotipruk, A. 2007. Microwave-assisted extraction of antioxidative anthraquinones from roots of Morinda citrifolia. *Separation and Purification Technology.* 54(1): pp.44–50.

Hernández-Martín, E. and Otero, C. 2008. Different enzyme requirements for the synthesis of biodiesel: Novozym® 435 and Lipozyme® TL IM. *Bioresource Technology.* 99(2): pp.277–286.

Hernando, J., Leton, P., Matia, M.P., Novella, J.L. and Alvarez-Builla, J. 2007. Biodiesel and FAME synthesis assisted by microwaves: homogeneous batch and flow processes. *Fuel.* 86(10): pp.1641–1644.

Hill, J.M. and Marchant, T.R. 1996. Modelling microwave heating. *Applied Mathematical Modelling.* 20(1): pp.3–15.

Hill, J., Nelson, E., Tilman, D., Polasky, S. and Tiffany, D. 2006. Environmental, economic, and energetic costs and benefits of biodiesel and ethanol biofuels. *Proceedings of the National Academy of Sciences.* 103(30): pp.11206–11210.

Hoekman, S.K. 2009. Biofuels in the US–challenges and opportunities. *Renewable Energy.* 34(1): pp.14–22.

Hong, N.I., Yaylayan, V.A., Vijaya Raghavan, G.S., Paré, J.J. and Bélanger, J.M. 2001. Microwave-assisted extraction of phenolic compounds from grape seed. *Natural Product Letters.* 15(3): pp.197–204.

Hsiao, M.C., Lin, C.C., Chang, Y.H. and Chen, L.C. 2010. Ultrasonic mixing and closed microwave irradiation-assisted transesterification of soybean oil. *Fuel.* 89(12): pp.3618–3622.

Hsiao, M.C., Lin, C.C. and Chang, Y.H. 2011. Microwave irradiation-assisted transesterification of soybean oil to biodiesel catalyzed by nanopowder calcium oxide. *Fuel.* 90(5): pp.1963–1967.

Hsu, A.F., Jones, K.C., Foglia, T.A. and Marmer, W.N. 2004. Continuous production of ethyl esters of grease using an immobilized lipase. *Journal of the American Oil Chemists' Society.* 81(8): pp.749–752.

Huffer, J.W., Westcott, J.E., Miller, L.V. and Krebs, N.F. 1998. Microwave method for preparing erythrocytes for measurement of zinc concentration and zinc stable isotope enrichment. *Analytical Chemistry.* 70(11): pp.2218–2220.

Illman, A.M., Scragg, A.H. and Shales, S.W. 2000. Increase in Chlorella strains calorific values when grown in low nitrogen medium. *Enzyme and Microbial Technology.* 27(8): pp.631–635.

Iqbal, J. and Theegala, C. 2013. Microwave assisted lipid extraction from microalgae using biodiesel as co-solvent. *Algal Research.* 2(1): pp.34–42.

Ito, T., Nakashimada, Y., Kakizono, T. and Nishio, N. 2004. High-yield production of hydrogen by Enterobacter aerogenes mutants with decreased α-acetolactate synthase activity. *Journal of Bioscience and Bioengineering.* 97(4): pp.227–232.

Janulis, P. 2004. Reduction of energy consumption in biodiesel fuel life cycle. *Renewable Energy.* 29(6): pp.861–871.

Jermolovicius, L.A., Schneiderman, B. and Senise, J.T. 2006. Alteration of esterification kinetics under microwave irradiation. *Advances in Microwave and Radio Frequency Processing.* pp.377–385. Springer Berlin Heidelberg.

Jin, L., Zhang, Y., Dombrowski, J.P., Chen, C.H., Pravatas, A., Xu, L., Perkins, C. and Suib, S.L. 2011. ZnO/La2O2CO3 layered composite: a new heterogeneous catalyst for the efficient ultra-fast microwave biofuel production. *Applied Catalysis B: Environmental.* 103(1): pp.200–205.

Johnson, M.B. and Wen, Z. 2009. Production of biodiesel fuel from the microalga Schizochytrium limacinum by direct transesterification of algal biomass. *Energy & Fuels.* 23(10): pp.5179–5183.

Jorquera, O., Kiperstok, A., Sales, E.A., Embiruçu, M. and Ghirardi, M.L. 2010. Comparative energy life-cycle analyses of microalgal biomass production in open ponds and photobioreactors. *Bioresource Technology.* 101(4): pp.1406–1413.

Kaieda, M., Samukawa, T., Kondo, A. and Fukuda, H. 2001. Effect of methanol and water contents on production of biodiesel fuel from plant oil catalyzed by various lipases in a solvent-free system. *Journal of Bioscience and Bioengineering*, 91(1): pp.12–15.

Kamath, H.V., Regupathi, I. and Saidutta, M.B. 2011. Optimization of two step karanja biodiesel synthesis under microwave irradiation. *Fuel Processing Technology.* 92(1): pp.100–105.

Kanitkar, A.V. 2010. *Parameterization of Microwave Assisted Oil Extraction and its Transesterification to Biodiesel* (Doctoral dissertation, Louisiana State University).

Kappe, C.O. 2004. Controlled microwave heating in modern organic synthesis. *Angewandte Chemie International Edition.* 43(46): pp.6250–6284.

Kappe, C.O. 2008. Microwave dielectric heating in synthetic organic chemistry. *Chemical Society Reviews.* 37(6): pp.1127–1139.

Kappe, C.O., Dallinger, D. and Murphree, S.S. 2009. Practical Microwave Synthesis for Organic Chemists: Strategies, Instruments, and Protocols Wiley.

Kappe, C.O., Stadler, A. and Dallinger, D. 2012. *Microwaves in Organic and Medicinal Chemistry.* John Wiley & Sons.

Kim, D., Choi, J., Kim, G.J., Seol, S.K., Ha, Y.C., Vijayan, M., Jung, S., Kim, B.H., Lee, G.D. and Park, S.S. 2011. Microwave-accelerated energy-efficient esterification of free fatty acid with a heterogeneous catalyst. *Bioresource Technology.* 102(3): pp.3639–3641.

King, J., Holliday, R. and List, G. 1999. Hydrolysis of soybean oil in a subcritical water flow reactor. *Green Chemistry.* 1(6): pp.261–264.

Kiss, G.A.C., Forgacs, E., Cserhati, T., Mota, T., Morais, H. and Ramos, A. 2000. Optimisation of the microwave-assisted extraction of pigments from paprika (Capsicum annuum L.) powders. *Journal of Chromatography A.* 889(1): pp.41–49.

Knothe, G. 2005. Dependence of biodiesel fuel properties on the structure of fatty acid alkyl esters. *Fuel Processing Technology.* 86(10): pp.1059–1070.

Knothe, G., Bagby, M.O. and Ryan III, T.W. 1998. Precombustion of fatty acids and esters of biodiesel. A possible explanation for differing cetane numbers. *Journal of the American Oil Chemists' Society.* 75(8): pp.1007–1013.

Koberg, M., Cohen, M., Ben-Amotz, A. and Gedanken, A. 2011. Bio-diesel production directly from the microalgae biomass of Nannochloropsis by microwave and ultrasound radiation. *Bioresource Technology.* 102(5): pp.4265–4269.

Kong, J., Han, X.F., Chen, L.P. and Huo, J.L. 2009. Preparation of biodiesel under microwave irradiation from sunflower Oil by solid super acid TiO2/SO4. *Guangzhou Chemical Industry.* 2: p.042.

Krammer, P. and Vogel, H. 2000. Hydrolysis of esters in subcritical and supercritical water. *The Journal of Supercritical Fluids.* 16(3): pp.189–206.

Krohn, B.J., McNeff, C.V., Yan, B. and Nowlan, D. 2011. Production of algae-based biodiesel using the continuous catalytic Mcgyan® process. *Bioresource Technology.* 102(1): pp.94–100.

Kulkarni, M.G., Dalai, A.K. and Bakhshi, N.N. 2007. Transesterification of canola oil in mixed methanol/ethanol system and use of esters as lubricity additive. *Bioresource Technology.* 98(10): pp.2027–2033.

Kumar, R., Kumar, G.R. and Chandrashekar, N. 2011. Microwave assisted alkali-catalyzed transesterification of Pongamia pinnata seed oil for biodiesel production. *Bioresource Technology.* 102(11): pp.6617–6620.

Kumari, V., Shah, S. and Gupta, M.N. 2007. Preparation of biodiesel by lipase-catalyzed transesterification of high free fatty acid containing oil from Madhuca indica. *Energy & Fuels.* 21(1): pp.368–372.

Kuramochi, H., Maeda, K., Kato, S., Osako, M., Nakamura, K. and Sakai, S.I. 2009. Application of UNIFAC models for prediction of vapor–liquid and liquid–liquid equilibria relevant to separation and purification processes of crude biodiesel fuel. *Fuel.* 88(8): pp.1472–1477.

Kusdiana, D. and Saka, S. 2001. Kinetics of transesterification in rapeseed oil to biodiesel fuel as treated in supercritical methanol. *Fuel.* 80(5): pp.693–698.

Kusdiana, D. and Saka, S. 2004a. Effects of water on biodiesel fuel production by supercritical methanol treatment. *Bioresource Technology.* 91(3): pp.289–295.

Kusdiana, D. and Saka, S. 2004b. Two-step preparation for catalyst-free biodiesel fuel production. *Applied Biochemistry and Biotechnology.* 115(1-3): pp.781–791.

Lagha, A., Chemat, S., Bartels, P.V. and Chemat, F. 1999. Microwave-ultrasound combined reactor suitable for atmospheric sample preparation procedure of biological and chemical products. *Analusis.* 27(5): pp.452–457.

Lancrenon, X. and Fedders, J. 2008. An innovation in glycerin purification. *Biodiesel Magazine.*

Lardon, L., Helias, A., Sialve, B., Steyer, J.P. and Bernard, O. 2009. Life-cycle assessment of biodiesel production from microalgae. *Environmental Science & Technology.* 43(17): pp.6475–6481.

Leadbeater, N.E. and Stencel, L.M. 2006. Fast, easy preparation of biodiesel using microwave heating. *Energy & Fuels.* 20(5): pp.2281–2283.

Leadbeater, N.E., Barnard, T.M. and Stencel, L.M. 2008. Batch and continuous-flow preparation of biodiesel derived from butanol and facilitated by microwave heating. *Energy & Fuels.* 22(3): pp.2005–2008.

Leadbeater, N.E. ed. 2010. *Microwave Heating as a Tool for Sustainable Chemistry.* CRC Press.

Lee, J.Y., Yoo, C., Jun, S.Y., Ahn, C.Y. and Oh, H.M. 2010. Comparison of several methods for effective lipid extraction from microalgae. *Bioresource Technology.* 101(1): pp.S75–S77.

Lee, A.K., Lewis, D.M. and Ashman, P.J. 2012. Disruption of microalgal cells for the extraction of lipids for biofuels: processes and specific energy requirements. *Biomass and Bioenergy.* 46: pp.89–101.

Leonelli, C. and Mason, T.J. 2010. Microwave and ultrasonic processing: now a realistic option for industry. *Chemical Engineering and Processing: Process Intensification.* 49(9): pp.885–900.

Lertsathapornsuk, V., Pairintra, R., Krisnangkura, K. and Chindaruksa, S. 2003. Direct conversion of used vegetable oil to biodiesel and its use as an alternative fuel for compression ignition engine. *Proc. First Int. Conf. Sustainable Energy and Green Architecture* (Vol. 2003, pp.SE091–SE096).

Lertsathapornsuk, V., Ruangying, P., Pairintra, R. and Krisnangkura, K. 2005. Continuous transethylation of vegetable oils by microwave irradiation. *Proceedings of the 1st conference on energy network* (pp.1–4).

Lertsathapornsuk, V., Pairintra, R., Aryusuk, K. and Krisnangkura, K. 2008. Microwave assisted in continuous biodiesel production from waste frying palm oil and its performance in a 100 kW diesel generator. *Fuel Processing Technology.* 89(12): pp.1330–1336.

Letellier, M. and Budzinski, H. 1999. Microwave assisted extraction of organic compounds. *Analusis.* 27(3): pp.259–270.

Letellier, M., Budzinski, H., Garrigues, P. and Wise, S. 1997. Focused microwave-assisted extraction of polycyclic aromatic hydrocarbons in open cell from reference materials (sediment, soil, air particulates). *Spectroscopy-ottawa-.* 13: pp.71–80.

Leung, D.Y.C. and Guo, Y. 2006. Transesterification of neat and used frying oil: optimization for biodiesel production. *Fuel Processing Technology.* 87(10): pp.883–890.

Li, Z., Gu, T., Kelder, B. and Kopchick, J.J. 2001. Analysis of fatty acids in mouse cells using reversed-phase high-performance liquid chromatography. *Chromatographia.* 54(7-8): pp.463–467.

Li, H., Pordesimo, L.O., Weiss, J. and Wilhelm, L.R. 2004. Microwave and ultrasound assisted extraction of soybean oil. *Transactions of the ASAE.* 47(4): p.1187.

Li, L., Du, W., Liu, D., Wang, L. and Li, Z. 2006. Lipase-catalyzed transesterification of rapeseed oils for biodiesel production with a novel organic solvent as the reaction medium. *Journal of Molecular Catalysis B: Enzymatic.* 43(1): pp.58–62.

Lidström, P., Tierney, J., Wathey, B. and Westman, J. 2001. Microwave assisted organic synthesis—a review. *Tetrahedron.* 57(45): pp.9225–9283.

Liu, S.Y., Wang, Y.F., McDonald, T. and Taylor, S.E. 2008. Efficient production of biodiesel using radio frequency heating. *Energy & Fuels.* 22(3): pp.2116–2120.

Lotero, E., Liu, Y., Lopez, D.E., Suwannakarn, K., Bruce, D.A. and Goodwin, J.G. 2005. Synthesis of biodiesel via acid catalysis. *Industrial & Engineering Chemistry Research.* 44(14): pp.5353–5363.

Loupy, A., Petit, A., Ramdani, M., Yvanaeff, C., Majdoub, M., Labiad, B. and Villemin, D. 1993. The synthesis of esters under microwave irradiation using dry-media conditions. *Canadian Journal of Chemistry.* 71(1): pp.90–95.

Lucchesi, M.E., Chemat, F. and Smadja, J. 2004. Solvent-free microwave extraction of essential oil from aromatic herbs: comparison with conventional hydro-distillation. *Journal of Chromatography A.* 1043(2): pp.323–327.

Machek, J. and Skopal, F. 2001. Biodiesel from rapeseed oil, methanol and KOH 3. Analysis of composition of actual reaction mixture. *Eur. J. Lipid Sci. Technol.* 103: pp.363–371.

Madras, G., Kolluru, C. and Kumar, R. 2004. Synthesis of biodiesel in supercritical fluids. *Fuel.* 83(14): pp.2029–2033.

Maher, K.D. and Bressler, D.C. 2007. Pyrolysis of triglyceride materials for the production of renewable fuels and chemicals. *Bioresource Technology.* 98(12): pp.2351–2368.

Meher, L.C., Sagar, D.V. and Naik, S.N. 2006. Technical aspects of biodiesel production by transesterification—a review. *Renewable and Sustainable Energy Reviews.* 10(3): pp.248–268.

Majewski, M.W., Pollack, S.A. and Curtis-Palmer, V.A. 2009. Diphenylammonium salt catalysts for microwave assisted triglyceride transesterification of corn and soybean oil for biodiesel production. *Tetrahedron Letters.* 50(37): pp.5175–5177.

Mandal, V., Mohan, Y. and Hemalatha, S. 2007. Microwave assisted extraction—an innovative and promising extraction tool for medicinal plant research. *Pharmacognosy Reviews.* 1(1): pp.7–18.

Marcato, B. and Vianello, M. 2000. Microwave-assisted extraction by fast sample preparation for the systematic analysis of additives in polyolefins by high-performance liquid chromatography. *Journal of Chromatography A.* 869(1): pp.285–300.

Marchetti, J.M. and Errazu, A.F. 2008. Technoeconomic study of supercritical biodiesel production plant. *Energy Conversion and Management.* 49(8): pp.2160–2164.
Martin, F. 1985. Monitoring plant metabolism by 13C, 15N and 14N nuclear magnetic resonance spectroscopy. A review of the applications to algae, fungi and higher plants. *Physiologie Végétale.* 23(4): pp.463–490.
Martinez-Guerra, E., Gude, V.G., Mondala, A., Holmes, W. and Hernandez, R. 2014a. Microwave and ultrasound enhanced extractive-transesterification of algal lipids. *Applied Energy.* 129: pp.354–363.
Martinez-Guerra, E., Gude, V.G., Mondala, A., Holmes, W. and Hernandez, R. 2014b. Extractive-transesterification of algal lipids under microwave irradiation with hexane as solvent. *Bioresource Technology.* 156: pp.240–247.
Mazzocchia, C., Modica, G., Martini, F., Nannicini, R. and Venegoni, D. 2004. Biodiesel and FAME from triglicerides over acid and basic catalysts assisted by microwave. *Proceedings of Secondo Convegno Nazionale delle Microonde nell'Ingegneria e nelle Scienze Applicate, MISA.* 6(8.10): p.2004.
Mazzocchia, C., Kaddouri, A., Modica, G. and Nannicini, R. 2006. Fast synthesis of biodiesel from triglycerides in presence of microwaves. In *Advances in Microwave and Radio Frequency Processing* (pp.370–376). Springer Berlin Heidelberg.
McNichol, J., MacDougall, K.M., Melanson, J.E. and McGinn, P.J. 2012. Suitability of soxhlet extraction to quantify microalgal fatty acids as determined by comparison with *in situ* transesterification. *Lipids.* 47(2): pp.195–207.
Melo-Júnior, C.A., Albuquerque, C.E., Fortuny, M., Dariva, C., Egues, S., Santos, A.F. and Ramos, A.L. 2009. Use of microwave irradiation in the noncatalytic esterification of C18 fatty acids. *Energy & Fuels.* 23(1): pp.580–585.
Metaxas, A.A. and Meredith, R.J. 1983. *Industrial Microwave Heating* (No. 4). IET.
METRE, A. and NATH, K. 2013. Revisiting Trans-esterification to Produce Biodiesel: Novel Reactors, Separation Strategies and Economic Outlook.
Miao, X. and Wu, Q. 2006. Biodiesel production from heterotrophic microalgal oil. *Bioresource Technology.* 97(6): pp.841–846.
Miranda, J.R., Passarinho, P.C. and Gouveia, L. 2012. Pre-treatment optimization of Scenedesmus obliquus microalga for bioethanol production. *Bioresource Technology.* 104: pp.342–348.
Mittelbach, M. 2005. Biodiesel: production technologies and perspectives. *Working Group Renewable Energy Resources.*
Moseley, J.D. and Woodman, E.K. 2009. Energy efficiency of microwave-and conventionally heated reactors compared at meso scale for organic reactions. *Energy & Fuels.* 23(11): pp.5438–5447.
Moseley, J.D., Lenden, P., Lockwood, M., Ruda, K., Sherlock, J.P., Thomson, A.D. and Gilday, J.P. 2007. A comparison of commercial microwave reactors for scale-up within process chemistry. *Organic Process Research & Development.* 12(1): pp.30–40.
Moser, B.R. 2009. Biodiesel production, properties, and feedstocks. *In Vitro Cellular & Developmental Biology-Plant.* 45(3): pp.229–266.
Motasemi, F. and Ani, F.N. 2012. A review on microwave-assisted production of biodiesel. *Renewable and Sustainable Energy Reviews.* 16(7): pp.4719–4733.
Natalello, A., Sasso, F. and Secundo, F. 2013. Enzymatic transesterification monitored by an easy-to-use Fourier transform infrared spectroscopy method. *Biotechnology Journal.* 8(1): pp.133–138.
Nelson, R.G., Howell, S.A. and Weber, J.A. 1994. *Potential Feedstock Supply and Costs for Biodiesel Production* (No. CONF-9410176—). Western Regional Biomass Energy Program, Reno, NV (United States).
Nelson, S.O. 1994. Measurement of microwave dielectric properties of particulate materials. *Journal of Food Engineering.* 21(3): pp.365–384.
Nogueira, B.M., Carretoni, C., Cruz, R., Freitas, S., Melo, P.A., Costa-Félix, R., Pinto, J.C. and Nele, M. 2010. Microwave activation of enzymatic catalysts for biodiesel production. *Journal of Molecular Catalysis B: Enzymatic.* 67(1): pp.117–121.
Noordam, M. and Withers, R. 1996. Producing biodiesel from canola in the inland northwest: an economic feasibility study. *Idaho Agricultural Experiment Station Bulletin.* 785: p.12.
Nüchter, M., Ondruschka, B., Bonrath, W. and Gum, A. 2004. Microwave assisted synthesis–a critical technology overview. *Green Chemistry.* 6(3): pp.128–141.
Ozturk, G., Kafadar, A., Duz, M., Saydut, A. and Hamamci, C. 2010. Microwave assisted transesterification of maize (Zea mays L.) oil as a biodiesel fuel. *Energy, Exploration & Exploitation.* 28(1): pp.47–58.
Pan, X., Niu, G. and Liu, H. 2003. Microwave-assisted extraction of tea polyphenols and tea caffeine from green tea leaves. *Chemical Engineering and Processing: Process Intensification.* 42(2): pp.129–133.
Paré, J.J., Bélanger, J.M. and Stafford, S.S. 1994. Microwave-assisted process (MAP™): a new tool for the analytical laboratory. *TrAC Trends in Analytical Chemistry.* 13(4): pp.176–184.
Patil, P.D. and Deng, S. 2009. Transesterification of *camelina sativa* oil using heterogeneous metal oxide catalysts. *Energy & Fuels.* 23(9): pp.4619–4624.
Patil, P.D., Gude, V.G. and Deng, S. 2009a. Transesterification of *Camelina sativa* oil using supercritical and subcritical methanol with cosolvents. *Energy & Fuels.* 24(2): pp.746–751.
Patil, P.D., Gude, V.G., Camacho, L.M. and Deng, S. 2009b. Microwave-assisted catalytic transesterification of *Camelina sativa* oil. *Energy & Fuels.* 24(2): pp.1298–1304.
Patil, P., Deng, S., Rhodes, J.I. and Lammers, P.J. 2010. Conversion of waste cooking oil to biodiesel using ferric sulfate and supercritical methanol processes. *Fuel.* 89(2): pp.360–364.
Patil, P., Gude, V.G., Pinappu, S. and Deng, S. 2011a. Transesterification kinetics of Camelina sativa oil on metal oxide catalysts under conventional and microwave heating conditions. *Chemical Engineering Journal.* 168(3): pp.1296–1300.

Patil, P.D., Gude, V.G., Mannarswamy, A., Cooke, P., Munson-McGee, S., Nirmalakhandan, N., Lammers, P. and Deng, S. 2011b. Optimization of microwave-assisted transesterification of dry algal biomass using response surface methodology. *Bioresource Technology*. 102(2): pp.1399–1405.

Patil, P.D., Gude, V.G., Mannarswamy, A., Deng, S., Cooke, P., Munson-McGee, S., Rhodes, I., Lammers, P. and Nirmalakhandan, N. 2011c. Optimization of direct conversion of wet algae to biodiesel under supercritical methanol conditions. *Bioresource Technology*. 102(1): pp.118–122.

Patil, P.D., Gude, V.G., Mannarswamy, A., Cooke, P., Nirmalakhandan, N., Lammers, P. and Deng, S. 2012. Comparison of direct transesterification of algal biomass under supercritical methanol and microwave irradiation conditions. *Fuel*. 97: pp.822–831.

Patil, P.D., Reddy, H., Muppaneni, T., Schaub, T., Holguin, F.O., Cooke, P., Lammers, P., Nirmalakhandan, N., Li, Y., Lu, X. and Deng, S. 2013. *In situ* ethyl ester production from wet algal biomass under microwave-mediated supercritical ethanol conditions. *Bioresource Technology*. 139: pp.308–315.

Pawar, R.R., Jadhav, S.V. and Bajaj, H.C. 2014. Microwave-assisted rapid valorization of glycerol towards acetals and ketals. *Chemical Engineering Journal*. 235: pp.61–66.

Perin, G., Álvaro, G., Westphal, E., Viana, L.H., Jacob, R.G., Lenardão, E.J. and D'Oca, M.G.M. 2008. Transesterification of castor oil assisted by microwave irradiation. *Fuel*. 87(12): pp.2838–2841.

Perreux, L. and Loupy, A. 2001. A tentative rationalization of microwave effects in organic synthesis according to the reaction medium, and mechanistic considerations. *Tetrahedron*. 57(45): pp.9199–9223.

Peters, M.S., Timmerhaus, K.D., West, R.E., Timmerhaus, K. and West, R. 1968. *Plant Design and Economics for Chemical Engineers* (Vol. 4). New York: McGraw-Hill.

Peterson, E.R. 1994. Microwave chemical processing. *Research on Chemical Intermediates*. 20(1): pp.93–96.

Pienkos, P.T. and Darzins, A.L. 2009. The promise and challenges of microalgal-derived biofuels. *Biofuels, Bioproducts and Biorefining*. 3(4): pp.431–440.

Pimentel, D. and Patzek, T.W. 2005. Ethanol production using corn, switchgrass, and wood; biodiesel production using soybean and sunflower. *Natural Resources Research*. 14(1): pp.65–76.

Pinto, A.C., Guarieiro, L.L., Rezende, M.J., Ribeiro, N.M., Torres, E.A., Lopes, W.A., Pereira, P.A.D.P. and Andrade, J.B.D. 2005. Biodiesel: an overview. *Journal of the Brazilian Chemical Society*. 16(6B): pp.1313–1330.

Pollesello, P., Toffanin, R., Murano, E., Paoletti, S., Rizzo, R. and Kvam, B.J. 1992. Lipid extracts from different algal species: 1H and 13C-NMR spectroscopic studies as a new tool to screen differences in the composition of fatty acids, sterols and carotenoids. *Journal of Applied Phycology*. 4(4): pp.315–322.

Porizka, P., Prochazková, P., Prochazka, D., Sládková, L., Novotný, J., Petrilak, M., Brada, M., Samek, O., Pilát, Z., Zemánek, P. and Adam, V. 2014. Algal biomass analysis by laser-based analytical techniques—A review. *Sensors*. 14(9): pp.17725–17752.

Potthast, R., Chung, C. and Mathur, I. 2009. Purification of glycerin obtained as a bioproduct from the transesterification of triglycerides in the synthesis of biofuel. *US Patent No. AC07C2980FI*.

Prabakaran, P. and Ravindran, A.D. 2011. A comparative study on effective cell disruption methods for lipid extraction from microalgae. *Letters in Applied Microbiology*. 53(2): pp.150–154.

Pradhan, A., Shrestha, D.S., McAloon, A., Yee, W., Haas, M., Duffield, J.A. and Shapouri, H. 2009. Energy life-cycle assessment of soybean biodiesel. *US Department of Agriculture: Washington, DC*.

Prafulla, D.P., Veera Gnaneswar, G., Harvind, K.R., Tapaswy, M. and Shuguang, D. 2012. Biodiesel production from waste cooking oil using sulfuric acid and microwave irradiation processes. *Journal of Environmental Protection*.

Quispe, C.A., Coronado, C.J. and Carvalho Jr, J.A. 2013. Glycerol: Production, consumption, prices, characterization and new trends in combustion. *Renewable and Sustainable Energy Reviews*. 27: pp.475–493.

Rahmanlar, İ., Yücel, S. and Özçimen, D. 2012. The Production of methyl esters from waste frying oil by microwave method. *Asia-Pacific Journal of Chemical Engineering*. 7(5): pp.698–704.

Ramos, M.J., Fernández, C.M., Casas, A., Rodríguez, L. and Pérez, Á. 2009. Influence of fatty acid composition of raw materials on biodiesel properties. *Bioresource Technology*. 100(1): pp.261–268.

Ranjan, A., Patil, C. and Moholkar, V.S. 2010. Mechanistic assessment of microalgal lipid extraction. *Industrial & Engineering Chemistry Research*. 49(6): pp.2979–2985.

Ranjith Kumar, R., Hanumantha Rao, P. and Arumugam, M. 2015. Lipid extraction methods from microalgae: a comprehensive review. *Frontiers in Energy Research*. 2: p.61.

Rathore, V. and Madras, G. 2007. Synthesis of biodiesel from edible and non-edible oils in supercritical alcohols and enzymatic synthesis in supercritical carbon dioxide. *Fuel*. 86(17): pp.2650–2659.

Ratledge, C. and Wynn, J.P. 2002. The biochemistry and molecular biology of lipid accumulation in oleaginous microorganisms. *Advances in Applied Microbiology*. 51: pp.1–52.

Rawat, I., Kumar, R.R., Mutanda, T. and Bux, F. 2013. Biodiesel from microalgae: a critical evaluation from laboratory to large scale production. *Applied Energy*. 103: pp.444–467.

Refaat, A.A., Attia, N.K., Sibak, H.A., El Sheltawy, S.T. and El Diwani, G.I. 2008a. Production optimization and quality assessment of biodiesel from waste vegetable oil. *International Journal of Environmental Science & Technology*. 5(1): pp.75–82.

Refaat, A.A., El Sheltawy, S.T. and Sadek, K.U. 2008b. Optimum reaction time, performance and exhaust emissions of biodiesel produced by microwave irradiation. *International Journal of Environmental Science & Technology*. 5(3): pp.315–322.

Reijnders, L. and Huijbregts, M.A.J. 2008. Biogenic greenhouse gas emissions linked to the life cycles of biodiesel derived from European rapeseed and Brazilian soybeans. *Journal of Cleaner Production.* 16(18): pp.1943–1948.

Reimbert, C.G., Minzoni, A.A. and Smyth, N.F. 1996. Effect of radiation losses on hotspot formation and propagation in microwave heating. *IMA Journal of Applied Mathematics.* 57(2): pp.165–179.

Revellame, E., Hernandez, R., French, W., Holmes, W. and Alley, E. 2010. Biodiesel from activated sludge through *in situ* transesterification. *Journal of Chemical Technology and Biotechnology.* 85(5): pp.614–620.

Rittmann, B.E. 2008. Opportunities for renewable bioenergy using microorganisms. *Biotechnology and Bioengineering.* 100(2): pp.203–212.

Romano, S.A.A.D. 1982. *Vegetable oils: A new alternative* (No. CONF-820860-).

Rostagno, M.A., Palma, M. and Barroso, C.G. 2007. Microwave assisted extraction of soy isoflavones. *Analytica Chimica Acta.* 588(2): pp.274–282.

Rosenberg, U. and Bogl, W. 1987. Microwave thawing, drying, and baking in the food industry. *Food Technology.*

Roy, I. and Gupta, M.N. 2003. Applications of microwaves in biological sciences. *Current Science-Bangalore.* 85(12): pp.1685–1692.

Saifuddin, N. and Chua, K.H. 2004. Production of ethyl ester (biodiesel) from used frying oils: Optimization transesterification process using microwave irradiation.

Sajjadi, B., Aziz, A.A. and Ibrahim, S. 2014. Investigation, modelling and reviewing the effective parameters in microwave-assisted transesterification. *Renewable and Sustainable Energy Reviews.* 37: pp.762–777.

Saka, S. and Kusdiana, D. 2001. Biodiesel fuel from rapeseed oil as prepared in supercritical methanol. *Fuel.* 80(2): pp.225–231.

Sakai, T., Kawashima, A. and Koshikawa, T. 2009. Economic assessment of batch biodiesel production processes using homogeneous and heterogeneous alkali catalysts. *Bioresource Technology.* 100(13): pp.3268–3276.

Salvi, D., Boldor, D., Ortego, J., Aita, G.M. and Sabliov, C.M. 2010. Numerical modeling of continuous flow microwave heating: a critical comparison of COMSOL and ANSYS. *Journal of Microwave Power and Electromagnetic Energy.* 44(4): pp.187–197.

Samek, O., Jonáš, A., Pilát, Z., Zemánek, P., Nedbal, L., Tříska, J., Kotas, P. and Trtílek, M. 2010. Raman microspectroscopy of individual algal cells: sensing unsaturation of storage lipids *in vivo*. *Sensors.* 10(9): pp.8635–8651.

Samukawa, T., Kaieda, M., Matsumoto, T., Ban, K., Kondo, A., Shimada, Y., Noda, H. and Fukuda, H. 2000. Pretreatment of immobilized Candida antarctica lipase for biodiesel fuel production from plant oil. *Journal of Bioscience and Bioengineering.* 90(2): pp.180–183.

Sanches-Silva, A., de Quirós, A.R.B., López-Hernández, J. and Paseiro-Losada, P. 2004. Comparison between high-performance liquid chromatography and gas chromatography methods for fatty acid identification and quantification in potato crisps. *Journal of Chromatography A.* 1032(1): pp.7–15.

Santana, G.C.S., Martins, P.F., Da Silva, N.D.L., Batistella, C.B., Maciel Filho, R. and Maciel, M.W. 2010. Simulation and cost estimate for biodiesel production using castor oil. *Chemical Engineering Research and Design.* 88(5): pp.626–632.

Satyanarayana, K.G., Mariano, A.B. and Vargas, J.V.C. 2011. A review on microalgae, a versatile source for sustainable energy and materials. *International Journal of Energy Research.* 35(4): pp.291–311.

Schenk, P.M., Thomas-Hall, S.R., Stephens, E., Marx, U.C., Mussgnug, J.H., Posten, C., Kruse, O. and Hankamer, B. 2008. Second generation biofuels: high-efficiency microalgae for biodiesel production. *Bioenergy Research.* 1(1): pp.20–43.

Schneider, B. 1997. *In-vivo* nuclear magnetic resonance spectroscopy of low-molecular-weight compounds in plant cells. *Planta.* 203(1): pp.1–8.

Schober, S. and Mittelbach, M. 2007. Iodine value and biodiesel: Is limitation still appropriate?. *Lipid Technology.* 19(12): pp.281–284.

Schuchardt, U., Sercheli, R. and Vargas, R.M. 1998. Transesterification of vegetable oils: a review. *Journal of the Brazilian Chemical Society.* 9(3): pp.199–210.

Schumann, R., Häubner, N., Klausch, S. and Karsten, U. 2005. Chlorophyll extraction methods for the quantification of green microalgae colonizing building facades. *International Biodeterioration & Biodegradation.* 55(3): pp.213–222.

Schwede, S., Kowalczyk, A., Gerber, M. and Span, R. 2011, May. Influence of different cell disruption techniques on mono digestion of algal biomass. *World Renewable Energy Congress, Linkoping, Sweden.* pp.8–13.

Scott, S.A., Davey, M.P., Dennis, J.S., Horst, I., Howe, C.J., Lea-Smith, D.J. and Smith, A.G. 2010. Biodiesel from algae: challenges and prospects. *Current Opinion in Biotechnology.* 21(3): pp.277–286.

Seider, W.D., Seader, J.D. and Lewin, D.R. 2009. *Product & Process Design Principles: Synthesis, Analysis and Evaluation, (With CD).* John Wiley & Sons.

Shah, S., Sharma, S. and Gupta, M.N. 2004. Biodiesel preparation by lipase-catalyzed transesterification of Jatropha oil. *Energy & Fuels.* 18(1): pp.154–159.

Sharma, Y.C., Singh, B. and Upadhyay, S.N. 2008. Advancements in development and characterization of biodiesel: a review. *Fuel.* 87(12): pp.2355–2373.

Sheehan, J., Dunahay, T., Benemann, J. and Roessler, P. 1998. A look back at the US Department of Energy's aquatic species program: biodiesel from algae. *National Renewable Energy Laboratory.* 328.

Sheehan, J., Dunahay, T., Benemann, J. and Roessler, P. 1998. A look back at the US Department of Energy's Aquatic Species Program-Biodiesel from Algae. National Renweable Energy Laboratory, www.nrel.gov/docs/fy04osti/34796.pdf.

Shen, Y., Pei, Z., Yuan, W. and Mao, E. 2009. Effect of nitrogen and extraction method on algae lipid yield. *International Journal of Agricultural and Biological Engineering.* 2(1): pp.51–57.

Sheng, J., Vannela, R. and Rittmann, B.E. 2012. Disruption of Synechocystis PCC 6803 for lipid extraction. *Water Science & Technology.* 65(3).

Shibasaki-Kitakawa, N., Honda, H., Kuribayashi, H., Toda, T., Fukumura, T. and Yonemoto, T. 2007. Biodiesel production using anionic ion-exchange resin as heterogeneous catalyst. *Bioresource Technology.* 98(2): pp.416–421.

Shimada, Y., Watanabe, Y., Samukawa, T., Sugihara, A., Noda, H., Fukuda, H. and Tominaga, Y. 1999. Conversion of vegetable oil to biodiesel using immobilized Candida antarctica lipase. *Journal of the American Oil Chemists' Society.* 76(7): pp.789–793.

Sim, T.S., Goh, A. and Becker, E.W. 1988. Comparison of centrifugation, dissolved air flotation and drum filtration techniques for harvesting sewage-grown algae. *Biomass.* 16(1): pp.51–62.

Singh, A.K. and Fernando, S.D. 2008. Transesterification of soybean oil using heterogeneous catalysts. *Energy & Fuels.* 22(3): pp.2067–2069.

Singh, J. and Gu, S. 2010. Commercialization potential of microalgae for biofuels production. *Renewable and Sustainable Energy Reviews.* 14(9): pp.2596–2610.

Sinnwell, S. and Ritter, H. 2007. Recent advances in microwave-assisted polymer synthesis. *Australian Journal of Chemistry.* 60(10): pp.729–743.

Solomon, B.O., Zeng, A.P., Biebl, H., Ejiofor, A.O., Posten, C. and Deckwer, W.D. 1994. Effects of substrate limitation on product distribution and H2/CO2 ratio inKlebsiella pneumoniae during anaerobic fermentation of glycerol. *Applied Microbiology and Biotechnology.* 42(2-3), pp.222–226.

Song, E.S., Lim, J.W., Lee, H.S. and Lee, Y.W. 2008. Transesterification of RBD palm oil using supercritical methanol. *The Journal of Supercritical Fluids.* 44(3): pp.356–363.

Šoštarič, M., Klinar, D., Bricelj, M., Golob, J., Berovič, M. and Likozar, B. 2012. Growth, lipid extraction and thermal degradation of the microalga Chlorella vulgaris. *New Biotechnology.* 29(3): pp.325–331.

Spigno, G. and De Faveri, D.M. 2009. Microwave-assisted extraction of tea phenols: A phenomenological study. *Journal of Food Engineering.* 93(2): pp.210–217.

Sridhara, R. and Mathai, I.M. 1974. Transesterification reactions. *Journal of Scientific & Industrial Research.* 33(4): pp.178–187.

Srivastava, A. and Prasad, R. 2000. Triglycerides-based diesel fuels. *Renewable and Sustainable Energy Reviews.* 4(2): pp.111–133.

Stavarache, C., Vinatoru, M., Nishimura, R. and Maeda, Y. 2005. Fatty acids methyl esters from vegetable oil by means of ultrasonic energy. *Ultrasonics Sonochemistry.* 12(5): pp.367–372.

Suppalakpanya, K., Ratanawilai, S.B. and Tongurai, C. 2010. Production of ethyl ester from esterified crude palm oil by microwave with dry washing by bleaching earth. *Applied Energy.* 87(7): pp.2356–2359.

Suppes, G.J., Dasari, M.A., Doskocil, E.J., Mankidy, P.J. and Goff, M.J. 2004. Transesterification of soybean oil with zeolite and metal catalysts. *Applied Catalysis A: General.* 257(2): pp.213–223.

Terigar, B.G. 2009. Advanced microwave technology for biodiesel feedstock processing (Doctoral dissertation, Faculty of the Louisiana State University and Agricultural and Mechanical College in partial fulfillment of the requirements for the degree of Master of Science in Biological and Agricultural Engineering in The Department of Biological and Agricultural Engineering by Beatrice G. Terigar BS, Aurel Vlaicu University of Arad, Romania, 2007).

Terigar, B.G., Balasubramanian, S. and Boldor, D. 2010a. Effect of storage conditions on the oil quality of Chinese tallow tree seeds. *Journal of the American Oil Chemists' Society.* 87(5): pp.573–582.

Terigar, B.G., Balasubramanian, S., Boldor, D., Xu, Z., Lima, M. and Sabliov, C.M. 2010b. Continuous microwave-assisted isoflavone extraction system: Design and performance evaluation. *Bioresource Technology.* 101(7): pp.2466–2471.

Terigar, B.G., Balasubramanian, S., Lima, M. and Boldor, D. 2010c. Transesterification of soybean and rice bran oil with ethanol in a continuous-flow microwave-assisted system: yields, quality, and reaction kinetics. *Energy & Fuels.* 24(12): pp.6609–6615.

Terigar, B.G., Balasubramanian, S., Sabliov, C.M., Lima, M. and Boldor, D. 2011. Soybean and rice bran oil extraction in a continuous microwave system: From laboratory-to pilot-scale. *Journal of Food Engineering.* 104(2): pp.208–217.

Tian, Y., Zuo, W., Ren, Z. and Chen, D., 2011. Estimation of a novel method to produce bio-oil from sewage sludge by microwave pyrolysis with the consideration of efficiency and safety. *Bioresource Technology.* 102(2): pp.2053–2061.

Tierney, J.P. and Lidstrom, P. 2009. Microwave Assisted Organic Synthesis.

Tomasevic, A.V. and Siler-Marinkovic, S.S. 2003. Methanolysis of used frying oil. *Fuel Processing Technology.* 81(1): pp.1–6.

Toukoniitty, B., Mikkola, J.P., Murzin, D.Y. and Salmi, T. 2005. Utilization of electromagnetic and acoustic irradiation in enhancing heterogeneous catalytic reactions. *Applied Catalysis A: General.* 279(1): pp.1–22.

Umdu, E.S., Tuncer, M. and Seker, E. 2009. Transesterification of Nannochloropsis oculata microalga's lipid to biodiesel on Al2O3 supported CaO and MgO catalysts. *Bioresource Technology.* 100(11): pp.2828–2831.

Van Kasteren, J.M.N. and Nisworo, A.P. 2007. A process model to estimate the cost of industrial scale biodiesel production from waste cooking oil by supercritical transesterification. *Resources, Conservation and Recycling.* 50(4): pp.442–458.

Varma, R.S. 1999. Solvent-free organic syntheses. using supported reagents and microwave irradiation. *Green Chemistry.* 1(1): pp.43–55.

Varma, M.N. and Madras, G. 2007. Synthesis of biodiesel from castor oil and linseed oil in supercritical fluids. *Industrial & Engineering Chemistry Research.* 46(1): pp.1–6.

Vieitez, I., da Silva, C., Alckmin, I., Borges, G.R., Corazza, F.C., Oliveira, J.V., Grompone, M.A. and Jachmanián, I. 2008. Effect of temperature on the continuous synthesis of soybean esters under supercritical ethanol. *Energy & Fuels.* 23(1): pp.558–563.

Vijayaraghavan, K. and Hemanathan, K. 2009. Biodiesel production from freshwater algae. *Energy & Fuels.* 23(11): pp.5448–5453.

Virot, M., Tomao, V., Colnagui, G., Visinoni, F. and Chemat, F. 2007. New microwave-integrated Soxhlet extraction: an advantageous tool for the extraction of lipids from food products. *Journal of Chromatography A.* 1174(1): pp.138–144.

Virot, M., Tomao, V., Ginies, C., Visinoni, F. and Chemat, F. 2008. Microwave-integrated extraction of total fats and oils. *Journal of Chromatography A.* 1196: pp.57–64.

Wahlen, B.D., Willis, R.M. and Seefeldt, L.C. 2011. Biodiesel production by simultaneous extraction and conversion of total lipids from microalgae, cyanobacteria, and wild mixed-cultures. *Bioresource Technology.* 102(3): pp.2724–2730.

West, A.H., Posarac, D. and Ellis, N. 2008. Assessment of four biodiesel production processes using HYSYS. Plant. *Bioresource Technology.* 99(14): pp.6587–6601.

Wiesbrock, F., Hoogenboom, R. and Schubert, U.S. 2004. Microwave-assisted polymer synthesis: state-of-the-art and future perspectives. *Macromolecular Rapid Communications.* 25(20): pp.1739–1764.

Xu, L., Brilman, D.W.W., Withag, J.A., Brem, G. and Kersten, S. 2011. Assessment of a dry and a wet route for the production of biofuels from microalgae: energy balance analysis. *Bioresource Technology.* 102(8): pp.5113–5122.

Yaakob, Z., Sukarman, I.S., Kamarudin, S.K., Abdullah, S.R.S. and Mohamed, F. 2008. Production of biodiesel from jatropha curcas by microwave irradiation. In *International conference on Renewable Energy Sources.*

Yang, J., Xu, M., Zhang, X., Hu, Q., Sommerfeld, M. and Chen, Y. 2011. Life-cycle analysis on biodiesel production from microalgae: water footprint and nutrients balance. *Bioresource Technology.* 102(1): pp.159–165.

Yin, J.Z., Xiao, M. and Song, J.B. 2008. Biodiesel from soybean oil in supercritical methanol with co-solvent. *Energy Conversion and Management.* 49(5): pp.908–912.

You, Y.D., Shie, J.L., Chang, C.Y., Huang, S.H., Pai, C.Y., Yu, Y.H. and Chang, C.H. 2007. Economic cost analysis of biodiesel production: Case in soybean oil. *Energy & Fuels.* 22(1): pp.182–189.

Young, G., Nippgen, F., Titterbrandt, S. and Cooney, M.J. 2010. Lipid extraction from biomass using co-solvent mixtures of ionic liquids and polar covalent molecules. *Separation and Purification Technology.* 72(1): pp.118–121.

Yuan, H., Yang, B.L. and Zhu, G.L. 2008. Synthesis of biodiesel using microwave absorption catalysts. *Energy & Fuels.* 23(1): pp.548–552.

Zabeti, M., Daud, W.M.A.W. and Aroua, M.K. 2009. Activity of solid catalysts for biodiesel production: a review. *Fuel Processing Technology.* 90(6): pp.770–777.

Zanin, F.G., Macedo, A., Archilha, M.V.L., Wendler, E.P. and Dos Santos, A.A. 2013. A one-pot glycerol-based additive-blended ethyl biodiesel production: A green process. *Bioresource Technology.* 143: pp.126–130.

Zhang, Y., Dube, M.A., McLean, D. and Kates, M. 2003a. Biodiesel production from waste cooking oil: 1. Process design and technological assessment. *Bioresource Technology.* 89(1): pp.1–16.

Zhang, Y., Dube, M.A., McLean, D.D. and Kates, M. 2003b. Biodiesel production from waste cooking oil: 2. Economic assessment and sensitivity analysis. *Bioresource Technology.* 90(3): pp.229–240.

Zhang, S., Zu, Y.G., Fu, Y.J., Luo, M., Zhang, D.Y. and Efferth, T. 2010. Rapid microwave-assisted transesterification of yellow horn oil to biodiesel using a heteropolyacid solid catalyst. *Bioresource Technology.* 101(3): pp.931–936.

Zhang, Y., Hu, L. and Ramström, O. 2013. Double parallel dynamic resolution through lipase-catalyzed asymmetric transformation. *Chemical Communications.* 49(18): pp.1805–1807.

Zheng, H., Yin, J., Gao, Z., Huang, H., Ji, X. and Dou, C. 2011. Disruption of Chlorella vulgaris cells for the release of biodiesel-producing lipids: a comparison of grinding, ultrasonication, bead milling, enzymatic lysis, and microwaves. *Applied Biochemistry and Biotechnology.* 164(7): pp.1215–1224.

Zigoneanu, I.G., Williams, L., Xu, Z. and Sabliov, C.M. 2008. Determination of antioxidant components in rice bran oil extracted by microwave-assisted method. *Bioresource Technology.* 99(11): pp.4910–4918.

Zuo, W., Tian, Y. and Ren, N. 2011. The important role of microwave receptors in bio-fuel production by microwave-induced pyrolysis of sewage sludge. *Waste Management.* 31(6): pp.1321–1326.

5
Microwave Pretreatment of Feedstock for Bioethanol Production

Bioethanol production has been receiving increasing interest at the international, national and regional levels. Increasing concerns over the greenhouse gas emissions and depletion of non-renewable fossil fuels have forced many nations to shift from conventional energy sources to renewable energy production. As a result, bioethanol production has experienced an unprecedented growth at global levels (Figure 1). Ethanol is already used as a replacement transportation fuel in many parts of the world for many years. The world bioethanol production increased from 31 billion liters per year in 2001 (Berg 2001) to 93 billion liters in 2015 indicating a 300% increase in production (EERE 2017). Brazil and USA are the major producers of the worldwide production (Kim and Dale 2004). Large scale production of fuel ethanol is mainly based on sucrose from sugarcane in Brazil or starch, mainly from corn, in the USA.

Current ethanol production based on corn, starch and sugarsubstances is not a sustainable method due to their food and feed value. Bioethanol production is mainly hindered by high feedstock costs. Lignocellulosic biomass can be used to overcome the feedstock cost issues. This feedstock eliminates the

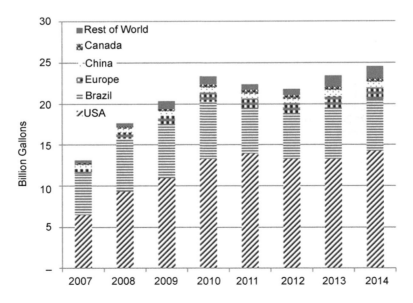

Figure 1. Bioethanol production around the world between 2007 and 2014.

concerns and conflicts related to food vs. fuel or energy vs. environment priorities thereby serving as an excellent source for bioethanol production. It has been estimated that 442 billion liters of bioethanol can be produced from lignocellulosic biomass and that total crop residues and wasted crops can produce 491 billion liters of bioethanol per year, about 4.5 times higher than the actual world bioethanol production (Kim and Dale 2004). Lignocellulosic materials are renewable, low cost and are abundantly available. It includes crop residues, grasses, sawdust, wood chips, etc. Extensive research has been carried out on ethanol production from lignocellulosics in the past two decades (Binod et al. 2010, Duff and Murray 1996). Hence bioethanol production could be the route to the effective utilization of agricultural wastes. Rice straw, wheat straw, corn straw, and sugarcane bagasse are the major agricultural wastes in terms of quantity of biomass available. This review aims to present a brief overview of the available and accessible technologies for bioethanol production using these major agricultural wastes.

Microwaves in Bioethanol Chemistry

Ethanol (ethyl alcohol) can be derived from both renewable and non-renewable feedstock. It can be used as a fuel or as a gasoline enhancer due its high oxygen content. The increased percentage of oxygen allows a better oxidation of the gasoline hydrocarbons with the consequent reduction in the emission of CO and aromatic compounds. Ethanol has better octane booster properties, it is non-toxic, and does not contaminate water sources. Many countries have implemented or are implementing programs for addition of ethanol to gasoline. Fuel ethanol production has increased remarkably because many countries look for reducing oil imports, boosting rural economies and improving air quality. The world ethyl alcohol production has reached about 26,600 million liters (Renewable Fuels Association 2017), USA and Brazil being the first producers. About 73% of produced ethanol worldwide is used as fuel, beverage and other industries use about 17% and 10% ethanol respectively.

There are several pathways for producing ethanol from biomass (Figure 2). Two major platforms can be identified based on the nature of the conversion process: the gasification platform and the carbohydrate platform. The latter can be sub-divided into three pathways based on the carbohydrate feedstock: lignocellulose pathway, starch pathway and glucose (pure sugar) pathway. Bioethanol from the starch and sucrose pathways is considered a first generation biofuel since the feedstocks are typically food crops and the conversion process uses conventional technology. Bioethanol from the gasification platform and the lignocellulose pathway is considered a second-generation biofuel since the feedstocks are non-food based and the conversion process is typically unconventional and still in development (Keshwani 2009).

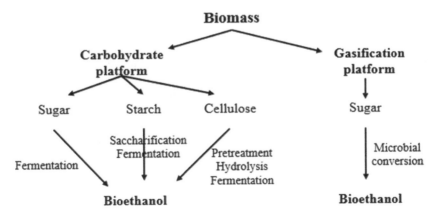

Figure 2. Different pathways for the production of bioethanol from lignocellulosic biomass.

Gasification platform

The gasification platform involves the production of syngas from biomass. Syngas is a mixture of carbon dioxide, carbon monoxide and hydrogen. The resulting gaseous mixture is then fermented into ethanol by microorganisms. Although syngas can be converted into ethanol and other value-added products using non-biological catalysts, microbial fermentation offers specific advantages. Some of these include high specificity of the biocatalyst, lower energy costs, and the ability to handle varying ratios of components in syngas (Bredwell et al. 1999; Klasson et al. 1992).

Synthesis gas, consisting of CO, H_2 and CO_2, may be produced from biomass according to the approximate reaction (Alden et al. 1988):

$$8CH_2O + 4N_2 \rightarrow 6CO + 8H_2 + 2CO_2 + 4N_2 \qquad [1]$$

Biomass

There are many gasifier designs that have been demonstrated and are commercially available. Gas composition is a function of the amount of air or oxygen necessary to generate the heat for pyrolysis of the biomass, as well as the type of gasifier and the moisture content of the biomass. Steam may also be added to adjust the hydrogen concentration. If oxygen is used, nitrogen is eliminated.

The components of synthesis gas may be converted into ethanol by certain anaerobic bacteria according to the equations [2–4]. Henstra et al. (2007) compiled a list of mesophilic microorganisms capable of producing ethanol from syngas. Examples include *Clostridium autoethanogenum*, *Clostridium ljungdahlii* and *Butyribacterium methylotrophicum*. Nearly all microorganisms capable of fermenting syngas into ethanol are mesophilic. This is problematic since it requires significant cooling of the syngas produced in gasifiers that operate at temperatures up to 1000°C. Research is ongoing to identify thermophilic microorganisms and to optimize fermentation conditions.

$$6CO + 3H_2O \rightarrow CH_3CH_2OH + 4CO_2 \qquad [2]$$

or

$$6H_2 + 2CO_2 \rightarrow CH_3CH_2OH + 3H_2O \qquad [3]$$

Equations [2] and [3] may be combined with Equation [1] to give

$$8CH_2O + 4N_2 + O_2 \rightarrow 2.33CH_3CH_2OH + 3.33CO_2 + H_2O + 4N_2 \qquad [4]$$

Biomass

Since nearly all the biomass (including the lignin, but not the ash) can be converted into gas by Equation [1], yields of ethanol of about 50 percent of the total biomass are possible (135 gal per ton). This compares to yields of only about 30 percent for enzymatic or acid hydrolysis/fermentation processes. Synthesis gas compositions from biomass are variable, depending upon the raw material temperature and process used. Typical compositions show a $CO:H_2$ ratio of about one with CO and H_2 compositions of about 35 percent (Alden et al. 1988). Yields of gas of about 90 percent are common. Gas composition can also be tailored for fuels production (Duraiswamy et al. 1989).

Carbohydrate platform

The carbohydrate platform involves the extraction of carbohydrates from biomass and the subsequent fermentation into ethanol. The sucrose (glucose) pathway is the simplest approach and involves extraction of readily fermentable six-carbon sugars present in biomass such as sugarcane, sugar beets and sweet sorghum. The ethanol industry in Brazil relies primarily on sugarcane as a feedstock. However, these feedstocks are geographically limited as they require specific soil and climate conditions. The starch pathway is currently the primary means of ethanol production in the US. Carbohydrates in the form of starch are present in large quantities in biomass such as corn, potato and sweet potato. Corn is the dominant starch feedstock for ethanol production in the US and the industry is quite mature. Unlike the sucrose pathway, starch needs to be first saccharified into simple sugars using hydrolytic enzymes (amylases). These sugars are then fermented into ethanol. Starch is a homopolymer of α-D-glucose, a six-carbon sugar and exists in two forms: amylose and amylopectin. Individual glucose units are linked via α-1-4 and α-1-6 glycosidic bonds. The nature of these bonds creates polymeric structures that have low crystallinity and are easily hydrolyzed by low cost amylase enzymes. The ease of saccharification coupled with government subsidies makes corn an attractive feedstock for bioethanol production in the US.

Bioethanol produced from starch and sucrose pathways is considered a first-generation biofuel derived from food crops. These food crops absorb carbon from the atmosphere over their growth cycle and have the potential for significant carbon sequestration to mitigate the effects of global warming. For example, the annual carbon sequestration estimate for corn is 595 kg per hectare (West and Marland 2003). However, this potential for sequestration has to be tempered by carbon dioxide emissions from intensive agricultural practices that require significant fossil fuel inputs. In fact, net carbon emissions from corn ethanol are only around 10% lower than net carbon emissions from gasoline (Farrell et al. 2006). The amount of arable land that can be allocated to crops such as corn and sugarcane for energy production is limited by the overall availability of cropland. According to the agricultural census conducted by the USDA, the total amount of cropland available in the lower 48 United States is 434 million acres (USDA 2002). The amount of cropland has been declining over the past 50 years and the trend is likely to continue as population pressures force urban development on agricultural land (Outlook 2007). Additionally, the use of food and feed crops for energy production will impact their availability for traditional uses. For example, increased use of corn for ethanol production will result in higher corn prices and will negatively impact the food and feed industries and could result in reduced exports of animal products.

The lignocellulose pathway is considered a viable long-term option for bioethanol production. Lignocellulosic biomass includes woody biomass, logging residues, dedicated energy crops like switchgrass, miscanthus and poplar, agricultural residues such as wheat straw, corn stover and bagasse, residual pulp from paper mills, municipal solid waste and wastes from food processing industries. The richness in diversity of lignocellulosic biomass minimizes geographic constraints and since most lignocellulosic biomass is considered waste material, it is readily available at low cost. Dedicated energy crops like switchgrass and poplar do not require high quality arable land. These attributes make them an attractive feedstock for bioethanol production. Switchgrass, in particular, shows promise due to high productivity across a wide geographic range, suitability for marginal land quality, low water and nutritional requirements and positive environmental benefits. Ethanol yield from switchgrass is estimated to be 5000–6000 liters per hectare while the same from corn starch is estimated around 4000 liters per hectare (Gulati et al. 1996, Parrish and Fike 2005).

Lignocellulosic biomass primarily consists of cellulose, hemicellulose, and lignin. The representative composition is 40%, 30% and 25% respectively as shown in Figure 3. Cellulose is the most abundant and readily renewable source of biomass on the planet with tremendous potential for production of sugars for biofuel production. Cellulose and hemicellulose can be used to produce bioethanol and other fuel intermediates and platform chemicals such as levulinic acid (LA) and hydroxymethylfurfural (HMF). Lignin is the second most abundant biopolymer besides cellulose, consisting primarily of three units: guaiacyl, sinapyl, and p-hydroxyphenyl units linked by aryl ether or C–C bonds. Lignin is difficult to treat and process which is preferable through combustion methods to produce heat and power. Cellulose and hemicellulose require pretreatment to hydrolyze and to increase the solubility. Pretreatment is also necessary to extract digestible sugars which can be fermented to produce bioethanol (Figure 4). Main goals of pretreatment include (1) production of highly digestible solids that enhance sugar yields during enzyme hydrolysis, (2) avoiding the degradation of sugars (mainly pentoses) including those derived from hemicellulose, (3) minimizing the formation of inhibitors for subsequent fermentation steps, (4) recovery of lignin for conversion into valuable coproducts, and (5) to be cost effective by operating in reactors of moderate size and by minimizing heat and power requirements (Brodeur et al. 2011).

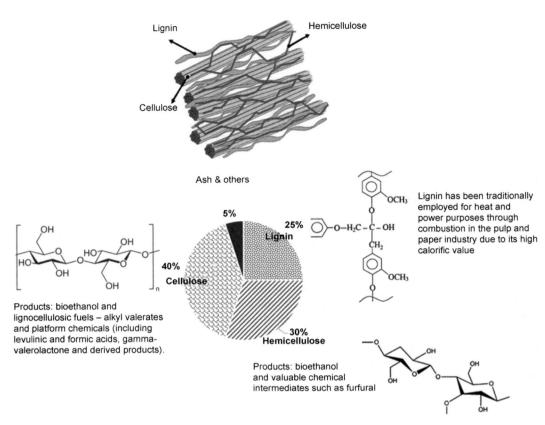

Figure 3. Lignocellulosic biomass composition and the desired products.

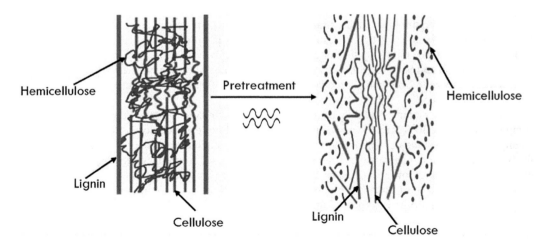

Figure 4. Microwave pretreatment effect on lignocellulosic biomass (Kumar et al. 2009).

Chemical bonds in lignocellulosic complex

There are four main types of bonds identified in the lignocellulose complex (Table 1). Those are ether bonds, ester bonds, carbon-to-carbon bonds and hydrogen bonds. These four bonds are the main types of bonds that provide linkages within the individual components of lignocellulose (intra-polymer linkages), and connect the different components to form the complex (inter-polymer linkages).

Table 1. Overview of linkages between the monomer units that form the individual polymers lignin, cellulose and hemicellulose, and between the polymers to form lignocellulose.

Bonds within different components (intra-polymer linkages)	
Ether bond	Lignin, hemicellulose
Carbon to carbon Lignin	Carbon to carbon Lignin
Hydrogen bond Cellulose	Hydrogen bond Cellulose
Ester bond Hemicellulose	Ester bond Hemicellulose
Bonds connecting different components (inter-polymer linkages)	
Ether bond	Cellulose-Lignin Hemicellulose-Lignin
Ester bond	Hemicellulose-Lignin
Hydrogen bond	Cellulose-Hemicellulose Hemicellulose-Lignin Cellulose-Lignin

Intra-polymer linkages

The main types of bonds that connect the building molecules within the lignin polymer are ether bonds and carbon-to-carbon bonds (Harmsen et al. 2010). Ether bonds may appear between allylic and aryl carbon atoms, or between aryl and aryl carbon atoms, or even between two allylic carbon atoms. The total fraction of ether type bonds in the lignin molecule is around 70% of the total bonds between the monomer units. The carbon-to-carbon linkages form the remaining 30% of the total bonds between the units. They can also appear between two aryl carbon atoms or two allylic carbon atoms, or between one aryl and one allylic carbon atom (Lebo et al. 2001).

Two main linkages form cellulose polymer: (1) the glucosidic linkage is the one that forms the initial polymer chain. More specifically, it is a 1-4 β D-glucosidic bond that connects the glucose units together. The glucosidic bond can also be considered as an ether bond, since it is in fact the connection of two carbon atoms with an elementary oxygen interfering (Solomon 1988); and (2) the hydrogen bond is considered to be responsible for the crystalline fibrous structure of cellulose. The arrangement of the polymer in long straight parallel chains together with the fact that the hydroxyl groups are evenly distributed on both sides of the glucose monomer, allowing the formation of hydrogen bond between two hydroxyl groups of different polymer chains (Faulon et al. 1994). Carboxyl groups are also present in cellulose as 1 carboxyl per 100 or 1000 monomer units of glucose ratio (Krassig and Schurz 2002), although not very obvious in structure of cellulose.

Inter-polymer linkages

It is difficult to determine the inter-polymer linkages in a lignocellulosic biomass without altering the original structure due to its complexity (Harmsen et al. 2010). Therefore conclusions on the connecting linkages between the polymers may not be definite. However, it has been identified that there are hydrogen bonds connecting lignin with cellulose and with hemicellulose, respectively. In addition, the existence of covalent bonds between lignin and polysaccharides is identified. More specifically, it is certain that hemicellulose connects to lignin via ester bonds. It is also known that there are ether bonds between lignin and the polysaccharides. It is still not clear though whether the ether bonds are formed between lignin and cellulose, or hemicellulose. Hydrogen bonding between hemicellulose and cellulose is also identified. However, this linkage is not expected to be strong due to the fact that hemicellulose lacks of primary alcohol functional group external to the pyranoside ring (Faulon et al. 1994).

Bioethanol Production from Starch

Starch is one of the most studied biopolymers due to its widespread applications as well as due to the global interest around renewable, cheap, and easy to process resources. Starch granules are synthesized in a broad array of plant tissues and within many plant species having a very complex structure mainly composed of two types of alpha-glucan, amylose and amylopectin, which represent approximately 98–99% of the dry weight. Starch is frequently subjected to different treatments in order to modify its structure and to obtain functional properties suitable for industrial applications.

Microwave pretreatment of starch

Microwave interaction with starch depends on the type, water content, density, dielectric properties, penetration depth, load geometry and placement inside the oven temperature. Other factors that influence the pretreatment will be microwave frequency, power, reactor type (single or multimode) and geometry, and exposure time. The main mechanism of microwave absorption in a polymer is the reorientation of dipoles in the imposed electric field (National Research Council 1994). The efficiency of microwave coupling with polymer materials is dependent on the dipole strength, its mobility and mass, and the matrix state of the dipole (Metaxas and Meredith 1983). The response of dipoles to an oscillating field is an increase in rotational and vibrational energies, depending on the degree of symmetry of the molecule, with a resulting frictional generation of heat (Venkatesh and Raghavan 2004). The most responsible dipole for heating is water and consequently amount and mobility of water influence the dielectric properties of a material. Dry starches are not polarized at microwave frequencies, but according to Roebuck et al. (1972) the dielectric properties depend primarily upon the free water content. Several investigations related to the influence of water content, temperature, salt addition or microwave frequency on the dielectric properties of different starches were reported (Ryynanen et al. 1996, Ndife et al. 1998, Piyasena et al. 2003, Motwani et al. 2007).

Microwave interactions with starch

Dielectric constant and loss factor of starches generally depend on moisture content, temperature and type of starch. When measurements were conducted by a cavity perturbation technique at 2.75 GHz for corn, wheat, potato and waxy corn starch in excess water systems (5–30% starch), the differences in dielectric properties due to starch type were not significant, but the most important factor affecting the dielectric properties was the water content, both ε' and ε'' decreasing with increasing temperature and starch concentration (Ryynanen et al. 1996). The dielectric loss factors of different starch suspensions were found to be a function of starch type when measured at 2.45 GHz (Ndife et al. 1998). A comprehensive investigation on the effects of temperature, concentration, microwave frequency and salt addition on dielectric properties of starch solutions (1–4% w/w) was conducted by Piyasena et al. (2003) where correlations were developed for estimation of dielectric properties of starch solutions with or without salt. The dielectric loss factor ε'' increased with increasing temperature and salt concentration and penetration depths associated with salt-enriched samples were low compared to samples without salt. Penetration depth relates inversely to the loss factor. The impact of salt on the loss factor was attributed to the high associated conductivity as salt concentration increased. Recently, Motwani et al. (2007) provided new information related to dielectric properties of corn starch slurries of 10–50% w/w. Starch concentration were measured at temperatures between 40 and 90°C and in the frequency range from 15 MHz to 3 GHz. Accordingly, the dielectric constant ε'' decreased with increasing starch concentration as a result of the reduction in the polarizability of water molecules. The dielectric loss factor ε'' increased with starch concentration indicating the increase in ability to dissipate the energy in the form of heat. The dielectric loss factor of the starch slurries decreased at lower frequency range (15–450 MHz) and increased at higher frequency range (450 MHz–3 GHz) showing a U-shaped curve behavior with the frequency. A significant effect of frequency on dielectric constant was evident only at higher starch concentrations (> 40%). The dielectric constant of starch slurries decreased with increasing temperature throughout the frequency range (15 MHz–3 GHz) due to decrease in the relaxation time of the molecules in the system as the temperature increased. The dielectric loss factor increased with the temperature in the frequency range of 15–450 MHz and decreased with the temperature in the higher frequency range of 450 MHz–3 GHz. Therefore, the dielectric properties are key factors in microwave processing of granular starch and are strongly dependent on type of starch, moisture content, and temperature and microwave frequency.

Mechanism of microwave irradiation on granular starch

Recent investigations have explained the mechanism of microwave effect on starch either as an aqueous dispersion (Sjöqvist and Gatenholm 2005, Fan et al. 2013a, Fan et al. 2013b) or as a primary component of cereal pellets (Boischot et al. 2003, Moraru and Kokini 2003). However, no mechanism for microwave irradiation of native starch under normal atmospheric conditions was clearly formulated. Native starches contain 10–20% moisture under normal atmospheric conditions (Swinkels 1985). This moisture can be considered the driving force in the microwave heating of granular starch. The action of microwaves on granular starch could be hypothetically divided in the following stages: (i) dielectric relaxation phenomenon of water molecules is responsible for the initial heating of starch, followed by (ii) rapid rise of temperature corresponding to the starch loss moisture due to the water vaporization, which creates a local high pressure inside granules and further causes (iii) granule expansion which occurs from the center, where the temperature is highest, and then it advances in the whole volume leading to (iv) granule degradation (Brasoveanu and Nemtanu 2014).

Microwave effects on structural changes of starch

Structure changes of starch under microwave radiation can be evaluated using spectroscopic methods. Near Infra-Red (NIR) analysis of the most characteristic carbohydrate regions (2080–2130 and 2270–2290 nm) highlighted structural alterations of the native corn and wheat starches treated by microwave irradiation at 300 and 600 W (Hodsagi et al. 2012). The peak around 2130 nm belonging to the amylopectin disappeared with increasing microwave treatment while the amylose peak (around 2096 nm) decreased with increase in the applied microwave energy when investigating the dispersive or the FT-NIR spectra. The peak minima decrease means more intense vibrations of O–H bending and C–O stretching (2080–2130 nm), indicating structural changes of the starches. But the changes were not as intensive in case of high amylose containing feedstock as reported by Lewandowicz et al. (2000). Higher amylose content suffered greater changes in crystallinity from the microwave treatment. Also, in the region 2270–2290 nm, increases in the peak minima corresponding to the intensity of microwave treatment revealed a possible cause of the vibrations of O–H stretching and C–C stretching decreased in accordance with the structural changes.

Microwave treatment also causes swelling in starches and increases viscosity of suspensions. These effects vary from feedstock to feedstock. For example, Lewandowicz et al. (2000) studied the effect of microwave treatment on wheat, corn and waxy corn starches. The results from this study are quite diverse for each of the starch. Microscope investigations revealed some interesting structural changes that explain the causes for gelatinization and viscosity changes. Native wheat starch suspension heated at 75°C (Figure 5a) gave an image typical of early stage of gelatinization. The granules were slightly swollen and a small amylose leakage from starch granules was observed. When the temperature rose to 95°C (Figure 5b), the solubilization process was advanced, starch granules were considerably swollen and an extensive amylose leakage was observed. In the case of microwaved wheat starch, the images were different. At 75°C (Figure 5c) there were almost no symptoms of solubilization which began at 95°C

(Figure 5d), mainly due to amylose leakage. Normal corn starch revealed a similar solubilization behavior (Figures 5e and 5f). In both wheat and corn starches microwave treatment caused some structural changes which make solubilization in water more difficult. Unlike wheat and normal corn starches, native waxy corn gave a different microscope image on heating. At temperatures as low as 75°C, the solubilization process was already advanced (Figures 5g and 5h) and starch and water became an almost homogenous mixture. Only small remnants of starch granules could be observed. There were no significant differences between native and microwaved waxy corn starches in terms of their microscope appearance on heating.

Wheat and corn starches of an intermediate moisture content (30%) subjected to microwave processing showed alteration of their physico-chemical properties and structure, while waxy corn starch remained

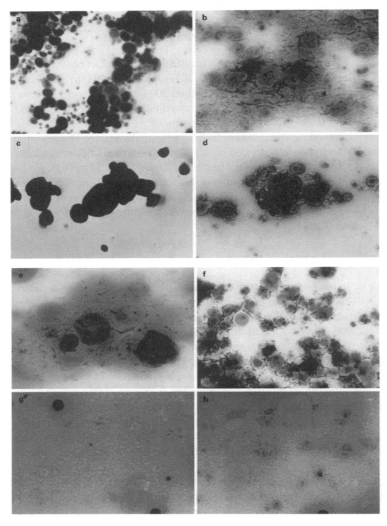

Figure 5. Light microphotographs of the native and irradiated starch samples. Native wheat at 75°C (a) and 95°C (b), microwaved wheat at 75°C (c) and 95°C (d), native corn at 95°C (e), microwaved corn at 95°C (f), native waxy corn at 75°C (g), microwaved corn waxy at 75°C (h).

almost unchanged. Microwave irradiation was evidenced to reduce the crystallinity, solubility, and swelling characteristics of wheat and corn starches as well as to increase the gelatinization temperature of all the investigated starches. The extent of the changes induced by microwave treatment depended not only on the crystal structure of starch, but also on its amylose content.

Microwave effect on chemical bonds and bond energies

Differences in chemical bond types, bond energy, and skeleton connection system between rice starch—water system heated in microwaves and rice starch—water system heated rapidly using a traditional method were lately examined by using FTIR and Raman spectroscopy methods (Fan et al. 2012). According to this study, microwave treatment neither changed the type of chemical groups nor produced new chemical groups in starch molecules (Figure 6). Moreover, the vibrational energy of the chemical bond of microwave-treated rice starch changed gradually during the heating process. The assignment of wave numbers and corresponding chemical compounds are shown in Table 2.

Table 2. FTIR wavenumber characterization (Pandey 2005, Wang et al. 2007, Fan et al. 2012).

FTIR		Raman	
Wavenumber (cm^{-1})	Assignment	Wavenumber (cm^{-1})	Assignment
3336	O-H stretching		
2880–2940	C-H stretching in methyl and methylene groups	2907	C-H stretching
1734	C=O stretching in unconjugated ketones, carbonyls and in ester groups, frequently of carbohydrate origin		
1598	Aromatic skeletal vibration plus C=O stretch		
1502	Aromatic skeletal vibration plus C=O stretch		
1316	C-H vibration in cellulose		
1372	C-H deformation in cellulose and hemicellulose	1380	CH$_2$ scissoring, C–H and C–O–H deformation
1270	C-O stretch in lignin; C-O linkages in guaiacyl aromatic methoxy groups		
1235	C = stretch in lignin and xylan		
1157	C-O-C vibration in cellulose and hemicellulose	1159	C–O, C–C, C–H relate mode
1034	Aromatic C-H in plane deformation, C-O deformation; primary alcohol	1083	C–O–H bending (in polymers)
897	β-glycosidic linkages	1050	C–C stretching

Table 3 shows the effect of microwave treatment on various starches reported in the literature. Changes in solubility, swelling, gelatinization and digestion characteristics are discussed for different starch feedstock as observed in experimental studies (Brasoveanu and Nemtanu 2014). It can be noted that microwave energy parameters were reported in terms of W or W/g of biomass and the exposure time also varied significantly in these studies. Therefore the observed effects are diverse which also depends on the type of feedstock.

168 *Microwave-Mediated Biofuel Production*

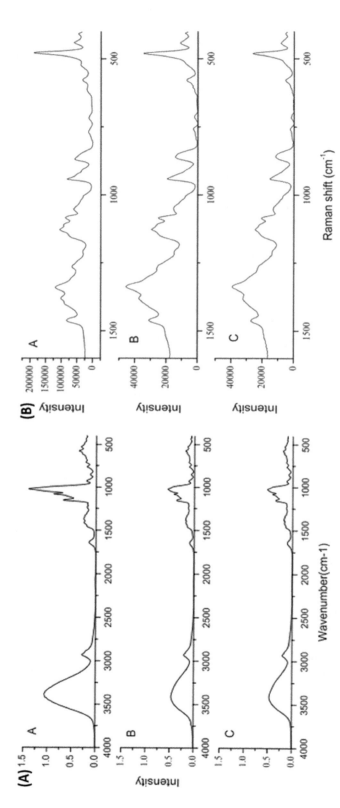

Figure 6. (a) The FTIR spectrum of native starch (NS). (b) The FTIR spectrum of starch of conducting rapid heating treatment (RS). (c) The FTIR spectrum of starch of microwave treatment (MS). Figure 5. (a) The Raman spectrum of native starch (NS). (b) The Raman spectrum of starch of conducting rapid heating treatment (RS). (c) The Raman spectrum of starch of microwave treatment (MS).

Table 3. Microwave effect on different starch feedstock (Brasoveanu and Nemtanu 2014).

Type of starch	Moisture content (%)	Microwave parameters	Exposure time	Observed effects	References
Corn	30	0.5 W/g	60 min	Decrease of solubility, swelling power, rheological properties, gelatinization enthalpy, crystallinity degree	Lewandowicz et al. 2000
	15–40	0.17 and 0.5 W/g	60 min	Increase of gelatinization onset temperature Decrease of paste viscosity parameters Change of gelatinization properties	Stevenson et al. 2005
	6.8	450 W	15 min	Drastic reduction of crystalline fraction Deformation of granule	Szepes et al. 2005
		30.1 W/g	20 min	Decrease of solubility, swelling power, freeze-thaw stability, rheological properties, enthalpy of gelatinization	Luo et al. 2006
		440 and 800 W	10–15 min	Appearance of free radicals	Dyrek et al. 2007 Fortuna et al. 2008 Bidzinska et al. 2010
		300 and 600 W	30–150 s	Decrease of rheological properties No change of digestibility Slight change starch vibrations (2080–2130 and 2270–2290 nm)	Luo et al. 2006
Wheat		300 W/g		Decrease of solubility, swelling power, rheological properties, gelatinization enthalpy, crystallinity degree	Lewandowicz et al. 2000
		300 and 600 W	30–150 s	Decrease of rheological properties No change of digestibility Slight change in starch vibrations (2080–2130 and 2270–2290 nm)	Hodsagi et al. 2012
Potato	9.6	450 W	15 min	No change of granule size and granule size distribution Increase of crystallinity Crystal structure changed from B pattern to A pattern	Szepes et al. 2005
		440 and 800 W	10–15 min	Appearance of free radicals	Dyrek et al. 2007 Fortuna et al. 2008 Bidzinska et al. 2010
	18.1	460, 500, 750 W	30–300 s	Increase of water absorption	Treppe et al. 2011
	18.6	1000 W	320 s	Appearance of cracks on granule surface Increase of bulk density Decrease of granule size	Treppe et al. 2011
Rice	20	270–1350 W	60 min	Increase of pasting parameters, digestibility No change of granule morphology	Anderson et al. 2006
Barley	42–45	220 W–1.1 kW		Increase of digestibility	Emami et al. 2012

Bioethanol from Lignocellulosic Feedstock

Bioethanol production from pure sugar and starch feedstock is not a sustainable process. Their use as feedstock raises many food vs. fuel conflicts. These feedstock were considered first generation feedstock. Sugar containing non-edible feedstock would be ideal for bioethanol production as it eliminates the human interests for consumption. Lignocellulosic materials abundantly available in nature can serve as excellent feedstock for bioethanol production if sugars trapped in these tough to treat feedstock can be released. Lignocellulosics comprise three markedly different profitable fractions (Table 4). Hemicellulose is composed of C_5 and C_6 polysaccharides, having important applications for biofuel production (e.g., bioethanol) and

Table 4. Composition of different feedstock for bioethanol production (Kumar et al. 2009, Brosse et al. 2012, David and Ragauskas 2010, Sannigrahi et al. 2010, NÚÑez et al. 2011, Huang and Ragauskas 2013, Saha et al. 2013, Ragauskas et al. 2014).

Plant resource	Hemicellulose %	Cellulose %	Lignin %	Reference
Miscanthus	24–33	45–52	9–13	Kumar et al. 2009
Switchgrass	26–33	37–32	17–18	Brosse et al. 2012
Corn stover	31	37	18	David and Ragauskas 2010
Poplar	16–22	42–48	21–27	David and Ragauskas 2010
Eucalyptus	24–28	39–46	29–32	Sannigrahi et al. 2010
Pine	23	46	28	Nunes et al. 2011
Hardwood stems	24–40	40–55	18–25	Huang and Ragauskas 2013
Softwood stems	25–35	45–50	25–35	Huang and Ragauskas 2013
Nut shells	25–30	25–30	30–40	Saha et al. 2013
Wheat straw	50	30	15	Ragauskas et al. 2014
Leaves	80–85	15–20	0	Ragauskas et al. 2014
Newspaper	25–40	40–55	18–30	Ragauskas et al. 2014
Primary wastewater solids	-	8–15	-	Ragauskas et al. 2014
Solid cattle manure	1.4–3.3	1.6–4.7	2.7–5.7	Ragauskas et al. 2014
Coastal Bermuda grass	35.7	25	6.4	Ragauskas et al. 2014
Corn cobs	33–36.4	38.8–44	13.1–18	Liu et al. 2010, Wang et al. 2011
Wheat straw	20–33.8	33–40	15–26.8	Ruiz et al. 2011, Talebnia et al. 2010
Corn stover	20.11–31.3	34.32–36.5	11.9–13.55	Weiss et al. 2010, Liu and Cheng 2010
Rice straw	16.1–22	35–36.6	12–14.9	Hsu et al. 2010, Yadav et al. 2011
Sugar cane bagasse	15.79–29.6	34.1–49	19.4–27.2	Mesa-Pérez, et al. 2013, Maeda et al. 2011
Barley straw	25.1–37.1	37.5	15.8–16.9	Sun et al. 2011, García-Aparicio et al. 2011
Rice husk	20.99	33.43	18.25	Garrote et al. 2007, Abbas and Ansumali 2010
Rye straw	23.8–24.4	41.1–42.1	19.5–22.9	Ingram et al. 2011, Gullón et al. 2010
Rapeseed straw	19.6–24.22	36.59–37	15.55–18	Díaz et al. 2010, Lu et al. 2009
Sunflower stalks	20.2–24.27	33.8	14.6–19.9	Ruiz et al. 2008, Caparrós et al. 2008
Sweet sorghum bagasse	22.01–26.3	41.33–45.3	15.2–16.47	Zhang et al. 2011, Goshadrou et al. 2011
Herbaceous Switchgrass	25.95–31.1	41.2–32.97	17.34–19.1	Keshwani and Cheng 2009, Hu et al. 2011
Alfalfa stems	14.7	24.7	14.9	Ai and Tschirner 2010
Coastal Bermuda grass	19.29	25.59	19.33	Wang et al. 2010
Hardwood Aspen	18	43.8	20.8	Tian et al. 2011
Hybrid Poplar	21.73	48.95	23.25	Pan et al. 2006
Eucalyptus	21.4	44.6	30.1	Gonzalez et al. 2011
Softwood Pinus radiata	22.5	45.3	26.8	Araque et al. 2008

Table 4. contd....

Table 4. contd.

Plant resource	Hemicellulose %	Cellulose %	Lignin %	Reference
Eucalyptus globulus	21.8	44.4	27.7	Romaní et al. 2011
Spruce	20.8	43.8	28.83	Shafiei et al. 2010
Cellulose wastes Newspapers	16.4	60.3	12.4	Lee et al. 2010
Recycled paper sludge Industry co-products	14.2	60.8	8.4	Peng and Chen 2011
Distiller's grains	16.9	12.63	–	Kim et al. 2010
Brewer's spent grain	15.18–32.8	18.8–20.97	21.7–25.62	Carvalheiro et al. 2005, Pires et al. 2012

for the generation of valuable chemical intermediates (e.g., furfural). Cellulose is comparably considered as one of the most abundant biopolymers on earth comprising linear β(1-4) glucose chain links. Cellulose can also be deconstructed into valuable products such as biofuels (e.g., bioethanol and lignocellulosic fuels—alkyl valerates) and platform chemicals (including levulinic and formic acids, gamma-valerolactone and derived products). Lignin is the third major component of lignocellulosics, being the most underutilized fraction. Lignin has been traditionally employed for heat and power purposes through combustion in the pulp and paper industry due to its high calorific value. Lignin is a complex and recalcitrant phenolic macromolecule comprising phenylpropane type units. Due to its highly irregular polymeric structure, it is resistant to microbial attack and can prevent water from destroying the polysaccharide–protein matrix of plant cells.

Pretreatment of Lignocellulosic Feedstock

Pretreatment is an essential process for enhancing bioethanol production from lignocellulosic biomass. The benefits of pretreatment are many fold. Pretreatment removes the lignin and hemicellulose content, reduces crystallinity of cellulose and increases porosity of the lignocellulosic materials. However, pretreatment methods should meet the goals of improving the formation of sugars or the ability to subsequently form sugars by hydrolysis, avoid degradation or loss of carbohydrates, avoid the formation of byproducts that are inhibitory to the subsequent hydrolysis and fermentation processes and be cost-effective. Pretreatment methods can be classified as physical (milling and grinding), physico-chemical (steam pretreatment/autohydrolysis, hydrothermolysis, and wet oxidation), chemical (alkali, dilute acid, oxidizing agents, and organic solvents), biological, electrical, or a combination of these (Kumar et al. 2009). There are many ways to produce bioethanol from various biomass feedstock and through different strategies. These include carbohydrate platform in which sugars, starch, and cellulose containing feedstock can be processed through fermentation, saccharification—fermentation, pretreatment-hydrolysis-fermentation process schemes respectively (Keshwani 2009). The hydrolysis rate of amorphous cellulose is 30 times faster than crystalline cellulose, making pre-treatment essential for any increase in enzyme hydrolysis, as this decrystallizes cellulose in plant biomass (Zhu et al. 2011).

Lignin plays a major role in recalcitrance of plant cell walls for saccharification during enzyme hydrolysis (Chen and Dixon 2007). The problem with lignin is the presence of phenylpropanoid in the polymer, which is in the vascular tissues and fibres, and this inhibits enzymes and yeast in the process. There are numerous requirements for a successful pretreatment method such as low cost, non-corrosive and safe chemicals and solvents, minimal waste production, fast reactions and high yields and compatibility with enzymes (Huber et al. 2006). Based on these requirements there are various methods employed for plant biomass pre-treatment including; (1) mechanical; (2) thermal; (3) acid; (4) alkaline; (5) oxidative; (6) ammonia expansion; (7) CO_2; (8) biological; and (9) ionic liquids (Hendriks and Zeeman 2009, Zhu and

Pan 2010, Wyman 1994). Figure 7 shows the chemical pretreatments on selective parts of lignocellulosic biomass (Lee et al. 2014). Acid, alkaline and oxidation treatments have different selective pathways to detach or solubilize cellulose while ionic liquids attach to the OH bonds.

Mechanical pretreatment

Milling is the most common mechanical pre-treatment method which allows for reducing the size of the lignocellulosic biomass. Size reduction can reduce the crystallinity, reduce the degree of polymerization and increase the available surface area for hydrolysis. Hydrolysis can be increased by up to 25% while reducing the hydrolysis time by 23–59% (Hendriks and Zeeman 2009). No additional solvent is required for milling which can be considered an advantage. Because addition of solvents can lead to inhibitors like furfural and others (Yabushita et al. 2014). However, milling is an energy-intensive process and for this reason it is not economically feasible on large scale production. In addition, a 40 mesh particle size is required for improvement of the hydrolysis reaction. Combination with an acid catalyst could be beneficial. Addition of a solid weak-acidic catalyst to milling can increase the rate constant by 13 times in comparison with milling process alone (Benoit et al. 2012). Ultrasound processing can be considered a mechanical process as well. Sonicated cellulose has been reported to be more accessible. This technique can be used in combination with other methods as it creates stable colloidal suspensions of cellulose with a significant reduction in particle size (Zhang et al. 2013).

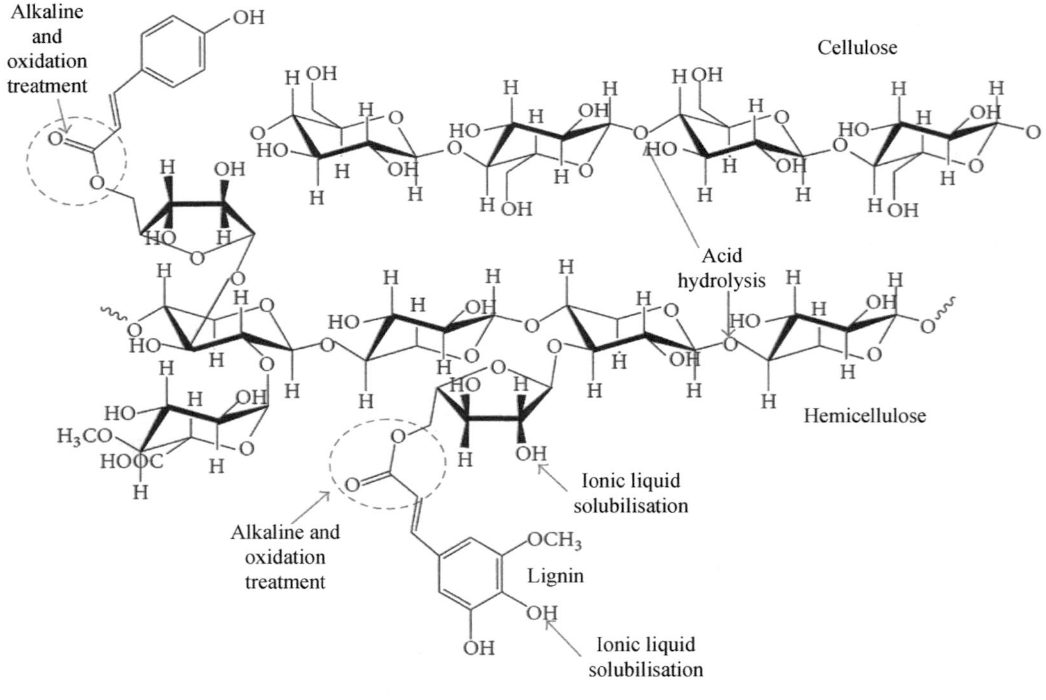

Figure 7. Selectivity of chemical treatments for fractionation of lignocellulosic biomass (Lee et al. 2014).

Thermal pretreatment

Thermal pre-treatment includes the use of hot water and steam explosion to break down the plant biomass. If the temperature starts to increase above 150–180°C, hemicellulose shortly followed by lignin, will start to dissolve in water. The exact temperature depends on the exact composition of the polymers based on the biosynthetic pathways. The hemicellulose forms acids, which can hydrolyze the rest of the biomass polymers. The lignin compounds formed are phenolic, like vanillin, which in most cases have the inhibitory effect on enzymes and yeast. If the temperature is raised to 220°C, ethanol production can be completely inhibited due to formation of furfurals and other soluble lignin compounds (Hendriks and Zeeman 2009).

Steam pretreatment/explosion involves biomass heating up to 240°C under high pressures for a few seconds, to minutes, in flow reactors. Steam can be released and the biomass cooled down quickly after the process, this is known as steam explosion. Both processes solubilize hemicellulose and give access to cellulose without the formation of inhibitors, however, the benefits of the explosion method are disputed in literature and high temperatures and energy input are required. At high pressures, there is also a considerable safety risk. Liquid hot water is also used as a method to remove hemicellulose and the pH must be controlled between 4 and 7 due to acid release. However, although the risk is lower than steam explosion, inhibitors can still be produced in the process.

Acid pretreatment

Both dilute and concentrated acids are used to solubilize the hemicellulose component. Sulfuric, nitric and hydrochloric acids are commonly used for the pre-treatment. Acid pretreatment is performed over a temperature range of 120–210°C, with 4 wt.% acid loading and pretreated for minutes to hours, depending on batch or flow-through reactors (Sannigrahi et al. 2011). Sometimes the acid insoluble lignin can increase during the pre-treatment due to modification of the carbohydrates present. With concentrated acids, this is highly toxic, corrosive and difficult to recover after pre-treatment, as well as producing inhibitors including furfurals and humins (Sannigrahi et al. 2011). Acid treatment can alter the crystalline structure, swelling the material, as the acid facilitates water in penetrating the cellulose crystals, expanding the surface area (Camacho et al. 1996). This expansion promotes single polymer chains to be formed, followed by the breakup of these molecules. However, hydrolysis of the cellulose at the reducing ends occurs, forming glucose which is washed away after pre-treatment and reduces the effective yield. It is known with dilute acid pre-treatments, a recovery of only 50% of the sugars is possible. This is due to conversion of sugars to other chemical products (Badger 2002). In hemicellulose, mostly xylan is hydrolyzed by the acid and hence can be removed from the biomass. Monomers, furfural and HMF are produced in acidic environments, which inhibit enzymes and yeast and over time these inhibitors increase (Tong et al. 2010). Lignin can precipitate in acid pre-treatment as all effects are more profound in concentrated acids compared to dilute acids. A selective oxalic acid-catalyzed hydrolysis has shown promise at removing hemicellulose at mild temperatures (vom Stein et al. 2011). This is a single process to separate the wood polymers, which leaves a xylose rich hemicellulose aqueous solution, a lignin fraction in an organic phase and the cellulose pulp (vom Stein et al. 2011). In acid treatments, some new methods using superacids are also in development (Martin-Mingot et al. 2012). Comparisons to current acid pre-treatments including HCl, H_2SO_4 and HF, with $HFSbF_5$ are being used to depolymerize cellulose to glucose at just 0°C (Martin-Mingot et al. 2012).

Alkaline pretreatment

Alkaline pretreatment involves the use of bases, such as sodium, potassium, calcium, and ammonium hydroxide, for the pretreatment of lignocellulosic biomass. The use of an alkali causes the degradation of ester and glycosidic side chains resulting in structural alteration of lignin, cellulose swelling, partial decrystallization of cellulose (Cheng et al. 2010, McIntosh and Vancov 2010) and partial solvation of hemicellulose (McIntosh and Vancov 2010, Sills and Gossett 2011). In essence, the biomass material is swollen, allowing enzymes to access the cellulose content. With strong alkali pre-treatments, the polysaccharides are 'peeled' as the end groups are hydrolyzed and removed. However, this 'peeling' can result in loss of carbon from the polymers and CO_2 is produced. If lower temperatures and aqueous potassium hydroxide is used, then xylan can be removed without this degradation as explained above.

Sodium hydroxide has been extensively studied for many years, and it has been shown to disrupt the lignin structure of the biomass, increasing the accessibility of enzymes to cellulose and hemicellulose (MacDonald et al. 1983, Soto et al. 1994, Zhu et al. 2010). Lime is also commonly used for biomass pretreatment. Corn Stover, switchgrass, bagasse, wheat, and rice straw are among the lignocellulosic feedstocks pretreated by this method (Sun and Cheng 2002, Hu et al. 2008, Liang et al. 2010, Hendricks and Zeeman 2009). Lime is more efficient than sodium hydroxide (Sills and Gossett 2011). However, the drawback of alkaline pretreatment combined with usual inhibitors is that it is consumed by the biomass. This effect causes modification to the cellulose, but the cellulose form produced can be denser and more stable than the native form.

Acid vs. Alkali Pretreatment

While most pretreatments increase the porosity of the biomass and reduce crystallinity of the cellulose fiber, fractionation of the biomass can differ based on the type of pretreatment process. Figure 8 illustrates this fact for acid and alkali pretreatments. The primary mode of action for acid pretreatment is the solubilization of hemicellulose, which is accompanied by a reduction in cellulose crystallinity and fracture of the lignin seal. However, the majority of the lignin stays in the solid phase. In contrast, alkali pretreatments almost exclusively target the removal of lignin. However, some hemicelluloses are also removed during alkali

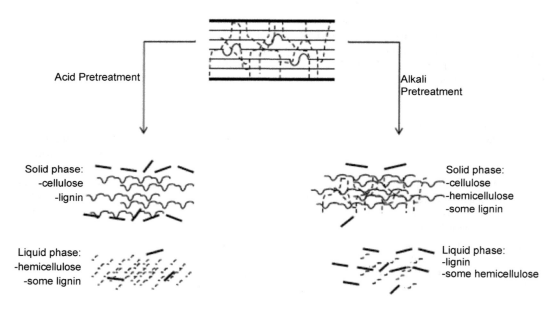

Figure 8. Fractionation of lignocellulosic biomass under acid and alkali pretreatment conditions.

pretreatment (especially at high temperatures) because of their random amorphous structure (Keshwani 2009).

Cellulose hydrolysis

Hydrogen bonding is a critical factor in polysaccharide structures, and it is well-known that the depolymerization of cellulose depends strongly on the structure of the hydrogen bond network (Fan et al. 2013a). More specifically the structure of cellulose involves intersheet, interchain, and intrachain hydrogen bonds (Figure 9). These impart rigidity and stability to the cellulose structure, but can be broken at elevated temperatures.

The mechanism of a cellulose hydrolysis reaction is shown in Figure 10. The scientific basis for hydrolysis reaction is that it involves splitting of the water molecule into hydrogen cations (H⁺) and hydroxide anions (OH⁻). In the presence of cellulose, the H⁺ attacks the oxygen atom in the 1,4'-β-glycosidic linkage. Thus the 1,4'-β-glycosidic linkage is broken to form a cyclic carbonium cation with a chair shape. This step is considered as the rate determining step. Finally, glucose is formed by rapid ion transfer of OH⁻ from the dissociation of water molecules to the glucose unit-based carbonium cation.

Based on this reaction mechanism, a solid acid catalyst and associated reaction conditions are more favorable for splitting of water in cellulose hydrolysis. The splitting of a water molecule into hydrogen cations (H⁺) and hydroxide anions (OH⁻) can be regarded as deprotonation. The catalytic activity of a catalyst in cellulose hydrolysis increases with a decrease in the deprotonation enthalpy of water on the surface of the solid acid catalyst. These findings suggested that stronger Brønsted acidity is more favorable for catalytic hydrolysis of cellulose. In addition, to some extent, an increase in the amount of water had a positive effect on the breakage of 1,4'-β-glycosidic bonds and intramolecular hydrogen bonds in insoluble

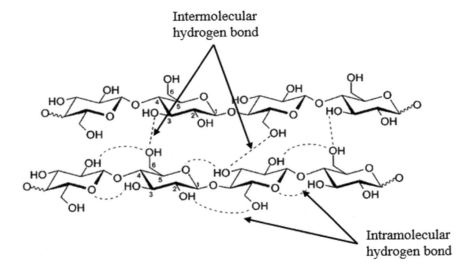

Figure 9. Schematic representation of cellulose and its hydrogen bond network (Fan et al. 2013b).

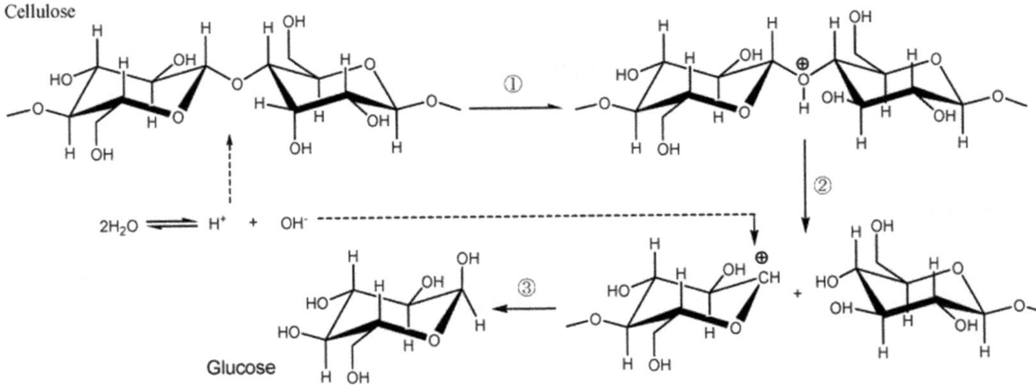

Figure 10. Schematic mechanism of breakage of 1,4'-β-glycosidic bonds and formation of glucose in the hydrolysis of cellulose.

cellulose, thereby leading to more hydrolytic products. Therefore, the formation of glucose requires more water than the formation of β-1,4-glucan. Yamaguchi and co-workers revealed that the amount of water comparable to the solid catalyst weight could lead to a maximum yield of glucose in the heterogeneous catalytic hydrolysis reaction of cellulose (Zhou et al. 2011).

Oxidative pretreatment

Oxidative pretreatment involves utilization of hydrogen peroxide or peracetic acid in water. These compounds can remove hemicellulose and lignin due to multiple reactions, including cleavage of alkyl or ether linkages, electrophilic substitution or oxidative cleavage of aromatic nuclei. However, since the oxidant is not selective, losses of cellulose can occur and some compounds formed from lignin are inhibitors to yeast and enzymes. Peracetic acid, however, is more selective and only oxidizes the lignin content as shown on sugarcane bagasse at ambient temperatures. The yields improved from 6.8% to 98% with this pretreatment (Hendriks and Zeeman 2009, Gammons 2014). Other drawbacks include loss of sugars due to non-selective oxidation. In addition, the pH control of the materials is not compatible with enzyme hydrolysis.

Ammonia and carbon dioxide pretreatment

Ammonia pretreatment (referred to as ammonia fiber expansion, AFEX), is conducted with 1:1 biomass loadings at ambient temperatures, lasting for 10–60 days or if completed at 120°C, then for several minutes to an hour in pressured vessels (Hendriks and Zeeman 2009, Gammons 2014). Pressure is released causing ammonia to expand quickly, which causes disruption to the biomass network, aiding hydrolysis. This can cause mechanical disruption but also modification of the hemicellulose and lignin polymers but not cellulose. A six-fold increase in enzyme activity was reported due to swelling of cellulose and delignification. However, this ammonia work is in very early days and any consequences and problems have not been fully studied or reported yet. CO_2 pre-treatment is used at high pressures and temperatures up to 200°C for several minutes. The effect causes an acidic liquid which can hydrolyze the hemicellulose. Supercritical CO_2 can also be used which has been reported to increase yields of glucose in bagasse from 50 to 70%.

Biological pretreatment

Biological pre-treatment can be conducted using white rot fungi, because they can completely degrade parts of the plant cell wall. Fungi can degrade lignin components to small molecule products. This method can be considered an environmentally friendly approach. *Phanerochaete chrysosporium* is most commonly used as it contains extracellular oxidative ligninolytic secreted enzymes. Microorganisms can be used as well to treat the lignocellulose and enhance enzymatic hydrolysis. These microorganisms usually degrade lignin and hemicellulose but cellulose to a small extent, since cellulose is more resistant than the other parts of lignocellulose to the biological attack.

Several fungi, e.g., brown-, white- and soft-rot fungi, have been used for this purpose. White-rot fungi are among the most effective microorganisms for biological pretreatment of lignocellulosic biomass. Taniguchi et al. (2005) evaluated biological pretreatment of rice straw using four white-rot fungi (*Phanerochaete chrysosporium, Trametes versicolor, Ceriporiopsis subvermispora,* and *Pleurotus ostreatus*) on the basis of quantitative and structural changes in the components of the pretreated rice straw as well as susceptibility to enzymatic hydrolysis. Pretreatment with *P. ostreatus* resulted in selective degradation of lignin rather than the holocellulose component, and increased the susceptibility of rice straw to enzymatic hydrolysis. Some bacteria can also be used for biological pretreatment of lignocellulosic materials. Kurakake et al. (2007) studied the biological pretreatment of office paper with two bacterial strains, *Sphingomonas paucimobilis* and *Bacillus circulans*, for enzymatic hydrolysis. Biological pretreatment with the combined strains improved the enzymatic hydrolysis of office paper from municipal wastes. Under optimum conditions, the sugar recovery was enhanced up to 94% for office paper.

Biological pretreatment with microorganisms or enzymes is also investigated to improve digestion in biogas production. The biological pretreatment might be used not only for lignin removal, but also for biological removal of specific components such as antimicrobial substances. Solid-state fermentation of orange peels by fungal strains of *Sporotrichum, Aspergillus, Fusarium* and *Penicillum* enhanced the availability of feed constituents and reduced the level of the antimicrobial substances (Srilatha et al. 1995). In a similar work, cultivation of white-rot fungi was used to detoxify olive mill wastewater and improve its digestion (Dhouib et al. 2006). Low energy requirement, no chemical requirement, and mild environmental conditions are the main advantages of biological pretreatment. However, the main drawback to develop biological methods is the low hydrolysis rate obtained in most biological materials compared to other technologies (Sun and Cheng 2002). To develop cost-competitive biological pretreatment of lignocellulose, and improve the hydrolysis to eventually improve ethanol yields, *basidiomycetes* and other fungi should be evaluated for their ability to delignify the plant material quickly and efficiently.

Combination pretreatment

The yields in steam explosion or hot water extraction pretreatment methods can be improved by adding an acid, as this catalyzes the solubilization of hemicellulose. This improves the solubility as the soaked biomass contains SO_2 which is converted to H_2SO_4, however, after a while all this is removed, which the acid replenishes. Another combination method is the steam explosion under alkaline condition, often involving lime addition. This improves the digestibility on low lignin content biomass but not high lignin content. Additionally a benefit of this combination is that lime is relatively cheap and safe and that the calcium can be regained afterwards. Finally, steam pre-treatment has been combined with oxidative pretreatment. The production of furfural content was low, a benefit for enzyme and yeast treatments down the line. However, parts of the hemicellulose were converted to water and CO_2, which can easily be removed, but decreases sugar production. A summary of advantages and disadvantages for various pretreatment methods is shown in Table 5.

Challenges in Lignocellulosic Bioethanol Production

There are several challenges that need to be addressed before the bioethanol production potential of lignocellulosic biomass can be fully realized (Keshwani 2009). Unlike the starch platform, the carbohydrates in lignocelluloses are not easily accessible for enzymatic hydrolysis. This recalcitrance is primarily due to the composition of lignocellulosic biomass and the way specific components interact with each other. Therefore, the first challenge is the pretreatment of lignocellulosic biomass to reduce biomass recalcitrance, thereby improving the yield of fermentable sugars as discussed in the previous sections. The second challenge is reducing the cost of hydrolytic enzymes (cellulases) used in the lignocellulose pathway. The cost of cellulases is approximately six times the cost of amylases used in the starch pathway (Schubert 2006). Overall enzyme costs for bioethanol from the lignocellulose pathway are estimated to be 30 to 50 cents per gallon (Keshwani 2009). The final challenge is the fermentation of five-carbon sugars, which can account for 20–30% of the carbohydrate fraction of lignocelluloses. To make the overall process economically feasible, these five-carbon sugars must be utilized. Most commercially available yeasts *Saccharomyces cerevisiae* and *Zymomonas mobilis* cannot ferment five-carbon sugars. Yeasts such as *Pachysolen tannophilus, Pichia stipitis* and *Candida shehatae* can ferment, but are limited by low ethanol tolerance and slow rate of fermentation (Saha 2003).

Key Factors for an Effective Pretreatment of Lignocellulosic Biomass

There are several important factors that must be taken into consideration for low-cost and advanced pretreatment process (Yang and Wyman 2008, Alvira et al. 2010). Some of the factors are discussed below.

Table 5. Summary of pretreatment methods (Kumar et al. 2009, Alvira et al. 2010, Brodeur et al. 2011).

Pretreatment method	Advantages	Disadvantages
Alkali	Efficient removal of lignin Low inhibitor formation Crystallinity reduction Delignification	High cost of alkaline catalyst Alteration of lignin structure
Acid	High glucose yield Solubilizes hemicellulose	High costs of acids and need for recovery High costs of corrosive resistant equipment Formation of inhibitors
Concentrated acid	High glucose yield Ambient temperatures	High cost of acid and need to be recovered Reactor corrosion problems Formation of inhibitors
Diluted acid	Less corrosion problems than concentrated acid Less formation of inhibitors	Generation of degradation products Low sugar concentration in exit stream
Green solvents	Lignin and hemicellulose hydrolysis Ability to dissolve high loadings of different biomass types Mild processing conditions (low temperatures)	High solvent costs Need for solvent recovery and recycle
Steam	Cost effective Lignin transformation and hemicellulose solubilization High yield of glucose and hemicellulose in two-step process Crystallinity reduction	Partial hemicellulose degradation Acid catalyst needed to make process efficient with high lignin content material Toxic compound generation
LHW–Liquid hot water	Separation of nearly pure hemicellulose from rest of feedstock No need for catalyst Hydrolysis of hemicellulose	High energy/water input Solid mass left over needs to be dealt with (cellulose/lignin)
AFEX–Ammonia fiber explosion	High effectiveness for herbaceous material and low lignin content biomass Cellulose becomes more accessible Causes inactivity between lignin and enzymes Low formation of inhibitors	Recycling of ammonia is needed Less effective process with increasing lignin content Alters lignin structure High cost of ammonia
ARP–Ammonia recycling percolation	Removes majority of lignin High cellulose content after pretreatment Herbaceous materials are most affected	High energy costs and liquid loading
Supercritical fluid	Low degradation of sugars Cost effective Increases cellulose accessible area	High pressure requirements Lignin and hemicelluloses unaffected
CO_2 Explosion	Increases accessible surface area; cost-effective; does not cause formation of inhibitory compounds	Does not modify lignin or hemicelluloses
Mechanical comminution	Reduces cellulose crystallinity	Power consumption usually higher than inherent biomass energy
Biological	Degrades lignin and hemicelluloses; low energy requirements	Rate of hydrolysis is very low
Ozonolysis	Reduces lignin content; does not produce toxic residues	Large amount of ozone required; expensive
Organosolv	Hydrolyzes lignin and hemicelluloses	Solvents need to be drained from the reactor, evaporated, condensed, and recycled; high cost

High yields for multiple crops, sites ages, harvesting times—Various pretreatments have shown to be better suited for specific feedstocks. For example, alkaline-based pretreatment methods such as lime, ammonia fiber explosion (AFEX), and ammonia recycling percolation (ARP), can effectively reduce the lignin content of agricultural residues but are less satisfactory for processing recalcitrant substrate as softwoods (Chandra et al. 2007). Acid based pretreatment processes have been shown to be effective on a wide range of lignocellulose substrates, but are relatively expensive (Mosier et al. 2005).

Highly digestible pretreated solid—Cellulose from pretreatment should be highly digestible with yields higher than 90% in less than five and preferably less than 3 days with enzyme loading lower than 10 FPU/g cellulose (Yang and Wyman 2008).

No significant sugars degradation—High yields close to 100% of fermentable cellulosic and hemicellulosic sugars should be achieved through pretreatment step.

Minimum amount of toxic compounds—The liquid hydrolyzate from pretreatment must be fermentable following a low-cost, high yield conditioning step. Harsh conditions during pretreatment lead to a partial hemicellulose degradation and generation of toxic compounds derived from sugar decomposition that could affect the proceeding hydrolysis and fermentation steps (Oliva et al. 2003). Toxic compounds generated and their amounts depend on raw material and harshness of pretreatment.

Degradation products from pretreatment of lignocellulose materials can be divided into the following classes: carboxylic acids, furan derivatives, and phenolic compounds. Main furan derivates are furfural and 5-hydroxymethyl-furfural (HMF) derived from pentoses and hexoses degradation, respectively (Palmqvist and Hahn-Hägerdal 2000). Weak acids are mostly acetic and formic and levulinic acids. Phenolic compounds include alcohols, aldehydes, ketones and acids (Klinke et al. 2002).

Biomass size reduction not required—Milling or grinding the raw material to small particle sizes before pretreatment are energy-intensive and costly technologies.

Operation in reasonable size and moderate cost reactors—Pretreatment reactors should be low in cost through minimizing their volume, employing appropriate materials of construction for highly corrosive chemical environments, and keeping operating pressures reasonable.

Non-production of solid-waste residues—The chemicals formed during hydrolyzate conditioning in preparation for subsequent steps should not present processing or disposal challenges.

Effectiveness at low moisture content—The use of raw materials at high dry matter content would reduce energy consumption during pretreatment.

Obtaining high sugar concentration—The concentration of sugars from the coupled operation of pretreatment and enzymatic hydrolysis should be above 10% to ensure an adequate ethanol concentration and to keep recovery and other downstream costs manageable.

Fermentation compatibility—The distribution of sugar recovery between pretreatment and subsequent enzymatic hydrolysis should be compatible with the choice of an organism able to ferment pentoses (arabinose and xylose) in hemicellulose.

Lignin recovery—Lignin and other constituents should be recovered to simplify downstream processing and for conversion into valuable coproducts (Yang and Wyman 2008).

Minimum heat and power requirements—Heat and power demands for pretreatment should be low and/or compatible with the thermally integrated process.

Microwaves in Cellulose Pretreatment

Microwave pretreatment of cellulose

The structure of cellulose within cell wall consists of amorphous and crystalline regions which alternate at distances of 15 nm approximately (see Figure 11). Crystalline cellulose contains a very ordered hydrogen bonded network within which a proton transport network is possible in the presence of an electromagnetic field. As the number of groups capable of rotating increases particularly, but not exclusively, in the amorphous region, the rate of decomposition increases.

While attempting to study the microwave degradation process of cellulose, Budarin et al. (2010) found that a reaction temperature of 180°C was ideal for the degradation. It follows the explanation that the microwave interaction through dipolar polarization in cellulose polar groups was higher at this temperature due to the special effects caused by microwave heating. The polar groups in cellulose had less freedom to oscillate freely in microwave field at low temperatures and therefore less interaction. This finding is critical in developing energy-efficient processes. This reaction can be carried out at much milder conditions than the severe conditions demanded by conventional methods. It could be considered that the softer the material, the faster will be the decomposition reaction. If this were the case, the same increase in rate of decomposition would be seen in the conventional pyrolysis chars after 180°C. The fact that there is no increase in rate at this temperature under conventional conditions strengthens the argument for a specific non-thermal MW effect. For these reasons cellulose fibers could be represented as alternating ionic conducting (crystalline) and non-conducting (amorphous) regions. When the crystalline region is placed within an electromagnetic field it will polarize, generating a charge on the crystalline interface (overall charge will be zero). The MW treatment of cellulose produces chars at significantly lower temperatures than conventional pyrolysis. The presence of acid has been found to promote char formation (Mamleev et al. 2009).

Acid catalyzed pyrolysis is also reported to increase the yield of 1,6-anhydrosugars in bio-oil (Dobele et al. 2003). Analysis of liquid fractions produced from our microwave treatment has shown significant levels of such products along with the furans and substituted phenols associated with conventional pyrolysis (Budarin et al. 2009). The production of these components will further plasticize the amorphous region

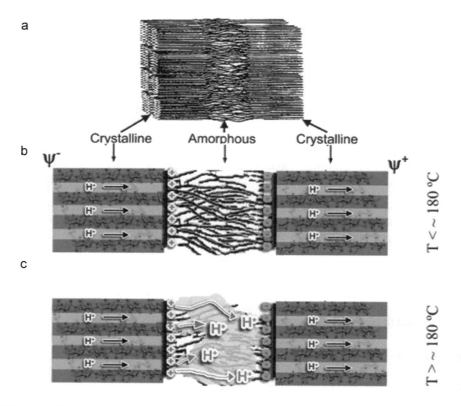

Figure 11. Scheme of interaction of microwave irradiation with cellulose. (a) Microstructure of cellulose fibre (Bhattacharya et al. 2008). (b) Schematic of crystalline and amorphous regions of cellulose below 180°C in the presence of microwave electromagnetic field.

exacerbating the enhanced MW effect. Increasing temperature and acidity will eventually disrupt the crystalline regions of cellulose aiding its degradation to char and volatile matter. In terms of an industrial process, this procedure can be easily adapted to a variety of biomass, producing a uniform char which can be handled with ease by the end user. In order to achieve sustainable development mankind needs to move away from petrochemical feedstocks to renewable alternatives. Microwave processing of biomass may be the route to solid and liquid fuel for the development of a society which is no longer reliant on fossil fuels (Budarin et al. 2010).

Microwaves can provide superior pretreatment while being energy-efficient. There are many ways through which microwaves can be utilized in the pretreatment process. A variety of combinations such as chemical, physical, mechanical and thermal combinations with microwaves are possible. A systematic investigation of the interaction of microwave irradiation with microcrystalline cellulose covering a broad temperature range (150 –> 270°C) was conducted using a variety of analytical techniques (e.g., HPLC 13C NMR, FTIR, hydrogen–deuterium exchange) to evaluate the liquid and solid products. At temperatures below 180°C, the CH_2OH groups are hindered from interacting with microwaves while they are strongly involved in hydrogen bonding within both the amorphous and crystalline regions (Figure 12a). Above the softening temperature (180°C), these CH_2OH groups could be involved in a localized rotation in the presence of microwaves. As such they could act similarly to "molecular radiators" allowing for the transfer of microwave energy to their surrounding environment. In view of the limited presence of water inside the rigid cellulose framework, this is likely to involve collisions between the CH_2OH groups and the anomeric C1 of the same glucose ring thus forming levoglucosan. The latter can easily hydrolyze to glucose (Figure 12b). The data obtained from both conventional and microwave based experiments

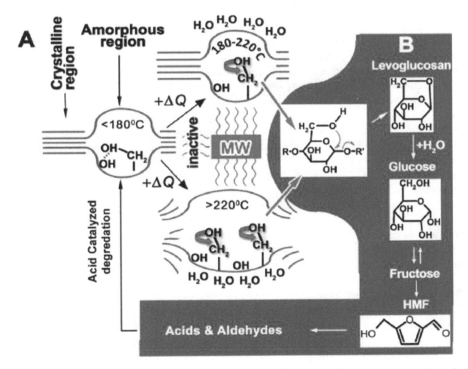

Figure 12. Schematic representation of the cellulose-microwave interaction as a function of temperature: (a) mechanism of CH_2OH group activation; (b) scheme of cellulose degradation toward acids and aldehydes (Fan et al. 2013c).

confirmed the molecular explanation of cellulose decomposition and its direct relationship to microwave activation of the CH_2OH pendant groups. At higher temperatures 190°C onward, microwave heating is found to be markedly more efficient toward the hydrolysis of cellulose than conventional heating (Fan et al. 2013c). Further, low microwave density heating at 220°C resulted in 12% release of fermentable sugars while a high microwave density heating at 250°C resulted in 21% release of fermentable sugars with the highest selectivity towards glucose of 75%, both of which are substantially higher than conventional heating method which released only 1.2% of fermentable sugars.

Microwave Applications in Pretreatment of Lignocellulosic Biomass

Microwave-chemical pretreatment

The combined microwave-chemical pretreatment of different feedstock may result in higher sugar recovery. Alkaline solution removes lignin, while the acidic solution removes hemicellulose. Various chemical agents can be used in microwave-chemical pretreatment such as dilute ammonia (Chen et al. 2012), microwave-assisted $FeCl_3$ (Lu and Zhou 2011) and the two most commonly studied chemical methods in the pretreatment of lignocellulosic biomass, i.e., microwave assisted-alkaline and microwave-assisted acid pretreatments. One-stage microwave-chemical pretreatment is generally carried out by addition of one of the chemicals to the lignocellulosic matter during microwave. NaOH aqueous solution and sulfuric acids are the most common chemicals used in pre-treatment.

Acid-Microwave Pretreatment

Microwave heating combined with dilute sulfuric acid solution to pretreat starch-free wheat fibers was investigated by Palmarola-Adrados et al. (2004). It was shown that the biomass pretreatment using microwave heating was able to produce a higher sugar yield than the steam explosion pretreatment. The impact of dilute acid concentration (up to 2% sulfuric acid) coupled with microwave heating reduced the hemicellulose content from 29.9 to 10.1% when compared to neutral solution (distilled water). An increase in acid concentration intensifies the depletion of hemicellulose so that the relative content of hemicellulose drops to around 1% once the acid concentration is larger than or equal to 0.015 M. Accordingly, it is realized that 79.8 and 97.8 wt% of hemicellulose contained in the raw bagasse were hydrolyzed using distilled water and 0.02 M sulfuric acid, respectively. It was also observed that the relative content of cellulose declines slightly by increasing acid concentration. With regards to lignin, similar behavior was also exhibited when the acid concentration was lifted from 0.015 to 0.02 M (Chen et al. 2012 a,b,c). It is known that dilute acid pretreatment is able to convert hemicellulose to soluble sugars and facilitates the subsequent enzymatic hydrolysis of cellulose. Binod et al. (2012) found the amount of reducing sugar yield was 0.091 g/g pretreated sugarcane bagasse with microwave assisted 1% sulfuric acid at 600 W for 4 min followed by enzymatic hydrolysis. The sugar yield was increased to 0.665 g/g pretreated biomass by changing H_2SO_4 with 1% NaOH at the same conditions. It was outlined that the difference with acid pretreatment, is that the lignin present in the solid fraction blocks enzymatic accessibility which in turn reduces the hydrolysis efficiency (Sun and Cheng 2002). Alkali pretreatment, in contrast, can remove lignin from the lignocellulosic materials.

Chen et al. (2011) evaluated the Scanning electron microscope (SEM) images of the raw bagasse and the pretreated bagasse at the three reaction temperatures with the heating time of 5 min and is shown in Figure 13. For the raw bagasse, a complete and compact lignocellulosic structure is clearly observed (Figure 13a). After undergoing the pretreatment with the reaction temperature of 130°C, the structure of bagasse has been damaged to a certain extent so that some cracks are seen on the bagasse surface

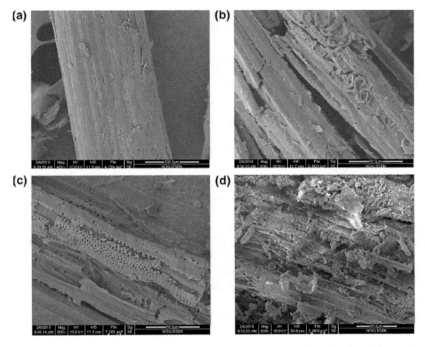

Figure 13. SEM images of (a) raw bagasse and pretreated bagasse with the heating time of 5 min and the reaction temperatures of (b) 130°C, (c) 160°C and (d) 190°C (Chen et al. 2011).

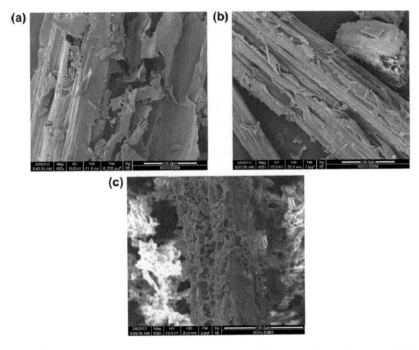

Figure 14. SEM images of pretreated bagasse with the heating time of 10 min and the reaction temperatures of (a) 130°C, (b) 160°C and (c) 190°C (Chen et al. 2011).

(Figure 13b). When the reaction temperature is 160°C, the disruption of the lignocellulosic structure becomes more pronounced (Figure 13c) and some tiny holes are exhibited on the surface. With the solution heated to 190°C, the lignocellulosic structure of bagasse has been destroyed in a significant way. As a result, a lot of debris are obtained (Figure 13d). The exposure time plays an important role in deconstruction of lignocellulosic biomass. A longer heating time of 10 min increased the fragmentation of the lignocellulosic structure because the impact of microwave irradiation is more pronounced at longer exposure times as shown in Figure 14. This is particularly noticeable for the reaction temperature of 190°C, where some fragments have flaked off from the biomass surface (Figure 14c).

Alkali-Microwave Pretreatment

The microwave/chemical pretreatment resulted in a more effective pretreatment than the conventional heating chemical pretreatment by accelerating reactions during the pretreatment process (Zhu et al. 2005, 2006). Three microwave/chemical processes for pretreatment of rice straw - microwave/alkali, microwave/acid/alkali and microwave/acid/alkali/H_2O_2- for its enzymatic hydrolysis and for xylose recovery from the pretreatment liquid were evaluated (Zhu and Yu 2005). It was found that xylose could not be recovered during the microwave/alkali pretreatment process, but it could be recovered as crystalline xylose during the microwave/acid/alkali and microwave/acid/alkali/H_2O_2 pretreatment. The enzymatic hydrolysis of pretreated rice straw showed that the pretreatment by microwave/acid/alkali/H_2O_2 had the highest hydrolysis rate and glucose content in the hydrolyzate.

Komolwanich et al. (2014) reported that the microwave-assisted NaOH pretreatment of *S. spontaneum* and *A. donax* positively affected the cellulose content while hemicellulose and lignin contents declined and the maximum yields of monomeric sugar was 6.8 g/100 g at 5% NaOH. Pretreatment of oil palm empty fruit bunch fiber (EFB) under the combination of MW and NaOH, it was found that the enzymatic saccharification of EFB was significantly improved by the removal of more lignin and hemicellulose and enhancing cellulose accessibility during the pretreatment. About 3% NaOH at a microwave power of 180 W for 12 minutes was reported as optimum pretreatment condition, achieving 74% of lignin and 24.5% holocellulose removal (Nomanbhay et al. 2013). Microwave-alkali pretreatment of sugarcane bagasse using 1% NaOH solution at 600 W for 4 min followed by enzymatic hydrolysis produced reducing sugar yield of 0.665 g/g dry matter (Binod et al. 2012). Vani et al. (2012) compared the pretreatment of cotton plant residue by alkali assisted microwave at 300W for 6 min with high pressure reactor pretreatment (180°C, 100 rpm for 45 min). They found that hydrolysis of solid fractions resulted in maximum reducing sugar yield of 0.495 g/g, requiring energy of 108 kJ for microwave-assisted alkali pretreatment while, 540 kJ of energy was need in the high pressure reactor pretreatment, a 5-fold reduction. Zhu et al. (2005) attributed that to solubilization of hemicellulose and lignin concentration in NaOH solution. Consequently, dilute NaOH pretreatment improves the enzymatic digestibility because it effectively removes lignin and enlarges the surface area and pore size of the substrate, as a result of decreased crystallinity of cellulose, and cleaved the structural bonds between lignin and other carbohydrates. Zhu et al. (2006) also reported that using microwave-assisted alkali to pretreat wheat straw proved that removing more lignin and hemicellulose from biomass took less pretreatment time than the alkali/conventional heating. This was confirmed by Zhao et al. in 2010. The pretreatment of rice hulls by NaOH solution in a microwave environment increased accessibility of the substrates due the rupture of the rigid structure of rice hulls. The reducing sugar content was increased by 13% compared to the samples without microwave pretreatment.

Microwave heating combined with sulfuric acid, hydrogen peroxide and water is also sometimes referred to as co-pretreatment method. Srimachai et al. (2014) investigated production of ethanol and methane from oil palm frond (OPF) using a microwave heating power of 500 W for 20 min. The composition of cellulose, hemicellulose of pretreated OPF was shown in Figure 15. The composition of OPF before co-pretreatment was 37.68% of cellulose, 35.34% of hemicellulose and 25.18% of lignin. Co-pretreatment of OPF with sulphuric acid and microwave could increase the amount of cellulose from 35–37 to 52–66% which was higher than co-pretreatment with hydrogen peroxide and microwave, water and microwave. Cellulose increased by 9–36% in comparison with the initial cellulose from the OPF (untreated). The SUL (3%) + microwave gave the maximum amount of cellulose as 62.79% and the lowest as 41.28% of PER

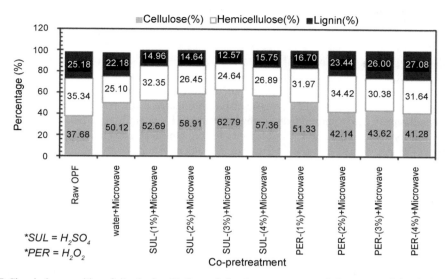

Figure 15. Chemical composition of oil palm frond before and after the co-pretreatment (microwave and chemical pretreatment, Srimachai et al. 2014).

(4%) + microwave. These results are similar to the experimental values of Goh et al. (2010), the largest amount of cellulose (62.50%) from OPF was achieved by hot compressed water pretreatment. Zhu et al. (2006), also found that high amount of cellulose (69.3%) from rice straw was achieved with NaOH (4%) + microwave (300 W) for 60 min pretreatment.

Two-Stage Microwave-Chemical Pretreatment

Table 6 shows summary of some recent research done on the use of microwave—chemical pretreatment of biomass for bioethanol production for two-stage. As shown previously, one-stage pretreatment methods have some limitations. For example, combination microwave-alkali pretreatment removes most of the lignin from biomass and the remaining solid residue on enzymatic hydrolysis results in the production of a mixture of hexose and pentose sugars, which need complex co-fermentation methods for comprehensive sugar utilization. With acid pretreatment, the lignin present in the solid fraction constrains enzymatic hydrolysis which decreases the hydrolysis efficiency. Two-stage microwave chemical pretreatment is one of the solutions to overcome some limitation of single-stage microwave–chemical pretreatment (Pedersen and Meyer 2010). Binod et al. (2012) reported that the pretreatment of sugarcane bagasse via sequential microwave-alkali-acid treatment with 1% NaOH and 1% sulfuric acid. The recovery of fermentable sugars from sugarcane bagasse was improved with an overall yield of 0.83 g/g dry biomass, as compared to microwave/alkali only pretreatment with reducing sugar yield of 0.665 g/g dry biomass, at microwave power of 600 W and 4 min pretreatment time. Obviously, these results indicate a combined microwave alkali-acid pretreatment for short duration improved the fermentable sugar yield. Zhu et al. (2006) subjected rice straw to microwave/alkali and microwave/acid/alkali for its enzymatic hydrolysis for xylose recovery from the pretreatment liquid. They observed that xylose could not be recovered by the microwave/alkali pretreatment process, but could be recovered as crystalline xylose by the microwave/acid/alkali. This is probably, due to the high lignin-derived impurities in the pretreatment liquors.

Mechanical-Microwave Pretreatment

Mechanical pretreatment is an important way to reduce cellulose crystallinity and increase delignification. Milling or grounding the biomass to reduce particle size positively reflects on the pretreatment process. It leads to increased available specific surface and enhanced enzyme accessibility for the subsequent enzymatic hydrolysis step. Vidal et al. (2011) suggested that necessary particle sizing needs to be determined in the context of thermochemical pretreatment employed for lignocellulose conversion because mechanical pretreatment by itself is insufficient to attain economically feasible biomass conversion. Chen et al. (2012b, c) report that the optimum lignin removal occurred with particle size 1–2 mm with highest delignification and lowest sugar loss. Hu (et al. 2008, Hu and Wen 2008) observed lignin removal ratio of 63–70% with various particle size ranges of 1.0–2.0, 0–0.5, 0.5–0.25, and < 0.25 mm of switchgrass under microwave-assisted alkali pretreatment but the energy was intensive, thus increased the pretreatment cost. Karunanithy and Muthukumarappan (2010) grounded switchgrass and big bluestem to a particle size range between 0.3 and 1.2 mm for switchgrass and 0.4 and 0.8 mm for big bluestem using a single screw extruder. The combination of extrusion and microwave pretreatment has improved recovery of glucose, xylose and total sugar by 27.0%, 16.7%, and 21.4%, respectively for switchgrass, while the same for big bluestem were 17.3%, 24.9%, and 19.7% respectively. Although microwave heating increased recovery of sugar, the energy costs were high. Optimizing the microwave pretreatment will reduce the energy consumption and increase product recovery due to the selective heating.

Microwave-Steam Explosion Pretreatment

Steam explosion (SE) can be used for biomass fractionation. In this method, biomass is exposed to pressurized steam and then the pressure is suddenly reduced. Due to pressure fluctuation, the materials undergo an explosive decompression. The effects of steam treatment have been attributed to several factors such as (de Castro 1994): complete hydrolysis of hemicellulose (Grohmann et al. 1986), lignin depolymerization (Karina et al. 1992, Shimizu et al. 1998) and redistribution within the cell wall (Michalowicz et al. 1991, Toussaint et al. 1991) and swelling of cell walls (Wong et al. 1988). There is evidence from auto-hydrolyzed sugar-cane bagasse that overall lignin content is not altered (de Castro 1989). However, lignin structure and location are modified by melting and agglomeration of the depolymerized lignin in different parts of the cell wall (Toussaint et al. 1991). As steam-treated material is hydrolyzed to a greater extent by enzymes compared to untreated material, this suggests that the presence of lignin in the cell wall tissue per se has little negative effect on cell wall susceptibility to enzyme attack (Saddler et al. 1982, Wong et al. 1988), but that its spatial distribution in unaltered secondary cell walls creates steric blocks to specific enzyme substrates.

Steam explosion under microwave irradiation (SE–MI) was studied by Pang et al. (2013). They compared SE-MI with steam explosion only treatment. The corn stover was irradiated under 540 W microwave power for 3 min. Both pretreatment methods were conducted at 170–210°C range for 3–15 min. SE–MI process improved the enzymatic hydrolysis yields of glucose and xylose, and slightly enhanced the total sugar yield when compared with SE alone process. The maximum glucose, xylose and total sugar yields were 57.4%, 17.8% and 75.2% (corresponding to 28.0 g glucose, 8.7 g xylose and 36.7 g total sugar generated per 100 g raw feedstock), respectively at 200°C for 5 min. SE–MI pretreatment showed clear advantage in inhibiting the increase of biomass crystallinity. The crystallinity after SE–MI pretreatment was 19% less than that from SE pretreatment at 190°C for 5 min.

Factors (Operating Parameters) that affect Sugar Recovery in Microwave Pretreatment

Biomass loading

Biomass loading is an important factor affecting microwave pretreatment efficiency. Selecting the suitable substrate concentration can enhance the pretreatment output by accelerating biomass dissolution. Lu et al. (2011) examined rape straw with various solid loadings of 30 (LTS), 50 (MTS), and 90% (HTS) at 900 W microwave power for a min. They found that the maximum glucose yield of 53.5% was attained at the solid loading of LTS, but this value is similar to that of MTS (48.6%). Thus at the expense of a small decrease in sugar yield, more straw can be pretreated and less energy is needed overall. Furthermore, both cellulose and hemicellulose saccharification was shown to improve with a simultaneous increase in substrate concentration at the middle levels where the optimum substrate concentration was 75 g/L achieving maximum saccharification under 680 W microwave power for 24 min (Ma et al. 2009). A further rise in the level of substrate concentration results in a gradual decrease in saccharifications. It might be caused by the different "energy effect" with different substrate concentrations. In other words, the samples with a high biomass loading (and thus relatively low water loading) will receive less energy absorbed by water due to oscillation of water molecules (Yang and Wyman 2004) which is not beneficial to the structure disruption of rice straw. As a result, a higher straw digestibility was obtained for samples at a relatively low biomass loading.

Microwave power

Unlike conventional heating sources, microwave irradiation produces higher power densities which increase production rates and shrink production costs. Microwave power level positively affects glucose recovery from alkali assisted pretreatment of wheat straw (Xu et al. 2011). On the other hand, Binod et al. (2012) reported that microwave-acid pretreatment of sugarcane bagasse had no increase in sugar yields even with varying microwave power levels. The reducing sugar yields for microwave assisted acid (MA), microwave-assisted alkali (MAL) and microwave-assisted alkali/acid (MAA) pretreatments at different microwave power levels were compared. An inverse relationship between the increasing microwave power and reducing sugar yield was observed for MA pretreatment. The highest reducing sugar yield was obtained at 100 W microwave power. For MAL and MAA pretreatment, the highest reducing sugar was produced at 600 W power level. In addition, pretreatment time can be reduced by increasing the microwave power level with some limitations. As the temperature rises, the degree of polymerization of cellulose declines, long cellulose chains collapse to shorter groups of molecules, it releases glucose that degrade to hydroxymethyl furfural (McParland et al. 1982). Therefore, it is essential to optimize microwave power to avoid excess energy consumption and/or degradation of useful components (Nomanbhay et al. 2013).

Microwave exposure time

One of the main factors that influence the pretreatment severity is the exposure time (Chum et al. 1990, Schell et al. 2003). The interactive effect between irradiation time and microwave power level significantly influences the biomass digestibility (Ma et al. 2009) by enhancing hemicellulose removal and cellulose digestibility (Kabel et al. 2007). An extended exposure time with a higher microwave power may lead to a decline in biomass digestibility, as increase in irradiation time and microwave power lead to a high temperature within the sample which could initiate decomposition of released sugar in the pretreatment process (Kabel et al. 2007). For instance, dissolved xylose can be decomposed into furfural at high temperature, consequently reducing the total xylose yield. Furfural is an undesirable byproduct which plays a negative role by inhibiting fermentation processes (Eggeman and Elander 2005).

Formation of inhibitors

Microwave pretreatment, particularly with chemicals, necessitate feedstock washing after pretreatment to eliminate fermentation inhibitors. The major types of fermentation inhibitors are furfural and hydroxymethyl furfural (HMF), weak acids, and phenolic compounds. Maiorella et al. (1983) reported the inhibitory effects at concentrations of acetic acid 0.5–9 g/L, formic acid 0.5–2.7 g/L and lactic acid 10–40 g/L on *Saccharomyces cerevisiae* yeast growth in the fermentation process due to interference of the acids with functions involved in cell maintenance. Meanwhile, glycerol at a concentration of 450 g/L alters the cell's osmotic pressure (Maiorella et al. 1983), and furfurals at concentrations of 3 g/L are considered antagonistic to cell growth (Palmqvist and Hahn-Hágerdal 2000). Karunanithy et al. (2010) reported 0.12 g/L of acetic acid was the only fermentation inhibitor found after pretreatment switchgrass by extrusion-microwave pretreatment. Higher acetic acid concentrations were found after pretreatment of Kans grass and Gaint reed via microwave-assisted NaOH and two-stage microwave/NaOH/H_2SO_4 but the furfural was untracable (Lu and Zhou 2011). There were 0.7 g/L acetic acid, 0.04 g/L lactic acid, 2.3 g/L glycerol,

0.14 g/L formic acid, less than 0.1 g/L HMF and furfurals in pretreated sorghum bagasse by microwave-assisted dilute ammonia (Chen et al. 2012a). However, the quantities of these chemicals were insufficient to generate any inhibitory effect. Steam explosion-microwave pretreatment of corn stover aggravated furfural, HMF, formic and acetic acids formation when compared to steam explosion pretreatment alone (Pang et al. 2013). An increase in acetic acid concentration from 2.14 to 2.94 g/100 g matter was observed with microwave power and exposure time for steam explosion combined with microwave pretreatment of corn stover (Pang et al. 2012). Pretreatment optimization should be considered to prevent the formation of inhibitory compounds.

Summary of Recent Studies

Microwave mediated hydrothermal processing is a technique that should be considered for the extraction of seaweeds' polysaccharides since the main sugars present in macro-algae (laminarin and fucoidans) are the watersoluble compounds (Zvyagintseva et al. 2000). Chen (2005) reported the use of microwave-assisted method to obtain polysaccharides from *Solanum nigrum*. Navarro and Stortz (2005) produced 3,6-anhydrogalactose units from galactose 6-sulfated residues of red seaweed galactans utilizing microwave irradiation to carry out the alkaline modification. The experiments were carried out in a domestic microwave oven heating the samples for 1 min at 1200 W using Teflon closed-vessels. Furthermore, Yang et al. (2008) and Rodriguez-Jasso et al. (2011), respectively evaluated the hydrothermal extraction of sulfated polysaccharides of *Undaria pinnatifida* and *Fucus vesiculosus* using a digestion microwave oven with a maximum delivered power of 630 W (at 172°C). Results showed that microwave heating around 30–60 s was more effective in improving polymer dissolution without a noticeable structural degradation. A preliminary development of a microwave prototype at industrial scale was reported by Uy et al. (2005). Carrageenan extraction of *Eucheumacottonii* and *Eucheumaspinosum* was carried out using an industrial single-mode cavity continuous microwave, at 38% of full power (800 W) and with a residence time of 30 min. Since these hydrocolloids are not water soluble, hydrothermal processing extraction was tested with aqueous mixtures of organic solvents. The extracted carrageenans showed high purity, without the need for further purification procedures. González-López et al. (2012), produced compounds with antioxidant activity using non-isothermal autohydrolysis process and Sargassum muticum as raw material. In a recent works, Anastyuk et al. (Anastyuk et al. 2012a,b) used autohydrolysis as an alternative strategy for fucoidan depolymerization from brown algae *Silvetia babingtonii* and *Fucus evanescens*. The term autohydrolysis is referred to the acid polysaccharide hydrolysis under very mild conditions using—SO_3H groups as source of catalyst in substrate reaction. Rodriguez-Jasso et al. (2013) reported the extraction of sulfated polysaccharides by autohydrolysis from *Fucus vesiculosus*. The results showed that the pH decrease in the reaction media at high temperatures and times, possibly due to the polysaccharides hydrolysis using the "*in situ*" –SO_3H groups as source of catalyst. Table 6 shows a summary of some recent research done on the use of microwave—chemical pretreatment of biomass for bioethanol production for one-stage and two-stage process configurations.

Energy Aspects of Microwave Enhanced Bioethanol Production

Few studies reported on the energy efficiency of microwave pretreatment process in bioethanol production. Energy consumption in pretreatment step influences the overall footprint of the bioethanol production process. The following relationships (equations [5–11]) can be used to determine the product yields and energy consumption in microwave enhanced bioethanol production (Thangavelu et al. 2014).

$$Glucose\ yield = \frac{Glucose\ obtained\ in\ hydrolysis\ (g)}{initial\ glucose\ in\ biomass\ (g)} \times 100\% \qquad [5]$$

$$Ethanol\ yield = \frac{Ethanol\ obtained\ in\ fermentation\ (g)}{theoretical\ ethanol\ in\ biomass\ (g)} \times 100\% \qquad [6]$$

$$\text{Ethanol yield (yield coefficient)} = \frac{\text{GlucoEthanol obtained in fermentation (g)}}{\text{Glucose in biomass hydrolysate (g)}} \times 100\% \qquad [7]$$

$$\text{Ethanol productivity} = \frac{\text{Ethanol obtained in fermentation (g)}}{\text{Fermentation time (g)}} \times 100\% \qquad [8]$$

Table 6. Recent studies on bioethanol production using microwave chemical pretreatment.

Raw Material	Pretreatment condition	Results	References
S. spontaneum and A. donax	One-stage 5% NaOH, 80°C, 5 min for S. spontaneum and 120°C, 5 min for A. donax	One-stage Max. yield of monomeric 6.8 g/100 g for both feedstocks	Komolwanich et al. (2014)
	Two-stage 5% NaOH, 200°C, 10 min for S. spontaneum 0.5% H_2SO_4, 180°C, 30 min for A. donax	Two-stage 33.8 g/100 g S. spontaneum 31.9 g/100 g A. donax	
Empty Fruit Bunch	One-stage 3% NaOH, 180 W, 12 min	One-stage reducing sugar 411 mg/g	Nomanbhay et al. (2013)
Sugarcane Bagasse	One-stage 1% NaOH, 600 W, 4 min	One-stage 0.665 g/g dry biomass	Binod et al. 2012
Miscanthus Sinensis	One-stage 1.0% NH_4OH, 300 W, 15 min	One-stage Monomeric sugar yields 2.93 g/100 g	Boonmanumsin et al. 2012
	One-stage 1.63% H_3PO_4, 300 W, 30 min	One-stage 62.28 g/100 g biomass	
	Two-stage 1.0% NH_4OH, 300 W, 15 min 1.78% (v/v) H_3PO_4, 300 W, 30 min	Two-stage 71.64 g/100 g biomass	
Switchgrass	One-stage 3% NaOH, 250 W, 10 min	One-stage highest yield of reducing sugars 30 mg/mL	Keshwani 2009
Wheat straw, 8.6%, w/v, in water	Temperature (160–240°C, 5 min) and duration (5–20 min at 200°C)	MW pretreatment in dilute acid (0.5% H_2SO_4, w/v) at 160°C for 10 min.	Saha et al. 2008
	Enzymatic saccharification (45°C, pH 5.0, 120 h) using a mix of 3 commercial enzyme preparations (cellulase, -glucosidase, and hemicellulase) at the dose level of 0.15 mL of each	Hydrolyzate with 29 ± 16 mg furfural, 3 ± 00 mg hydroxymethyl furfural, and 60 ± 8 mg acetic acid per g of straw. Maximum release of sugars - 651 ± 7 mg/g straw, 84% yield MW pretreatment with lime (0.1 g/g straw) at 160°C for 10 min released 604 ± 30 mg total sugars/g straw (78% yield) after enzymatic hydrolysis.	
		Ethanol concentration was 167 ± 15 g/L with a yield of 0.21 g/g straw in the case of simultaneous saccharification and fermentation	
Rape straw	10 g immersed in 90 g of 2% (v/v) H_2SO_4. Pretreatments at 900, 700 and 550 W for 1, 3, 6, and 10 min, and treatments at 900 W/1 min, 900 W/10 min, 700 W/3 min, and 700 W/6 min	Untreated straw gave a very low ethanol yield of around 13% of the theoretical yield corresponding to 2.7 g ethanol/l. The highest ethanol yield after MW pretreatment reached 65% of the theoretical ethanol yield with ethanol concentration of 13.6 g/l	Lu et al. 2011

Table 6. contd....

Table 6. contd.

Raw Material	Pretreatment condition	Results	References
Saccharomyces cerevisiae	Oil palm frond (OPF) soaked in dilute sulfuric acid, hydrogen peroxide and water consequently pretreated by MW for simultaneous saccharification and fermentation followed by anaerobic digestion of effluent	OPF soaking in water gave a maximal ethanol yield of 0.32 g-ethanol/g-glucose which was 62.75% of the ethanol theoretical yield (0.51 g-ethanol/g-glucose). The effluent from the ethanol production process was used to produce methane with the yield of 514 ml CH_4/g VS added.	Srimachai et al. 2014
Sweet sorghum bagasse	MW pretreatment at 121°C for 30 to 120 min. Steam-assisted and MW-assisted processes were evaluated	Steam-assisted autohydrolysis extracted 72.69 (±0.08)% by weight of the hemicelluloses while the MW-assisted autohydrolysis extracted 70.83 (±0.49)% of the hemicelluloses from the sweet sorghum bagasse. Steam-assisted lime treatment produced 69.67 (±1.26)% of the lignin extraction while MW-assisted lime treatment resulted in 68.27 (±1.19)% of the lignin extraction. The MW-assisted process increased the total crystallinity index (TCI) of the cellulose in the treated SSB and also increased the concentration of guaiacyl lignin content in the recovered lignin which was thermally more stable than the lignin produced in the steam-assisted process	Kurian et al. 2015
Sugarcane bagasse	Pretreatments with a white rot fungus, *Ceriporiopsis subvermispora*, and microwave hydrothermolysis of bagasse on enzymatic saccharification and fermentation	Sugar yield, 44.9 g per 100 g of bagasse with fungal treatments followed by microwave hydrothermolysis at 180°C for 20 min MW pretreatments with and without the fungal cultivation resulted in similar levels of cellulose exposure, but the combined treatment caused more defibration and thinning of the plant tissues. Simultaneous saccharification and fermentation of the pulp obtained by MW hydrothermolysis with and without fungal treatment, gave ethanol yields of 35.8% and 27.0%, respectively, based on the holocellulose content in the pulp	Sasaki et al. 2011
Sago bark waste	Domestic microwave oven with a frequency setting at 2450 MHz and power outputs of 700, 900 and 1100 W was used	The maximum sugar yield of 62.6% at the biomass loading of 20% (w/w). Highest ethanol yield of 60.2% theoretical yield, ethanol	Thangavelu et al. 2014

Table 6. contd....

Table 6. contd.

Raw Material	Pretreatment condition	Results	References
	10 grams of dried SBW submerged in 40 g of 3% (v/v) H_2SO_4 in a 250-mL conical flask. The pretreatments at 700 W/2 min, 900 W/30 s, 900 W/1 min, and 1100 W/ 30 s were selected	concentration 30.67 g/L was achieved by diluted sulfuric acid supported microwave irradiation with 40% (w/w) biomass loading at 60 h fermentation	
Switchgrass	MW for 10 minutes at 250 W in 3% NaOH	MW pretreatment with 3% NaOH produced highest yields of reducing sugar from switchgrass	Keshwani et al. 2007
Sugarcane bagasse	MW-alkali pretreatment in 1% NaOH with 10% biomass loading at varying power consumption 850 W, 600 W, 450 W, 300 W, 180 W and 100 W for different residence times between 1 min and 30 min. MW in 1% H_2SO_4 with 10% solid loading	Sugarcane bagasse with 1% NaOH at 600 W (MW) for 4 min followed by enzymatic hydrolysis gave reducing sugar yield of 0.665 g/g dry biomass, while combined microwave-alkali-acid treatment with 1% NaOH followed by 1% sulfuric acid, the reducing sugar yield increased to 0.83 g/g dry biomass. Microwave-alkali treatment at 450 W for 5 min resulted almost 90% of lignin removal from the bagasse	Binod et al. 2012
Rice straw (41% cellulose; 17.8% lignin; 20% hemicellulose; 12% ash; and 9.2% others)	Domestic microwave with output power of 119–700 W in acetic acid or propionic acid between 2 and 5 min	MW intensity > solid–liquid ratio > acetic acid concentration > microwave irradiating time. Optimal conditions (25% acid concentration, 1:15 solid–liquid ratio, 230 W microwave intensity, and 5 min irradiating time), the removal ratios of lignin are 46.1 and 51.5%, and the sugar yields are 71.4 and 80.1% when acetic acid and propionic acid are used as solvents, respectively. Sugar yield was only 35.3% in blank	Gong et al. 2010
Rice straw	MW and alkali and its enzymic hydrolysis 20 g of rape straw in 160 mL of 1% NaOH aqueous solution and boiled from 15 min to 2 h. MW power set at 700, 500 and 300 W	Rice straw weight loss of 44.6% and composition of cellulose 69.2%, lignin 4.9% and hemicellulose 10.2% after 30-min MW-alkali pretreatment at 700 W. Weight loss of 41.5% and composition cellulose 65.4%, lignin 6.0% and hemicellulose 14.3% after 70-min alkali-alone pretreatment. MW-alkali pretreatment removes more lignin and hemicellulose from rice straw with shorter pretreatment time compared with the alkali-alone	Zhu et al. 2005
Switchgrass and Coastal Bermudagrass	Pretreatments in dilute alkali reagents with MW at 250 W for residence times ranging from 5 to 20 min	82% glucose and 63% xylose yields were achieved for switchgrass and 87% glucose and 59% xylose yields were achieved for coastal bermudagrass following enzymatic hydrolysis of biomass pretreated under optimal conditions	Keshwani and Cheng 2010

Table 6. contd....

Table 6. contd.

Raw Material	Pretreatment condition	Results	References
Garden biomass (GB)	Hot plate (HP), Autoclave (AC) & Microwave (MW) were evaluated	MW was better than AC and HP. A maximum of 53.95% of cellulose for MW heating and 46.97% of reducing sugar yield.	Gabhane et al. 2011
Barley husk	MLS-1200 Mega Microwave workstation, Milestone, 0.5% H_2SO_4 or water only, 200°C for 5 min or 210°C for 10 min N/A	Glucose yield of 85% and Xylose yield of 46%	Palmarola-Adrados et al. 2005
Rice straw	WD700 (MG-5062T) type domestic microwave, 1% NaOH, Temperature uncontrolled, 15 min to 2 h Fixed at 300–700 W, Temperature uncontrolled, 10 min Fixed at 240 W	Glucose yield of 65%; total sugar conversion: 78%	Zhu et al. 2005
Rice straw and sugar cane bagasse	Turbora Model TRX-1963 domestic microwave, Glycerin in water solution		
Switchgrass	Customized Sharp/R-21 HT domestic microwave, 0.1 g NaOH/g biomass, 190°C, 30 min Up to 1 kW	Glucose + Xylose yield of 99%	Hu and Wen 2008
Softwoods	Toshiba Model TMB 3210, Water-only, 219–226°C s for (estimated with an infra-red thermometer) up to 2.4 kW	88–93% total sugar	Azuma et al. 1985
Switchgrass	Panasonic Corporation, model NN-S954, 1–3% NaOH, Temperature uncontrolled, 5–20 min	Fixed at 250 W, 80–85% total sugar	Keshwani and Cheng 2010

$$Fermentation\ efficiency = \frac{Ethanol\ obtained\ in\ fermentation\ (g)}{0.51 \times Glucose\ in\ hydrolysate\ (g)} \times 100\% \qquad [9]$$

$$Energy\ consumption\ (\frac{kJ}{g}Glucose) = \frac{Microwave\ power\ (kW) \times time\ (s)}{Highest\ glucose\ in\ hydrolysis\ (g)} \qquad [10]$$

$$Energy\ consumption\ (\frac{kJ}{g}Ethanol) = \frac{Microwave\ power\ (kW) \times time\ (s)}{Highest\ ethanol\ during\ fermentation\ (g)} \qquad [11]$$

Lu et al. (2011) reported on the energy inputs for different solids loading rates in pretreatment of rape straw. The pretreatment times were 1, 3, 6, and 10 min at microwave output of 700 W or 900 W. The lowest energy consumption was observed when solid loading and energy input were fixed at 50% (w/w) and 54 kJ (900 W for 1 min), respectively, and amounted to 5.5 and 10.9 kJ to produce 1 g of glucose after enzymatic hydrolysis and 1 g ethanol after fermentation, respectively. In general, 1 g ethanol can produce about 30 kJ of energy, and therefore, the energy input for the pretreatment was only 35% of the energy output. The approach developed in this study resulted in 92.9% higher energy savings for producing 1 g ethanol when compared with the results of microwave pretreatments previously reported (Lu et al. 2011). Untreated straw gave a very low ethanol yield of around 13% of the theoretical yield corresponding to 2.7 g ethanol/L. The highest ethanol yield after microwave-pretreated rape straw at fixed solid loading of 10% was achieved with 900 W at 1 min (65% of the theoretical ethanol yield, ethanol concentration of 13.6 g/L). They also reported that longer pretreatment time has no significant effect on ethanol production.

Unlike conventional heating, microwaves generate higher power densities, enabling higher production rates and lower production costs. At 10% solids loading, the energy consumption of 700 W for 3 min, 700 W for 6 min and 900 W for 10 min were 67.2, 153.9 and 169.2 kJ for producing 1 g of glucose and

105.4, 230.7 and 235.2 kJ for producing 1 g of ethanol respectively. In general, 1 g ethanol can produce 30 kJ energy, while 1 g glucose can only produce 15.8 kJ energy. The energy input is several times higher than the energy output in the above three processes because most of the energy was used for water evaporation during pretreatment. Evaporation of 1 g of water requires 2.5 kJ of energy at an initial temperature of 25°C. For the three schemes, the water loss was 10.3, 37.1, 68.9 and 85.5 g, which means 25.8, 92.8, 172.3 and 213.8 kJ of energy used for the water evaporation. Thus, 48–65% of the energy input was used for water evaporation. The water losses for 30%, 50% and 90% solids loadings were 9.2, 7.9, and 5.1 g respectively corresponding to energy inputs of 23.0, 19.8 and 12.8 kJ of energy for evaporation. The energy consumption of these samples was thus much lower than those of samples with less solid loading; however, glucose and ethanol yields decline at high solids loading. The energy consumption under optimal conditions in this study was much lower than that observed in other studies.

Zhu et al. (2006a,b in Table 7) used 700 W for 25 min and achieved an ethanol yield of 69.3%, but with an energy input of 153.1 kJ/g ethanol. Finally, it was reported that microwave pretreatment can

Table 7. Comparison of energy consumption in microwave enhanced bioethanol production.

Feedstock	Purpose	Pretreatment conditions	Energy consumption (kJ/g glucose)	Energy consumption (kJ/g ethanol)	Water lost (g)	Reference
Rice Straw	Pretreatment	300 W	83.1	-	-	Zhu et al. 2006b
Wheat Straw	Pretreatment	700 W/25 min	-	153.1	148.5	Zhu et al. 2006a,c
	Pretreatment	250 W/10 min	11.8–23.6	-	-	Keshwani et al. 2007
Rape Straw	Pretreatment	900 W/1 min	23.5	40.3	10.3	Lu et al. 2011
Rape Straw	Pretreatment	700 W/3 min	67.2	105.4	37.1	Lu et al. 2011
Rape Straw	Pretreatment	700 W/6 min	153.9	230.7	68.9	Lu et al. 2011
Rape Straw	Pretreatment	900 W/10 min	169.2	235.2	85.5	Lu et al. 2011
Rape Straw	Pretreatment	30% solids loading	8.2	13.5	9.2	Lu et al. 2011
Rape Straw	Pretreatment	50% solids loading	5.5	10.9	7.9	Lu et al. 2011
Rape Straw	Pretreatment	90% solids loading	10.0	14.5	5.1	Lu et al. 2011
Sorghum	Pretreatment		-	29	-	Chen et al. 2012a
Sago Bark waste, 20%	Pretreatment	700 W/2 min	2.18	4.74	8.2	Kannan et al. 2013
20% SBW	Pretreatment	900 W/30 s	1.77	3.52	5.6	Kannan et al. 2013
20% SBW	Pretreatment	900 W/1 min	1.72	3.87	11.6	Kannan et al. 2013
20% SBW	Pretreatment	1100 W/30 s	1.09	2.37	7.8	Kannan et al. 2013
Sago Bark waste	Pretreatment	20% SBW	1.15	2.01	10.2	Kannan et al. 2013
Sago Bark waste	Pretreatment	40% SBW	1.27	1.76	8.5	Kannan et al. 2013
Sago Bark waste	Pretreatment	60% SBW	1.72	2.13	6.5	Kannan et al. 2013
Sago Bark waste	Pretreatment	80% SBW	3.35	3	4.5	Kannan et al. 2013
Sago pith waste	Pretreatment & hydrolysis	700 W/3 min; H_2O	188	525	16.3	Kannan et al. 2013
Sago pith waste	Pretreatment & hydrolysis	900 W/2 min; H_2O	128	348	12.4	Kannan et al. 2013
Sago pith waste	Pretreatment & hydrolysis	700 W/3 min; $H_2O + CO_2$	44	98	14.8	Kannan et al. 2013
Sago pith waste	Pretreatment & hydrolysis	900 W/2 min; $H_2O + CO_2$	33	69	7.2	Kannan et al. 2013

improve enzymatic hydrolysis yields four fold over that of untreated and results in hydrolysates suitable for fermentation. During microwave pretreatment process, with the time increase, a lot of water evaporates and much energy is lost which proved that longer time had no good effect on microwave pretreatment process. The results in this study indicated that pretreatment with high power and short time was more energy-saving technology and more practical in the future.

Thangavelu et al. (2014) studied the energy efficiency of microwave irradiation for bioethanol production from sago bark waste (SBW). A maximum sugar yield of 62.6% was reached at the biomass loading of 20% (w/w). An ethanol yield of 60.2% (theoretical) with a concentration of 30.67 g/L was obtained under diluted sulfuric acid supported microwave irradiation with 40% (w/w) biomass loading at 60 h fermentation time. Specific microwave energy consumption for producing 1 g sugar (after enzymatic hydrolysis) and 1 g ethanol (after enzymatic hydrolysis) were reported as 1.27 and 1.76 kJ to produce 1 g of sugar after enzymatic hydrolysis and 1 g ethanol after fermentation respectively when biomass loading and energy input were fixed at 40% (w/w) and 33 kJ (1100 W for 30 s) respectively. In general, 1 g ethanol can produce approximately 27 kJ of energy, and therefore, the energy input for the microwave pretreatment was only 7% of the energy output. The microwave heating application resulted in 80% energy savings for producing 1 g ethanol compared to rape straw by microwave pretreatment previously reported by Lu et al. (2011).

Energy requirements for microwave pretreatment are shown in Table 7. At 20% SBW loading, the energy consumption of 700 W/2 min, 900 W/30 s, 900 W/1 min and 1100 W/30 s was 2.18, 1.77, 1.72 and 1.09 kJ for producing 1 g of sugar and 4.74, 3.52, 3.87 and 2.37 kJ for producing 1 g of ethanol respectively. In common, 1 g ethanol can produce 27 kJ of energy; however 1 g sugar can only produce 16 kJ of energy. The biomass loadings of 40% SBW with 1100 W/30 s only utilizes 1.76 kJ for producing 1 g of ethanol which is more suitable for commercial scale ethanol production. It was observed that the energy input is several times higher than the energy output in the above four processes because most of the energy was utilized for evaporation of water during pretreatment. Evaporation of 1 g of water requires 2.5 kJ of energy at an initial temperature of 25°C. For, the water loss of 700 W/2 min, 900 W/30 s, 900 W/1 min and 1100 W/30 s was 8.2, 5.6, 11.2 and 7.8 which means 20.5, 14, 28 and 19.5 kJ of energy used for the water evaporation. Hence, 30–60% of the energy input was used for water evaporation. The water losses for 20, 40, 60 and 80% SBW were 10.2, 8.5, 6.5 and 4.5 g respectively matching to energy inputs of 25.5, 21.3, 16.3 and 11.3 kJ of energy for evaporation. The energy consumption of samples with higher biomass loadings was thus much lower than those of samples with less biomass loading; however, sugar and ethanol yields drop at high biomass loadings.

Other bioproducts from biomass feedstock

5-Hydroxymethylfurfural, HMF is a polyfunctional chemical intermediate and a possible substitute to petroleum based building blocks used in the commercial production of high potential synthetic chemicals. Numerous studies have been reported on the production of HMF from different carbohydrate feedstocks (e.g., cellulose, cellobiose, fructose, glucose, hemicellulose, inulin, sucrose, and xylose) at various efficiencies and reaction conditions. Levulinic acid is another key compound used as raw material for various fuel additive and other platform chemical production. Figure 16 shows the production routes for production of valuable chemicals using HMF and LA as reactants. Table 8 summarizes experimental studies utilizing microwave irradiation to produce HMF and LA (summarized from Mukherjee et al. 2015). These can be produced from starch as well as lignocellulosic feedstock such as fructose, glucose, starch, sucrose and cellulosic waste. Microwave process conditions vary depending on the catalysts and solvents in the reaction.

Concluding Remarks

Microwave heating offers several advantages in bioethanol production. Enzymatic hydrolysis and fermentation yields are enhanced under microwave mediated pretreatment. Higher and uniform power densities generated by microwaves enable for faster sugar recovery, higher ethanol production and lower

Microwave Pretreatment of Feedstock for Bioethanol Production 195

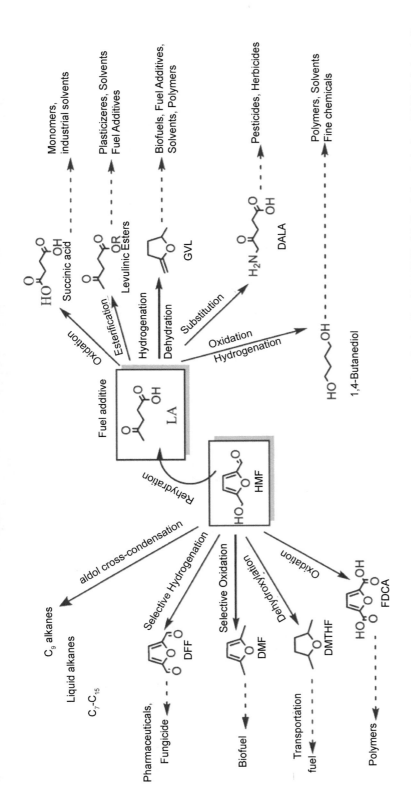

Figure 16. Application domains of HMF and LA (HMF - 5-Hydroxymethylfurfural; FDCA - 2, 5-furandicarboxylic acid; DMTHF - 2,5-dimethyltetrahydrofuran; DMF - 2, 5-dimethylfuran; DFF - 2, 5-diformylfuran; GVL - γ-valerolactone; DALA - 5-aminolevulinic acid) (Roy Goswami et al. 2015).

Table 8. HMF and LA synthesis using microwave irradiation (Mukherjee et al. 2015).

Substrate & conc. (wt%)	Solvent	Catalyst	MW power (W)	Primary final products	Temp. (°C)	Reaction/ residence time	Conversion (%)	Yield (wt.%)	Reference
Cellulose (8)	[C$_3$SO$_3$H$_{mim}$]HSO$_4^-$ Water (1:2 w/w)	-	800	LA	160	30 min	N/A	32	Ren et al. 2013
Fructose (10)	Acetone-DMSO (7:3 w/w)	DOWEX 50 WX8–100 strong acidic ion-exchange resin (4 wt.%)	700	HMF	150	20 min	99	58	Qi et al. 2008a
Fructose (2)	Water	TiO$_2$ (0.4 wt.%)	700	HMF	200	10 min	90.4	29	Qi et al. 2008b
Fructose (2)	Water	H$_2$SO$_4$ (1 wt.%)	700	HMF	200	5 min	97.3	33	Qi et al. 2008b
Glucose (2)	Water	TiO$_2$ (0.4 wt.%)	700	HMF	200	5 min	63.8	13	Qi et al. 2008b
Cellulose (5)	[BMIM]Cl	CrCl$_3$·6H$_2$O (0.5 wt.%)	400	HMF	-	2 min 30 s	-	48	Zhang and Zhao 2010
Fructose	Water	HCl (2 M)	250	LA	170	30 min	-	31.8	Szabolcs et al. 2013
Glucose (5)	Water	HCl (2 M)	250	LA	170	30 min	-	31.4	Szabolcs et al. 2013
Cellobiose	Water	HCl (2 M)	250	LA	170	30 min	-	29.9	Szabolcs et al. 2013
Chitosan	Water	HCl (2 M)	250	LA	170	20 min	-	26.3	Szabolcs et al. 2013
Sweet sorghum juice (25)	-	H$_2$SO$_4$ (2 M)	135	LA	160	30 min	-	31.4	Novodárszki et al. 2014
Fructose (2.5)	[BMIM]Cl	Lignin-derived carbonaceous catalyst (10 wt.% of substrate)	100	HMF	110	10 min	100	59	Guo et al. 2012
Fructose (2)	Acetone-DMSO (7:3 w/w)	Sulphated zirconia (20 wt% of substrate)	700	HMF	180	5 min	91.3	46	Qi et al. 2009
Starch (5)	[BMIM]Cl-MIBK (1:1 v/v)	Zr(O)Cl$_2$ (10 mol%)	300	HMF	120	5 min	-	34	Saha et al. 2013
Glucose (5)	DMSO	AlCl$_3$ (50 mol%)	300	HMF	140	5 min	-	37	De et al. 2011
Sucrose (5)	DMSO	AlCl$_3$ (50 mol%)	300	HMF	140	5 min	-	33	De et al. 2011
Starch (5)	DMSO	AlCl$_3$ (50 mol%)	300	HMF	140	5 min	-	24	De et al. 2011

production costs when compared with conventional pretreatment. Other benefits such as energy efficiency (savings) due to shorter reaction times and selective recovery of target compounds can be expected due to unique capacities of microwave pretreatment. While these features are attractive, overheating and formation of inhibitory products should be avoided under microwave pretreatment. Microwave power output rate and reaction time optimization should be the key to maximize the sugar recovery and to minimize the byproduct formation. Pilot-scale and large-scale demonstrations are required prior to their commercialization.

References

Abbas, A. and Ansumali, S. 2010. Global potential of rice husk as a renewable feedstock for ethanol biofuel production. *Bioenergy Research*. 3(4): pp.328–334.

Ai, J. and Tschirner, U. 2010. Fiber length and pulping characteristics of switchgrass, alfalfa stems, hybrid poplar and willow biomasses. *Bioresource Technology*. 101(1): pp.215–221.

Alden, H., Espenls, B.-G. and Rensfelt. E. 1988. pp. 987–1001. *In*: A. V. Bridgwater and J. L. Kuester (eds.). Research in Thermochemical Biomass Conversion. Elsevier. New York.

Alvira, P., Tomás-Pejó, E., Ballesteros, M. and Negro, M.J. 2010. Pretreatment technologies for an efficient bioethanol production process based on enzymatic hydrolysis: a review. *Bioresource Technology*. 101(13): pp.4851–4861.

Anastyuk, S.D., Shevchenko, N.M., Dmitrenok, P.S. and Zvyagintseva, T.N. 2012a. Structural similarities of fucoidans from brown algae Silvetia babingtonii and Fucus evanescens, determined by tandem MALDI-TOF mass spectrometry. *Carbohydrate Research*. 358: pp.78–81.

Anastyuk, S.D., Shevchenko, N.M., Ermakova, S.P., Vishchuk, O.S., Nazarenko, E.L., Dmitrenok, P.S. and Zvyagintseva, T.N. 2012b. Anticancer activity *in vitro* of a fucoidan from the brown alga Fucus evanescens and its low-molecular fragments, structurally characterized by tandem mass-spectrometry. *Carbohydrate Polymers*. 87(1): pp.186–194.

Anderson, A.K. and Guraya, H.S. 2006. Effects of microwave heat-moisture treatment on properties of waxy and non-waxy rice starches. *Food Chemistry*. 97(2): pp.318–323.

Araque, E., Parra, C., Freer, J., Contreras, D., Rodríguez, J., Mendonça, R. and Baeza, J. 2008. Evaluation of organosolv pretreatment for the conversion of Pinus radiata D. Don to ethanol. *Enzyme and Microbial Technology*. 43(2): pp.214–219.

Azuma, J.I., Higashino, J., Isaka, M. and Koshijima, T. 1985. <Original> Microwave Irradiation of Lignocellulosic Materials: IV. *Enhancement of Enzymatic Susceptibility of Microwave-irradiated Softwoods*.

Badger, P.C. 2002. Ethanol from cellulose: A general review. *Trends in New Crops and New Uses*. 14: pp.17–21.

Balat, M. and Balat, H. 2009. Recent trends in global production and utilization of bio-ethanol fuel. *Applied Energy*. 86(11): pp.2273–2282.

Benoit, M., Rodrigues, A., Vigier, K.D.O., Fourré, E., Barrault, J., Tatibouët, J.M. and Jérôme, F. 2012. Combination of ball-milling and non-thermal atmospheric plasma as physical treatments for the saccharification of microcrystalline cellulose. *Green Chemistry*. 14(8): pp.2212–2215.

Berg, C. 2001. World bioethanol production, the distillery and bioethanol network. *línea: http://www. distill. com/worldethanolproduction. ht m (Mayo 2012)*.

Bhattacharya, D., Germinario, L.T. and Winter, W.T. 2008. Isolation, preparation and characterization of cellulose microfibers obtained from bagasse. *Carbohydrate Polymers*. 73(3): pp.371–377.

Bidzińska, E., Dyrek, K.R.Y.S.T.Y.N.A. and Wenda, E.L.Ż.B.I.E.T.A. 2010. Electron paramagnetic resonance study of thermally generated radicals in native and modified starches. *Current Topics in Biophysics*. 33(Suppl. A): pp.21–25.

Binod, P., Sindhu, R., Singhania, R.R., Vikram, S., Devi, L., Nagalakshmi, S., Kurien, N., Sukumaran, R.K. and Pandey, A. 2010. Bioethanol production from rice straw: an overview. *Bioresource Technology*. 101(13): pp.4767–4774.

Binod, P., Satyanagalakshmi, K., Sindhu, R., Janu, K.U., Sukumaran, R.K. and Pandey, A. 2012. Short duration microwave assisted pretreatment enhances the enzymatic saccharification and fermentable sugar yield from sugarcane bagasse. *Renewable Energy*. 37(1): pp.109–116.

Bjerre, A.B., Olesen, A.B., Fernqvist, T., Plöger, A. and Schmidt, A.S. 1996. Pretreatment of wheat straw using combined wet oxidation and alkaline hydrolysis resulting in convertible cellulose and hemicellulose. *Biotechnology and Bioengineering*. 49(5): pp.568–577.

Bobleter, O. 1998. Hydrothermal degradation and fractionation of saccharides and polysaccharides. *ChemInform*. 29(52).

Bobleter, O., Niesner, R. and Röhr, M. 1976. The hydrothermal degradation of cellulosic matter to sugars and their fermentative conversion to protein. *Journal of Applied Polymer Science*. 20(8): pp.2083–2093.

Boischot, C., Moraru, C.I. and Kokini, J.L. 2003. Factors that influence the microwave expansion of glassy amylopectin extrudates. *Cereal Chemistry*. 80(1): pp.56–61.

Boonmanumsin, P., Treeboobpha, S., Jeamjumnunja, K., Luengnaruemitchai, A., Chaisuwan, T. and Wongkasemjit, S. 2012. Release of monomeric sugars from Miscanthus sinensis by microwave-assisted ammonia and phosphoric acid treatments. *Bioresource Technology*. 103(1): pp.425–431.

Brașoveanu, M. and Nemțanu, M.R. 2014. Behaviour of starch exposed to microwave radiation treatment. *Starch-Stärke*. 66(1-2): pp.3–14.

Bredwell, M.D., Srivastava, P. and Worden, R.M., 1999. Reactor design issues for synthesis gas fermentations. Biotechnology progress, 15(5), pp.834-844.

Brodeur, G., Yau, E., Badal, K., Collier, J., Ramachandran, K.B. and Ramakrishnan, S. 2011. Chemical and physicochemical pretreatment of lignocellulosic biomass: a review. *Enzyme Research*.

Brosse, N., Dufour, A., Meng, X., Sun, Q. and Ragauskas, A. 2012. Miscanthus: a fast-growing crop for biofuels and chemicals production. *Biofuels, Bioproducts and Biorefining*. 6(5): pp.580–598.

Budarin, V.L., Clark, J.H., Lanigan, B.A., Shuttleworth, P., Breeden, S.W., Wilson, A.J., Macquarrie, D.J., Milkowski, K., Jones, J., Bridgeman, T. and Ross, A. 2009. The preparation of high-grade bio-oils through the controlled, low temperature microwave activation of wheat straw. *Bioresource Technology*. 100(23): pp.6064–6068.

Budarin, V.L., Clark, J.H., Lanigan, B.A., Shuttleworth, P. and Macquarrie, D.J. 2010. Microwave assisted decomposition of cellulose: a new thermochemical route for biomass exploitation. *Bioresource Technology*. 101(10): pp.3776–3779.

Cadoche, L. and López, G.D. 1989. Assessment of size reduction as a preliminary step in the production of ethanol from lignocellulosic wastes. *Biological Wastes*. 30(2): pp.153–157.

Camacho, F., Gonzalez-Tello, P., Jurado, E. and Robles, A. 1996. Microcrystalline-cellulose hydrolysis with concentrated sulphuric acid. *Journal of Chemical Technology and Biotechnology*. 67(4): pp.350–356.

Caparrós, S., Ariza, J., López, F., Nacimiento, J.A., Garrote, G. and Jiménez, L. 2008. Hydrothermal treatment and ethanol pulping of sunflower stalks. *Bioresource Technology*. 99(5): pp.1368–1372.

Carvalheiro, F., Garrote, G., Parajó, J.C., Pereira, H. and Gírio, F.M. 2005. Kinetic modeling of breweryapos; spent grain autohydrolysis. *Biotechnology Progress*. 21(1): pp.233–243.

Castro, F.B. and Machado, P.F. 1989. Digestive process evaluation of steam and pressure treated sugarcane bagasse. *Bol. Ind. Anim.* 46: pp.213–217.

Chandra, R.P., Bura, R., Mabee, W.E., Berlin, D.A., Pan, X. and Saddler, J.N. 2007. Substrate pretreatment: The key to effective enzymatic hydrolysis of lignocellulosics?. In *Biofuels*. pp.67–93. Springer Berlin Heidelberg.

Chen, X.Q. 2005. Microwave-assisted extraction of polysaccharides from Solanum nigrum. *Journal of Central South University of Technology*. 12(5): pp.556–560.

Chen, X.Q. 2005. Microwave-assisted extraction of polysaccharides from Solanum nigrum. *Journal of Central South University of Technology*. 12(5): pp.556–560.

Chen, F. and Dixon, R.A. 2007. Lignin modification improves fermentable sugar yields for biofuel production. *Nature Biotechnology*. 25(7): pp.759–761.

Chen, W.H., Tu, Y.J. and Sheen, H.K. 2011. Disruption of sugarcane bagasse lignocellulosic structure by means of dilute sulfuric acid pretreatment with microwave-assisted heating. *Applied Energy*. 88(8): pp.2726–2734.

Chen, C., Boldor, D., Aita, G. and Walker, M. 2012a. Ethanol production from sorghum by a microwave-assisted dilute ammonia pretreatment. *Bioresource Technology*. 110: pp.190–197.

Chen, W.H., Ye, S.C. and Sheen, H.K. 2012b. Hydrolysis characteristics of sugarcane bagasse pretreated by dilute acid solution in a microwave irradiation environment. *Applied Energy*. 93: pp.237–244.

Chen, X., Tao, L., Shekiro, J., Mohaghaghi, A., Decker, S., Wang, W., Smith, H., Park, S., Himmel, M.E. and Tucker, M. 2012c. Improved ethanol yield and reduced Minimum Ethanol Selling Price (MESP) by modifying low severity dilute acid pretreatment with deacetylation and mechanical refining: 1 Experimental. *Biotechnology for Biofuels*. 5(1): p.1.

Cheng, Y.S., Zheng, Y., Yu, C.W., Dooley, T.M., Jenkins, B.M. and VanderGheynst, J.S. 2010. Evaluation of high solids alkaline pretreatment of rice straw. *Applied Biochemistry and Biotechnology*. 162(6): pp.1768–1784.

Cherubini, F. and Ulgiati, S. 2010. Crop residues as raw materials for biorefinery systems–A LCA case study. *Applied Energy*. 87(1): pp.47–57.

Chisti, Y. 2007. Biodiesel from microalgae. *Biotechnology Advances*. 25(3): pp.294–306.

Chum, H.L., Johnson, D.K., Black, S.K. and Overend, R.P. 1990. Pretreatment-catalyst effects and the combined severity parameter. *Applied Biochemistry and Biotechnology*. 24(1): pp.1–14.

Cybulska, I., Lei, H. and Julson, J. 2009. Hydrothermal pretreatment and enzymatic hydrolysis of prairie cord grass. *Energy & Fuels*. 24(1): pp.718–727.

David, K. and Ragauskas, A.J. 2010. Switchgrass as an energy crop for biofuel production: a review of its ligno-cellulosic chemical properties. *Energy & Environmental Science*. 3(9): pp.1182–1190.

de Castro, F.B. 1994. *The Use of Steam Treatment to Upgrade Lignocellulosic Materials for Animal Feed* (Doctoral dissertation, University of Aberdeen).

de Castro, F.B. 1989. Evaluation of the digestion process of auto-hydrolyzed sugarcane bagasse in bovines.

De, S., Dutta, S. and Saha, B. 2011. Microwave assisted conversion of carbohydrates and biopolymers to 5-hydroxymethylfurfural with aluminium chloride catalyst in water. *Green Chemistry*. 13(10): pp.2859–2868.

Dhouib, A., Ellouz, M., Aloui, F. and Sayadi, S. 2006. Effect of bioaugmentation of activated sludge with white-rot fungi on olive mill wastewater detoxification. *Letters in Applied Microbiology*. 42(4): pp.405–411.

Díaz, M.J., Cara, C., Ruiz, E., Romero, I., Moya, M. and Castro, E. 2010. Hydrothermal pre-treatment of rapeseed straw. *Bioresource Technology*. 101(7): pp.2428–2435.

Dobele, G., Dizhbite, T., Rossinskaja, G., Telysheva, G., Meier, D., Radtke, S. and Faix, O. 2003. Pre-treatment of biomass with phosphoric acid prior to fast pyrolysis: a promising method for obtaining 1, 6-anhydrosaccharides in high yields. *Journal of Analytical and Applied Pyrolysis*. 68: pp.197–211.

Duff, S.J. and Murray, W.D. 1996. Bioconversion of forest products industry waste cellulosics to fuel ethanol: a review. *Bioresource Technology*. 55(1): pp.1–33.

Dumitriu, S. 2004. *Polysaccharides: Structural Diversity and Functional Versatility*. CRC press.

Duraiswamy, K., Colaninno, J. and Mansour, M.N. 1989. Energy from Biomass and Wastes XII, DL Klass ed.. Inst. Gas Tech., Chicago, IL, p.833.

Dyrek, K., Bidzińska, E., Łabanowska, M., Fortuna, T., Przetaczek, I. and Pietrzyk, S. 2007. EPR study of radicals generated in starch by microwaves or by conventional heating. *Starch-Stärke.* 59(7): pp.318–325.

Eggeman, T. and Elander, R.T. 2005. Process and economic analysis of pretreatment technologies. *Bioresource Technology.* 96(18): pp.2019–2025.

Elliott, D.C. 2004. *Biomass, chemicals from* (No. PNNL-SA-36685). Pacific Northwest National Laboratory (PNNL), Richland, WA (US).

Emami, S., Perera, A., Meda, V. and Tyler, R.T. 2012. Effect of microwave treatment on starch digestibility and physicochemical properties of three barley types. *Food and Bioprocess Technology.* 5(6): pp.2266–2274.

Faga, B.A., Wilkins, M.R. and Banat, I.M. 2010. Ethanol production through simultaneous saccharification and fermentation of switchgrass using Saccharomyces cerevisiae D 5 A and thermotolerant Kluyveromyces marxianus IMB strains. *Bioresource Technology.* 101(7): pp.2273–2279.

Fan, D., Ma, W., Wang, L., Huang, J., Zhao, J., Zhang, H. and Chen, W. 2012. Determination of structural changes in microwaved rice starch using Fourier transform infrared and Raman spectroscopy. *Starch-Stärke.* 64(8): pp.598–606.

Fan, D., Ma, W., Wang, L., Huang, J., Zhang, F., Zhao, J., Zhang, H. and Chen, W. 2013a. Determining the effects of microwave heating on the ordered structures of rice starch by NMR. *Carbohydrate Polymers.* 92(2): pp.1395–1401.

Fan, D., Wang, L., Ma, S., Ma, W., Liu, X., Huang, J., Zhao, J., Zhang, H. and Chen, W. 2013b. Structural variation of rice starch in response to temperature during microwave heating before gelatinisation. *Carbohydrate Polymers.* 92(2): pp.1249–1255.

Fan, J., De Bruyn, M., Budarin, V.L., Gronnow, M.J., Shuttleworth, P.S., Breeden, S., Macquarrie, D.J. and Clark, J.H. 2013c. Direct microwave-assisted hydrothermal depolymerization of cellulose. *Journal of the American Chemical Society.* 135(32): pp.11728–11731.

Farrell, A.E., Plevin, R.J., Turner, B.T., Jones, A.D., O'hare, M. and Kammen, D.M. 2006. Ethanol can contribute to energy and environmental goals. *Science.* 311(5760): pp.506–508.

Faulon, J.L., Carlson, G.A. and Hatcher, P.G. 1994. A three-dimensional model for lignocellulose from gymnospermous wood. *Organic Geochemistry.* 21(12): pp.1169–1179.

Feria, M.J., López, F., García, J.C., Pérez, A., Zamudio, M.A. and Alfaro, A. 2011. Valorization of Leucaena leucocephala for energy and chemicals from autohydrolysis. *Biomass and Bioenergy.* 35(5): pp.2224–2233.

Fortuna, T., Przetaczek, I., Dyrek, K., Bidzińska, E. and Łabanowska, M. 2008. Some physicochemical properties of commercial modified starches irradiated with microwaves. *EJPAU.* 11(4): p.20.

Gabhane, J., William, S.P., Vaidya, A.N., Mahapatra, K. and Chakrabarti, T. 2011. Influence of heating source on the efficacy of lignocellulosic pretreatment–a cellulosic ethanol perspective. *Biomass and Bioenergy.* 35(1): pp.96–102.

García-Aparicio, M.P., Oliva, J.M., Manzanares, P., Ballesteros, M., Ballesteros, I., González, A. and Negro, M.J. 2011. Second-generation ethanol production from steam exploded barley straw by Kluyveromyces marxianus CECT 10875. *Fuel.* 90(4): pp.1624–1630.

Gammons, R. 2014. Optimising the pre-treatment effects of protic ionic liquids on lignocellulosic materials (Doctoral dissertation, University of York), York, UK.

Garrote, G., Dominguez, H. and Parajo, J.C. 1999. Hydrothermal processing of lignocellulosic materials. *European Journal of Wood and Wood Products.* 57(3): pp.191–202.

Garrote, G., Cruz, J.M., Domínguez, H. and Parajó, J.C. 2003. Valorisation of waste fractions from autohydrolysis of selected lignocellulosic materials. *Journal of Chemical Technology and Biotechnology.* 78(4): pp.392–398.

Garrote, G., Falqué, E., Domínguez, H. and Parajó, J.C. 2007. Autohydrolysis of agricultural residues: Study of reaction byproducts. *Bioresource Technology.* 98(10): pp.1951–1957.

Garrote, G., Cruz, J.M., Domínguez, H. and Parajó, J.C. 2008. Non-isothermal autohydrolysis of barley husks: Product distribution and antioxidant activity of ethyl acetate soluble fractions. *Journal of Food Engineering.* 84(4): pp.544–552.

Goh, C.S., Lee, K.T. and Bhatia, S. 2010. Hot compressed water pretreatment of oil palm fronds to enhance glucose recovery for production of second generation bio-ethanol. *Bioresource Technology.* 101(19): pp.7362–7367.

Gong, G., Liu, D. and Huang, Y. 2010. Microwave-assisted organic acid pretreatment for enzymatic hydrolysis of rice straw. *Biosystems Engineering.* 107(2): pp.67–73.

Gonzalez, R., Treasure, T., Phillips, R., Jameel, H., Saloni, D., Abt, R. and Wright, J. 2011. Converting Eucalyptus biomass into ethanol: Financial and sensitivity analysis in a co-current dilute acid process. Part II. *Biomass and Bioenergy.* 35(2): pp.767–772.

González-López, N., Moure, A. and Domínguez, H. 2012. Hydrothermal fractionation of Sargassum muticum biomass. *Journal of Applied Phycology.* 24(6): pp.1569–1578.

Goshadrou, A., Karimi, K. and Taherzadeh, M.J. 2011. Bioethanol production from sweet sorghum bagasse by Mucor hiemalis. *Industrial Crops and Products.* 34(1): pp.1219–1225.

Grohmann, K., Torget, R., Himmel, M. and Scott, C.D. 1986, January. Dilute acid pretreatment of biomass at high acid concentrations. In *Biotechnol. Bioeng. Symp.*;(United States) (Vol. 17, No. CONF-860508-). Solar Energy Research Institute, Golden, CO 80401, USA. Solar Fuels Research Div.

Gulati, M., Kohlmann, K., Ladisch, M.R., Hespell, R. and Bothast, R.J. 1996. Assessment of ethanol production options for corn products. *Bioresource Technology.* 58(3): pp.253–264.

Gullón, B., Yáñez, R., Alonso, J.L. and Parajó, J.C. 2010. Production of oligosaccharides and sugars from rye straw: a kinetic approach. *Bioresource Technology.* 101(17): pp.6676–6684.

Guo, F., Fang, Z. and Zhou, T.J. 2012. Conversion of fructose and glucose into 5-hydroxymethylfurfural with lignin-derived carbonaceous catalyst under microwave irradiation in dimethyl sulfoxide–ionic liquid mixtures. *Bioresource Technology.* 112: pp.313–318.

Harmsen, P., Huijgen, W., Bermudez, L. and Bakker, R. 2010. Literature review of physical and chemical pretreatment processes for lignocellulosic. *Biomass.* pp.1–49.

Hendriks, A.T.W.M. and Zeeman, G. 2009. Pretreatments to enhance the digestibility of lignocellulosic biomass. *Bioresource Technology.* 100(1): pp.10–18.

Henstra, A.M., Sipma, J., Rinzema, A. and Stams, A.J. 2007. Microbiology of synthesis gas fermentation for biofuel production. *Current Opinion in Biotechnology.* 18(3): pp.200–206.

Hodsagi, M., Jámbor, A., Juhász, E., Gergely, S., Gelencsér, T. and Salgó, A. 2012. Effects of microwave heating on native and resistant starches. *Acta Alimentaria.* 41(2): pp.233–247.

Hsu, T.C., Guo, G.L., Chen, W.H. and Hwang, W.S. 2010. Effect of dilute acid pretreatment of rice straw on structural properties and enzymatic hydrolysis. *Bioresource Technology.* 101(13): pp.4907–4913.

http://www.afdc.energy.gov/fuels/ethanol.html

http://www.ethanolrfa.org/resources/industry/statistics/

Hu, G., Heitmann, J.A. and Rojas, O.J. 2008. Feedstock pretreatment strategies for producing ethanol from wood, bark, and forest residues. *BioResources.* 3(1): pp.270–294.

Hu, Z. and Wen, Z. 2008. Enhancing enzymatic digestibility of switchgrass by microwave-assisted alkali pretreatment. *Biochemical Engineering Journal.* 38(3): pp.369–378.

Hu, Z., Wang, Y. and Wen, Z. 2008. Alkali (NaOH) pretreatment of switchgrass by radio frequency-based dielectric heating. *Applied Biochemistry and Biotechnology.* 148(1-3): pp.71–81.

Hu, Z., Foston, M. and Ragauskas, A.J. 2011. Comparative studies on hydrothermal pretreatment and enzymatic saccharification of leaves and internodes of alamo switchgrass. *Bioresource Technology.* 102(14): pp.7224–7228.

Huang, F. and Ragauskas, A. 2013. Extraction of hemicellulose from loblolly pine woodchips and subsequent kraft pulping. *Industrial & Engineering Chemistry Research.* 52(4): pp.1743–1749.

Huber, G.W., Iborra, S. and Corma, A. 2006. Synthesis of transportation fuels from biomass: chemistry, catalysts, and engineering. *Chemical Reviews.* 106(9): pp.4044–4098.

Huber, G.W. and Corma, A. 2007. Synergies between bio-and oil refineries for the production of fuels from biomass. *Angewandte Chemie International Edition.* 46(38): pp.7184–7201.

Ingram, T., Wörmeyer, K., Lima, J.C.I., Bockemühl, V., Antranikian, G., Brunner, G. and Smirnova, I. 2011. Comparison of different pretreatment methods for lignocellulosic materials. Part I: conversion of rye straw to valuable products. *Bioresource Technology.* 102(8): pp.5221–5228.

Jackowiak, D., Frigon, J.C., Ribeiro, T., Pauss, A. and Guiot, S. 2011. Enhancing solubilisation and methane production kinetic of switchgrass by microwave pretreatment. *Bioresource Technology.* 102(3): pp.3535–3540.

Kabel, M.A., Bos, G., Zeevalking, J., Voragen, A.G. and Schols, H.A. 2007. Effect of pretreatment severity on xylan solubility and enzymatic breakdown of the remaining cellulose from wheat straw. *Bioresource Technology.* 98(10): pp.2034–2042.

Kannan, T.S., Ahmed, A.S. and Farid Nasir, A. 2013, July. Energy efficient microwave irradiation of sago bark waste (SBW) for bioethanol production. *Advanced Materials Research.* 701: pp.249–253.

Karina, M., Tanahashi, M. and Higuchi, T. 1992. Degradation mechanism of lignin by a steam explosion, 4: Steam treatment of a dehydrogenative polymer of coniferyl alcohol. *Journal of the Japan Wood Research Society.*

Karunanithy, C. and Muthukumarappan, K. 2010. Effect of extruder parameters and moisture content of corn stover and big bluestem on sugar recovery from enzymatic hydrolysis. *Biological Engineering Transactions.* 2(2): pp.91–113.

Keshwani, D.R., Cheng, J.J., Burns, J.C., Li, L. and Chiang, V. 2007. Microwave pretreatment of switchgrass to enhance enzymatic hydrolysis. In *2007 ASAE Annual Meeting* (p. 1). American Society of Agricultural and Biological Engineers.

Keshwani, D.R. 2009. *Microwave Pretreatment of Switchgrass for Bioethanol Production.* ProQuest. Ph.D. Dissertation, North Carolina State University, Raleigh, North Carolina.

Keshwani, D.R. and Cheng, J.J. 2009. Switchgrass for bioethanol and other value-added applications: a review. *Bioresource Technology.* 100(4): pp.1515–1523.

Keshwani, D.R. and Cheng, J.J. 2010. Microwave-based alkali pretreatment of switchgrass and coastal bermudagrass for bioethanol production. *Biotechnology Progress.* 26(3): pp.644–652.

Kim, S. and Dale, B.E. 2004. Global potential bioethanol production from wasted crops and crop residues. *Biomass and Bioenergy.* 26(4): pp.361–375.

Kim, Y., Mosier, N.S. and Ladisch, M.R. 2009. Enzymatic digestion of liquid hot water pretreated hybrid poplar. *Biotechnology Progress.* 25(2): pp.340–348.

Kim, Y., Hendrickson, R., Mosier, N.S., Ladisch, M.R., Bals, B., Balan, V., Dale, B.E., Dien, B.S. and Cotta, M.A. 2010. Effect of compositional variability of distillers' grains on cellulosic ethanol production. *Bioresource Technology.* 101(14): pp.5385–5393.

Klasson, K.T., Ackerson, M.D., Clausen, E.C. and Gaddy, J.L., 1992. Bioconversion of synthesis gas into liquid or gaseous fuels. Enzyme and Microbial Technology, 14(8), pp.602-608.

Klinke, H.B., Ahring, B.K., Schmidt, A.S. and Thomsen, A.B. 2002. Characterization of degradation products from alkaline wet oxidation of wheat straw. *Bioresource Technology.* 82(1): pp.15–26.

Komolwanich, T., Tatijarern, P., Prasertwasu, S., Khumsupan, D., Chaisuwan, T., Luengnaruemitchai, A. and Wongkasemjit, S. 2014. Comparative potentiality of Kans grass (Saccharum spontaneum) and Giant reed (Arundo donax) as lignocellulosic feedstocks for the release of monomeric sugars by microwave/chemical pretreatment. *Cellulose.* 21(3): pp.1327–1340.

Krassig, H. and Schurz, J. 2002. Cellulose, Ullmann's Encyclopedia of Industrial Chemistry.

Kumar, P., Barrett, D.M., Delwiche, M.J. and Stroeve, P. 2009. Methods for pretreatment of lignocellulosic biomass for efficient hydrolysis and biofuel production. *Industrial & Engineering Chemistry Research.* 48(8): pp.3713–3729.

Kurakake, M., Ide, N. and Komaki, T. 2007. Biological pretreatment with two bacterial strains for enzymatic hydrolysis of office paper. *Current Microbiology.* 54(6): pp.424–428.

Kurian, J.K., Gariepy, Y., Orsat, V. and Raghavan, G.V. 2015. Comparison of steam-assisted versus microwave-assisted treatments for the fractionation of sweet sorghum bagasse. *Bioresources and Bioprocessing.* 2(1): pp.1–16.

Lebo, S.E., Gargulak, J.D. and McNally, T.J. 2001. *Lignin, Kirk-Othmer Encyclopedia of Chemical Technology.* John Wiley & Sons, Inc. doi: 10(1002), p.0471238961.

Lee, D.H., Cho, E.Y., Kim, C.J. and Kim, S.B. 2010. Pretreatment of waste newspaper using ethylene glycol for bioethanol production. *Biotechnology and Bioprocess Engineering.* 15(6): pp.1094–1101.

Lee, H.V., Hamid, S.B.A. and Zain, S.K. 2014. Conversion of lignocellulosic biomass to nanocellulose: structure and chemical process. *The Scientific World Journal, 2014.*

Lewandowicz, G., Fornal, J. and Walkowski, A. 1997. Effect of microwave radiation on physico-chemical properties and structure of potato and tapioca starches. *Carbohydrate Polymers.* 34(4): pp.213–220.

Lewandowicz, G., Jankowski, T. and Fornal, J. 2000. Effect of microwave radiation on physico-chemical properties and structure of cereal starches. *Carbohydrate Polymers.* 42(2): pp.193–199.

Liang, Y., Siddaramu, T., Yesuf, J. and Sarkany, N. 2010. Fermentable sugar release from Jatropha seed cakes following lime pretreatment and enzymatic hydrolysis. *Bioresource Technology.* 101(16): pp.6417–6424.

Liu, C. and Wyman, C.E. 2005. Partial flow of compressed-hot water through corn stover to enhance hemicellulose sugar recovery and enzymatic digestibility of cellulose. *Bioresource Technology.* 96(18): pp.1978–1985.

Liu, C.Z. and Cheng, X.Y. 2010. Improved hydrogen production via thermophilic fermentation of corn stover by microwave-assisted acid pretreatment. *International Journal of Hydrogen Energy.* 35(17): pp.8945–8952.

Liu, K., Lin, X., Yue, J., Li, X., Fang, X., Zhu, M., Lin, J., Qu, Y. and Xiao, L. 2010. High concentration ethanol production from corncob residues by fed-batch strategy. *Bioresource Technology.* 101(13): pp.4952–4958.

Lü, J. and Zhou, P. 2011. Optimization of microwave-assisted FeCl 3 pretreatment conditions of rice straw and utilization of Trichoderma viride and Bacillus pumilus for production of reducing sugars. *Bioresource Technology.* 102(13): pp.6966–6971.

Lu, X., Zhang, Y. and Angelidaki, I. 2009. Optimization of H_2SO_4-catalyzed hydrothermal pretreatment of rapeseed straw for bioconversion to ethanol: focusing on pretreatment at high solids content. *Bioresource Technology.* 100(12): pp.3048–3053.

Lu, X., Xi, B., Zhang, Y. and Angelidaki, I. 2011. Microwave pretreatment of rape straw for bioethanol production: focus on energy efficiency. *Bioresource Technology.* 102(17): pp.7937–7940.

Luo, Z., He, X., Fu, X., Luo, F. and Gao, Q. 2006. Effect of microwave radiation on the physicochemical properties of normal maize, waxy maize and amylomaize V starches. *Starch-Stärke.* 58(9): pp.468–474.

Ma, H., Liu, W.W., Chen, X., Wu, Y.J. and Yu, Z.L. 2009. Enhanced enzymatic saccharification of rice straw by microwave pretreatment. *Bioresource Technology.* 100(3): pp.1279–1284.

MacDonald, D.G., Bakhshi, N.N., Mathews, J.F., Roychowdhury, A., Bajpai, P. and Moo-Young, M. 1983. Alkali treatment of corn stover to improve sugar production by enzymatic hydrolysis. *Biotechnology and Bioengineering.* 25(8): pp.2067–2076.

Maeda, R.N., Serpa, V.I., Rocha, V.A.L., Mesquita, R.A.A., Santa Anna, L.M.M., De Castro, A.M., Driemeier, C.E., Pereira, N. and Polikarpov, I. 2011. Enzymatic hydrolysis of pretreated sugar cane bagasse using Penicillium funiculosum and Trichoderma harzianum cellulases. *Process Biochemistry.* 46(5): pp.1196–1201.

Maiorella, B., Blanch, H.W. and Wilke, C.R. 1983. By-product inhibition effects on ethanolic fermentation by saccharomyces cerevisiae. *Biotechnology and Bioengineering.* 25(1): pp.103–121.

Mamleev, V., Bourbigot, S., Le Bras, M. and Yvon, J. 2009. The facts and hypotheses relating to the phenomenological model of cellulose pyrolysis: Interdependence of the steps. *Journal of Analytical and Applied Pyrolysis.* 84(1): pp.1–17.

Martin-Mingot, A., Compain, G., Liu, F., Jouannetaud, M.P., Bachmann, C., Frapper, G. and Thibaudeau, S. 2012. Dications in superacid HF/SbF 5: When superelectrophilic activation makes possible fluorination and/or C–H bond activation. *Journal of Fluorine Chemistry.* 134: pp.56–62.

McIntosh, S. and Vancov, T. 2010. Enhanced enzyme saccharification of Sorghum bicolor straw using dilute alkali pretreatment. *Bioresource Technology.* 101(17): pp.6718–6727.

McParland, J.J., Grethlein, H.E. and Converse, A.O. 1982. Kinetics of acid hydrolysis of corn stover. *Solar Energy.* 28(1): pp.55–63.

Menon, V. and Rao, M. 2012. Trends in bioconversion of lignocellulose: biofuels, platform chemicals & biorefinery concept. *Progress in Energy and Combustion Science.* 38(4): pp.522–550.

Mesa, L., González, E., Ruiz, E., Romero, I., Cara, C., Felissia, F. and Castro, E. 2010. Preliminary evaluation of organosolv pre-treatment of sugar cane bagasse for glucose production: Application of 2 3 experimental design. *Applied Energy.* 87(1): pp.109–114.

Mesa-Pérez, J.M., Rocha, J.D., Barbosa-Cortez, L.A., Penedo-Medina, M., Luengo, C.A. and Cascarosa, E. 2013. Fast oxidative pyrolysis of sugar cane straw in a fluidized bed reactor. *Applied Thermal Engineering.* 56(1): pp.167–175.

Metaxas, A.C. and Meredith, R.J. 1983. Industrial Microwave Heating"-Peter Peregrinus LTD (IEE). *London (UK).*

Michalowicz, G., Toussaint, B. and Vignon, M.R. 1991. Ultrastructural changes in poplar cell wall during steam explosion treatment. *Holzforschung.* 45: pp.175–179.

Mollekopf, N., Treppe, K., Fiala, P. and Dixit, O. 2011. Vacuum microwave treatment of potato starch and the resultant modification of properties. *Chemie Ingenieur Technik.* 83(3): pp.262–272.

Moraru, C.I. and Kokini, J.L. 2003. Nucleation and expansion during extrusion and microwave heating of cereal foods. *Comprehensive Reviews in Food Science and Food Safety.* 2(4): pp.147–165.

Mosier, N., Hendrickson, R., Ho, N., Sedlak, M. and Ladisch, M.R. 2005. Optimization of pH controlled liquid hot water pretreatment of corn stover. *Bioresource Technology.* 96(18): pp.1986–1993.

Motwani, T., Seetharaman, K. and Anantheswaran, R.C. 2007. Dielectric properties of starch slurries as influenced by starch concentration and gelatinization. *Carbohydrate Polymers.* 67(1): pp.73–79.

Mukherjee, A., Dumont, M.J. and Raghavan, V. 2015. Review: Sustainable production of hydroxymethylfurfural and levulinic acid: Challenges and opportunities. *Biomass and Bioenergy.* 72: pp.143–183.

National Research Council (US). Committee on Microwave Processing of Materials, An Emerging Industrial Technology, National Research Council (US). National Materials Advisory Board, National Research Council (US). Commission on Engineering and Technical Systems, 1994. *Microwave processing of Materials* (Vol. 473). National Academies Press.

Navarro, D.A. and Stortz, C.A. 2005. Microwave-assisted alkaline modification of red seaweed galactans. *Carbohydrate Polymers.* 62(2): pp.187–191.

Navarro, D.A. and Stortz, C.A. 2005. Microwave-assisted alkaline modification of red seaweed galactans. *Carbohydrate polymers.* 62(2): pp.187–191.

Ndife, M.K., Şumnu, G. and Bayindirli, L. 1998. Dielectric properties of six different species of starch at 2450 MHz. *Food Research International.* 31(1): pp.43–52.

Nomanbhay, S.M., Hussain, R. and Palanisamy, K. 2013. Microwave-assisted alkaline pretreatment and microwave assisted enzymatic saccharification of oil palm empty fruit bunch fiber for enhanced fermentable sugar yield. DOI:10.4236/jsbs.2013.31002.

Novodárszki, G., Rétfalvi, N., Dibó, G., Mizsey, P., Cséfalvay, E. and Mika, L.T. 2014. Production of platform molecules from sweet sorghum. *RSC Advances.* 4(4): pp.2081–2088.

NÚÑez, H.M., Rodriguez, L.F. and Khanna, M. 2011. Agave for tequila and biofuels: an economic assessment and potential opportunities. *Gcb Bioenergy.* 3(1): pp.43–57.

Octave, S. and Thomas, D. 2009. Biorefinery: toward an industrial metabolism. *Biochimie.* 91(6): pp.659–664.

Oliva, J.M., Sáez, F., Ballesteros, I., González, A., Negro, M.J., Manzanares, P. and Ballesteros, M. 2003. Effect of lignocellulosic degradation compounds from steam explosion pretreatment on ethanol fermentation by thermotolerant yeast Kluyveromyces marxianus. *Biotechnology for Fuels and Chemicals.* pp.141–153. Humana Press.

Outlook, A.E. 2007. With Projections to 2030. US Energy Information Administration.

Palmarola-Adrados, B., Galbe, M. and Zacchi, G. 2004. Combined steam pretreatment and enzymatic hydrolysis of starch-free wheat fibers. *Proceedings of the Twenty-Fifth Symposium on Biotechnology for Fuels and Chemicals Held May 4–7, 2003, in Breckenridge, CO.* pp.989–1002. Humana Press.

Palmarola-Adrados, B., Galbe, M. and Zacchi, G. 2005. Pretreatment of barley husk for bioethanol production. *Journal of Chemical Technology and Biotechnology.* 80(1): pp.85–91.

Palmqvist, E. and Hahn-Hägerdal, B. 2000. Fermentation of lignocellulosic hydrolysates. II: inhibitors and mechanisms of inhibition. *Bioresource Technology.* 74(1): pp.25–33.

Pan, X., Gilkes, N., Kadla, J., Pye, K., Saka, S., Gregg, D., Ehara, K., Xie, D., Lam, D. and Saddler, J. 2006. Bioconversion of hybrid poplar to ethanol and co-products using an organosolv fractionation process: Optimization of process yields. *Biotechnology and Bioengineering.* 94(5): pp.851–861.

Pandey, K.K. 2005. Study of the effect of photo-irradiation on the surface chemistry of wood. *Polymer Degradation and Stability.* 90(1): pp.9–20.

Pang, F., Xue, S., Yu, S., Zhang, C., Li, B. and Kang, Y. 2012. Effects of microwave power and microwave irradiation time on pretreatment efficiency and characteristics of corn stover using combination of steam explosion and microwave irradiation (SE–MI) pretreatment. *Bioresource Technology.* 118: pp.111–119.

Pang, F., Xue, S., Yu, S., Zhang, C., Li, B. and Kang, Y. 2013. Effects of combination of steam explosion and microwave irradiation (SE–MI) pretreatment on enzymatic hydrolysis, sugar yields and structural properties of corn stover. *Industrial Crops and Products.* 42: pp.402–408.

Parrish, D.J. and Fike, J.H. 2005. The biology and agronomy of switchgrass for biofuels. *BPTS.* 24(5-6): pp.423–459.

Pedersen, M. and Meyer, A.S. 2010. Lignocellulose pretreatment severity–relating pH to biomatrix opening. *New Biotechnology.* 27(6): pp.739–750.

Peng, L. and Chen, Y. 2011. Conversion of paper sludge to ethanol by separate hydrolysis and fermentation (SHF) using Saccharomyces cerevisiae. *Biomass and Bioenergy.* 35(4): pp.1600–1606.

Pérez, J.A., Ballesteros, I., Ballesteros, M., Sáez, F., Negro, M.J. and Manzanares, P. 2008. Optimizing liquid hot water pretreatment conditions to enhance sugar recovery from wheat straw for fuel-ethanol production. *Fuel.* 87(17): pp.3640–3647.

Petersen, M.Ø., Larsen, J. and Thomsen, M.H. 2009. Optimization of hydrothermal pretreatment of wheat straw for production of bioethanol at low water consumption without addition of chemicals. *Biomass and Bioenergy.* 33(5): pp.834–840.

Pires, E.J., Ruiz, H.A., Teixeira, J.A. and Vicente, A.A. 2012. A new approach on Brewer's spent grains treatment and potential use as lignocellulosic yeast cells carriers. *Journal of Agricultural and Food Chemistry.* 60(23): pp.5994–5999.

Piyasena, P., Ramaswamy, H.S., Awuah, G.B. and Defelice, C. 2003. Dielectric properties of starch solutions as influenced by temperature, concentration, frequency and salt. *Journal of Food Process Engineering.* 26(1): pp.93–119.

Pronyk, C., Mazza, G. and Tamaki, Y. 2011. Production of carbohydrates, lignins, and minor components from triticale straw by hydrothermal treatment. *Journal of Agricultural and Food Chemistry.* 59(8): pp.3788–3796.

Qi, X., Watanabe, M., Aida, T.M. and Smith, Jr, R.L. 2008a. Selective conversion of D-fructose to 5-hydroxymethylfurfural by ion-exchange resin in acetone/dimethyl sulfoxide solvent mixtures. *Industrial & Engineering Chemistry Research.* 47(23): pp.9234–9239.

Qi, X., Watanabe, M., Aida, T.M. and Smith, R.L. 2008b. Catalytical conversion of fructose and glucose into 5-hydroxymethylfurfural in hot compressed water by microwave heating. *Catalysis Communications.* 9(13): pp.2244–2249.
Qi, X., Watanabe, M., Aida, T.M. and Smith, R.L. 2009. Sulfated zirconia as a solid acid catalyst for the dehydration of fructose to 5-hydroxymethylfurfural. *Catalysis Communications.* 10(13): pp.1771–1775.
Ragauskas, A.J., Beckham, G.T., Biddy, M.J., Chandra, R., Chen, F., Davis, M.F., Davison, B.H., Dixon, R.A., Gilna, P., Keller, M. and Langan, P. 2014. Lignin valorization: improving lignin processing in the biorefinery. *Science.* 344(6185): p.1246843.
Ramos, L.P. 2003. The chemistry involved in the steam treatment of lignocellulosic materials. *Química Nova.* 26(6): pp.863–871.
Ren, H., Zhou, Y. and Liu, L. 2013. Selective conversion of cellulose to levulinic acid via microwave-assisted synthesis in ionic liquids. *Bioresource Technology.* 129: pp.616–619.
Rodriguez-Jasso, R.M., Mussatto, S.I., Pastrana, L., Aguilar, C.N. and Teixeira, J.A. 2011. Microwave-assisted extraction of sulfated polysaccharides (fucoidan) from brown seaweed. *Carbohydrate Polymers.* 86(3): pp.1137–1144.
Rodríguez-Jasso, R.M., Mussatto, S.I., Pastrana, L., Aguilar, C.N. and Teixeira, J.A. 2013. Extraction of sulfated polysaccharides by autohydrolysis of brown seaweed Fucus vesiculosus. *Journal of Applied Phycology.* 25(1): pp.31–39.
Roebuck, B.D., Goldblith, S.A. and Westphal, W.B. 1972. Dielectric properties of carbohydrate-water mixtures at microwave frequencies. *Journal of Food Science.* 37(2): pp.199–204.
Roesijadi, G., Jones, S.B., Snowden-Swan, L.J. and Zhu, Y. 2010. Macroalgae as a biomass feedstock: a preliminary analysis, PNNL 19944. *Pacific Northwest National Laboratory, Richland.*
Romaní, A., Garrote, G., Alonso, J.L. and Parajó, J.C. 2010. Bioethanol production from hydrothermally pretreated Eucalyptus globulus wood. *Bioresource Technology.* 101(22): pp.8706–8712.
Romaní, A., Garrote, G., López, F. and Parajó, J.C. 2011. Eucalyptus globulus wood fractionation by autohydrolysis and organosolv delignification. *Bioresource Technology.* 102(10): pp.5896–5904.
Romaní, A., Garrote, G. and Parajó, J.C. 2012. Bioethanol production from autohydrolyzed Eucalyptus globulus by Simultaneous Saccharification and Fermentation operating at high solids loading. *Fuel.* 94: pp.305–312.
Roy Goswami, S., Dumont, M.J. and Raghavan, V. 2015. Starch to value added biochemicals. *Starch-Stärke.*
Ruiz, E., Cara, C., Manzanares, P., Ballesteros, M. and Castro, E. 2008. Evaluation of steam explosion pre-treatment for enzymatic hydrolysis of sunflower stalks. *Enzyme and Microbial Technology.* 42(2): pp.160–166.
Ruiz, H.A., Ruzene, D.S., Silva, D.P., da Silva, F.F.M., Vicente, A.A. and Teixeira, J.A. 2011. Development and characterization of an environmentally friendly process sequence (autohydrolysis and organosolv) for wheat straw delignification. *Applied Biochemistry and Biotechnology.* 164(5): pp.629–641.
Ruiz, H.A., Ruzene, D.S., Silva, D.P., Quintas, M.A., Vicente, A.A. and Teixeira, J.A. 2011. Evaluation of a hydrothermal process for pretreatment of wheat straw—effect of particle size and process conditions. *Journal of Chemical Technology and Biotechnology.* 86(1): pp.88–94.
Ruiz, H.A., Rodríguez-Jasso, R.M., Fernandes, B.D., Vicente, A.A. and Teixeira, J.A. 2013. Hydrothermal processing, as an alternative for upgrading agriculture residues and marine biomass according to the biorefinery concept: a review. *Renewable and Sustainable Energy Reviews.* 21: pp.35–51.
Ryynanen, S., Risman, P.O. and Ohlsson, T. 1996. The dielectric properties of native starch solutions - A research note.
Saddler, J.N., Brownell, H.H., Clermont, L.P. and Levitin, N. 1982. Enzymatic hydrolysis of cellulose and various pretreated wood fractions. *Biotechnology and Bioengineering.* 24(6): pp.1389–1402.
Saha, B., De, S. and Fan, M. 2013. Zr(O)Cl$_2$ catalyst for selective conversion of biorenewable carbohydrates and biopolymers to biofuel precursor 5-hydroxymethylfurfural in aqueous medium. *Fuel.* 111: pp.598–605.
Saha, B.C. 2003. Hemicellulose bioconversion. *Journal of Industrial Microbiology and Biotechnology.* 30(5): pp.279–291.
Saha, B.C., Biswas, A. and Cotta, M.A. 2008. Microwave pretreatment, enzymatic saccharification and fermentation of wheat straw to ethanol. *Journal of Biobased Materials and Bioenergy.* 2(3): pp.210–217.
Saha, B.C., Yoshida, T., Cotta, M.A. and Sonomoto, K. 2013. Hydrothermal pretreatment and enzymatic saccharification of corn stover for efficient ethanol production. *Industrial Crops and Products.* 44: pp.367–372.
Sanchez, O.J. and Cardona, C.A. 2008. Trends in biotechnological production of fuel ethanol from different feedstocks. *Bioresource Technology.* 99(13): pp.5270–5295.
Sannigrahi, P., Ragauskas, A.J. and Tuskan, G.A. 2010. Poplar as a feedstock for biofuels: a review of compositional characteristics. *Biofuels, Bioproducts and Biorefining.* 4(2): pp.209–226.
Sannigrahi, P., Kim, D.H., Jung, S. and Ragauskas, A. 2011. Pseudo-lignin and pretreatment chemistry. *Energy & Environmental Science.* 4(4): pp.1306–1310.
Sasaki, C., Takada, R., Watanabe, T., Honda, Y., Karita, S., Nakamura, Y. and Watanabe, T. 2011. Surface carbohydrate analysis and bioethanol production of sugarcane bagasse pretreated with the white rot fungus, Ceriporiopsis subvermispora and microwave hydrothermolysis. *Bioresource Technology.* 102(21): pp.9942–9946.
Schell, D.J., Farmer, J., Newman, M. and Mcmillan, J.D. 2003. Dilute-sulfuric acid pretreatment of corn stover in pilot-scale reactor. *Biotechnology for Fuels and Chemicals.* pp.69–85. Humana Press.
Schubert, C. 2006. Can biofuels finally take center stage? *Nature Biotechnology.* 24(7): pp.777–784.
Serrano-Ruiz, J.C., Campelo, J.M., Francavilla, M., Romero, A.A., Luque, R., Menendez-Vazquez, C., García, A.B. and García-Suárez, E.J. 2012. Efficient microwave-assisted production of furfural from C 5 sugars in aqueous media catalysed by Brönsted acidic ionic liquids. *Catalysis Science & Technology.* 2(9): pp.1828–1832.
Shafiei, M., Karimi, K. and Taherzadeh, M.J. 2010. Pretreatment of spruce and oak by N-methylmorpholine-N-oxide (NMMO) for efficient conversion of their cellulose to ethanol. *Bioresource Technology.* 101(13): pp.4914–4918.

Shengdong, Z., Ziniu, Y., Yuanxin, W., Xia, Z., Hui, L. and Ming, G. 2005. Enhancing enzymatic hydrolysis of rice straw by microwave pretreatment. *Chemical Engineering Communications.* 192(12): pp.1559–1566.

Shimizu, K., Sudo, K., Ono, H., Ishihara, M., Fujii, T. and Hishiyama, S. 1998. Integrated process for total utilization of wood components by steam-explosion pretreatment. *Biomass and Bioenergy.* 14(3): pp.195–203.

Sills, D.L. and Gossett, J.M. 2011. Assessment of commercial hemicellulases for saccharification of alkaline pretreated perennial biomass. *Bioresource Technology.* 102(2): pp.1389–1398.

Sjöqvist, M. and Gatenholm, P. 2005. The effect of starch composition on structure of foams prepared by microwave treatment. *Journal of Polymers and the Environment.* 13(1): pp.29–37.

Solomon, T.W.G. 1988. Organic Chemistry, 4th Edition, John Wiley & Sons.

Soto, M.L., Dominguez, H., Nunez, M.J. and Lema, J.M. 1994. Enzymatic saccharification of alkali-treated sunflower hulls. *Bioresource Technology.* 49(1): pp.53–59.

Srilatha, H.R., Nand, K., Babu, K.S. and Madhukara, K. 1995. Fungal pretreatment of orange processing waste by solid-state fermentation for improved production of methane. *Process Biochemistry.* 30(4): pp.327–331.

Srimachai, T., Thonglimp, V. and Sompong, O. 2014. Ethanol and Methane Production from Oil Palm Frond by Two Stage SSF. *Energy Procedia.* 52: pp.352–361.

Stevenson, D.G., Biswas, A. and Inglett, G.E. 2005. Thermal and pasting properties of microwaved corn starch. *Starch-Stärke.* 57(8): pp.347–353.

Sun, Y. and Cheng, J. 2002. Hydrolysis of lignocellulosic materials for ethanol production: a review. *Bioresource Technology.* 83(1): pp.1–11.

Sun, X.F., Jing, Z., Fowler, P., Wu, Y. and Rajaratnam, M. 2011. Structural characterization and isolation of lignin and hemicelluloses from barley straw. *Industrial Crops and Products.* 33(3): pp.588–598.

Suryawati, L., Wilkins, M.R., Bellmer, D.D., Huhnke, R.L., Maness, N.O. and Banat, I.M. 2009. Effect of hydrothermolysis process conditions on pretreated switchgrass composition and ethanol yield by SSF with Kluyveromyces marxianus IMB4. *Process Biochemistry.* 44(5): pp.540–545.

Swinkels, J.J.M. 1985. Composition and properties of commercial native starches. *Starch-Stärke.* 37(1): pp.1–5.

Szabolcs, Á., Molnár, M., Dibó, G. and Mika, L.T. 2013. Microwave-assisted conversion of carbohydrates to levulinic acid: an essential step in biomass conversion. *Green Chemistry.* 15(2): pp.439–445.

Szepes, A., Hasznos-Nezdei, M., Kovács, J., Funke, Z., Ulrich, J. and Szabó-Révész, P. 2005. Microwave processing of natural biopolymers—studies on the properties of different starches. *International Journal of Pharmaceutics.* 302(1): pp.166–171.

Taherzadeh, M.J. and Karimi, K. 2007. Acid-based hydrolysis processes for ethanol from lignocellulosic materials: a review. *Bioresources.* 2(3): pp.472–499.

Talebnia, F., Karakashev, D. and Angelidaki, I. 2010. Production of bioethanol from wheat straw: an overview on pretreatment, hydrolysis and fermentation. *Bioresource Technology.* 101(13): pp.4744–4753.

Taniguchi, M., Suzuki, H., Watanabe, D., Sakai, K., Hoshino, K. and Tanaka, T. 2005. Evaluation of pretreatment with Pleurotus ostreatus for enzymatic hydrolysis of rice straw. *Journal of Bioscience and Bioengineering.* 100(6): pp.637–643.

Thangavelu, S.K., Ahmed, A.S. and Ani, F.N. 2014. Bioethanol production from sago pith waste using microwave hydrothermal hydrolysis accelerated by carbon dioxide. *Applied Energy.* 128: pp.277–283.

Tian, S., Zhu, W., Gleisner, R., Pan, X.J. and Zhu, J.Y. 2011. Comparisons of SPORL and dilute acid pretreatments for sugar and ethanol productions from aspen. *Biotechnology Progress.* 27(2): pp.419–427.

Tong, X., Ma, Y. and Li, Y., 2010. An efficient catalytic dehydration of fructose and sucrose to 5-hydroxymethylfurfural with protic ionic liquids. Carbohydrate research, 345(12), pp.1698-1701.

Toussaint, B., Excoffier, G. and Vignon, M.R. 1991. Effect of steam explosion treatment on the physico-chemical characteristics and enzymic hydrolysis of poplar cell wall components. *Animal Feed Science and Technology.* 32(1): pp.235–242.

Treppe, F., Hochmuth, C., Schneider, R., Junker, T., Schneider, J. and Stoll, A., 2011. Steigerung der Ressourceneffizienz durch hybride Prozesse. In Congress Sustainable Production, Proceedings (pp. 127-148).

U.S. Department of Agriculture. 2002. Census of Agriculture (Washington, DC, June 2004), Vol. 1, Chapter 1, U.S. National Level Data, Table 8, "Land: 2002 and 1997," web site www.nass.usda.gov/census/ census02/volume1/us/st99_1_008_008.pdf.

Uy, S.F., Easteal, A.J., Farid, M.M., Keam, R.B. and Conner, G.T. 2005. Seaweed processing using industrial single-mode cavity microwave heating: a preliminary investigation. *Carbohydrate Research.* 340(7): pp.1357–1364.

Venkatesh, M.S. and Raghavan, G.S.V. 2004. An overview of microwave processing and dielectric properties of agri-food materials. *Biosystems Engineering.* 88(1): pp.1–18.

Vidal Jr, B.C., Dien, B.S., Ting, K.C. and Singh, V. 2011. Influence of feedstock particle size on lignocellulose conversion—a review. *Applied Biochemistry and Biotechnology.* 164(8): pp.1405–1421.

vom Stein, T., Grande, P.M., Kayser, H., Sibilla, F., Leitner, W. and de María, P.D., 2011. From biomass to feedstock: one-step fractionation of lignocellulose components by the selective organic acid-catalyzed depolymerization of hemicellulose in a biphasic system. Green Chemistry, 13(7), pp.1772-1777.

Wang, L., Han, G. and Zhang, Y. 2007. Comparative study of composition, structure and properties of Apocynum venetum fibers under different pretreatments. *Carbohydrate Polymers.* 69(2): pp.391–397.

Wang, Z., Keshwani, D.R., Redding, A.P. and Cheng, J.J. 2010. Sodium hydroxide pretreatment and enzymatic hydrolysis of coastal Bermuda grass. *Bioresource Technology.* 101(10): pp.3583–3585.

Wang, L., Yang, M., Fan, X., Zhu, X., Xu, T. and Yuan, Q. 2011. An environmentally friendly and efficient method for xylitol bioconversion with high-temperature-steaming corncob hydrolysate by adapted Candida tropicalis. *Process Biochemistry.* 46(8): pp.1619–1626.

Weil, J., Sarikaya, A., Rau, S.L., Goetz, J., Ladisch, C.M., Brewer, M., Hendrickson, R. and Ladisch, M.R. 1997. Pretreatment of yellow poplar sawdust by pressure cooking in water. *Applied Biochemistry and Biotechnology.* 68(1-2): pp.21–40.

Weiss, N.D., Farmer, J.D. and Schell, D.J. 2010. Impact of corn stover composition on hemicellulose conversion during dilute acid pretreatment and enzymatic cellulose digestibility of the pretreated solids. *Bioresource Technology.* 101(2): pp.674–678.

West, T.O. and Marland, G. 2003. Net carbon flux from agriculture: Carbon emissions, carbon sequestration, crop yield, and land-use change. *Biogeochemistry.* 63(1): pp.73–83.

Wong, K.K., Deverell, K.F., Mackie, K.L., Clark, T.A. and Donaldson, L.A. 1988. The relationship between fiber-porosity and cellulose digestibility in steam-exploded Pinus radiata. *Biotechnology and Bioengineering.* 31(5): pp.447–456.

Wyman, C.E. 1994. Ethanol from lignocellulosic biomass: technology, economics, and opportunities. *Bioresource Technology.* 50(1): pp.3–15.

Xu, C., Arancon, R.A.D., Labidi, J. and Luque, R. 2014. Lignin depolymerisation strategies: towards valuable chemicals and fuels. *Chemical Society Reviews.* 43(22): pp.7485–7500.

Xu, J., Chen, H., Kádár, Z., Thomsen, A.B., Schmidt, J.E. and Peng, H. 2011. Optimization of microwave pretreatment on wheat straw for ethanol production. *Biomass and Bioenergy.* 35(9): pp.3859–3864.

Yabushita, M., Kobayashi, H., Hara, K. and Fukuoka, A. 2014. Quantitative evaluation of ball-milling effects on the hydrolysis of cellulose catalysed by activated carbon. *Catalysis Science & Technology.* 4(8): pp.2312–2317.

Yadav, K.S., Naseeruddin, S., Prashanthi, G.S., Sateesh, L. and Rao, L.V. 2011. Bioethanol fermentation of concentrated rice straw hydrolysate using co-culture of Saccharomyces cerevisiae and Pichia stipitis. *Bioresource Technology.* 102(11): pp.6473–6478.

Yang, B. and Wyman, C.E. 2004. Effect of xylan and lignin removal by batch and flowthrough pretreatment on the enzymatic digestibility of corn stover cellulose. *Biotechnology and Bioengineering.* 86(1): pp.88–98.

Yang, B. and Wyman, C.E. 2008. Pretreatment: the key to unlocking low-cost cellulosic ethanol. *Biofuels, Bioproducts and Biorefining.* 2(1): pp.26–40.

Yang, C., Chung, D. and You, S. 2008. Determination of physicochemical properties of sulphated fucans from sporophyll of Undaria pinnatifida using light scattering technique. *Food Chemistry.* 111(2): pp.503–507.

Yu, G., Yano, S., Inoue, H., Inoue, S., Endo, T. and Sawayama, S. 2010. Pretreatment of rice straw by a hot-compressed water process for enzymatic hydrolysis. *Applied Biochemistry and Biotechnology.* 160(2): pp.539–551.

Yu, Y. and Wu, H. 2010. Understanding the primary liquid products of cellulose hydrolysis in hot-compressed water at various reaction temperatures. *Energy & Fuels.* 24(3): pp.1963–1971.

Zhang, Y.H.P., Ding, S.Y., Mielenz, J.R., Cui, J.B., Elander, R.T., Laser, M., Himmel, M.E., McMillan, J.R. and Lynd, L.R. 2007. Fractionating recalcitrant lignocellulose at modest reaction conditions. *Biotechnology and Bioengineering.* 97(2): pp.214–223.

Zhang, Z. and Zhao, Z.K. 2010. Microwave-assisted conversion of lignocellulosic biomass into furans in ionic liquid. *Bioresource Technology.* 101(3): pp.1111–1114.

Zhang, J., Ma, X., Yu, J., Zhang, X. and Tan, T. 2011. The effects of four different pretreatments on enzymatic hydrolysis of sweet sorghum bagasse. *Bioresource Technology.* 102(6): pp.4585–4589.

Zhang, Q., Benoit, M., Vigier, K.D.O., Barrault, J., Jégou, G., Philippe, M. and Jérôme, F. 2013. Pretreatment of microcrystalline cellulose by ultrasounds: effect of particle size in the heterogeneously-catalyzed hydrolysis of cellulose to glucose. *Green Chemistry.* 15(4): pp.963–969.

Zhao, X., Zhou, Y., Zheng, G. and Liu, D. 2010. Microwave pretreatment of substrates for cellulase production by solid-state fermentation. *Applied Biochemistry and Biotechnology.* 160(5): pp.1557–1571.

Zhou, Q., Lv, X., Zhang, X., Meng, X., Chen, G. and Liu, W. 2011. Evaluation of swollenin from Trichoderma pseudokoningii as a potential synergistic factor in the enzymatic hydrolysis of cellulose with low cellulase loadings. *World Journal of Microbiology and Biotechnology.* 27(8): pp.1905–1910.

Zhu, S., Wu, Y., Yu, Z., Liao, J. and Zhang, Y. 2005. Pretreatment by microwave/alkali of rice straw and its enzymic hydrolysis. *Process Biochemistry.* 40(9): pp.3082–3086.

Zhu, S., Wu, Y., Yu, Z., Chen, Q., Wu, G., Yu, F., Wang, C. and Jin, S. 2006a. Microwave-assisted alkali pre-treatment of wheat straw and its enzymatic hydrolysis. *Biosystems Engineering.* 94(3): pp.437–442.

Zhu, S., Wu, Y., Yu, Z., Wang, C., Yu, F., Jin, S., Ding, Y., Chi, R.A., Liao, J. and Zhang, Y. 2006b. Comparison of three microwave/chemical pretreatment processes for enzymatic hydrolysis of rice straw. *Biosystems Engineering.* 93(3): pp.279–283.

Zhu, S., Wu, Y., Yu, Z., Zhang, X., Wang, C., Yu, F. and Jin, S. 2006c. Production of ethanol from microwave-assisted alkali pretreated wheat straw. *Process Biochemistry.* 41(4): pp.869–873.

Zhu, J., Wan, C. and Li, Y. 2010. Enhanced solid-state anaerobic digestion of corn stover by alkaline pretreatment. *Bioresource Technology.* 101(19): pp.7523–7528.

Zhu, J.Y. and Pan, X.J. 2010. Woody biomass pretreatment for cellulosic ethanol production: technology and energy consumption evaluation. *Bioresource Technology.* 101(13): pp.4992–5002.

Zhu, J.Y., Sabo, R. and Luo, X. 2011. Integrated production of nano-fibrillated cellulose and cellulosic biofuel (ethanol) by enzymatic fractionation of wood fibers. *Green Chemistry.* 13(5): pp.1339–1344.

Zvyagintseva, T.N., Shevchenko, N.M., Nazarova, I.V., Scobun, A.S., Luk'yanov, P.A. and Elyakova, L.A. 2000. Inhibition of complement activation by water-soluble polysaccharides of some far-eastern brown seaweeds. *Comparative Biochemistry and Physiology Part C: Pharmacology, Toxicology and Endocrinology.* 126(3): pp.209–215.

6

Microwave Enhanced Biogas Production

Biogas can be produced from a variety of organic compounds through a complex microbial process called anaerobic digestion (AD), and can be used for methane production following separation and purification steps (Zinoviev et al. 2010). Biogas is typically a mixture of 60–75% CH_4 and 25–40% CO_2 which can be produced from fermentation of various sources of biodegradable waste and this technology is relatively simple in design and operation, flexible in utilizing various feedstocks, and less demanding in terms of consumables and infrastructure when compared with other technologies. A wide range of waste biomass, such as kitchen waste, sewage sludge, organic effluents from food and dairy industries, agricultural or crop residues (e.g., maize silage), municipal solid waste, livestock manure, and others can be processed. The main applications of biogas include combined heat and power generation and cooking applications. The effluent byproduct (sludge), which is rich in nutrients such as ammonia, phosphorus, potassium, and other trace elements, has a potential use as a fertilizer and soil conditioner. The renewed interest in this technology is driven by the need for an affordable fuel in less-developed regions. Currently, about 25 million households worldwide receive energy for lighting and cooking from household-scale biogas plants, China and India being the leaders; a few thousand medium- and large-scale industrial biogas plants are also operating in these countries.

Anaerobic digestion proceeds through a series of decomposition phases: hydrolysis, acidogenesis, acetogenesis, and methanogenesis (Wang et al. 2009) as shown in Figure 1. In the first step, organic compounds, which are primarily composed of complex carbohydrates, lipids, proteins, and inorganic materials, are solubilized by extracellular enzymes such as cellulase, amylase, protease, and lipase into simpler molecules at this stage. This process is called hydrolysis. For example, complex polysaccharides (cellulose) are broken down to dimeric and monomeric sugar (glucose), proteins are split into amino acids and peptides, and fats to fatty acids and glycerol. The first step is followed by acidogenesis stage in which acid-producing fermentation bacteria, commonly named acidogens, convert the sugar monomers and other hydrolysates produced in the first stage into low-molecular-weight compounds, such as alcohols, acetic acid, and volatile fatty acids (VFA, e.g., propionic acids, butyric acids), amino acids, H_2, CO_2, H_2S, and CH_4 under anaerobic conditions. As an example, the main reactions of glucose fermentation to acetic and butyric acids are described by equations 1 and 2 shown below.

Glucose fermentation to acetic acid

$$C_6H_{12}O_6 + 2H_2O \rightarrow 2CH_3COOH + 4H_2 + 2CO_2 \qquad [1]$$

When the by-product is butyric acid, 2 moles of H_2 are produced:

$$C_6H_{12}O_6 \rightarrow CH_3CH_2CH_2COOH + 2H_2 + 2CO_2 \qquad [2]$$

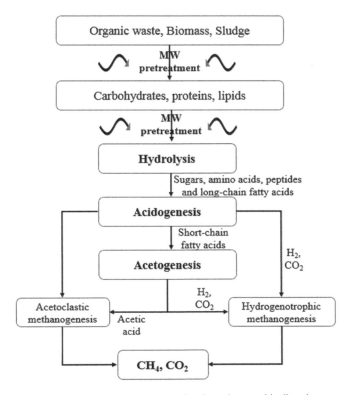

Figure 1. Steps involved in methane production through anaerobic digestion process.

In acetogenesis stage, VFA and other intermediates produced in acidogenesis are digested by acetogens to largely produce acetic acid, CO_2, and H_2. For the above mentioned propionic and butyric acids, the stoichiometry is described by the following reaction equations 3 and 4, respectively:

$$CH_3CH_2COOH + 2H_2O \rightarrow 3H_2 + CH_3COOH + CO_2 \qquad [3]$$

$$CH_3CH_2COOH + 2H_2O \rightarrow 2H_2 + 2CH_3COOH \qquad [4]$$

Finally, acetates and other intermediates are metabolized by methanogenic bacteria or methanogens to form biogas via hydrogenotrophic (hydrogen producing or utilizing as energy) or acetoclastic (acetic acid to methane conversion) pathways as shown in equations 5 and 6.

$$CO_2 + 4H_2 \rightarrow CH_4 + 2H_2O \qquad \text{(hydrogenotrophic)} \qquad [5]$$

$$CH_3COOH \rightarrow CH_4 + CO_2 \qquad \text{(acetoclastic)} \qquad [6]$$

Methanogens are obligatorily anaerobic microorganisms and are very sensitive to pH, temperature, and other environmental parameters. In contrast to the acidogens, the methanogens belong to a group of bacteria with a heterogeneous morphology and the main difference lies in the makeup of the cell walls of the bacteria. Methanogens and acidogens have a symbiotic relationship. Acidogens consume the dissolved oxygen in the organic media and thereby create ideal anaerobic conditions to facilitate the growth of methanogens. Methanogens, on the other hand, use the intermediate acids produced by acidogens to reduce the toxicity and pH of the medium.

When the composition of organic matter is known, it is possible to estimate theoretical methane and ammonium yields that can be generated from the anaerobic digestion. These yields can be calculated with the following formula adapted from Symons and Buswell (1933). However, it should be noted that this theoretical approach does not take into account the needs for cell maintenance and anabolism (Sialve et al. 2009).

$$C_aH_bO_cN_d + \left(\frac{4a-b-2c+3d}{4}\right)H_2O \Rightarrow \left(\frac{4a+b-2c-3d}{8}\right)CH_4 + \left(\frac{4a-b+2c+3d}{8}\right)CO_2 + dNH_3 \quad [7]$$

In this equation, the organic matter is converted to methane, carbon dioxide and ammonia. The specific methane yield (B) expressed in liters of CH_4 per gram of volatile solids (VS) can thus be calculated as:

$$B = \frac{4a+b-2c-3d}{12a+b+16c+14d} \times V_m \quad [8]$$

The ratio r_G of methane to carbon dioxide can therefore be computed from

$$n = \frac{-b+2c+3d}{a}$$

The average carbon oxidation state in the substrate (Harris and Adams 1979) is as follows:

$$r_G = \frac{4-n}{4+n}$$

The biogas composition, however, also depends on the amount of CO_2 which is dissolved in the liquid phase through the carbonate system, and is therefore strongly related to pH. The ammonium production yield in the digester can be evaluated using the relation given below:

$$Y_{N-NH_3}(mg/gVS) = \frac{d \times 17 \times 1000}{12a+b+16c+14d} \quad [9]$$

Equation [7] is a theoretical approach that allows estimation of the maximum potential yields.

Anaerobic Digestion Process Development

Several digester models have been developed to treat heterogeneous substrates such as fruit, vegetable wastes, municipal solid wastes, and others over the years. These are one- and two-stage systems, depending on the separation of the acidogenic phase and methanogenic phase in the reactor (Zinoviev et al. 2010). One-stage systems are one of two types: wet systems (total solids < 15%) and dry systems (total solid > 15 %). The most popular industrial designs under one-stage systems are: continuous stirred tank reactor, anaerobic sequencing batch reactor (ASBR), plug flow reactor, anaerobic filter (AF), and up-flow anaerobic sludge reactor (UASB).

The kinetics of the AD process are expressed in terms of hydraulic retention time (HRT), which is the ratio of the volume of reactor tank to the influent flow rate (V/Q). This is a measure of average retention time of the organic mass inside the digester tank for complete conversion. Three types of temperature conditions are maintained in industrial biogas generation.

Thermophylic system—This operates in a high temperature range (50–70°C), leading to a rapid break down of organic matter with a short HRT of 3–5 days to produce biogas. While these systems are more sensitive to the N_2 levels of substrate and temperature variations, they are more effective in pathogen removal.

Mesophylic system—This needs longer retention times (15–20 days or more) and a moderate temperature range (35–40°C) for the substrate to decompose. However, this system is more robust in terms of temperature variation and is most commonly used.

Psychrophylic system—This type operates more specifically in low-temperature conditions (15–25°C). Very high retention times (months) are required to achieve a high gas conversion efficiency and a high degree of pathogen removal. Most of the commercial processes rely on mesophilic systems, since thermophilic technologies became available more recently. Studies have shown that the average biogas yield in thermophilic process exceeds that of psychrophilic and mesophilic processes, by 144% and 41% respectively (De Bere 2000).

Pretreatment of Lignocellulosic Biomass Feedstock for Anaerobic Digestion

Pretreatment methods are seldom mentioned in literature when digesting sugar and starch crops, possibly because these types of crops are already well digestible after a simple size reduction step. The rate-limiting step in anaerobic digestion of solid feedstock such as lignocellulosic crops is the hydrolysis of complex polymeric substances and in particular, the cross-linking of lignin which is non-biodegradable with cellulose and hemicellulose. In addition, the crystalline structure of cellulose prevents penetration by microorganisms or extracellular enzymes. Pretreatment is necessary for partial or complete decomposition of this feedstock into soluble fermentable products.

Pretreatment of the feedstock improves hydrolysis yield and the total methane yield. The hydrolysis of lignocellulose and conversion to methane is, in general, done in one reactor by a consortium of microorganisms. The advantage of using a mixture of microorganisms is that almost all products, like pentoses, hexoses, volatile products and even sometimes inhibiting compounds like furfural and soluble lignin compounds (in not too high concentrations) can be converted to methane after a period of adaptation (Benjamin et al. 1984, Fox et al. 2003).

The choice of pretreatment should be made carefully to maximize hydrolysis, or liquefaction of the substrate, in order to maximize the methane potential from the target crop. Abundant literature is available on lignocellulosic pretreatment technologies such as enzymatic liquefaction and saccharification, solvent-based, dilute acid, ammonia fiber explosion, ammonia recycle percolation, lime, steam explosion, and OrganoSolv pretreatment, that are under intensive investigation at both laboratory scale and as pilot plants, mostly in order to generate ethanol in a biorefinery concept (Frigon and Guiot 2010).

Pretreatment methods directed more specifically to enhanced methane production from lignocellulosic biomass were reviewed recently by Hendriks and Zeeman (2009), who concluded that steam-, lime-, liquid-hot-water-, and ammonia-based pretreatment methods showed high potentials. Other physical pretreatment methods such as steam explosion, thermal hydrolysis, wet oxidation, pre-incubation in water, and treatment with ultrasound or microwave irradiation also show potential for improving methane yields from lignocellulosic materials. Energy-intensive pretreatment methods including steam explosion, wet explosion and ammonia fiber explosion (AFEX) have the advantage of practically solubilizing the whole substrates and achieve very high methane yields. However, energy costs should be taken into consideration, and the net energy gain of using these pretreatments has yet to be clearly demonstrated. The following sections discuss various types of pretreatment techniques.

Acid pretreatment

Acid pretreatment solubilizes hemicellulose which makes cellulose to be more easily accessible for the enzymes. Volatile degradation products can be formed in this step which can be converted to methane. Acid treatment favors methane production because methanogens can handle compounds like furfural and HMF to a certain concentration and with an acclimatization period. However, there is a risk of inhibition from soluble lignin components. Methanogens are, however, capable of adapting to such inhibiting compounds (Benjamin et al. 1984, Xiao and Clarkson 1997). When sulfuric or nitric acids are used in the acidic pretreatment, the methane production during anaerobic treatment will be reduced as a result of reduction of sulfate and nitrate to H_2S and N_2 respectively.

Alkaline pretreatment

During alkaline pretreatment the first reactions taking place are solvation and saphonication. This causes a swollen state of the biomass and makes it more accessible for enzymes and bacteria. An important aspect of alkali pretreatment is that the biomass itself consumes some of the alkali. The residual alkali concentration after the alkali consumption by the biomass is the alkali concentration available for the reaction (Gossett et al. 1982). Pavlostathis and Gossett (1985) for example found during their experiments an alkali consumption of approximately 3 g NaOH/100 g TS. Lime is better candidate for pretreatment than sodium hydroxide (Gossett et al. 1982). Alkali extraction can also cause solubilization, redistribution and condensation of lignin and modifications in the crystalline state of the cellulose. These effects can lower or counteract the positive effects of lignin removal and cellulose swelling (Gregg and Saddler 1996). Another important aspect of alkaline pretreatment is the change of the cellulose structure to a form that is denser and thermodynamically more stable than the native cellulose (Petersen 1984).

Mechanical pretreatment

Milling (cutting the lignocellulosic biomass into smaller pieces) is a mechanical pretreatment of the lignocellulosic biomass (Hendriks and Zeeman 2009). Milling reduces the particle size and crystallinity which leads to an increase of available specific surface and a reduction of the degree of polymerization (DP) (Palmowski and Muller 1999). Milling also shears the biomass increasing the cellular content availability. An increased methane production by 5–25% was reported for this pretreatment (Delgenés et al. 2002). Apart from milling, beating, blending, shaking and pressure application are other mechanical pretreatment options. Table 1 shows a summary of the effects of mechanical pretreatment techniques performed in algae biogas production.

Table 1. Mechanical pre-treatments for algae biogas production.

Pre-treatment	Reactor mode	Machine	Feedstock	Results	References
Beating	Batch	Hollander beater	*Pelvetia caniculata*	444.3 mL biogas/g TS	Tedesco et al. 2013
			Fucus serratus	181.2 mL biogas/g TS	
			Gracilaria gracilis	171.8 mL biogas/g TS	
			Fucus vesiculosus	230.5 mL biogas/g TS	
			Laminaria digitata	156.4 mL biogas/g TS	
Blending	Batch	Waring blender	*Rhizoclonium*	93–100 mL CH_4/g TS	Izumi et al. 2010
Shaking	Batch	Glass tubes with glass beads	*Isochrysis galbana*	61.7% increment in COD	Santos et al. 2014
Pressure	Batch	TK Energi A/S prototype	*Fucus vessiculosus*	90 NL CH_4/kg VS	Rodriguez et al. 2015
			Filamentous red algae	Filamentous red algae 223 NL CH_4/kg VS	
		French press	*Nannochloropsis salina*	460 mL biogas/g VS	Schwede et al. 2011

Steam pretreatment

Steam pretreatment is conducted by exposing the biomass to steam at high temperature (temperatures up to 240°C) and high pressure for a few minutes. The steam is released after the set time and the biomass is quickly cooled down. The objective of a steam pretreatment/steam explosion is to solubilize the hemicellulose to make the cellulose better accessible for enzymatic hydrolysis and to avoid the formation of inhibitors. The difference between 'steam' pretreatment and 'steam explosion' pretreatment is the quick depressurization and cooling down of the biomass at the end of the steam explosion pretreatment, which causes the water in the biomass to 'explode'. The positive effect of steam pretreatment is mostly due to the removal of a large part of the hemicellulose, causing an increase of cellulose fiber reactivity, probably because the cellulose is easily accessible for the enzymes (Converse et al. 1989, Grohmann et al. 1986, Laser et al. 2002). Steam pretreatment includes a risk on condensation and precipitation of soluble lignin components, making the biomass less digestible, and affecting methane production.

Thermal pretreatment

Thermal pretreatment involves temperature increases above 150–180°C in parts of the lignocellulosic biomass. First, the hemicelluloses start to solubilize followed by lignin shortly (Bobleter 1994, Garrote et al. 1999). The composition of the hemicellulose backbone and the branching groups determine the thermal, acid and alkali stability of the hemicellulose. From the two dominant components of hemicelluloses (xylan and glucomannan), xylans are thermally the least stable, but the difference with the glucomannans is only small. Above 180°C, an exothermal reaction (probably solubilization) of the hemicellulose starts (Beall and Eickner 1970). This temperature of 180°C is probably just an indication of the temperature at which an exothermal reaction of the hemicellulose starts, because the thermal reactivity of lignocellulosic biomass depends largely on its composition (Fengel and Wegener 1984). During thermal processes, a part of the hemicellulose is hydrolyzed to form acids. These acids are assumed to catalyze further the hydrolysis of the hemicellulose (Gregg and Saddler 1996). There may be some unknown factors, other than the catalyzing effect of *in situ* formed acids, which play a role in the solubilization of hemicellulose (Liu and Wyman 2003, Zhu et al. 2004, 2005).

Thermal pretreatment with temperatures of 160°C and higher, causes, besides the solubilization of hemicellulose, also the solubilization of lignin. The produced compounds are almost always phenolic compounds and have in many cases an inhibitory or toxic effect on bacteria, yeast and methanogens/archaea (Gossett et al. 1982). These soluble lignin compounds are very reactive and will, if not removed quickly, recondensate and precipitate on the biomass (Liu and Wyman 2003). Severe pretreatment conditions promote the condensation and precipitation of soluble lignin compounds, sometimes even with soluble hemicellulosic compounds like furfural and HMF (Negro et al. 2003). Low thermal pretreatment at temperatures less than 100°C also proved to be beneficial in increasing biogas production. The yields seem to depend on the pretreatment time and the temperature. Table 2 summarizes the conditions for low thermal pretreatment and anaerobic digestion of microalgae.

Microwave Pretreatment

Microwaves increase the kinetic energy of the water leading to a boiling state; the process polarizes macro-molecules, causing changes in the structure of biomass and a rapid generation of heat and pressure in the biological system that produces cell hydrolysis, forcing out compounds from the biological matrix. The mechanism of microwave irradiation may include a thermal effect and a non-thermal effect. The non-thermal effect refers to the effect that is not associated with a temperature increase, and it is caused by the polarized parts of macro-molecules aligning with the poles of the electromagnetic field, resulting in the possible breaking of hydrogen bonds. Thermal effect refers to the part of the process that generates heat as a result of the microwave energy absorption by water or by organic complexes. Thermal and non-thermal effects of the microwave (MW) irradiation play a role in the "hot-spot" phenomena, and the different dielectric parameter of cell components led to selective heating manifested in the different thermal stress, which contributes in the intensive degradation of cell wall components such as cellulose and pectin (Banik et al. 2003). MW pretreatment has verified positive effects on cell wall destruction and releasing of organic matter into the soluble phase, but combining it with addition of chemicals such as alkali, acid and oxidizer agents cause synergetic mechanism to accelerate the decomposition under aerobic and anaerobic condition as well (Beszédes et al. 2011).

Table 2. Low temperature pretreatment of microalgae (Passos et al. 2013).

Microalgae	Pretreatment conditions	Anaerobic digestion conditions	Results	References
Microalgal biomass grown in wastewater	100°C, 8 h	Batch	Methane production increase by 33%*	Chen and Oswald 1998
Scenedesmus biomass	90°C, 3 h	Batch 35°C	Methane production increase by 220%	González-Fernández et al. 2012a
Scenedesmus biomass	80°C, 25 min	Batch 35°C	Methane production increase by 57%	González-Fernández et al. 2012b
Scenedesmus and *Chlamydomonas* biomass	55°C, 12 and 24 h	Batch 35°C	Methane production decrease by 4–8%	Alzate et al. 2012
Acutodesmus obliquus and *Oocystis* sp. biomass	55°C, 12 and 24 h	Batch 35°C	Methane production decrease by 3–13%	Alzate et al. 2012
Microspora biomass	55°C, 12 and 24 h	Batch 35°C	Methane production increase by 4–5%	Alzate et al. 2012
Microalgal biomass grown in wastewater	21.8, 43.6, 65.4 MJ/kg TS		Methane production increase by 12–78%	Passos et al. 2013a
Microalgal biomass grown in wastewater	900 W, 3 min, 70 MJ/kg VS		Methane production increase by 60%	Passos et al. 2014
Methanosarcina ceae	Loadings of 1–2 kg COD/(m^3-d)	35 and 55°C	Highest production of biogas with methane content of ca. 67% at mesophilic reactors	Zielinska et al. 2013
Nannochloropsis salina	5 times until boiling at 600 W		Biogas yield increase of 40%	Schwede et al. 2011
Microalgae from a high rate algal ponds	CSTR 20 days HRT 110.2 MJ/kg VS		Methane yield improvement of 58%	Passos et al. 2014
Microalgae from a high rate algal ponds	Batch 65.4 MJ/kg TS		Biogas yield increase of 78%	Passos et al. 2013b
Algae blooms	Batch 15 min + H$_2$SO$_4$ (1% v/v)		253.5 mL CH$_4$/g TVS	Cheng et al. 2014

* Compared to control.

Anaerobic Digestion of Sewage Sludge

Conventional wastewater treatment by activated sludge process generates large quantities of sludge called excess sewage sludge (ESS). The treatment and management scheme for excess sewage sludge involves several steps, such as anaerobic digestion, chemical conditioning, and mechanical dewatering to 70–80% (wt.) water content, followed by disposal at landfill, application to cropland or incineration when faced with land shortage (Tang et al. 2010). Volume reduction is a critical step to reduce the costs related to transportation and disposal of this excess sludge.

Municipal sewage sludge contains mainly water and various organic and inorganic solids. Anaerobic digestion is a suitable mechanism for solids reduction through biodegradation. However, the slow degradation of sewage sludge by anaerobic digestion is a major problem, with a retention time of about 20 days in conventional digesters, which has significant space requirements in a wastewater treatment plant, making it is very necessary to enhance the efficiency of anaerobic digestion.

Among the four major steps; hydrolysis, acidogenesis, acetogenesis and methanogenesis in anaerobic digestion, hydrolysis is the rate-limiting step and prolongs the overall process. Cell membrane and extracellular polymeric substances (EPS) present physical and chemical barriers to direct anaerobic degradation, respectively (Higgins and Novak 1997, Novak et al. 2003, Sheng et al. 2008). The solid components in ESS are very complex, and include biomass produced by the biological conversion of organics, oil and grease, nutrients (nitrogen and phosphorus), heavy metals, synthetic organic compounds and pathogens. Different groups of microorganisms, organic and inorganic matter are agglomerated into a polymeric network formed by EPS or divalent cations. Because of this special structure and the hydrophilic characteristics of sewage sludge, it increases the difficulty to achieve effective volume reduction with anaerobic digestion and mechanical dehydration. A pre-treatment step is necessary to disrupt the network and increase the biodegradability of sewage sludge. Thermal pretreatment is often applied for sewage sludge disintegration at temperatures from 50 to 270°C (Carrère et al. 2010). However, temperatures above 180°C may lead to the production of recalcitrant and/or inhibitory compounds, which reduce biomass digestibility (Wilson and Novak 2009). Pretreatments at low temperatures (< 100°C) reduce the energy demand, which can improve the energy balance and profitability of the system (Ferrer et al. 2008).

Application of Microwaves in Anaerobic Digestion of Sewage Sludge

Conventional thermal or thermo-chemical treatment is a time-consuming process (Shahriari 2011). Microwave irradiation can be a better alternative due to its ability to simultaneously improve digestion and decrease the pathogen content. Sludge is a multi-phase medium containing water, mineral and organic substances, proteins, and cells of microorganisms. Due to its high water content, sewage sludge absorbs microwave irradiation effectively.

Microwave pretreatment may increase the solubility of organic matter (Park et al. 2004). The ratio of soluble COD/total COD (sCOD/tCOD) increased from 2 to 22%, when domestic wastewater sludge was irradiated with microwave to its boiling temperature. Anaerobic digestion of pretreated sludge resulted in almost 2.5% increase of VS reduction under mesophilic conditions. COD removal and methane production rates from the pretreated sludge were 64% and 79% higher than those from the control system respectively. Microwave pretreatment also resulted in a decrease in the HRT of the AD from 15 to 8 days, since the biogas production and VS removal rates were still high and the concentration of VFAs in the effluent was low.

In another study, microwave pretreatment of primary sludge, waste activated sludge and anaerobic digester sludge at 70°C increased the sCOD by 16%, 125% and 45%, respectively (Hong et al. 2006). Microwave heating of primary sludge at 85°C and 100°C resulted in an increase of 11.9% and 22.7% respectively for biogas production while waste activated sludge irradiated to 85°C and 100°C produced 11.4% and 15% more biogas, respectively. Microwave pretreatment of primary sludge with 3 to 4% TS to 65°C and 90°C at various microwave intensities resulted in 2–3 fold increase of sCOD and a 15%–30% improvement in the biogas production (Zheng 2006). Similar results were obtained at low temperature (50–96°C) microwave irradiation of thickened waste activated sludge (Eskicioglu et al. 2007a,b). Microwave pretreated waste activated sludge had 3.6 and 3.2 fold increase in sCOD/tCOD ratio and 13% and 17% increase in biomethane potential at concentrations of 1.4 and 3% TS respectively. Solubilization was slightly higher at 50% than at 100% microwave intensities for both sludge concentrations at the same pretreatment temperatures, possibly due to longer exposure time to the microwave field at low microwave intensities.

The most important factors affecting waste activated sludge solubilization are temperature, intensity, and sludge concentration and hydraulic retention time of sludge. High hydraulic retention times seemed to improve the anaerobic biodegradability of combined aerobic sequencing batch reactors (SBR) sludge (Kennedy et al. 2007). The sCOD/tCOD ratio of control samples was approximately 1.4%. Increasing microwave temperature from 45 to 85°C had a positive effect on the solubilization, as the sCOD/tCOD ratio approximately doubled when sludge was pretreated with microwaves to 85°C. Overall, temperature had the most significant effect on solubility in comparison to sludge concentration and microwave intensity. The maximum increase in mesophilic biogas production was 16.2% when 100% of SBR sludge received microwave pretreatment to 85°C. Multiple microwave irradiation cycles and maintaining sludge at 85°C showed no increase in biodegradation. In a continuous flow system, Eskicioglu et al. (2007) observed that the tCOD removal efficiency of pretreated digesters increased as the HRT was gradually shortened from 20 days to 5 days, and as the pretreatment temperature was increased from 50°C to 96°C. Thickened waste activated sludge pretreated at 96°C by microwave and conventional heating achieved 29% and 32% higher total solids and 23% and 26% volatile solids removal efficiencies compared to control at HRT of 5 days, respectively. Microwave and CH digesters fed by pretreated sludge at 96°C had 80% and 96%

higher sCOD and 120% and 56% higher effluent NH3-N (ammonia-Nitrogen) at HRT of 5 days compared to control respectively.

Microwaves can be combined with acid or alkali medium to improve the outcomes. In a recent study, Qiao et al. (2008) investigated the treatment of sludge by microwave irradiation combined with alkali addition. Around 50–70% of volatile suspended solids (VSS) was dissolved into solution at 120°C to 170°C with a dose of 0.2 g NaOH/g-DS (dried solids) within 5 minutes, and 80% of the tCOD was transformed to sCOD. For the fresh sludge, microwave heating alone reduced VSS by an additional 40% at 170°C within 1 minute. Adding 0.05 g NaOH/g-DS increased the VSS dissolution ratio to 50%. In addition, the settleability improved after 1 minute of microwave pretreatment combined with alkali addition. In another study, microwave irradiation was applied along with NaOH at a final temperature of 171°C and duration of 16 minutes with pH adjustment (Doğan and Sanin 2009). The ratio of sCOD/tCOD increased from 0.005 to 0.18, 0.27, 0.34 and 0.37 with MW/pH of 10, MW/pH of 11, MW/pH of 12 and MW/pH of 12.5, respectively which were higher than the effect of microwave or NaOH alone. Biomethane potential (BMP) tests were also carried out for pH of 10, pH of 12, MW, MW/pH of 10 and MW/pH of 12. A combination of microwave heating and a pH of 12 had the greatest methane production, which was 18.9% higher than the control. In semi continuous study of WAS with MW/pH of 12 at an HRT of 15 days, 55% more methane was produced compared to the control reactor. Furthermore, the VS and TCOD removal were increased by 35.4% and 30.3%, respectively.

Microwave power levels may have a significant effect on the pretreatment and the outcomes. Different combinations of microwave power (500, 750, and 900 W) and contact times in the range of 0–140 seconds were evaluated (Yu et al. 2010). The turbidity for treated sludge supernatant was the highest at the highest energy input, probably due to the release of biopolymers. With the increase in the microwave energy and contact time, VSS solubilization increased. After 140 seconds of pretreatment, VSS solubilization was 24.7%, 25.7% and 29.6% at 500 W, 750 W and 900 W, respectively. This might result from disruption of the sludge floc and cells causing the release of organic matter into the soluble phase. The sCOD/tCOD ratio increased from 0.0622 (raw sludge) to 0.1571, 0.1581 and 0.1611 at energies of 500 W, 750 W and 900 W, respectively, after 140 seconds irradiation. The results showed that the SCOD increased gradually, becoming asymptotic as the maximum solubilization was approached and no further increases in solubilization occurred. They concluded 900 W and 60 seconds is a good option for optimizing sludge digestibility and energy consumption. A summary of studies reporting the microwave pretreatment effects on biomethane production potential from activated sludge is shown in Table 3.

Factors Affecting Microwave Digestion of Biomass

The main factors affecting the microwave pretreatment for biomass digestion are namely temperature, microwave power and the time of exposure. Many studies reported on the effects of microwave wastewater sludge treatment with the aim of solubilizing organic matter or enhancing anaerobic digestion. In some studies, microwave treatment was compared with other thermal or ultrasonic processes in order to assess its potential as a sludge disintegration method (Beszédes et al. 2009, 2011a, Climent et al. 2007, Eskicioglu et al. 2006a, 2007a,b). However, the results reported in these studies lead to conflicting results in terms of organic matter solubilization, power effect on biogas production and the non-thermal effects of microwaves which are summarized below.

Effect of power on organic matter solubilization

Organic matter solubility is calculated from the COD (chemical oxygen demand) of the soluble phase. According to some authors, the use of higher power during microwave treatment leads to lower solubilization (Eskicioglu et al. 2007, Park et al. 2010, Toreci et al. 2009), while others report the opposite effect (Climent et al. 2007).

Table 3. Effect of microwave pretreatment on activated sludge solubilization and biochemical methane potential tests.

Pretreatment conditions	Batch test	Results	References
14.3 min; 400 W; 102°C; 2.3% TS		Increase of 17.9% in the CODs/COD ratio*	Park et al. 2010
5 min; 800 W; 13,000 kJ/kg SS	55°C; 32 d	Increase of 311% in the VSs/VS ratio and no difference in biogas production and production rate	Climent et al. 2007
5 min; 96°C; 5.5% TS	33°C; 23 d	Increase of 143% in the CODs/COD ratio and 211% in the cumulative biogas production*	Eskicioglu et al. 2006a
Progressive heating 1.2–1.4°C/min;175°C	33°C; 18 d	Increase of 74.3% in COD solubilization and 34% in biogas production	Eskicioglu et al. 2009
0.83 kJ/ml; 1000 W; 7–8% TS	35°C; 20–25 d	Decreased solubilization (CODs/COD) and increase of 15.4% in methane production*	Sólyom et al. 2011
1168 W; 90°C; 4% TS	35°C; 22 d	Increase of 2.5% in the CODs/COD ratio, 37% in the digestion rate and no impact on methane production*	Zheng et al. 2009
Organic fraction of municipal solid waste Microwave pretreatment at 115–145°C for 40 min	Mesophilic batch	4–7% Higher biogas produced than untreated	Shahriari et al. 2012
Food waste -Microwave with intensity of 7.8°C/min	Mesophilic batch	24% Higher COD solubilization and 6% higher biogas production	Marin et al. 2010
91.2°C, 7 min	35°C, SRT 15 days	25.9% VS reduction, 23.6% TCOD removal 64% and 79% improvement in TCOD removal and in methane production. Anaerobic digestion of MW pretreated sludge reduced the reactor SRT from 15 days to 8 days	Park et al. 2004
96°C, 3% TS	Semi-continuous, 33–71°C, 5 day SRT	20% Higher biogas production over control reactor and 26% higher VS removal	Eskicioglu 2007b
85°C	Batch 35°C, 25 days SRT	12% and 16% improvement in VS destruction and in methane production, respectively	Kennedy et al. 2007
175°C, 3% TS	Mesophilic batch, 35–71°C	31% Higher biogas production than the control	Eskicioglu 2008, 2009
170°C, 30 min	Batch, 35°C, 30 d SRT	25.9% Higher biogas production, 12% higher VS removal over control	Qiao et al. 2008
30–100°C		900 W, temperature: 60–70°C gave 35% more methane	Kuglarz et al. 2013

* TS: total solids, SS: suspended solids, VS: volatile solids, VSs: soluble volatile solids, COD: chemical oxygen demand, CODs: soluble chemical oxygen demand.

Effect of power on biogas production

Increase in microwave power output rate increases biogas production (Toreci et al. 2009). However, some studies report that microwave power output rate did not significantly increase the biogas production rates (Eskicioglu et al. 2007a).

Non-thermal effect

Microwave non-thermal effects are usually expressed in terms of the pre-exponential factor and the activation energy required for the reaction. Non-thermal effect can be quantitatively identified by comparing the activation energies for conventional and microwave pretreatment methods. The activation energy (E_a) for waste activated sludge hydrolysis was compared between microwave irradiation (MW) and conventional heating (CH) methods to evaluate the non-thermal effect of microwave pretreatment (Byun et al. 2014). The microwave-assisted hydrolysis of waste activated sludge was assumed to follow the first-order kinetics on the basis of (VSS) conversion to soluble chemical oxygen demand (sCOD) (sCOD) for different initial VSS concentrations. The relationship between the VSS reduction and the sCOD increase between microwave and conventional pretreatments at different absolute temperatures of 323, 348 and 373 K was presented as the average ratio of VSS conversion to sCOD which was determined to range from 1.42 to 1.64 g sCOD/g VSS. The activation energy of the microwave assisted waste activated sludge hydrolysis was much lower than that of conventional heating for the same temperature conditions. This confirmed the non-thermal effect of microwaves as a special effect on the hydrolysis of waste activated sludge over conventional heating.

Non-thermal effects are usually determined through a comparison of microwave heating and conventional heating processes and their yields. Biogas production reported in these studies, where microwave instead of conventional heating was used, was in some cases lower (Climent et al. 2007, Eskicioglu et al. 2006a), in another case higher (Beszédes et al. 2011a), and in another study it was the same in both conventional and microwave heating (Eskicioglu et al. 2007a). The use of different raw materials can lead to different quantitative results. However, the reason for these contradictory conclusions cannot be explained due to the difficulty in comparing results using different experimental conditions in microwave treatments. Only a fraction of the power applied in microwave heating is usually absorbed by the sample material (Meredith 1998). This fraction depends on factors such as dielectric properties, the size and geometry of the sample, microwave frequency and intensity, process time, and oven cavity characteristics (Swain and James 2005, Campanone and Zaritzky 2005, Gunasekaran and Yang 2007, Zhu et al. 2007). In order to analyze and compare results from several experiments and other research studies it is necessary to determine the fraction of energy absorbed by the sample.

Relationship Between the Solubility of Organic Matter and Dielectric Properties

The dielectric properties, namely the dielectric constant (ε') and the dielectric loss factor (ε'') are important for predicting the behavior of materials during microwave processing, because both of them determine the interaction between the molecules with the oscillating electromagnetic field (Beszédes et al. 2014). Dielectric constant measures the ability of material to store the irradiated energy, the value loss factor related to the ability of material to convert electric energy into heat (Zheng et al. 2009).

Although the thermal and non-thermal effects of microwave irradiation may alter the structure of biomass, the dielectric properties of processed feedstock (biomass) may also have pronounced effects on the heat transfer simultaneously. The relationship between the dielectric parameters and the disintegration degree or the biodegradability was investigated for wastewater sludge recently. A study focused on the measurement of dielectric properties of sludge originated from meat processing wastewater to determine the relationship and correlation between the parameters related to the organic matter solubility, biodegradability and the dielectric parameters, such as the dielectric constant and dielectric loss factor (Beszedes et al. 2014). Dielectric constant (ε') and dielectric loss factor (ε'') were determined in a tailor made dielectrometer equipped with a dual channel NRVD power meter (Rohde and Schwarz). Magnetron of dielectrometer operates at a frequency of 2450 MHz. ε' and ε'' were calculated from the reflection coefficient phase shift, incident and reflected power.

Experimental results have verified that during microwave pretreatment, the structure of wastewater sludge was efficiently disintegrated, and this effect was manifested in an increased migration ability of ions liberated from the polymeric matrix of sludge and from intracellular components. These structural changes have led to increased value of dielectric constant and loss factor. The temperature dependent disintegration degree of sludge has a good correlation with dielectric loss factor and biodegradability, characterized by the ratio of biochemical to chemical oxygen demand as shown in Figure 2 and Figure 3.

ε' and ε'' decreased with increasing temperature, but over a certain value of temperature (depending on the dry matter content of sludge), they start to increase. This behavior of dielectric parameters is in a relationship with the structural change of sludge, which was characterized by the sCOD/tCOD ratio and

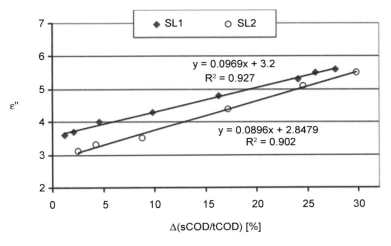

Figure 2. Relationship between solubility of organic matter and ε'' for solids content of 9.7 wt% (SL1) and 14.8 wt% (SL2), respectively.

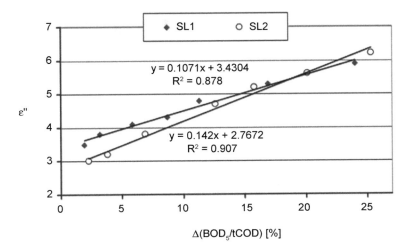

Figure 3. Correlation between the biodegradability and dielectric loss factor for solids content of 9.7 wt% (SL1) and 14.8 wt% (SL2).

the biodegradability of organic matter content of sludge, which was given by the ratio of BOD_5 to tCOD. Correlation between the electrical parameters and biodegradability indicators enable to develop a real-time and on-line measuring and control system for batch and continuous flow microwave sludge pretreatment (Beszedes et al. 2014).

Potential Microwave Pretreatment Applications in Sewage Sludge Treatment

Wastewater treatment produces different forms of sludge with varying characteristics. Sludge generated from the primary and secondary processes is usually concentrated in a thickener unit, and then often pumped to an anaerobic digester. Following anaerobic digestion, the sludge is dewatered. A flow schematic of a wastewater treatment process, along with potential locations for microwave pretreatment application, is shown in Figure 4 (NYSERDA 2011). Combined, primary and secondary sludge (Figure 4a and 4b) generally contain no more than 4% solids. Thickened sludge (Figure 4c) typically contains between 4% to 6% solids. Post-digestion biosolids (Figure 4d) typically contain between 3% and 8% solids. Polymer-treated, mechanically-dewatered solids (Figure 4e) typically contain between 20% and 45% solids. The energy requirements and corresponding size of a microwave applicator will be heavily dependent on the quantity of sludge requiring treatment, the solids/moisture content of the sludge, and the design temperature of the system. Sludge having excess moisture will require additional energy, and if temperatures are raised above the boiling point of water, the heat of vaporization will impose a significant energy penalty on the process. While pressurization is an option, the introduction of high-pressure microwave applicator would be an extremely costly process. Considering the above, location e is the most practical design location for several reasons:

1. The solids contents at this location are well over five times higher than any other location, which would significantly reduce the applicator energy requirements and subsequently, the size of the microwave required to process the sludge.
2. Disinfection was selected as the most beneficial process for microwave application, for which the back-of-the plant location is the most practical.
3. The limited apparent benefits of using microwave energy for dewatering or for the dissolution of organic matter to enhance biogas recovery did not justify consideration of other locations.

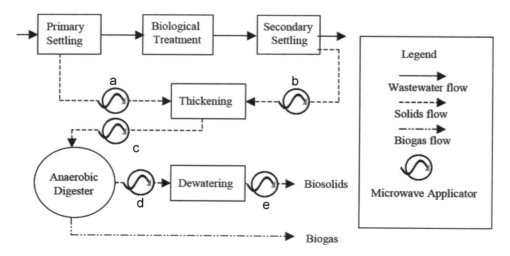

Figure 4. Potential stream locations for microwave applications in wastewater treatment plants (adapted from NYSERDA 2011).

The benefits of microwave pretreatment reflected in better sludge dewaterability and higher organic matter dissolution than the fixed film sludge. Microwave application was less effective on polymer pretreated sludge compared to non-pretreated sludge. Microwave disinfection was effective but equivalent to that could be achieved by conventional heating of the sludge at equivalent temperatures. From a commercial perspective, the primary benefit associated with microwave treatment of sludge is the potential to produce a Class A biosolids product for use as a fertilizer, soil conditioner or nutrient supplement (for land application), which could be most strategically employed at small-scale facilities that have high disposal costs or at a central processing station for treating sludge from multiple small-scale facilities.

Algae Biomass as Feedstock for Anaerobic Digestion

Theory and practice of anaerobic digestion of various organic feedstock has been well known for almost two centuries. The strong interest for anaerobic digestion is mainly due to its ability to treat and to convert a wide range of organic wastes into renewable energy. Due to abundance and the biological composition (carbohydrates, proteins and lipids), microalgae have been identified as an ideal feedstock for anaerobic digestion. The first study on anaerobic digestion of algal biomass was reported in 1960 by Golueke et al. (Oswald and Golueke 1960). An integrated approach to treat wastewater in open raceway ponds followed by energy extraction from algal biomass via anaerobic digestion process was studied previously.

Table 4 shows potential specific methane yields from carbohydrates, lipids and proteins for various microalgae species (Angelidaki and Sanders 2004, Sialve et al. 2009). The theoretical methane potential increases with the lipid content of the algal cell. High energetic content of lipids makes them attractive substrates for anaerobic digestion due to their higher gas production potential compared with carbohydrates. However, lipid hydrolysis is considered to be slower than protein and carbohydrate hydrolysis. The anaerobic treatment of various substrates requires around 0.18, 0.43 and 3.2 days for carbohydrates, proteins and lipids respectively. Therefore, lipid extraction followed by anaerobic digestion could be a better process scheme to enhance the productivity.

Microalgae biomass harvested from wastewater treatment high rate algal ponds can be valorized through anaerobic digestion for biogas production. González-Fernández et al. (2012a) reported a specific methane production of 0.1–0.5 L CH_4/g VS, with 60–80% CH_4 in biogas, depending on process temperature (15–52°C) and hydraulic retention time (HRT) (3–64 days). This value is commensurate with biogas production from other substrates. For instance, the specific methane production of waste activated sludge ranges between 0.15 and 0.3 L CH_4/g VS (Ferrer et al. 2011); and that of lignocellulosic agricultural crops, such as maize, wheat, rice and sugarcane wastes, between 0.28 and 0.34 L CH_4/g VS (Chandra et al. 2012).

Methane production from microalgae anaerobic digestion was investigated through experimental studies by many researchers. Table 5 summarizes these studies for various microalgae species (Ward et al. 2014). It can be noted from this table that there is no clear correlation between the methane yields and the organic loading rates, at least from the data reported in the literature. A wide range of biomethane

Table 4. Gross composition of several microalgae species (Becker 2004) and calculated (using Eq. [7]) theoretical methane potential and theoretical ammonia release during the anaerobic digestion of the total biomass (Sialve et al. 2009).

Species	Proteins (%)	Lipids (%)	Carbohydrates (%)	CH_4 (L CH_4 g VS^{-1})	N–NH_3 (mg g VS^{-1})
Euglena gracilis	39–61	14–20	14–18	0.53–0.8	54.3–84.9
Chlamydomonas reinhardtii	48	21	17	0.69	44.7
Chlorella pyrenoidosa	57	2	26	0.8	53.1
Chlorella vulgaris	51–58	14–22	12–17	0.63–0.79	47.5–54.0
Dunaliella salina	57	6	32	0.68	53.1
Spirulina maxima	60–71	6–7	13–16	0.63–0.74	55.9–66.1
Spirulina platensis	46–63	4–9	8–14	0.47–0.69	42.8–58.7
Scenedesmus obliquus	50–56	12–14	10–17	0.59–0.69	46.6–42.2

Table 5. Methane production from the anaerobic digestion of microalgae biomass reported in scientific literature (adapted from Ward et al. 2014).

Microalgae species	Methane yield	Loading rate	Reference
Arthrospira maxima	173 mL/g VS	500 mgTS/L	Inglesby and Fisher 2012
Arthrospira platensis	481 mL/g VS	2000 mgTS/L	Mussgnug et al. 2010
Blue green algae	366 mL/g VS	281.96 mgVS/L	Rui et al. 2008
Chlamydomonas reinhardtii	587 mL/g VS	2000 mgTS/L	Mussgnug et al. 2010
Chlorella kessleri	335 mL/g VS	2000 mgTS/L	Mussgnug et al. 2010
Chlorella sp., *Pseudokirchneriella* sp. and *Chlaqmydomas* sp.	280–600 mL/g VS	402 mg VS	De Schamphelaire and Verstraete 2009
Chlorella sp., *Scenedesmus, Euglena* and *Oscillatoria*	300–800 mL/g VS	-	Golueke and Oswald 1959
Chlorella sp., *Scenedesmus*	170–320 mL/g VS	1.44–2.89 g VS/L	Golueke et al. 1957
Chlorella sorokiniana	212 mL/g VS	-	Polakovicova et al. 2012
Chlorella vulgaris	403 mL/g VS	2000 mg VS/L	Lu et al. 2013
Chlorella vulgaris	286 mL/g VS	5000 mgVS/L	Lakaniemi et al. 2011
Chlorella vulgaris	240 mL/g VS	1000 mg VS/L	Ras et al. 2012
Chlorella vulgaris	189 mL/g VS	-	Polakovicova et al. 2012
Chlorella vulgaris	0.40–0.45 L/g VS	2677–6714 mg (COD)	Hernández and Córdoba 1993
Dunaliella	440 mL/g VS	910 mgVS/L	Chen 1987
Dunaliella salina	505 mL/g VS	2000 mgTS/L	Mussgnug et al. 2010
Dunaliella tertiolecta	24 mL/g VS	5000 mgVS/L	Lakaniemi et al. 2011
Durvillea Antarctica	492 mL/g VS	VS 3000 mgdryTS/d	Vergara-Fernandez et al. 2008
Euglena gracilis	485 mL/g VS	2000 mgTS/L	Mussgnug et al. 2010
Lake Chaohu natural population consortium	295 mL/g VS	-	Shuchuan et al. 2012
Macroystis pyrifera and *Durvillea Antartica* (50% blend)	540 mL/g VS	3000 mgdryTS/d	Vergara-Fernández et al. 2008
Macroystis pyrifera	545 mL/g VS	3000 mgdryTS/day	Vergara-Fernández et al. 2008
Microcystis sp.	70–153.5 mL/g VS	1500–6000 mgVS/L	Zeng et al. 2010
Nannochloropsis oculata	204 mL/g VS	-	Buxy et al. 2013
Nannochloropsis salina (lipid extracted biomass)	130 mL/g VS	2000 mg VS/L	Park and Li 2012
Phaeodactylum tricornutun	0.35 L/g VS	1.3 ± 0.4–5.8 ± 0.9	Zamalloa et al. 2012
Scenedesmus obliquus	287 mL/g VS	2000 mg TS/L	Mussgnug et al. 2010
Scenedesmus obliquus	240 mL/g VS	2000 mg VS/L	Zamalloa et al. 2012
Scenedesmus sp.	170 mL/g COD	1000 mgCOD/L	Gonzalez-Fernandez et al. 2012a
Scenedesmus sp. (single stage)	290 mL/g VS	18,000 mg VS/L	Yang et al. 2011
Scenedesmus sp. (two stage) Note: 46 mL/g/VS Hydrogen	354 mL/g VS	18,000 mg VS/L	Yang et al. 2011
Scenedesmus sp. and *Chlorella* sp.	16.3–15.8 ft^3	7.8–9.2 ft^3 /lb (VS)	Golueke et al. 1957
Scenedesmus sp. and *Chlorella* sp.	143 mL/g VS	4000 mg/VS/L	Yen and Brune 2007
Spirulina Leb 18	0.79 g/L	72,000 mg/L/TS	Costa et al. 2008
Spirulina maxima	0.35–0.80 m^3	20–100 kg/m^3 (VS)	Samson and Leduyt 1986
Spirulina maxima	320 mL/g VS	910 mg/VS/L	Chen 1987
Spirulina maxima	330 mL/g VS	22,500 mg/VS/L	Varel et al. 1988
Spirulina platensis UTEX1926	0.40 m^3 kg	-	Converti et al. 2009
Tetraselmis	0.25–0.31 L/g VS	2000 mg/VS	Marzano et al. 1982
Wastewater grown community	497 mL/g VS	2.16 g/L/TS	Salerno et al. 2009
Zygogonium sp.	344 mL/g VS	-	Ramamoorthy and Sulochana 1989

yields were reported in the literature. This variation is due to the fact that the biomethane yield is affected by various factors including the type of species (freshwater vs. saline water), difference among species of same type such as cell wall thickness, cell composition of carbohydrates, lipids and proteins, and the pretreatment conditions (temperature, pressure, exposure time) and the type of pretreatment (conventional thermal treatment, microwave or ultrasound, chemical or mechanical and biological treatment).

The focus of the study is another factor influencing the variations in biomethane production. For example, the lowest gas production recorded from freshwater microalgae biomass was ~70 mL g^{-1} VS (volatile solids) for untreated *Microcystis* sp. (Zeng et al. 2010). The gas production reported in this experiment was low due to the variations in inoculum start up volumes during the bio-methane potential assays rather than maximizing gas productivity. A gas production of 153 mL g^{-1} VS was recorded later for the same microalgae species utilizing an optimized inoculum ratio.

In another study, Lakaniemi et al. (2011) reported a low production rate of 24 mL g^{-1} VS for the saline microalgae species *Dunaliellater tiolecta*. This low production rate was attributed to the effects of salinity. The highest methane production recorded was by De Schamphelaire and Verstraete et al. (2009) with a recorded gas production of 600 mL g^{-1} VS for a mixed undetermined freshwater microalgae consortium. The data shown in Table 5 also highlights the difference in units and terminology used to report gas production from microalgae. Units range from gas production per grams of chemical oxygen demand (COD) destroyed, gas produced per gram of volatile solids loaded and gas produced per gram of total solids loaded. The standardization of terminology and standard units to report biogas productivities are essential for comparing microalgae and other digestible substrates.

Microwave Pretreatment of Microalgae Biomass

Microalgae biomass is more difficult to decompose when compared with other wastewater sludge and other biomass feedstock. The rigid cell wall structure makes it a challenge for efficient extraction of energy embedded in algal biomass. Similar to other feedstock, hydrolysis is the key step to achieve higher biogas yields. Pretreatment increases the solubility of the intracellular compounds of biomass and enhances accessibility of carbohydrates, proteins and lipids compounds in the feedstock. Pretreatment enhances biogas yield by cell disruption after extracting a specific sub-product such as lipids for biofuel (Chisti 2008) and/or molecules of other interests (Spoalore et al. 2006). In this case, the remaining intracellular components become accessible for the anaerobic bacteria.

Physical, chemical and biological processes have proven successful at improving the disintegration and anaerobic biodegradability of lignocellulosic biomass. Some studies reported an increased methane yield after thermal and ultrasound pretreatments (Chen and Oswald 1998, Gonzalez-Fernandez 2012a, Gonzalez-Fernandez 2012b, Alzate et al. 2012). Thermal pretreatment even at low temperatures of 75–95°C can improve microalgae methane yield. Microalgae anaerobic biodegradability increases which in turn increases biogas (methane) yield by 70% in respect to non-pretreated biomass.

Microwaves were also investigated as a pretreatment process for microalgae biomass pretreatment (Marin et al. 2010). The quantum energy applied by microwave irradiation is not capable of breaking down chemical bonds, however hydrogen bonds are or can be broken. Induction heating and dielectric polarization result in changes in the secondary and tertiary structure of proteins and cause cell hydrolysis. The polarization of macromolecules occurs by a consistent rotation through an alternating electric field. This process is influenced by microwave frequency, radiation time, biomass concentration and penetration depth (Park et al. 2010).

When a mixture of *Scenedesmus* and *Chlorella* biomasses was pretreated with microwave, the solubilization degree increased linearly with the applied specific energy. Final biogas yield was reported to have increased by 78% after a microalga grown in high rate algal ponds was pretreated with microwaves at 900 W for 3 min (Passos et al. 2013a). For the same pretreatment conditions (900 W for 3 min), at 15 days of hydraulic retention time (HRT) the methane production rate increased by 33% (from 0.12 to 0.16 L CH_4/L·day) and the methane yields by 30% (from 0.13 to 0.17 L CH_4/g VS) and at 20 day HRT, the methane production rate increased by 43% (from 0.14 to 0.20 L CH_4/L·day) and the methane yield by 58% (from 0.17 to 0.27 L CH_4/g VS) compared with the control reactor (Passos et al. 2014). This was

222 *Microwave-Mediated Biofuel Production*

the only test performed in continuous mode. Although continuous mode is the proper way to quantify the biogas production, the majority of the studies were carried out in batch mode, as it is easier and faster to develop (Rodriguez et al. 2015).

Microwave Effects on Microalgae Cell Wall Structure

The impact of microwave irradiation on the microalgae cell wall structure can be analyzed using microscopic imaging. It is hypothesized, in general, that pretreatment techniques such as microwave irradiation allow for cell disruption to increase the readily available organic matter and hydrolysis rate. Figure 5 shows images of different microalgae species before (Figures 5a, c, e, and g) and after (Figures 5b, d, f and h) microwave

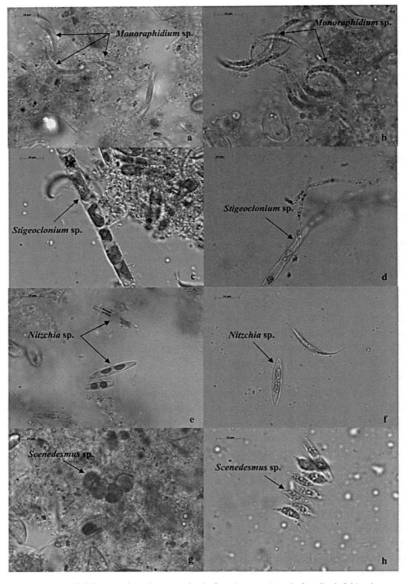

Figure 5. Microscopic images of different microalgae species before (a, c, e, g) and after (b, d, f, h) microwave pretreatment (Passos and Ferrer 2014).

pretreatment (Passos and Ferrer 2014). Observation in the optic microscope showed that microalgae were affected by microwave irradiation, as indicated by the decrease in chlorophyll pigmentation. However, it did not seem to induce complete microalgae cell wall lysis, since some cells appeared to be still intact after the pretreatment step.

Transmission electron microscopy (TEM) is another technology to study the individual cell structure integrity in biomass. Figure 6 shows the TEM images of *Monoraphidium* sp. which displays the cell wall integrity and contents (Passos and Ferrer 2014). It can be evidenced that cells were stressed by microwave irradiation, as organelles were affected and damaged beyond repair. It is also possible that the pretreatment was not responsible for cell wall lysis, stressed cells may have been more susceptible to bacteria attack, enhancing the anaerobic biodegradability of pretreated microalgae. To further evaluate the effect of pretreatment on microalgal biomass solubilization and anaerobic digestion, microscopic images of digested biomass ought to be analyzed. In this manner, it would be possible to elucidate whether pretreated cells were more accessible to methanogens, even if cell walls were not lysed after the pretreatment step.

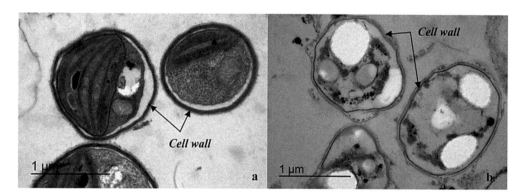

Figure 6. TEM images of *Monoraphidium* sp. before (a) and after (b) microwave pretreatment. Note that cell walls seem to be intact, although organelles are damaged (Passos and Ferrer 2014).

Energy Recovery Options from Microalgae

Microalgae can be used as feedstock for biodiesel production as discussed in Chapter 4. Lipid extraction process for biodiesel production leaves a biomass residue which accounts for approximately 65% of the harvested biomass rich in carbohydrates and proteins. This is considered waste with disposal costs that further increase the unfavorable economics of biodiesel production from microalgae. However, the significant quantities of proteins and carbohydrates in algal biomass residues can be used as feedstock for an anaerobic digestion process to yield biogas, methane. As shown in Figure 7, microalgae biomass can be used directly for biogas production via anaerobic digestion or the disrupted biomass after microwave treatment or the lipid extracted biomass can be fed to the anaerobic digestion system. The nutrient-rich supernatant can be recycled for reuse in microalgae cultivation.

There are two schemes for energy extraction from microalgae that involve biogas (methane) production (Figure 8). In scheme 1, biodiesel is produced from microalgal lipids followed by anaerobic digestion of residual biomass. Considering microalgae feedstock equivalent of 1 kg of volatile solids (VS), for a cell composition of 30% lipids, 45% proteins and 25% carbohydrates, 11.7 MJ of energy can be extracted in the form of biodiesel while 17.3 MJ of thermal energy can be produced, of which 5.5 MJ can be converted to electricity and 11.8 MJ can be recovered in the form of thermal energy in a cogeneration scheme. A total of 29 MJ/kg VS is possible in this scheme. In scheme 2, sole methane production results in an energy recovery potential of 28.2 MJ/kg VS.

Sialve et al. (2009) have reported on the possible additional energetic benefits with lipid recovery. Table 6 shows a comparison of two production schemes, sole biogas production and anaerobic digestion of lipid extracted algal biomass residues for three different species under normal and nitrogen limiting

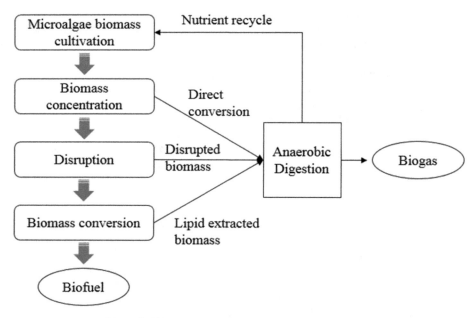

Figure 7. Biogas production routes from microalgae biomass.

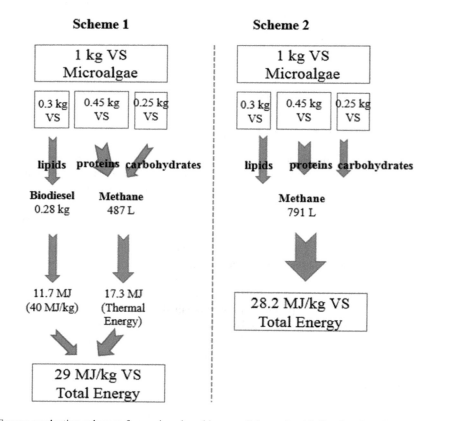

Figure 8. Energy production schemes from microalgae biomass: Scheme 1 – biodiesel and methane production; and scheme 2 – methane production only.

Table 6. Estimation of added energetic values from integrated biodiesel and methane production process from microalgae biomass (Sialve et al. 2009).

Species	Growth conditions	S1: anaerobic digestion of the whole algal biomass	S2: anaerobic digestion of algal biomass residues			Energetic added value with lipid recovery
		Methane[a] (kJ/g VS)	Methane[a] (kJ/g VS)	Lipids[b] (kJ/g VS)	Total energy (kJ/g VS)a	Additional energy (kJ/g VS)
C. vulgaris	–	23.0	20.1	6.6	26.7	3.7
C. vulgaris	Low N	24.9	17.2	14.7	32.0	7.1
C. emersonii	–	26.4	22.4	10.7	33.1	6.6
C. emersonii	Low N	33.1	27.6	23.2	50.8	17.7
C. protothecoides	–	23.4	21.8	4.1	25.8	2.4
C. protothecoides	Low N	25.5	22.2	8.5	30.7	5.2

[a] Computed with a methane calorific value of 35.6 kJ/L.
[b] Computed with the calorific value of rapeseed crude oil: 36.87 MJ/t

conditions. A maximum additional energetic value of 17.7 MJ/kg VS was estimated for *C. emersonii* under low nitrogen growth conditions. These estimates were made based on the reported experimental values from studies by Illman et al. (2000).

A few studies reported on the feasibility of biogas production from lipid extracted microalgal biomass residues and problems associated with the processes (Uggetti et al. 2014). A methane yield of 390 mLCH$_4$/gVS influent from residual *Scenedesmus* biomass derived from oil extraction processes by Yang et al. (2011). Co-digestion of algae biomass residue and lipid-rich fat, oil, and grease waste resulted in a specific methane production rate of 540 mL CH$_4$/gVS influent·d with regards to a rate of 150 mL CH$_4$/gVS influent·d when microalgae biomass was digested alone (Park and Li 2012). In another study, glycerol produced from the transesterification process was co-digested with the microalgae biomass residues (Ehimen et al. 2009).The effect of solvent used in oil extraction step on methane yields was reported in this study.

Problems associated with these approaches can be diverse. First, microalgae biomass residues generated after lipid extraction may cause more severe ammonia inhibition than the whole algae, due to their higher protein contents (Chisti 2008). However, this problem can be overcome by co-digestion of different biomass feedstock to increase the C/N ratio. Another important issue is with the presence of solvent in the biomass residues which can be toxic to the anaerobic bacteria in biogas production. Solvents are commonly used for lipid extraction from microalgae biomass. Chloroform, in particular, affects methane production adversely. Use and selection of solvent should be given proper consideration when developing integrated processes involving biodiesel and biogas production. The inhibitory effects of solvents can be reduced by washing off the residual solvent from the spent biomass. This process might create additional wastewater requiring further treatment. The energy and water requirements and the costs associated with washing and remediating the "wash water" should be considered as well. The washing process may also entrain the energy-rich polar molecules from the biomass. Altogether, the washing step may reduce the potential energy benefits from the entire process. There is a lack of literature in this area which is critical to energy and environmental sustainability of the solvents and their effect on the subsequent process steps and the overall process. More research is required in this area.

Agricultural and Other Wastes

Agricultural wastes such as rice straw and switchgrass and other lignocellulosic waste can be used as feedstock for biogas production. Similar to other wastes, these require a pretreatment for easy digestion by the anaerobic bacterial consortium. The microwave irradiation could lead to one or more changes in the features of the cellulosic biomass, including increased specific surface area, decreased polymerization and crystallinity of the cellulose, hydrolysis of the solubilized oligomers of hemicellulose and partial depolymerization of lignin (Sapci 2013). It can make the substrate more accessible to enzymes due to increased availability of contact surfaces and decreased crystalline structure. Therefore, thermal pretreatment using microwave radiation may be a good method for biomass pretreatment, as it disrupts the complex and rigid structure of the biomass that makes it resistant to mechanical stress and enzymatic attack.

A microwave-alkali pretreatment of rice straw was compared with alkali pretreatment alone (Zhu et al. 2005). The rice straw had a TSS weight loss of 44.6% composed of cellulose 69.2%, lignin 4.9%, and hemicellulose 10.2% after 30 minutes MW/alkali pretreatment at 700 W while it only had a weight loss of 41.5% and composition of cellulose 65.4%, lignin 6.0%, and hemicellulose 14.3% after 70 minutes for alkali pretreatment only. The rice straw pretreated by MW/alkali had a higher hydrolysis rate and glucose content in the hydrolysate in comparison with the alkali pretreatment.

In another study, the enhancement of the enzymatic digestibility of switchgrass by microwave assisted alkali pretreatment was investigated (Hu and Wen 2008). Here, microwave was used to pretreat switchgrass, which was then hydrolyzed by cellulase enzymes. When switchgrass was soaked in water and treated by microwave, the total sugar yield from the combined treatment and hydrolysis was 34.5 g/100 g biomass (58.5% of the maximum potential sugars released). This yield was 53% higher than the conventional heating of switchgrass. In addition, switchgrass was presoaked in alkali solution (0.1 g alkali/g biomass) and treated by microwave irradiation at 190°C and 5% TS for 30 minutes which resulted in a sugar yield of 99%. The advantage of microwave pretreatment over conventional heating pretreatment was attributed to the disruption of recalcitrant structures.

The effect of microwave pretreatment varies with the type of feedstock. It may have adverse or no effect on certain feedstock. For example, Sapci (2013) studied the effects of microwave pretreatment on the digestibility of Norwegian agricultural residual materials under anaerobic mesophilic conditions. Four different types of agricultural straws, winter wheat (WW), spring wheat (SW), oat straw (OS) and barley straw (B), were investigated for the biogas potential. Microwave pretreatment of biomass (100 g) was performed at 200 and 300°C with a hold time of 15 min in a laboratory-scale CEM Microwave Max asphalt oven (15 Amps, 50 Hz for 220–240 V). Sixty-six identical batch anaerobic reactors were run under mesophilic conditions for 60 days. Preliminary test results showed that the microwave pretreatment of the different straws did not improve their anaerobic digestion. An increase in the treatment temperature led to lower biogas production levels. An inverse relationship between the thermal conversion yield and cumulative biogas production was observed. Based on the elemental analysis of the pretreated straws, the content of carbon (C) and hydrogen (H) in the microwave pretreated biomass increased when the air temperature in the oven increased from 200 to 300°C. The C/N ratio analysis showed that when the temperature of the process was increased, the percentages of C and N in the pretreated MS decreased. In another study, conventional thermal and microwave methods were performed on *Pennisetum* hybrid grass, to analyze the effect of pretreatment on anaerobic digestion by the calculation of performance parameters using Logistic function, modified Gompertz equation, and transference function (Li et al. 2012). Results indicated that thermal pretreatment improved the biogas production of *Pennisetum* hybrid, whereas microwave method had an adverse effect on the performance.

Beszedes et al. (2011a) studied the biogas production potential of sludges originating from dairy and meat processing industries. Dairy waste had a total COD concentration of tCOD – 398 kg/m^3 and TS (wt.%) of 31.8% while the meat processing industry had 478.4 kg/m^3 and 27.2% of tCOD and TS respectively. The efficiency of microwave pretreatment was evaluated through the changes in the soluble fraction of the organic matter, the VS/TS ratio, the biogas yield, the methane content in the biogas, and the rate of batch mesophilic digestion. The microwave pretreatment proved to increase both the sCOD/tCOD and the VS/TS ratio. In addition, the biogas and methane yields increased during the digestion of the microwave pretreated food-industry sludge. A higher power level generally enhanced the biogas and methane production. Considering energy efficiency, the most economic pre-treatment of sludge from dairy and meat processing was at a power level of 1.5 Wg^{-1} and 2.5 Wg^{-1} microwave respectively; the surplus energy content of the enhanced biogas product could not compensate the extra energy demand of the stronger microwave pretreatment. Energy efficiency analysis for different microwave intensities are summarized in Table 8. The effect of microwave pretreatment at different energy intensities of 0.5, 2.5, and 5 W/g on the carbonaceous biochemical oxygen demand (cBOD$_5$), solubilization of organic matters (sCOD/tCOD), and the mesophilic anaerobic digestion of dairy sewage sludge was studied by Beszedes et al. (2011). Microwave pretreatment increased the specific biogas production from 220 mL g^{-1} to more than 600 mL g^{-1} because of the increased solubility (from 9.7% to more than 40%), and the enhanced accessibility of organic compounds for decomposing bacteria.

The effects of microwave irradiation at different intensities on solubilization, biodegradation and anaerobic digestion of sludge from the dairy sludge were studied by Rani et al. (2013). The changes in the soluble fractions of the organic matter, the biogas yield, the methane content in the biogas were used as control parameters for evaluating the efficiency of the microwave pretreatment. The most economical pretreatment of sludge was at 70% intensity for 12 min irradiation time for improving energy recovery. The COD solubilization, suspended solids reduction and biogas production were found to be 18.6%, 14% and 35% higher than the control, respectively. Combining microwave pretreatment with anaerobic digestion led to 67%, 64% and 57% of suspended solids reduction, VS reduction and biogas production higher than the control, respectively. A summary of effects of microwave pretreatment on wastewater sludge, organic wastes, energy crops and plant residues and food industry wastes and manure is presented in Table 7. Microwave cannot achieve size reduction as anticipated, however, solubilization and biodegradability can be improved which further enhance the biogas production.

Table 7. Effects of microwave pretreatment reported in various studies (Lagerkvist and Morgan-Sagastume 2012).

Feedstock/ pretreatment effect	Particle size reduction	Solubilization	Formation of Refractory compounds	Biodegradability enhancement	Loss of organic material	References
WWTP residues	na	+	0	0/+	na	Eskicioglu et al. 2007b, Toreci et al. 2009, Doğan and Sanin 2009, Tang et al. 2010
Organic waste from households	na	+	na	0/+	na	Marin et al. 2010
Energy crops/ plant residues	na	+	+	+	na	Jackowiak et al. 2011
Waste from food industry	na	+	na	+	na	Beszédes et al. 2011b
Manure	na	+	0	0	na	Jin et al. 2009

na – not applicable

Codigestion

Codigestion is a process in which two or more organic waste materials are digested together in a reactor, thereby improving anaerobic digestion. It may provide for a higher degradation of organics than the individual processes. Codigestion improves biogas yield due to the positive synergism established in the digestion medium and the supply of missing nutrients by the co-substrates. Codigestion of at least two suitable feedstocks enables a balanced nutrient composition, an appropriate C/N ratio and stable pH adjustment (Mata-Alvarez 2003, Mata-Alvarez et al. 2011).

Codigestion of different wastes can be accomplished to meet the requirements for mutual positive effects by optimizing the mixing ratio. For example, organic and inorganic nutrient composition can be balanced due to high amounts of fast degradable carbohydrates in sewage sludge and addition of slower degradable proteins and lipids in the algal biomass as well as essential trace elements and alkalinity for process stability. Equivalent amounts of carbohydrates, fats and proteins improve the process performance in anaerobic digestion (Nges et al. 2012). High C/N ratios cause nitrogen deficiency, whereas low ratios induce ammonia accumulation and inhibition. C/N ratios in the feedstock between 15 and 30 are considered to be optimal for anaerobic digestion (Weiland 2010). Feedstock composition can be optimized to adjust the C/N ratio in the biomass (Schwede et al. 2013).

Table 8. Characterization of anaerobic digestion of pretreated sludge. Energy content calculated from the methane component of biogas (Beszedes et al. 2011a).

Pretreatment	Time [min]	Specific MW energy cons. [J mL^{-1}]	Biogas production [mL g^{-1}TS] Dairy sludge	Biogas production [mL g^{-1}TS] Meat-proc. sludge	Methane/Biogas [%] Dairy sludge	Methane/Biogas [%] Meat-proc. sludge	Energy cont. [MJ m^{-3}] Dairy sludge	Energy cont. [MJ m^{-3}] Meat-proc. sludge	ΔE [J g^{-1}] Dairy sludge	ΔE [J g^{-1}] Meat-proc. sludge
Untreated	–	–	110.2 ± 5.3	211.4 ± 9.3	48.3 ± 1.4	48.1 ± 0.3	19.32	19.24	–	–
CH	6	–	160.1 ± 5.1	293.1 ± 9.1	50.1 ± 0,9	54.9 ± 0.8	20.04	21.96	-123.96	-52.09
	10	375	160.2 ± 3.8	261.8 ± 4.2	50.7 ± 0.6	52.1 ± 0.9	20.28	20.84	-136.08	-25.39
MW 0.5 Wg^{-1}	20	750	173.5 ± 3.5	266.3 ± 9.1	50.2 ± 0.4	54.7 ± 0.5	20.08	21.88	-189.59	-10.63
	30	1125	180.1 ± 4.3	278.5 ± 9.3	53.8 ± 0.7	57.1 ± 0.7	21.52	22.84	-294.82	-287.18
	40	1500	196.8 ± 3.6	314.5 ± 5.4	55.9 ± 1.1	56.3 ± 0.6	22.36	22.52	-692.11	-614.21
	10	1125	240.1 ± 4.3	291.9 ± 5.7	52.8 ± 1.1	58.7 ± 0.9	21.12	23.48	-12.71	191.85
MW 1.5 Wg^{-1}	20	2250	293.4 ± 3.9	318.8 ± 9.8	54.1 ± 0.7	60.1 ± 0.5	21.64	24.04	-22.43	20.49
	30	3375	370.3 ± 3.8	359.6 ± 7.1	55.4 ± 0.4	60.5 ± 0.5	22.16	24.20	-236.47	254.65
	40	4500	383.0 ± 4.6	392.7 ± 4.3	56.1 ± 1.3	60.2 ± 0.8	22.44	24.08	-498.23	103.51
	10	1875	300.2 ± 4.1	319.6 ± 5.1	57.7 ± 0.7	60.4 ± 0.8	23.08	24.16	18.63	-7.65
MW 2.5 Wg^{-1}	20	3750	334.6 ± 3.9	383.2 ± 5.4	59.6 ± 0.5	60.1 ± 0.5	23.84	24.04	332.96	-27.07
	30	5625	387.3 ± 4.4	434.5 ± 7.2	59.8 ± 0.2	60.8 ± 0.7	23.92	24.32	406.42	-335.87
	40	7500	397,9 ± 2.8	487.8 ± 4.7	60.6 ± 0.4	61.0 ± 1.3	24.24	24.40	391.40	-436.8
	10	3750	370.1 ± 5.1	405.2 ± 4.1	58.9 ± 0.5	60.5 ± 0.2	23.56	24.20	126.53	-269.32
MW 5 Wg^{-1}	20	7500	397.2 ± 2.9	481.3 ± 2.8	60.1 ± 0.5	61.5 ± 0.6	24.04	24.60	-257.87	-622.45
	30	11250	400.2 ± 6.1	493.2 ± 2.3	60.7 ± 0.7	61.2 ± 0.3	24.28	24.48	-615.03	-1598.63
	40	15000	406.4 ± 4.8	495.1 ± 3.8	60.5 ± 0.3	61.3 ± 0.6	24.20	24.52	-1037.12	-1921.86

A simple codigestion example is the anaerobic digestion of organic fraction of municipal solid waste, with thickened waste activated sludge and primary sludge to enhance biodegradation of solid waste, increase longevity of existing landfills and lead to more sustainable development by improving waste to energy production (Ara et al. 2014). Improvements in specific biogas production for batch assays, with concomitant improvements in total chemical oxygen demand and volatile solid removal, were obtained with organic fraction of municipal solid waste:thickened waste activated sludge:primary sludge mixtures at a ratio of 50:25:25 (with and without thickened waste activated sludge microwave pretreatment). This combination was used for continuous digester studies. At 15 d hydraulic retention times, the codigestion of organic fraction of municipal solid waste: organic fraction of municipal solid waste:primary sludge and organic fraction of municipal solid waste: thickened waste activated sludge microwave:primary sludge resulted in a 1.38- and 1.46-fold increase in biogas production and concomitant waste stabilization when compared with thickened waste activated sludge: primary sludge (50:50) and thickened waste activated sludge microwave:primary sludge (50:50) digestion at the same hydraulic retention times and volumetric volatile solid loading rate, respectively. The digestion of organic fraction of municipal solid waste with primary sludge and thickened waste activated sludge provides beneficial effects that could be implemented at municipal wastewater treatment plants that are operating at loading rates of less than design capacity.

Co-digestion with food waste and sewage sludge

The effects of microwave pretreatment on codigestion of food waste (FW) and sewage sludge (SS) were investigated through a series of mesophilic biochemical methane potential (BMP) tests to determine the optimum ratio of FW and SS based on microwave pretreatment (Zhang et al. 2016). An optimized ratio of 3:2 for co-digestion of FW and SS with microwave pretreatment resulted in methane production of 316.24 and 338.44 mL CH_4/gVS_{added} for MW-FW and MW-SS, respectively. The daily methane production trends over 35 day operation are shown Figure 9. The MW-SS was superior for methane production compared to MW-FW, in which accumulation of propionic acid led to the inhibition of methanogenesis.

The investigation of bacterial and archaeal community evolution by high-throughput sequencing method revealed that *Proteiniborus* and *Parabacteroides* were responsible for proteins and polysaccharides degradation for all, respectively, while *Bacteroides* only dominated in codigestion. *Methanosphaera* dominated in MW-FW at the active methane production phase, while it was *Methanosarcina* in MW-SS and mono-SS. Detailed evolution of bacterial community for MW-FW and MW-SS on anaerobic co-digestion of FW and SS through chimer analysis revealed a total of 499, 307 high quality reads. A total of 33 phylum were identified, and *Firmicutes, Proteobacteria, Bacteroidetes* and *Actinobacteria* were the predominant phylum in the bacterial community. *Firmicutes* was enriched by 5.05, 6.45 and 4.61 times for mono-SS, MW-SS and MW-FW, respectively. The abundance of *Firmicutes* increased significantly on day 5. After that, *Firmicutes* changed little for mono-SS and MW-SS while the abundance increased much further on day 12 for MW-FW and changed little to the end. The evolution of *Proteobacteria* was quite different for different treatments especially for mono-SS. The abundance of *Proteobacteria* decreased significantly for MW-SS and MW-FW in comparison with mono-SS, and it was the maximum contributor to the difference of mono-SS and co-digestion. The phylum *Actinobacteria* showed similar evolution to *Proteobacteria*. The abundance of *Bacteroidetes* decreased a little for mono-SS while increased for MW-SS and MW-FW. There was a common phenomenon for these four dominant phylum that the maximum or minimum abundance, that is, greatest changes, all appeared on day 5, which indicated that recombination of bacterial community occurred at acidification phase corresponding to VFAs accumulation.

Codigestion of olive pomace and wastewater sludges was investigated for enhancement of biogas production through microwave pretreatment (Alagöz et al. 2015). Codigestion with appropriate sludge pretreatment together resulted in improved balance of nutrients, efficient pathogen, organic removals and higher methane yields compared to mono-digestions of these substrates. Codigestion yielded 0.21 L CH_4/g VS added, whereas the maximum yields from mono-digestions of olive pomace and wastewater sludges were 0.18 and 0.16 L CH_4/g VS added, respectively. The applied codigestion led to 17–31% increase in methane production. Microwave pretreatment improved the methane yields by 24% and 52%, respectively,

230   *Microwave-Mediated Biofuel Production*

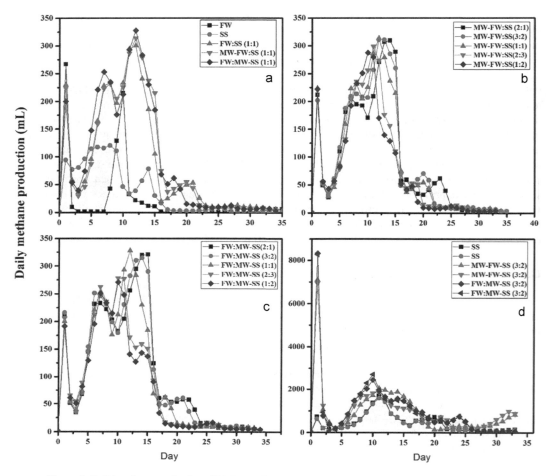

Figure 9. Daily methane production of BMP tests in different treatments (adapted from Zhang et al. 2016).

compared to codigestion without pretreatment. The highest yields were obtained at 30 min microwave pretreatment wastewater sludge as 0.46 L biogas/g VS added and 0.32 L CH_4/g VS added, respectively. In addition to the improved biogas generation, the codigestion of these substrates seemed to generate an energy efficient alternative solution for the disposal of agricultural product residuals and the wastewater sludges.

Microwaves in Syngas (Biogas to Syngas) Production

Microwave radiation can be used to upgrade biogas (CH_4 + CO_2) into syngas. The conversion of biogas into hydrogen is currently carried out via steam reforming and after the removal of CO_2, which is a highly energy demanding operation. Dry reforming of biogas (reaction between methane and carbon dioxide) is carried out without the need for any conditioning stage. High conversions and production of synthesis gas can be achieved by microwave-assisted dry reforming of biogas, even using low-cost carbon materials as catalysts. Low conversions of CH_4 and CO_2 have been obtained using similar catalysts and conditions under conventional heating. Moreover microwave-assisted dry reforming of biogas over a mixture of carbon and metal-based catalysts has exhibited similar conversions to those obtained under conventional heating using metal-based catalysts alone. Therefore, the application of microwaves has been proven to allow the replacement of part of metal catalyst by low cost carbonaceous catalysts without affecting conversions.

Energy Efficiency and Net Energy Ratio

The energy balance of pretreatment methods is crucial to determine their suitability for full-scale implementation. It can be realized that the heat and electricity demands for pretreatment and anaerobic digestion must be lower than the additional energy generated from biogas through these extra efforts. Very few studies focused on the net energy benefit analysis of the pretreatment techniques. Passos et al. (2015) compared the energy efficiency and net energy benefit ratios for various pretreatment techniques including microwave, thermal and hydrothermal pretreatment methods. Similar to results reported by others, thermal pretreatment returned favorable energy benefits followed by hydrothermal pretreatment. Although microwave pretreatment enhances the solubilization and biogas production for a variety of feedstock, the energy input requirements seem to be too high, offsetting the potential energy return benefits. These findings suggest that microwave pretreatment has potential to enhance the energy benefits in biogas production provided a suitable operating scheme combined with process optimization is necessary for improving the energy efficiency. This can be done by combining the microwave pretreatment with other chemical (acid or base conditions) treatments to reduce the energy requirements and enhance the pretreatment effects.

As a case study, Beszedes et al. (2009) compared the net energy benefits of three different pretreatment methods namely microwave irradiation at pH 2, microwave irradiation combined with ozonation, and ozonation alone at 30 and 60 minutes with the goal of enhancing the biogas production from maize production sludge. All the pretreatment methods enhanced the methane-production and there was no significant difference between the biogas production of the 30 min, the 60 min and the combined ozone/microwave pretreatments. The microwave irradiation at pH 2 resulted in significantly higher methane production compared with ozonation. The combined acidic and microwave treatment could significantly accelerate the hydrolysis of organic matter, which is the limiting step of the microbial degradation.

When combined ozone/microwave treatment was applied, the increment was 10-fold, however, the acidic microwave treatment enhanced the initial biogas production rate as high as 25-fold. There was a considerable difference between the change of the biodegradability—determined via aerobic microorganisms—and methane or biogas product, which was measured by anaerobic microorganisms in the case of 30-min ozone pretreatment comparing with control and other pretreatments. The 30-min ozone pretreatment caused an oxidizing effect, but aerobic microorganisms during 5 days were not able to degrade organic matter to the same extent as anaerobic bacteria during 30 days.

The net energy product (NEP) of processes involving microwave pretreatment can be calculated using equation [10] which includes the terms related energy production from combustion of methane and the energy consumption due to microwave pretreatment.

$$NEP = q_{comb} \times M_{methane} - P_{mw} \times \tau \qquad [10]$$

where NEP is the net energy product [J], q_{comb} is the combustion heat [J kg^{-1}] of methane, $M_{methane}$ is the mass of the methane produced [kg], P_{mw} is the power of the microwave magnetron [W], and t is the duration of treatment [s]. Table 10 shows the comparison of the net energy balance for four pretreatment methods. The net energy balance for the microwave pretreatment was due to the increased solubilization of organic matter which enhances the biogas production.

Table 10. Net energy balance comparison for different pretreatment methods.

Pretreatment	Treatment time [min]	BD5% (BOD5/COD) × 100	Initial biogas production rate [cm^3 g^{-1} day^{-1}] U	Net Energy Balance [J]
Untreated	-	26	1.04	500
Ozone	30	63	3.77	2000
Ozone	60	94	7.40	−900
Ozone/microwave	30 + 5	96	9.52	−950
Microwave (pH = 2)	5	95	25.75	2500

Microwave pretreatment increases the sugar solubility when compared with thermal pretreatment (3.2 vs. 2.7 times) (Park and Ahn 2011). However, the energy efficiency may not be positive. Some studies reported negative net energy balances for the microwave pretreatment (Tang et al. 2010, Jackowaik et al. 2011, Hu et al. 2012). The water content and the pretreatment time were identified as the most influencing parameters on the net energy balance (Tang et al. 2010). Jackowaik et al. (2011) reported a wide range of values for the ratio of energy consumption to energy production between 54 and 103. Hu et al. (2012) reported ratios of as low as 4.9 and up to 130.4 for microwave pretreatment between 200 W and 600 W. It should be noted that these values are based on the laboratory scale studies which implies that there is a possibility to improve the net energy benefits by considering optimum reaction conditions and microwave process design. It appears that high solid content and short treatment time may significantly increase the energy efficiency. Therefore, for large-sale applications a high solid content and a flash microwave device could be used to improve the energy efficiency of pretreatment.

Economic Analysis

Studies involving economic analysis of pretreatment methods are rare. Bordeleau and Droste (2011) developed a comprehensive model using Microsoft Excel and its Visual Basic Assistant to evaluate pretreatment permutations for conventional wastewater treatment plants. Most commonly used conventional heating, chemical pretreatment, microwave and ultrasound treatments were considered based on the fact that these techniques have the ability to increase the organic matter hydrolysis to improve biogas production by 30–50% and reduce solids by 20–60%. The model in conjunction with user defined values (UDVs) creates a treatment and cost output.

Well-established energy equations and wastewater characteristics, both average and high, were used. Table 11 shows a comparison of the daily costs and unit costs for the four different pretreatment methods. Average and high flows were 460×10^3 m³/d and 750×10^3 m³/d, respectively. Net costs per influent flow for ultrasound, chemical, conventional heating, and microwave were 0.0166, 0.0217, 0.0124, 0.0119 $/m³ and 0.0264, 0.0357, 0.0187, and 0.0162 $/m³ for average and high conditions, respectively. The average cost increase from results excluding pretreatment use for all processes was 0.003 and 0.0055 $/m³ for average and high conditions, respectively. No matter the permutation, pretreatments requiring more energy to achieve required hydrolysis levels were costlier. It was concluded that these pretreatment techniques are viable only if energy recovery can be maximized, dewaterability is increased, and solids meet environmental constraints for disposal at lower costs (Bordeleau and Droste 2011).

Table 11. Model results for base and advanced costs.

	Base Results		Advanced Results	
	Daily costs ($/d)	Costs ($/m³)	Daily costs ($/d)	Costs ($/m³)
UDV_{avg} (user defined value)				
Ultrasonic (ULT)	10293	0.0133	12669	0.0166
Alkali Treatment (CM)	10293	0.0133	15035	0.0217
Conventional heating (CH)	9695	0.0120	10755	0.0124
Microwave heating (MWH)	9695	0.0120	10576	0.0119
UDV_{high}				
Ultrasonic (ULT)	27627	0.0200	35091	0.0264
Alkali Treatment (CM)	27627	0.0200	42027	0.0357
Conventional heating (CH)	25779	0.0175	29319	0.0187
Microwave heating (MWH)	25779	0.0175	28788	0.0162

Challenges with Microwave Pretreatment

A comparison of the methane yields under different pretreatment conditions is shown in Figure 10 (Passos et al. 2014). These pretreatment methods include thermal, hydrothermal, steam explosion, ultrasound and microwave irradiations. It is difficult to make a proper comparison among the results reported by many researchers with different pretreatment conditions. This is because the feedstock type, process scheme and the pretreatment technique vary among the studies. It can be noted from Figure 10 that the microwave pretreatment is more effective in increasing the biogas yield (80%) even at very low solubilization rates (10%). This trend is somewhat similar to thermal pretreatment technique in terms of solubilization rate (20%) and biogas yields (60%). This confirms that both technologies are based on a temperature gradient as a driving force. On the other hand, hydrothermal and steam explosion pretreatment techniques seem to produce similar effects. Ultrasound pretreatment method increases the solubility of the organic matter (80–100%) but the biogas yields are slightly lower (20–30%).

The differences in the effects of pretreatment methods on solubilization and biogas yields can be explained as the varying impacts on the microalgae cell structure and the bioavailability of different organic compounds. When the pretreatment technique increases biomass solubilization, but methane yield remains similar to non-pretreated biomass, one possible explanation is the formation of recalcitrant compounds in the hydrolysis step, which hinders the anaerobic digestion process (Passos et al. 2014). On the other hand, when the pretreatment technique results in low solubilization degree, but high methane yield, non-soluble organic compounds may have been modified and possibly are more available to anaerobic bacteria. Finally, the comparison of pretreatment methods can be more reliable when using the same microalgal biomass, since pretreatment effects are species-specific and difficult to extrapolate. In addition, microscopic analysis of pretreated biomass would help understanding the effect of each pretreatment on the cell structure.

The energy return ratio (the ratio of energy output to the energy input) analysis of different pretreatment methods has shown that microwave pretreatment has the lowest return ratio (Passos et al. 2014). The biogas yields under microwave pretreatment are higher than any other pretreatment method albeit at a high energy input. This calls for several challenges which include improvements in microwave application, reactor design and optimization of power input and energy utilization. In addition, scaling-up of microwave pretreatment for biogas production will also pose several challenges.

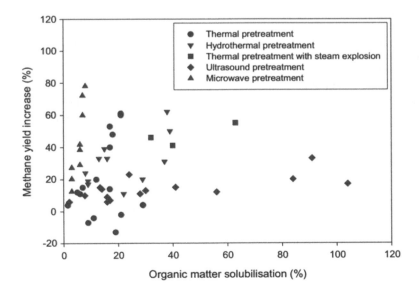

Figure 10. Organic matter solubilization vs. methane yield increase after mixed microalgae biomass pretreatment using different techniques (Passos et al. 2014).

Other considerations include that microwave irradiation has 1 mm^{-1} m wavelengths in the electromagnetic spectrum with equivalent frequencies of 300 GHz–300 MHz, respectively. For the heating or drying of thin substances, frequencies of 2,450 MHz with correspondingly short wavelengths (12.24 cm) are adequate. If deep penetration into materials is required, frequencies of 900 MHz with correspondingly long wavelengths (37.24 cm) and energy outputs of up to 100 kW are required. Microwave heating is absorbed selectively by substances containing more moisture, sugars or fats. The distribution of heat in microwave irradiated WW is thus not uniform because the fluid is two phase (solid and liquid) and heterogeneous in both phases.

Concluding Remarks

The key parameters in biogas production are the biodegradable organic matter; a pH between 6.8 to 7.3; and an optimal C/N ratio for the development of the microorganisms and minimal concentrations of inhibitory substances for microorganisms. Pretreatment is necessary for the easy and increased digestion and improved biogas production. Microwave irradiation effectively disintegrates the sludge which increases the readily biodegradable portion in liquid phase of sludge. Microwave pretreatment leads to enhanced volatile solids reduction, improvement in sludge dewaterability and biogas production during the anaerobic digestion and reduces the solids retention time remarkably. Higher biogas production potential have been reported for combination microwave pretreatment with alkaline, acidic and advances peroxidation methods at reduced energy consumption and reduced costs. In sludge treatment, microwave pretreatment produces pathogen-free (class A) sludge suitable for land applications. Microwave pretreatment offers the environmentally safe disposal of industrial sludge.

From the process point of view, co-digestion and microwave pretreatment together improves biogas yield by positive synergies established in the digestion and nutrients supply to the lack of co-sublayers. The ratio carbon and nitrogen (C/N) of the raw material is essential in the production of biogas. The optimal C/N ratio is expected to be in the range 15–25 when the anaerobic digestion process is carried out in a single stage, and for the situation when the process develops in two steps, the optimal C/N ratio will range: for step I: 10–45, for step II: 20–30. Temperature control is critical to the development of microwave anaerobic digestion process, having a strong influence over the quality and quantity of biogas production. Hydraulic retention time should be optimized in the process design to maximize the biogas yields.

For the microwave pretreatment step, several areas need improvement. The microwave pretreatment conditions should be optimized to enhance sludge solubilization, biogas generation and biosolids minimization. Scale-up of the lab-scale units to full-scale system needs to be carried out with careful experimental design and innovations. Investigation of fate of spore and cyst forming organisms and Enteroviruses under low and high MW thermal treatment conditions during sludge pretreatment and subsequent digestion is important. Potential synergistic benefits between microwave combination pretreatment schemes should be explored. The fundaments understanding of the dielectric properties of materials and their interaction with microwave irradiation should be pursued. Reuse of waste chemicals and heat should be practiced to achieve an efficient sludge treatment scheme and to minimize costs and lower carbon footprints.

References

Alagöz, B.A., Yenigün, O. and Erdinçler, A. 2015. Enhancement of anaerobic digestion efficiency of wastewater sludge and olive waste: Synergistic effect of co-digestion and ultrasonic/microwave sludge pre-treatment. *Waste Management.* 46: pp.182–188.

Alzate, M.E., Muñoz, R., Rogalla, F., Fdz-Polanco, F. and Pérez-Elvira, S.I. 2012. Biochemical methane potential of microalgae: influence of substrate to inoculum ratio, biomass concentration and pretreatment. *Bioresource Technology.* 123: pp.488–494.

Angelidaki, I. and Sanders, W. 2004. Assessment of the anaerobic biodegradability of macropollutants. *Re/Views in Environmental Science & Bio/Technology.* 3(2): pp.117–129.

Ara, E., Sartaj, M. and Kennedy, K. 2014. Effect of microwave pre-treatment of thickened waste activated sludge on biogas production from co-digestion of organic fraction of municipal solid waste, thickened waste activated sludge and municipal sludge. *Waste Management & Research.* 32(12): pp.1200–1209.

Ariunbaatar, J., Panico, A., Esposito, G., Pirozzi, F. and Lens, P.N. 2014. Pretreatment methods to enhance anaerobic digestion of organic solid waste. *Applied Energy.* 123: pp.143–156.

Banik, S., Bandyopadhyay, S. and Ganguly, S. 2003. Bioeffects of microwave—a brief review. *Bioresource Technology.* 87(2): pp.155–159.

Beall, F.C. and Eickner, H.W. 1970. *Thermal Degradation of Wood Components: A Review of the Literature* (No. FSRP-FPL-130). Forest Products Lab Madison Wis.

Becker, W. 2004. 18 Microalgae in Human and Animal Nutrition. *Handbook of Microalgal Culture: Biotechnology and Applied Phycology.* p.312.

Benjamin, M.M., Woods, S.L. and Ferguson, J.F. 1984. Anaerobic toxicity and biodegradability of pulp mill waste constituents. *Water Research.* 18(5): pp.601–607.

Beszédes, S., Kertész, S., László, Z., Szabo, G. and Hodur, C. 2009. Biogas production of ozone and/or microwave-pretreated canned maize production sludge. *Ozone: Science & Engineering.* 31(3): pp. 257–261.

Beszédes, S., László, Z., Horváth, Z.H., Szabó, G. and Hodúr, C. 2011a. Comparison of the effects of microwave irradiation with different intensities on the biodegradability of sludge from the dairy-and meat-industry. *Bioresource Technology.* 102(2): pp.814–821.

Beszédes, S., László, Z., Szabó, G. and Hodúr, C. 2011b. Effects of microwave pretreatments on the anaerobic digestion of food industrial sewage sludge. *Environmental Progress & Sustainable Energy.* 30(3): pp.486–492.

Beszédes, S., Veszelovszki, P., Ludányi, L., Keszthelyi-Szabó, G. and Hodúr, C. 2014. Correlation between dielectric parameters and biodegradability of meat processing sludge. *Annals of the Faculty of Engineering Hunedoara.* 12(3): p.207.

Bobleter, O. 1994. Hydrothermal degradation of polymers derived from plants. *Progress in Polymer Science.* 19(5): pp.797–841.

Bordeleau, E.L. and Droste, R.L. 2011. Comprehensive review and compilation of pretreatments for mesophilic and thermophilic anaerobic digestion. *Water Science & Technology.* 63(2): pp.291–296.

Bouallagui, H., Touhami, Y., Cheikh, R.B. and Hamdi, M. 2005. Bioreactor performance in anaerobic digestion of fruit and vegetable wastes. *Process Biochemistry.* 40(3): pp.989–995.

Buxy, S., Diltz, R. and Pullammanappallil, P. 2013. Biogasification of marine algae Nannochloropsisoculata. *Materials Challenges in Alternative and Renewable Energy II: Ceramic Transactions.* pp.59–67.

Byun, I.G., Lee, J.H., Lee, J.M., Lim, J.S. and Park, T.J. 2014. Evaluation of non-thermal effects by microwave irradiation in hydrolysis of waste-activated sludge. *Water Science & Technology.* 70(4): pp.742–749.

Campanone, L.A. and Zaritzky, N.E. 2005. Mathematical analysis of microwave heating process. *Journal of Food Engineering.* 69(3): pp.359–368.

Carrère, H., Dumas, C., Battimelli, A., Batstone, D.J., Delgenès, J.P., Steyer, J.P. and Ferrer, I. 2010. Pretreatment methods to improve sludge anaerobic degradability: a review. *Journal of Hazardous Materials.* 183(1): pp.1–15.

Chandra, R., Takeuchi, H. and Hasegawa, T. 2012. Methane production from lignocellulosic agricultural crop wastes: A review in context to second generation of biofuel production. *Renewable and Sustainable Energy Reviews.* 16(3): pp.1462–1476.

Chen, P.H. 1987. *Factors Influencing Methane Fermentation of Micro-algae.* California Univ., Berkeley (USA).

Chen, P.H. and Oswald, W.J. 1998. Thermochemical treatment for algal fermentation. *Environment International.* 24(8): pp.889–897.

Cheng, J., Liu, Y., Lin, R., Xia, A., Zhou, J. and Cen, K. 2014. Cogeneration of hydrogen and methane from the pretreated biomass of algae bloom in Taihu Lake. *International Journal of Hydrogen Energy.* 39(33): pp.18793–18802.

Chisti, Y. 2008. Biodiesel from microalgae beats bioethanol. *Trends in Biotechnology.* 26(3): pp.126–131.

Climent, M., Ferrer, I., del Mar Baeza, M., Artola, A., Vázquez, F. and Font, X. 2007. Effects of thermal and mechanical pretreatments of secondary sludge on biogas production under thermophilic conditions. *Chemical Engineering Journal.* 133(1): pp.335–342.

Converse, A.O., Kwarteng, I.K., Grethlein, H.E. and Ooshima, H. 1989. Kinetics of thermochemical pretreatment of lignocellulosic materials. *Applied Biochemistry and Biotechnology.* 20(1): pp.63–78.

Converti, A., Oliveira, R.P.S., Torres, B.R., Lodi, A. and Zilli, M. 2009. Biogas production and valorization by means of a two-step biological process. *Bioresource Technology.* 100(23): pp.5771–5776.

Costa, J.A.V., Santana, F.B., da Rosa Andrade, M., Lima, M.B. and Franck, D.T. 2008. Microalga biomass and biomethane production in the south of Brazil. *Journal of Biotechnology.* 136: p.S430.

De Bere, L. 2000. Anaerobic digestion of solid waste: state-of-the-art. *Water Science and Technology.* 41(3): pp.283–290.

De Schamphelaire, L. and Verstraete, W. 2009. Revival of the biological sunlight-to-biogas energy conversion system. *Biotechnology and Bioengineering.* 103(2): pp.296–304.

Delgenés, J.P., Penaud, V. and Moletta, R. 2002. Pretreatments for the enhancement of anaerobic digestion of solid wastes. In: Biomethanization of the Organic Fraction of Municipal Solid Wastes. IWA Publishing, pp.201–228.

Doğan, I. and Sanin, F.D. 2009. Alkaline solubilization and microwave irradiation as a combined sludge disintegration and minimization method. *Water Research.* 43(8): pp.2139–2148.

Domansky, R. and Rendos, F. 1962. On the Pyrolysis of Wood and its Components. *HolzRoh-Werkst.* 20: pp.473–476.

Ehimen, E.A., Connaughton, S., Sun, Z. and Carrington, G.C. 2009. Energy recovery from lipid extracted, transesterified and glycerol codigested microalgae biomass. *GCB Bioenergy.* 1(6): pp.371–381.

Eskicioglu, C., Droste, R.L. and Kennedy, K.J. 2006a. Performance of continuous flow anaerobic sludge digesters after microwave pretreatment. *Proceedings of the Water Environment Federation.* 2006(13): pp.526–540.

Eskicioglu, C., Kennedy, K.J. and Droste, R.L. 2006b. Characterization of soluble organic matter of waste activated sludge before and after thermal pretreatment. *Water Research.* 40(20): pp.3725–3736.

Eskicioglu, C., Droste, R.L. and Kennedy, K.J. 2007a. Performance of anaerobic waste activated sludge digesters after microwave pretreatment. *Water Environment Research.* 79(11): pp.2265–2273.

Eskicioglu, C., Terzian, N., Kennedy, K.J., Droste, R.L. and Hamoda, M. 2007b. A thermal microwave effects for enhancing digestibility of waste activated sludge. *Water Research.* 41(11): pp.2457–2466.

Eskicioglu, C., Kennedy, K.J. and Droste, R.L. 2008. Initial examination of microwave pretreatment on primary, secondary and mixed sludges before and after anaerobic digestion. *Water Science and Technology.* 57(3): pp.311–318.

Eskicioglu, C., Kennedy, K.J. and Droste, R.L. 2009. Enhanced disinfection and methane production from sewage sludge by microwave irradiation. *Desalination.* 248(1): pp.279–285.

Fengel, D. and Wegener, G. 1984. Wood: chemistry, ultrastructure, reactions. *Walter de Gruyter.* 613: pp.1960–82.

Ferrer, I., Ponsá, S., Vázquez, F. and Font, X. 2008. Increasing biogas production by thermal (70 C) sludge pre-treatment prior to thermophilic anaerobic digestion. *Biochemical Engineering Journal.* 42(2): pp.186–192.

Ferrer, I., Vázquez, F. and Font, X. 2011. Comparison of the mesophilic and thermophilic anaerobic sludge digestion from an energy perspective. *Journal of Residuals Science & Technology.* 8(2): pp.81–87.

Fox, M.H., Noike, T. and Ohki, T. 2003. Alkaline subcritical-water treatment and alkaline heat treatment for the increase in biodegradability of newsprint waste. *Water Science and Technology.* 48(4): pp.77–84.

Fox, M. and Noike, T. 2004. Wet oxidation pretreatment for the increase in anaerobic biodegradability of newspaper waste. *Bioresource Technology.* 91(3): pp.273–281.

Frigon, J.C. and Guiot, S.R. 2010. Biomethane production from starch and lignocellulosic crops: a comparative review. *Biofuels, Bioproducts and Biorefining.* 4(4): pp.447–458.

Garrote, G., Dominguez, H. and Parajo, J.C. 1999. Hydrothermal processing of lignocellulosic materials. *European Journal of Wood and Wood Products.* 57(3): pp.191–202.

Golueke, C.G., Oswald, W.J. and Gotaas, H.B. 1957. Anaerobic digestion of algae. *Applied Microbiology.* 5(1): p.47.

Golueke, C.G. and Oswald, W.J. 1959. Biological conversion of light energy to the chemical energy of methane. *Applied Microbiology.* 7(4): pp.219–227.

González-Fernández, C., Sialve, B., Bernet, N. and Steyer, J.P. 2012a. Thermal pretreatment to improve methane production of Scenedesmus biomass. *Biomass and Bioenergy.* 40: pp.105–111.

González-Fernández, C., Sialve, B., Bernet, N. and Steyer, J.P. 2012b. Comparison of ultrasound and thermal pretreatment of Scenedesmus biomass on methane production. *Bioresource Technology.* 110: pp.610–616.

Gossett, J.M., Stuckey, D.C., Owen, W.F. and McCarty, P.L. 1982. Heat treatment and anaerobic digestion of refuse. *J. Environ. Eng. Div., ASCE (United States)*, 108.

Gregg, D. and Saddler, J.N. 1996. A techno-economic assessment of the pretreatment and fractionation steps of a biomass-to-ethanol process. *Seventeenth Symposium on Biotechnology for Fuels and Chemicals.* pp.711–727. Humana Press.

Grohmann, K., Torget, R., Himmel, M. and Scott, C.D. 1986, January. Dilute acid pretreatment of biomass at high acid concentrations. In *Biotechnol. Bioeng. Symp.;(United States)* (Vol. 17, No. CONF-860508-). Solar Energy Research Institute, Golden, CO 80401, USA. Solar Fuels Research Div.

Gunasekaran, S. and Yang, H.W. 2007. Effect of experimental parameters on temperature distribution during continuous and pulsed microwave heating. *Journal of Food Engineering.* 78(4): pp.1452–1456.

Harris, R.F. and Adams, S.S. 1979. Determination of the carbon-bound electron composition of microbial cells and metabolites by dichromate oxidation. *Applied and Environmental Microbiology.* 37(2): pp.237–243.

Hendriks, A.T.W.M. and Zeeman, G. 2009. Pretreatments to enhance the digestibility of lignocellulosic biomass. *Bioresource Technology.* 100(1): pp.10–18.

Hernández, E.S. and Córdoba, L.T. 1993. Anaerobic digestion of Chlorella vulgaris for energy production. *Resources, Conservation and Recycling.* 9(1): pp.127–132.

Higgins, M.J. and Novak, J.T. 1997. Dewatering and settling of activated sludges: The case for using cation analysis. *Water Environment Research.* 69(2): pp.225–232.

Hong, S.M., Park, J.K., Teeradej, N., Lee, Y.O., Cho, Y.K. and Park, C.H. 2006. Pretreatment of sludge with microwaves for pathogen destruction and improved anaerobic digestion performance. *Water Environment Research.* pp.76–83.

https://www.rohde-schwarz.com/us/ accessed online on 04/19/2017.

Hu, Z. and Wen, Z. 2008. Enhancing enzymatic digestibility of switchgrass by microwave-assisted alkali pretreatment. *Biochemical Engineering Journal.* 38(3): pp.369–378.

Hu, Z.H., Yue, Z.B., Yu, H.Q., Liu, S.Y., Harada, H. and Li, Y.Y. 2012. Mechanisms of microwave irradiation pretreatment for enhancing anaerobic digestion of cattail by rumen microorganisms. *Applied Energy.* 93: pp.229–236.

Illman, A.M., Scragg, A.H. and Shales, S.W. 2000. Increase in Chlorella strains calorific values when grown in low nitrogen medium. *Enzyme and Microbial Technology.* 27(8): pp.631–635.

Inglesby, A.E. and Fisher, A.C. 2012. Enhanced methane yields from anaerobic digestion of Arthrospira maxima biomass in an advanced flow-through reactor with an integrated recirculation loop microbial fuel cell. *Energy & Environmental Science.* 5(7): pp.7996–8006.

Izumi, K., Okishio, Y.K., Nagao, N., Niwa, C., Yamamoto, S. and Toda, T. 2010. Effects of particle size on anaerobic digestion of food waste. *International Biodeterioration & Biodegradation.* 64(7): pp.601–608.

Jackowiak, D., Frigon, J.C., Ribeiro, T., Pauss, A. and Guiot, S. 2011. Enhancing solubilisation and methane production kinetic of switchgrass by microwave pretreatment. *Bioresource Technology.* 102(3): pp.3535–3540.

Jin, Y., Hu, Z. and Wen, Z. 2009. Enhancing anaerobic digestibility and phosphorus recovery of dairy manure through microwave-based thermochemical pretreatment. *Water Research.* 43(14): pp.3493–3502.

Kennedy, K.J., Thibault, G. and Droste, R.L. 2007. Microwave enhanced digestion of aerobic SBR sludge. *Water Sa.* 33(2).

Kuglarz, M., Karakashev, D. and Angelidaki, I. 2013. Microwave and thermal pretreatment as methods for increasing the biogas potential of secondary sludge from municipal wastewater treatment plants. *Bioresource Technology.* 134: pp.290–297.

Lagerkvist, A. and Morgan-Sagastume, F. 2012. The effects of substrate pre-treatment on anaerobic digestion systems: a review. *Waste Management.* 32(9): pp.1634–1650.

Lakaniemi, A.M., Hulatt, C.J., Thomas, D.N., Tuovinen, O.H. and Puhakka, J.A. 2011. Biogenic hydrogen and methane production from Chlorella vulgaris and Dunaliellatertiolecta biomass. *Biotechnology for Biofuels.* 4(1): p.1.

Laser, M., Schulman, D., Allen, S.G., Lichwa, J., Antal, M.J. and Lynd, L.R. 2002. A comparison of liquid hot water and steam pretreatments of sugar cane bagasse for bioconversion to ethanol. *Bioresource Technology.* 81(1): pp.33–44.

Li, L., Kong, X., Yang, F., Li, D., Yuan, Z. and Sun, Y. 2012. Biogas production potential and kinetics of microwave and conventional thermal pretreatment of grass. *Applied Biochemistry and Biotechnology.* 166(5): pp.1183–1191.

Liu, C. and Wyman, C.E. 2003. The effect of flow rate of compressed hot water on xylan, lignin, and total mass removal from corn stover. *Industrial & Engineering Chemistry Research.* 42(21): pp.5409–5416.

Marin, J., Kennedy, K.J. and Eskicioglu, C. 2010. Effect of microwave irradiation on anaerobic degradability of model kitchen waste. *Waste Management.* 30(10): pp.1772–1779.

Marzano, C.M.A.D.S., Legros, A., Naveau, H.P. and Nyns, E.J. 1982. Biomethanation of the marine algae Tetraselmis. *International Journal of Solar Energy.* 1(4): pp.263–272.

Mata-Alvarez, J. (ed.). 2003. Biomethanization of the organic fraction of municipal solid wastes. IWA publishing.

Mata-Alvarez, J., Dosta, J., Macé, S. and Astals, S. 2011. Codigestion of solid wastes: a review of its uses and perspectives including modeling. *Critical Reviews in Biotechnology.* 31(2): pp.99–111.

Mussgnug, J.H., Klassen, V., Schlüter, A. and Kruse, O. 2010. Microalgae as substrates for fermentative biogas production in a combined biorefinery concept. *Journal of Biotechnology.* 150(1): pp.51–56.

Negro, M.J., Manzanares, P., Oliva, J.M., Ballesteros, I. and Ballesteros, M. 2003. Changes in various physical/chemical parameters of Pinuspinaster wood after steam explosion pretreatment. *Biomass and Bioenergy.* 25(3): pp.301–308.

Nges, I.A. and Liu, J. 2010. Effects of solid retention time on anaerobic digestion of dewatered-sewage sludge in mesophilic and thermophilic conditions. *Renewable Energy.* 35(10): pp.2200–2206.

Nges, I.A., Escobar, F., Fu, X. and Björnsson, L. 2012. Benefits of supplementing an industrial waste anaerobic digester with energy crops for increased biogas production. *Waste Management.* 32(1): pp.53–59.

Nges, I.A., Mbatia, B. and Björnsson, L. 2012. Improved utilization of fish waste by anaerobic digestion following omega-3 fatty acids extraction. *Journal of Environmental Management.* 110: pp.159–165.

Novak, J.T., Sadler, M.E. and Murthy, S.N. 2003. Mechanisms of floc destruction during anaerobic and aerobic digestion and the effect on conditioning and dewatering of biosolids. *Water Research.* 37(13): pp.3136–3144.

NYSERDA. 2011. Feasibility of Using Microwave Radiation to Facilitate the Dewatering, Anaerobic Digestion and Disinfection of Wastewater Treatment Plant Sludge, Report 10881. 11-05, April 2011.

Oswald, J.W. and Golueke, C.G. 1960. Biological transformation of solar energy. *Advance Applied Microbiology.* 2: pp.223–261.

Palmowski, L. and Muller, J. 1999. Influence of comminution of biogenic materials on their bioavailability. *Muell Abfall.* 31(6): pp.368–372.

Palmowski, L.M. and Müller, J.A. 2000. Influence of the size reduction of organic waste on their anaerobic digestion. *Water Science and Technology.* 41(3): pp.155–162.

Park, B., Ahn, J.H., Kim, J. and Hwang, S. 2004. Use of microwave pretreatment for enhanced anaerobiosis of secondary sludge. *Water Science and Technology.* 50(9): pp.17–23.

Park, H.J., Heo, H.S., Park, Y.K., Yim, J.H., Jeon, J.K., Park, J., Ryu, C. and Kim, S.S. 2010. Clean bio-oil production from fast pyrolysis of sewage sludge: effects of reaction conditions and metal oxide catalysts. *Bioresource Technology.* 101(1): pp.S83–S85.

Park, W.J. and Ahn, J.H. 2011. Effects of microwave pretreatment on mesophilic anaerobic digestion for mixture of primary and secondary sludges compared with thermal pretreatment. *Environmental Engineering Research.*16(2): pp.103–109.

Park, S. and Li, Y. 2012. Evaluation of methane production and macronutrient degradation in the anaerobic co-digestion of algae biomass residue and lipid waste. *Bioresource Technology.* 111: pp.42–48.

Passos, F., García, J. and Ferrer, I. 2013b. Impact of low temperature pretreatment on the anaerobic digestion of microalgal biomass. *Bioresource Technology.* 138: pp.79–86.

Passos, F., Solé, M., García, J. and Ferrer, I. 2013a. Biogas production from microalgae grown in wastewater: effect of microwave pretreatment. *Applied Energy.* 108: pp.168–175.

Passos, F. and Ferrer, I. 2014. Microalgae conversion to biogas: thermal pretreatment contribution on net energy production. *Environmental Science & Technology.* 48(12): pp.7171–7178.

Passos, F., Hernandez-Marine, M., García, J. and Ferrer, I. 2014. Long-term anaerobic digestion of microalgae grown in HRAP for wastewater treatment. Effect of microwave pretreatment. *Water Research.* 49: pp. 351–359.

Passos, F., Carretero, J. and Ferrer, I. 2015. Comparing pretreatment methods for improving microalgae anaerobic digestion: thermal, hydrothermal, microwave and ultrasound. *Chemical Engineering Journal*. 279: pp.667–672.

Pavlostathis, S.G. and Gossett, J.M. 1985. Alkaline treatment of wheat straw for increasing anaerobic biodegradability. *Biotechnology and Bioengineering*. 27(3): pp.334–344.

Petersen, R.C. 1984. The chemical composition of wood (Chapter 2). In: Rowell, R.M. (ed.). The chemistry of solid wood, Advances in Chemistry Series, vol. 207. American Chemical Society, Washington, DC, p. 984

Polakovičová, G., Kušnír, P., Nagyová, S. and Mikulec, J. 2012. Process integration of algae production and anaerobic digestion. In 15th International Conference on Process Integration, Modelling and (Vol. 29).

Qdais, H.A., Hani, K.B. and Shatnawi, N. 2010. Modeling and optimization of biogas production from a waste digester using artificial neural network and genetic algorithm. *Resources, Conservation and Recycling*. 54(6): pp.359–363.

Qiao, W., Wang, W., Xun, R., Lu, W. and Yin, K. 2008. Sewage sludge hydrothermal treatment by microwave irradiation combined with alkali addition. *Journal of Materials Science*. 43(7): pp.2431–2436.

Ramamoorthy, S. and Sulochana, N. 1989. Enhancement of biogas production using algae. *Current Science*. 58(11): pp.646–647.

Rani, R.U., Kumar, S.A., Kaliappan, S., Yeom, I. and Banu, J.R. 2013. Impacts of microwave pretreatments on the semi-continuous anaerobic digestion of dairy waste activated sludge. *Waste Management*. 33(5): pp.1119–1127.

Ras, M., Lardon, L., Bruno, S., Bernet, N. and Steyer, J.P. 2011. Experimental study on a coupled process of production and anaerobic digestion of Chlorella vulgaris. *Bioresource Technology*. 102(1): pp.200–206.

Rodriguez, C., Alaswad, A., Mooney, J., Prescott, T. and Olabi, A.G. 2015. Pre-treatment techniques used for anaerobic digestion of algae. *Fuel Processing Technology*. 138: pp.765–779.

Rui, X., Pay, E., Tianrong, G., Fang, Y. and Wudi, Z. 2008. The potential of blue-green algae for producing methane in biogas fermentation. In *Proceedings of ISES World Congress 2007 (Vol. I–Vol. V)* (pp.2426–2429). Springer Berlin Heidelberg.

Salerno, M., Nurdogan, Y. and Lundquist, T.J. 2009. Biogas production from algae biomass harvested at wastewater treatment ponds (p.18). *American Society of Agricultural and Biological Engineers*.

Samson, R. and Leduyt, A. 1986. Detailed study of anaerobic digestion of Spirulina maxima algal biomass. *Biotechnology and Bioengineering*. 28(7): pp.1014–1023.

Santos, N.O., Oliveira, S.M., Alves, L.C. and Cammarota, M.C. 2014. Methane production from marine microalgae Isochrysisgalbana. *Bioresource Technology*. 157: pp.60–67.

Sapci, Z. 2013. The effect of microwave pretreatment on biogas production from agricultural straws. *Bioresource Technology*. 128: pp.487–494.

Schwede, S., Kowalczyk, A., Gerber, M. and Span, R. 2011. Influence of different cell disruption techniques on mono digestion of algal biomass. *World Renewable Energy Congress, Linkoping, Sweden*. pp.8–13.

Schwede, S., Kowalczyk, A., Gerber, M. and Span, R. 2013. Anaerobic co-digestion of the marine microalga Nannochloropsissalina with energy crops. *Bioresource Technology*. 148: pp.428–435.

Shahriari, S. 2011. Enhancement of Anaerobic Digestion of Organic Fraction of Municipal Solid Waste by Microwave Pretreatment. Thesis submitted to the Faculty of Graduate and Postdoctoral Studies In partial fulfillment of the requirements For the Ph.D. degree in Environmental Engineering University of Ottawa 2011.

Shahriari, H., Warith, M., Hamoda, M. and Kennedy, K.J. 2012. Effect of leachate recirculation on mesophilic anaerobic digestion of food waste. *Waste Management*. 32(3): pp.400–403.

Sheng, G.P., Zhang, M.L. and Yu, H.Q. 2008. Characterization of adsorption properties of extracellular polymeric substances (EPS) extracted from sludge. *Colloids and Surfaces B: Biointerfaces*. 62(1): pp.83–90.

Shuchuan, P., Chenghu, H., Jin, W., TianHu, C., XiaoMeng, L. and ZhengBo, Y. 2012. Performance of anaerobic co-digestion of corn straw and algae biomass from lake Chaohu. *Transactions of the Chinese Society of Agricultural Engineering*. 28(15): pp.173–178.

Sialve, B., Bernet, N. and Bernard, O. 2009. Anaerobic digestion of microalgae as a necessary step to make microalgal biodiesel sustainable. *Biotechnology Advances*. 27(4): pp.409–416.

Sólyom, K., Mato, R.B., Pérez-Elvira, S.I. and Cocero, M.J. 2011. The influence of the energy absorbed from microwave pretreatment on biogas production from secondary wastewater sludge. *Bioresource Technology*. 102(23): pp.10849–10854.

Spoalore, P., Joannis-Cassan, C., Duran, E. and Isambert, A. 2006. Commercial application of microalgae. *J. Biosci. Bioeng*. 101: pp.87–96.

Srimachai, T., Thonglimp, V. and Sompong, O. 2014. Ethanol and methane production from oil palm frond by two stage SSF. *Energy Procedia*. 52: pp.352–361.

Swain, M., James, S., Schubert, H. and Regier, M. 2005. Measuring the heating performance of microwave ovens. *The Microwave Processing of Foods*. pp.221–242.

Symons, G.E. and Buswell, A.M. 1933. The methane fermentation of carbohydrates 1, 2. *Journal of the American Chemical Society*. 55(5): pp.2028–2036.

Tang, B., Yu, L., Huang, S., Luo, J. and Zhuo, Y. 2010. Energy efficiency of pre-treating excess sewage sludge with microwave irradiation. *Bioresource Technology*. 101(14): pp.5092–5097.

Tedesco, S., Benyounis, K.Y. and Olabi, A.G. 2013. Mechanical pretreatment effects on macroalgae-derived biogas production in co-digestion with sludge in Ireland. *Energy*. 61: pp.27–33.

Toreci, I., Kennedy, K.J. and Droste, R.L. 2009. Evaluation of continuous mesophilic anaerobic sludge digestion after high temperature microwave pretreatment. *Water Research*. 43(5): pp.1273–1284.

Uggetti, E., Sialve, B., Trably, E. and Steyer, J.P. 2014. Integrating microalgae production with anaerobic digestion: a biorefinery approach. *Biofuels, Bioproducts and Biorefining*. 8(4): pp.516–529.

Varel, V.H., Chen, T.H. and Hashimoto, A.G. 1988. Thermophilic and mesophilic methane production from anaerobic degradation of the cyanobacterium Spirulina maxima. *Resources, Conservation and Recycling*. 1(1): pp.19–26.

Vergara-Fernández, A., Vargas, G., Alarcón, N. and Velasco, A. 2008. Evaluation of marine algae as a source of biogas in a two-stage anaerobic reactor system. *Biomass and Bioenergy*. 32(4): pp.338–344.

Wang, Y., Zhang, Y., Wang, J. and Meng, L. 2009. Effects of volatile fatty acid concentrations on methane yield and methanogenic bacteria. *Biomass and Bioenergy*. 33(5): pp.848–853.

Wang, N. and Wang, P. 2016. Study and application status of microwave in organic wastewater treatment–A review. *Chemical Engineering Journal*. 283: pp.193–214.

Ward, A.J., Lewis, D.M. and Green, F.B. 2014. Anaerobic digestion of algae biomass: a review. *Algal Research*. 5: pp.204–214.

Weiland, P. 2010. Biogas production: current state and perspectives. *Applied Microbiology and Biotechnology*. 85(4): pp.849–860.

Wilson, C.A. and Novak, J.T. 2009. Hydrolysis of macromolecular components of primary and secondary wastewater sludge by thermal hydrolytic pretreatment. *Water Research*. 43(18): pp.4489–4498.

Xiao, W. and Clarkson, W.W. 1997. Acid solubilization of lignin and bioconversion of treated newsprint to methane. *Biodegradation*. 8(1): pp.61–66.

Yang, Z., Guo, R., Xu, X., Fan, X. and Luo, S. 2011. Hydrogen and methane production from lipid-extracted microalgal biomass residues. *International Journal of Hydrogen Energy*. 36(5): pp.3465–3470.

Yen, H.W. and Brune, D.E. 2007. Anaerobic co-digestion of algal sludge and waste paper to produce methane. *Bioresource Technology*. 98(1): pp.130–134.

Yu, Y.Y., Chan, W.C.I., Lo, I.L.W., Liao, P.L.H. and Lo, K.L.V. 2010. Sewage sludge treatment by a continuous microwave enhanced advanced oxidation process. A paper submitted to the Journal of Environmental Engineering and Science. *Canadian Journal of Civil Engineering*.

Zamalloa, C., De Vrieze, J., Boon, N. and Verstraete, W. 2012. Anaerobic digestibility of marine microalgae Phaeodactylumtricornutum in a lab-scale anaerobic membrane bioreactor. *Applied Microbiology and Biotechnology*. 93(2): pp.859–869.

Zeng, S., Yuan, X., Shi, X. and Qiu, Y. 2010. Effect of inoculum/substrate ratio on methane yield and orthophosphate release from anaerobic digestion of Microcystis spp. *Journal of Hazardous Materials*. 178(1): pp.89–93.

Zhang, J., Lv, C., Tong, J., Liu, J., Liu, J., Yu, D., Wang, Y., Chen, M. and Wei, Y. 2016. Optimization and microbial community analysis of anaerobic co-digestion of food waste and sewage sludge based on microwave pretreatment. *Bioresource Technology*. 200: pp.253–261.

Zheng, J. 2006. *Effect of Mild Microwave Pretreatment on Characteristics and Mesophilic Digestion of Primary Sludge* (Doctoral dissertation, University of Ottawa (Canada)).

Zheng, J., Kennedy, K.J. and Eskicioglu, C. 2009. Effect of low temperature microwave pretreatment on characteristics and mesophilic digestion of primary sludge. *Environmental Technology*. 30(4): pp.319–327.

Zhu, J., Kuznetsov, A.V. and Sandeep, K.P. 2007. Mathematical modeling of continuous flow microwave heating of liquids (effects of dielectric properties and design parameters). *International Journal of Thermal Sciences*. 46(4): pp.328–341.

Zhu, Y., Lee, Y.Y. and Elander, R.T. 2004. Dilute-acid pretreatment of corn stover using a high-solids percolation reactor. *Applied Biochemistry and Biotechnology*. 117(2): pp.103–114.

Zhu, Y., Lee, Y.Y. and Elander, R.T. 2005. Optimization of dilute-acid pretreatment of corn stover using a high-solids percolation reactor. *Twenty-Sixth Symposium on Biotechnology for Fuels and Chemicals*. pp.1045–1054. Humana Press.

Zielińska, M., Cydzik-Kwiatkowska, A., Zieliński, M. and Dębowski, M. 2013. Impact of temperature, microwave radiation and organic loading rate on methanogenic community and biogas production during fermentation of dairy wastewater. *Bioresource Technology*. 129: pp.308–314.

Zinoviev, S., Müller-Langer, F., Das, P., Bertero, N., Fornasiero, P., Kaltschmitt, M., Centi, G. and Miertus, S. 2010. Next-generation biofuels: survey of emerging technologies and sustainability issues. *ChemSusChem*. 3(10): pp.1106–1133.

7

Microwave Mediated Thermochemical Conversion of Biomass

Introduction

Many biochemical and thermochemical processes have been developed in the past to produce valuable platform chemicals, intermediate products and transportation fuels from biomass. Thermochemical processes have received more attention due to their special advantages such as faster conversion and techno-economic feasibility. Three different thermochemical conversion processes can be described based on the reaction conditions especially related to oxygen utilization (see Figure 1). They are namely combustion (complete oxidation), gasification (partial oxidation) and pyrolysis (thermal degradation without oxygen). Combustion, also known as incineration, is the most commonly practiced and well-established process. Incineration of biomass and other waste substances generate environmental pollutants such as carbon oxides, sulfur, nitrogen, chlorine products (dioxins and furans), volatile organic compounds, polycyclic aromatic hydrocarbons, dust, etc. On the contrary, gasification and pyrolysis offer the potential for greater efficiencies in energy production and less pollution. Although pyrolysis is still under development in the waste industry, this process has received special attention, not only as a primary process of combustion and gasification, but also as independent process leading to the production of energy-dense products with numerous uses. This makes the pyrolysis treatment process self-sufficient in terms of energy use, and also significantly reduces operating costs.

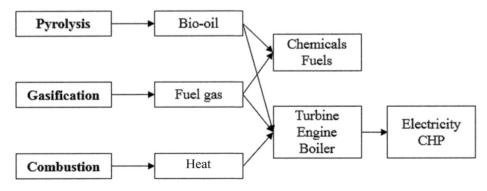

Figure 1. Products from thermochemical conversion of biomass (adapted from Bridgwater 2003).

Gasification

Fuel gas can be produced from biomass and related materials by either partial oxidation (by air or oxygen) or by steam or pyrolytic gasification to generate a mixture of carbon monoxide, carbon dioxide, hydrogen and methane which is presented in Table 1. Gasification occurs in a number of sequential steps:

- Drying (to evaporate moisture),
- Pyrolysis (to release gas, vaporized tars or oils and a solid char residue), and
- Gasification (or partial oxidation of the solid char, pyrolysis tars and pyrolysis gases).

Gasification is a thermochemical process similar to pyrolysis involving thermal treatment of biomass in a limited amount of oxygen or air at temperatures over 750°C. The heat of the reaction necessary for the conversion of biomass (from pyrolysis and cracking to drying and reduction) is provided by the partial combustion of biomass (see Figure 2). Usually, gasification of biomass yields over 80% of gas (see Table 1 for the composition) (Rajvanshi 1986), which can be cleaned and then directly used as a fuel (e.g., in heat and power generators) or be upgraded to a CO/H_2 synthesis gas mixture. Similar to the synthesis gas obtained via methane or coal gasification, bio-syngas is a valuable intermediate feedstock for the chemical industry because it can be converted into a range of useful products, such as biofuels and biohydrogen (Bain 2004) using various catalytic processes.

Table 1. Modes of thermal gasification (Bridgwater 2003).

	Partial oxidation with air	Partial oxidation with oxygen	Steam (pyrolytic) gasification
Main products	CO, CO_2, H_2, CH_4, N_2, tar	CO, CO_2, H_2, CH_4, tar (no N_2)	CO, CO_2, H_2, CH_4, tar
Heating value	~5 MJ/m³ gas	~10–12 MJ/m³ gas	~15–20 MJ/m³ gas
Considerations	Utilization problems can arise in combustion, particularly in gas turbines	Oxygen utilization costs compensated by a better quality fuel gas. The trade-off is finely balanced	Two stage process with a primary reactor producing gas and char, and a second reactor for char combustion to reheat sand which is recirculated. High heating value due to a higher methane and higher hydrocarbon gas content, but at the expense of lower overall efficiency due to loss of carbon in the second reactor

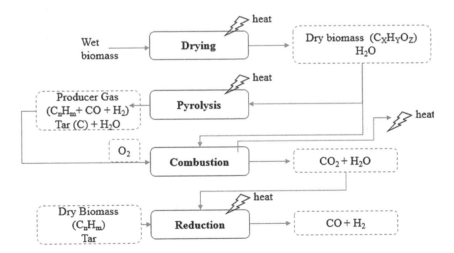

Figure 2. Chemical and physicochemical processes occurring in gasification of biomass, and their relationships.

Char gasification is the interactive combination of several gas–solid and gas–gas reactions in which solid carbon is oxidized to carbon monoxide and carbon dioxide, and hydrogen is generated through the water gas shift reaction. The gas–solid reactions of char oxidation are the slowest and limit the overall rate of the gasification process. Many of the reactions are catalyzed by the alkali metals present in wood ash, but still do not reach equilibrium. The gas composition is influenced by many factors such as feed composition, water content, reaction temperature, and the extent of oxidation of the pyrolysis products. Not all the liquid products from the pyrolysis step are completely converted due to the physical or geometrical limitations of the reactor and the chemical limitations of the reactions involved, and these give rise to contaminated tars in the final product gas. Due to the higher temperatures involved in gasification compared to pyrolysis, these tars tend to be refractory and are difficult to remove by thermal, catalytic or physical processes. This aspect of tar cracking or removal in gas clean-up is one of the most important technical uncertainties in implementation of gasification technologies and is discussed below. A recent survey of gasifier manufacturers found that 75% of gasifiers offered commercially were downdraft, 20% were fluid beds (including circulating fluid beds), 2.5% were updraft and 2.5% were other types. The fuel gas quality requirements, for turbines in particular, are very high.

Tar is a particular problem and remains the most significant technical barrier in gasification process. Tars can be destroyed either by catalytic cracking using dolomite or nickel or by thermal cracking, for example by partial oxidation or direct contact. The gas is very costly to store or transport so it has to be used immediately. Hot-gas efficiencies for the gasifier (total energy in raw product gas divided by energy in feed) can be as high as 95–97% for close-coupled turbine and boiler applications, and up to 85% for cold gas efficiencies. In power generation, using combined cycle operation, efficiencies of up to 50% for the largest installations have been proposed which reduces to 35% for smaller applications. Microwave mediated gasification process will be discussed in a later section followed by microwave mediated pyrolysis.

Pyrolysis

Pyrolysis is defined as a thermal degradation process in the absence of oxygen, which converts a raw material into different reactive intermediate products: solid (char), liquid (heavy molecular weight compounds that condense when cooled down) and gaseous products (light molecular weight gases). The pyrolysis process is complex because it is influenced by many factors related to feedstock composition and operating conditions.

The first stage mainly involves dehydration, dehydrogenation, decarboxylation or decarbonilation reactions. The second comprises of processes such as cracking (thermal or catalytic), where heavy compounds further break into gases, or char is also converted into gases such as CO, CO_2, CH_4 and H_2 by reactions with gasifying agents, as well as partial oxidation, polymerization and condensation reactions.

Solid fraction

Pyrolysis char is a carbonaceous residue mainly composed of elemental carbon originating from thermal decomposition of the organic components, unconverted organic compounds, e.g., solid additives, and even carbon nanoparticles produced in the gas phase secondary reactions. This carbonaceous residue plays an important role in the pyrolysis process since it contains the mineral content of the original feed material, relevant to specific catalytic processes (Raveendran et al. 1995). The importance of the char cannot be understated as it may be involved as a reactive in heterogeneous or catalytic heterogeneous reactions (Menéndez et al. 2007). The utilization of the char can vary considerably according to its characteristics. The main industrial uses of char can be summarized as follows: (i) as solid fuel for boilers which can be directly converted to pellets or mixed with other materials such as biomass, carbon, etc., to form the same, (ii) as feedstock for the production of activated carbon, (iii) as feedstock for making carbon nanofilaments, (iv) as feedstock for the gasification process to obtain hydrogen rich gas, (v) as feedstock for producing high surface area catalysts to be used in electrochemical capacitors, etc.

Liquid fraction

Pyrolysis oil (bio-oil) is a complex mixture of several organic compounds which may be accompanied by inorganic species. In the case of biomass, the liquid or oil fraction (bio-oils) is found to be highly oxygenated and complex, chemically unstable and less miscible in conventional fuels (Demirbaş 2002). As such, the liquid products need to be upgraded by lowering the oxygen content and removing residues. The oil obtained from pyrolysis can have the following industrial uses:

    i) combustion fuel,
    ii) used for power generation,
    iii) production of chemicals and resins,
    iv) transportation fuel,
    v) production of an hydro-sugars like levoglucosan,
    vi) as binder for pelletizing and briquetting combustible organic waste materials,
    vii) bio-oil can be used as preservatives, e.g., wood preservatives,
    viii) a suitable blend of a pyrolysis liquid with the diesel oil may be used as diesel engine fuels,
    ix) bio-oils can be used in making adhesives, etc. Moreover, the oil may be stored and transported, and hence need not be used at the production site.

Gas fraction

The gas produced in a pyrolysis process is mainly composed of combustible gases, such as H_2, CO, C_2H_2, CH_4, C_2H_4, C_2H_6, etc. Other gases, such as CO_2 and pollutants (SO_2, NOx) can also appear, although in lower concentrations. The gas produced from pyrolysis can be used directly as a fuel in various energy applications, such as: (i) direct firing in boilers without the need for flue gas treatment, and (ii) in gas turbines/engines associated with electricity generation. Pyrolysis gas containing significant amounts of hydrogen and carbon monoxide might be utilized in syngas applications. It is known that synthesis gas (H_2 + CO) having different H_2/CO molar ratios is suitable for different applications. For example, synthesis gas having a high H_2/CO molar ratio is desirable for producing hydrogen for ammonia synthesis. This ratio is increased further during the water-gas shift reaction for the removal of CO.

Three different pyrolysis methods are available: (i) slow; (ii) fast; and (iii) flash type pyrolysis. The differences among these methods can be expressed in terms of reaction temperature, heating speed, duration of reaction time or residence time. Table 2 shows the difference in the three methods.

Table 2. Main differences in pyrolysis methods.

Pyrolysis technology	Slow	Fast	Flash
Temperature (K)	550–950	850–1250	1050–1300
Heating rate (K/s)	0.1–1	10–200	> 1000
Residence time (s)	450–550	0.5–10	< 0.5

Microwave Mediated Pyrolysis

Microwave mediated thermochemical conversion of biomass and biofuels has been far more investigated than the biochemical processes. Microwave-assisted pyrolysis of a wide range of biomass feedstocks is also receiving increasing attention in recent years. Pyrolysis is the conversion of organic substances of complex structures (e.g., biomass) to smaller molecules, whereby cleavage of chemical bonds occurs by heating in the absence of oxygen (Mohan et al. 2006, Sanderson 2009, Laird et al. 2009). Pyrolysis of vegetable feedstock normally produces mixtures of solid char, oxygenated liquid products (pyrolysis oil or bio-oil), and gases of different composition depending on the feedstock, temperature regime, reaction time, and other parameters (Zinoviev et al. 2010).

Microwave mediated pyrolysis may offer several advantages over conventional heating (Robinson et al. 2015) which include: instantaneous volumetric heating that overcomes heat transfer constraints; smaller equipment size; pyrolysis of larger particles without the need for size reduction and downstream separation; low requirements for inert sweep gas; and potential to produce a different range or grade of products due to the unique thermal gradients that result from microwave heating.

Figure 3 depicts the possible mechanisms by which microwave pyrolysis will take place. Because the biomass contains significant portion of water molecules, microwave pyrolysis is highly influenced by their interaction with water molecules within the biomass. The microwave interaction occurs with free and bound water. Free water is expected to evaporate through the capillaries of the biomass even at low temperatures compared to bound water, which needs high temperatures to entrain from the material. Thus, free water would not provide enough temperature within the biomass material due to its early escape. On the other hand, bound water would enter the super-heated steam stage which may increase the temperature of the biomass material significantly. If the water within the biomass does not build up enough pressure, it would restrict the temperature to a lower value, which cannot contribute to the pyrolysis reaction (Robinson et al. 2009). In addition, at low temperatures it will be impossible to conduct pyrolysis of biomass (Salema and Ani 2012a).

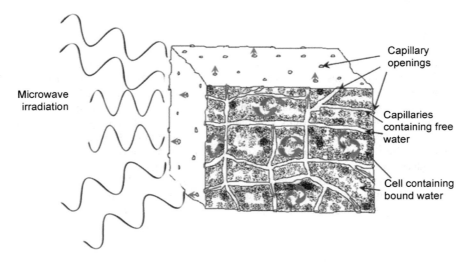

Figure 3. Schematic diagram of the MW interaction with materials and its effect on water evaporation from the sample (Salema and Ani 2012).

Method of microwave mediated pyrolysis

Microwave heating in pyrolysis eliminates a number of issues/limitations associated with conventional heating, the most important issue being char layer formation during conventional pyrolysis. In conventional heating, heat is transferred from a heating source to the outer surface of the heated material. Thus, surface temperature begins to rise, which results in heat transfer towards the core, primarily by thermal conduction. Once the temperature reaches the pyrolysis temperature, the heated material begins to decompose from the surface to the core. This forms a layer of char that grows in the same direction of the heat transfer, which behaves like a thermal insulator. This layer limits heat transfer through the heated material, which results in an outer surface hotter than the core, as shown in Figure 4a (Farag 2013). Consequently, volatile products would be affected quantitatively and/or qualitatively as a result of further thermal degradation during the flow out through this layer due to the pressure gradient at the pyrolysis zone.

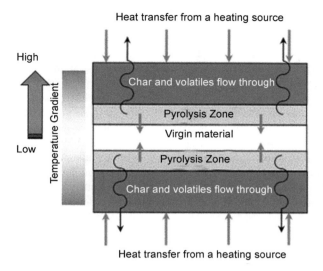

Figure 4a. Biomass pyrolysis under conventional heating conditions.

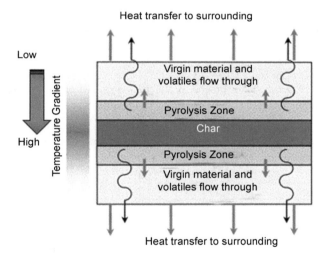

Figure 4b. Biomass pyrolysis under microwave heating process conditions (adapted from Farag 2013).

On the other hand, in microwave heating, electromagnetic waves penetrate the entire heated material at almost the same time, limited only by penetration limits. Therefore, theoretically, microwave heating leads to uniform temperature distribution within the heated material. However, in practice, it is likely to produce non-uniform temperature distribution, as the core is often hotter than the surface (Yang and Gunasekaran 2001, Cuccurullo et al. 2002, Pandit and Prasad 2003, Campanone and Zaritzky 2005, Rattanadecho 2006, Gunasekaran and Yang 2007, Ciacci et al. 2010) which depends on several factors. In microwave mediated pyrolysis, the char layer formation grows in the direction opposite to that of conventional pyrolysis, starting at the core and moving toward the outer surface, as shown in Figure 4b. In microwave pyrolysis, volatile products flow out through virgin material, i.e., through zones with a lower temperature, which preserves their chemical structure. Therefore, using electromagnetic irradiation rather than superficial heat transfer in pyrolysis would produce a better product, different in terms of quality and quantity compared to traditional pyrolysis (Farag 2013).

Benefits and reaction products reported in literature

Various studies have established the technical feasibility of microwave pyrolysis for bio-oil (liquid fraction) production, which is carried out at high heating rates and moderates temperatures. Since the operating conditions, the biomass feedstock and the reactor configuration have an effect on the yield and composition of the bio-oil produced, it is difficult to establish general conclusions. Nevertheless, most researchers have observed that the quantity and quality of the bio-oils produced under microwave heating are significantly better than those under conventional pyrolysis. Less polycyclic aromatic hydrocarbons, increasing amount of phenolic compounds and a larger degree of deoxygenation have been observed in the bio-oils produced under microwave heating. This improved composition of the bio-oil is actually related to the selective heating provided by the microwaves. The microwave process can be carried out flowing a cold sweep gas (which is barely heated by the microwaves) and so the released volatiles are cooled and condensed rapidly. The secondary reactions of the liquid precursors due to high temperatures are largely avoided and much more compounds are preserved in the final bio-oil. The implications of the enhanced composition of the bio-oil produced from microwave pyrolysis are significant since one of the main issues in the production of pyrolysis bio-oil is its high oxygen content. Moreover, few preliminary results have shown the effectiveness of microwave heating applied to *ex-situ* upgrading of bio-oil.

In general, raw biomass is not a good microwave absorbing material. It cannot be readily heated by microwave radiation. The dielectric properties for biomass improve as the pyrolysis proceeds and the carbon content of the remaining residue increases. In fact, the ability of char obtained from biomass pyrolysis to be heated by the microwaves is very good and it can be used as microwave receptor to indirectly heat the biomass up to the temperatures required for thermal decomposition. Other materials such as activated carbon or metal oxides have also been used as microwave receptors. Similar to conventional heating, microwave pyrolysis generates liquid, gas and solid fractions, and the proportion in which the three fractions are produced depends on the operating conditions. Table 3 shows the dielectric values of char obtained from various feedstock.

Microwave-assisted pyrolysis for biochar production is carried out at low temperatures (i.e., lower input microwave power). The solid residue produced by this process has been observed to exhibit a good

Table 3. Dielectric properties of carbon (char) materials obtained from various feedstock.

Carbon material	$\tan \delta = \varepsilon''/\varepsilon'$	Reference
Wet empty fruit bunch char	0.31	Omar et al. 2011
Dried empty fruit bunch char	0.34	
Oil palm shell activated carbon	0.40	
Dried oil palm empty fruit bunch char	0.13	
Empty fruit bunch sample with 18, 45 and 64 wt% moisture	0.30, 0.54, 0.32	
Oil palm fiber	0.08	Salema et al. 2013
Oil palm shell	0.12	
Oil palm shell char	0.08	
Wood	0.11	Torgovnikov 1993
Fir plywood	0.01–0.05	
Particle board	0.1–1.0	
Aspen bark	0.22	
Pine bark	0.18	

potential for gas adsorption and higher energy content than the char produced by conventionally-heated pyrolysis. However, lower energy efficiency has been reported in the case of microwave pyrolysis. It should be noted though that this efficiency depends on the suitability of microwave oven design for the process requirements, which may not be optimum when considering lab-scale rigs.

Microwave-assisted pyrolysis for gas production at high temperature favors secondary reactions between vapors and solid residue. The yield of the gas produced from microwave pyrolysis has been found to be larger than that obtained from conventional pyrolysis. In addition, the gas composition is reportedly higher in valuable synthesis gas (H_2 + CO) and lower in undesirable greenhouse gases (CH_4 and CO). The change in the gas composition is explained based on the enhancement of CO_2 gasification of the char and methane decomposition into H_2. Microwave radiation is known to have the potential to increase reaction rate, selectivity and yield of catalytic heterogeneous reactions (as in the case of CO_2 gasification and CH_4 decomposition) due to the random formation of microplasmas throughout the solid residue, in which temperature is much higher than the average temperature. In addition to the improved composition of the gas, it has been reported that less char and tar are produced under these gasification conditions. The lower amount of tar can potentially be simplified by subsequent gas cleaning operations, thus improving the overall process efficiency. Microwave-assisted steam and CO_2 gasification of biomass has also been investigated, although to a lesser extent than pyrolysis. Microwave gasification has been mainly carried out over biochar produced from conventional pyrolysis. High conversions have been observed at lower temperatures and shorter reaction times than those needed under conventional heating. These high conversions are attributed to the formation of microplasmas.

A summary of recent studies on conversion of lignocellulosic biomass shows that microwave assisted pyrolysis offers several benefits over conventional pyrolysis in terms of bio-oils and gas yields and enrichment of specialty chemicals such as levoglucosan, furfural and phenolic compounds. Gas enrichment refers to higher hydrogen or biogas production. Table 4 presents a summary of microwave-assisted thermochemical conversion of lignocellulosic materials into valuable chemical and fuel products (Richel and Jacquet 2015). Various products such as bio-oils, furfural, hydrogen, levoglucosan, phenols and phenol derivatives can be obtained by manipulating the process parameters.

Table 4. Microwave mediated thermochemical conversion of lignocellulosic materials (Richel and Jacquet 2015).

Substrate	Technology	Main products	Reference
Filter paper pulps	Domestic oven, 5 min at 130 W	Levoglucosan (+CO_2+H_2O+char)	Allan et al. 1980
Larch log, used papers and filter papers	Modified domestic oven, 10 min	Levoglucosan+others (levoglucosenone and anhydrosugars)	Miura et al. 2001
Cellulose	12 min, 620 W- acidic conditions	Levoglucosenone+2-furfural	Sarotti et al. 2007
Coffee hulls	15 min, 500–1,000°C + catalyst	Gas enriched in H_2 and biogas	Dominguez et al. 2007
Wheat straw	< 350°C, Bronsted acid/base and continuous removal of water	Bio-oils	Budarin et al. 2009
Corn stover, wood pellets	450–500°C, $MgCl_2$	Bio-oils enriched in furfural (> 80%)	Wan et al. 2009
Sewage sludge	Lab-scale equipment, 530 W, 1,000°C	Gas enriched in biogas (> 94%)	Dominguez et al. 2007
Wood pellets	Lab-scale equipment, 0.5–3.0 kW, 50–300 s	Bio-oils (40%), biogas (35%)	Robinson et al. 2009
Douglas fir	Douglas fir Large-scale reactor, 700 W, 12 min at 450°C + activated carbon	Bio-oils (48% with a high yield in phenols and phenolic derivatives) and biogas (40%)	Bu et al. 2012

Variables affecting microwave mediated pyrolysis

The yield and quality of produced value-added products are affected by some critical parameters in the microwave mediated pyrolysis process. To achieve best quality products at maximum conversion yield, these variables should be optimized. The most important variables in microwave mediated pyrolysis can be broadly classified into feedstock related, process related and microwave technology related variables which are listed below:

- Type and size of input biomass/materials
- Moisture and water content of input biomass/materials
- Reaction temperature
- Reaction time (residence time)
- Microwave output power
- Microwave type (multimode or single-mode)
- Reactor design/type
- Microwave receptor type, size, and amount/concentration, absorber
- Catalyst type and concentration
- Mixing intensity (stirring)
- Type and flow rate of carrier gas

Table 5 summarizes recent studies on microwave mediated pyrolysis of various feedstocks for bio-oil or valuable biochemical production. From the observations made in these studies, it can be noted that the microwave mediated pyrolysis process is influenced by various parameters which can be categorized into biomass-specific, microwave source-specific, and microwave process-specific. All these demand proper consideration for optimizing the benefits and techno-economics of the process.

Biomass (feedstock) related characteristics

An ideal feedstock is one which does not require any pretreatment. Because the feedstocks contain undesired elements, the product yield and quality will be compromised without proper pretreatment. The properties of biomass and the composition determine the heating rate and the reaction temperature. Particle size is critical in optimizing the energy consumption and product yields. Small particle size is typically required for easy absorption of microwave irradiation. Moisture content helps accelerate initial heating under microwave process. The concerns with feedstock physicochemical characteristics include the energy consumption for preparation of feedstock and understanding the degradation behavior during microwave heating.

Microwave source related characteristics

Microwave reactor configurations vary in mode (single vs. multimode), power output and control features. Addition of absorbers to maximize absorption of microwave irradiation is another important factor. Microwave power should be adjusted to match the dielectric constant of the absorber. For example, low microwave power is required for absorbers with high dielectric constant. Catalyst used under microwave heating also affects the heating rates, process yield and quality of the products. Hot spot formation and thermal runaway are other important issues to be considered in microwave heating. The particle size may influence the hot spot formation phenomenon. Smaller particles absorb microwave power quickly and distribute the heat evenly whereas the large particles can cause hot spots due to heat accumulation. As such the choice of absorber, catalyst and particle size determine the performance of the microwave pyrolysis process. Absorber loading rate and the penetration depths at various loading rates should be considered. Separation, regeneration and reuse of microwave absorber and catalysts is another important concern.

Microwave Mediated Thermochemical Conversion of Biomass   249

Table 5. Summary of recent studies on microwave pyrolysis of biomass (Yin 2012).

Feedstock and pretreatment and use of additives/catalysts/microwave absorbers	Reactor and process conditions	Pyrolysis products	References
Wheat straw pelletized (1 cm diameter, 1 cm high particles) predried; 46.48% C; 6.7% H, 33.71% O, 5.52% ash; density 930 kg/m³ No pretreatment	Batch MW reactor with Helium sweeping gas; Reaction time 1–5 min; High power levels (1–2 kW)	Pyrolysis products (wt.%): char 17.8, tar 27.9 (incl. 8.2 in cold trap), water 8.2, gases 46.1 ($H_2/CO/CO_2/CH_4/C_2H_2/C_2H_4/C_2H_6 = 1.3/24.4/14.5/2.6/1.2/1.5/0.6$), similar to those from conventional pyrolysis Tar contains phenols, guaiacols, cresols, condensed aromatics	Krieger-Brockett 1994
Aerobically digested sewage sludge with 71% moisture; 31.2% ash, 62.3% volatiles, 6.5% fixed carbon, heating value 16.68 MJ/kg (dry); C/H/O/N/S = 52.3/8.0/32.3/6.7/0.7 (wt.%, on dry ash-free (DAF) basis) Absorbers: pure graphite, 3 × 3 mm, and char from experiments were used as microwave absorbers	MW-1000 W for 6 min 15 mL/min Helium carrier gas Pyrolysis in electrical oven	When char used as absorber, yields of char/aqueous/oil/gas (wt.%): 10.7/68.8/4.0/16.5 for multimode MW; 10.1/62.7/3.0/24.2 for single-mode; 10.4/58.6/0.9/30.1 for electrical oven Quality of bio-oils from multi- and single-mode is similar	Dominguez et al. 2006
Larch: cylindrical blocks of different sizes (60, 80, 100 and 300 mm in diameter, moisture adjusted to 8–12%; main components 40% a-cellulose, 31% lignin, 12% pentosan No pretreatment	MW power 1.5 and 3 kW; up to 90 min reaction time. 5 L/min N_2 as sweeping gas and evacuated with an aspirator, vapor residence time > 10 min	For the 80 mm block (microwave power 1.5 kW): all moisture released in 3 min; yield of char decreases monotonically with time, ~20% after 15 min; yield of tar peaks at ~25% after 12 min (5% levoglucosan based on tar weight) and decreases with further irradiation	Miura et al. 2004
Coffee hulls (rich in cellulose): cylindrical pellets, 3 mm (diameter) × 20 mm (height); 8.2% moisture (as received); 5.6% ash, 77.0% volatiles, 17.4% fixed carbon (dry); HHV (higher heating value) 17.9 MJ/kg. C/H/N/S/O = 47.3/6.4/2.7/0.3/37.7 (wt.%, on dry basis) Pretreatment: Char (obtained from previous pyrolysis experiments at 1000°C) used as microwave absorber	60 mL/min N_2 as carrier gas, reaction temperature 500, 800 and 1000°C, total reaction time 20 min Pyrolysis in electrical oven under 500, 800 and 1000°C, reaction time 25 min	MW pyrolysis produces less oil and more gas than electrical pyrolysis for all the tested conditions MW pyrolysis at 500°C: 7.9% oil (34.4 MJ/kg), 30.2% char (24.3 MJ/kg), and 61.9% gas (incl. 61.4 vol.% $H_2 + CO$, 28.4 vol.% CO_2; 12.5 MJ/kg) Electrical pyrolysis at 500°C: 13.6% oil (31.2 MJ/kg), 29.2% char (24.2 MJ/kg), and 57.2% (incl. 29.9 vol.% $H_2 + CO$, 56.58 vol.% CO_2; 6.6 MJ/kg)	Dominguez et al. 2007

Table 5. cont....

Table 5. contd.

Feedstock and pretreatment and use of additives/catalysts/microwave absorbers	Reactor and process conditions	Pyrolysis products	References
Pine wood sawdust (dry basis): 81.98% volatile matter, 15.86% fixed carbon, 2.16% ash; 19.92 MJ/kg; 40.32% cellulose, 26.73% hemicellulose, 30.97% lignin SiC added as microwave absorber	15 g untreated sawdust or 16.5 g treated sawdust (incl. 1.5 g additive), blended homogeneously with 25 g SiC, are put in a quartz beaker-shaped reactor (12 cm i.d. × 13 cm height) in a MW cavity. Actual pyrolysis temperature 470°C, heating rate 5°C/s (lower, favoring coke formation), run time 12 min Purge gas 0.2 m^3/h N$_2$, vapor residence time ca. 2.7 s (relatively long, adequate for 2nd decomposition of some heavy components into light gases)	For untreated sawdust: 22.7% liquid (in which 30.1% water), 17.3% solid, 60% gas. max. oil yield is from Fe$_2$(SO$_4$)$_3$-treated sawdust: 26% liquid (27.7% water), 36.7% solid, 37.3% gas. Liquid from untreated sawdust (area%): acetol 26.6, furfura 15.72, 2-furanmethanol 4.81, 4-methyl-guaiacol 5.05, guaiacol 5.12, levoglucosan 1.04. Long residence time and low heating rate concluded to be the main reasons for the low oil yield	Chen et al. 2008
Pine wood sawdust, peanut shell, and maize stalk: their moisture/volatiles/fixed carbon/ash (air-dried, wt.%) and LHV (lower heating value, MJ/kg) are 8.1/77.3/14.3/0.3 and 17.2 for pine wood sawdust, 8.8/68.5/18.4/4.7 and 16.1 for peanut shell; 8.5/68.1/16.3/7.1 and 15.5 for maize stalk. All particle size 0.5–1.0 mm. Pretreatment: Microwave (600 W, 6 min) used only for drying	Bench-scale pyrolysis, conventional, electrically-heated fluidized-bed reactor. Pyrolysis temperature 500°C. Pure N$_2$ as carrier gas, < 1 s vapor residence time Quick moisture release induced by MW drying increases surface area and modifies surface/pore structure, leading to quick release of volatiles during biomass pyrolysis, reducing secondary cracking, yielding more liquid oil and solid char. Water content is reduced due to weakened secondary pyrolysis	Pine sawdust: 70% oil, 15% gas, 15% char; oil quality: 26% water, pH 2.5, viscosity 15.2 mPa s at 40°C, LHV 15 MJ/kg Peanut shell: 52% oil, 17% gas, 31% char; oil quality: 34% water, pH 2.6, viscosity 12.6 mPa s at 40°C, LHV 14 MJ/kg Maize stalk: 51% oil, 19% gas, 30% char. Oil quality: 35% water, pH 2.7, viscosity 9.8 mPa s at 40°C, LHV 12 MJ/kg Main oil compounds: alcohols, ketones, furfural, phenol, alkylated phenols, furan derivatives, etc.	Chen et al. 2008
Rice straw: 8.25% moisture (as received); volatiles/fixed carbon/ash (wt.%, dry): 72.2/14.4/13.4. C:H:O = 1:1.66:0.78 (molar ratio, dry ash-free). Heating value 15.26 MJ/kg. 30–35% hemicellulose, 21–31% cellulose, 4–19% lignin; particle size < 0.85 mm No pretreatment	Single-mode MW power 100–500 W, 50 mL/min N$_2$ as carrier gas	At 300 W: 28.07% solids (_19 MJ/kg); 49.37% gases (incl. H$_2$/CO$_2$/CO/CH$_4$ = 53/20/15/9 vol.%; _11 MJ/Nm^3); 22.56% liquids (43.68 area% alkanes from C$_{12}$ to C$_{32}$; 25.84 area% polars incl. phenol and its derivatives; 8.94 area% polycyclic aromatic hydrocarbons, 2–3 rings and their derivatives)	Huang et al. 2008, 2010

Wheat straw pellets: 10% moisture (as received); C/H/N = 40.21/5.34/0.7 wt.% in the starting materials The biomass sample was treated by: (1) H_2SO_4 solution, 10% w/w, then dried; (2) NH_3 (3% v/v) into the sample as a gas; (3) HCl (3% v/v) into the sample as a gas	150–200 g samples exposed to a max. microwave power of 1200 W, heated up to a max. temperature of 180°C Low temperature biomass pyrolysis	H_2SO_4 and NH_3 result in a sacrifice of oil yield for greater quantities of char; HCl has minimal effect on the oil yield Pyrolysis products (1000 W, 130°C, no additive): 29% char, 21% oil, 36% water, 14% gas. Oil composition: 11 area% 4-vinyl guaiacol/dihydrobenzofuran, 28 area% levoglucosan, 10 area% phenols, 1.3 area% desaspidinol. Oil heating value: 16–22 MJ/kg	Budarin et al. 2009
Air dried poplar sawdust (particle sizes: 0.7–1.1 mm) without bark particles. Wood chips (1.5–4.0 mm), bark (spruce and fir, 0.0–1.2 mm), cotton wool (fibers, diameter approx. 20 μm), and filter paper (thickness 0.15 mm, max. particle size 4 mm) used as liquefaction substrates	Liquefactions were carried out in a Milestone Mega 1200 microwave oven equipped with 100 mL sealed teflon reaction vessels with an internal temperature sensor (not used as a feedback control for regulating microwave intensity). Wood liquefaction with glycols using p-toluenesulfonic acid as the catalyst was carried out under microwave heating. With rapid heating and temperatures in the 190–210°C range complete liquefaction was achieved in 7 min	Liquefaction efficiency was dependent on the choice of glycol. Simple glycols such as ethylene glycol and propylene glycol were more effective than higher analogues. The use of glycerol in mixtures with glycols showed a synergistic effect. Size exclusion chromatography was used to follow the gradual emergence of liquefaction products in solution as well as the recondensation products that start forming early in the reaction and precipitate from solution when molar masses of approx. 1104 g/mol are reached	Kržan and Žagar 2009
Corn stover: air-dried at room temperature, mechanically pulverized into 0.5–4 mm	1L quartz flask in a MW reactor with 700 W, N_2 as purge gas; reaction temperatures 515–685°C, reaction time 4–22 min, biomass particle size 0.5–4 mm	Maximum volatile yield of 76.08% (33.72% biooil, 42.36% syngas) at 8 min reaction time, 650°C temperature, and 3 mm particle size. Maximum bio-oil yield (36.98%) for 18 min reaction time, 650°C temperature and 1 mm particle size. The bio-oil includes phenols (28–40%), aliphatic hydrocarbons (11–24%), aromatic hydrocarbons, furan derivatives, some acids	Lei et al. 2009
Corn stover pellets: 6.2 mm in diameter _ (10–20) mm in length (D _ L); 6.4% moisture, 24.5 MJ/kg; 18% lignin, 37% cellulose, 27% hemicellulose Aspen pellets: 4.8 _ (4–8) mm (D _ L); 5.9% moisture, 17.9 MJ/kg; 19% lignin, 53% cellulose, 27% hemicellulose Pretreatment: Inorganic compounds (e.g., $MgCl_2$, Na_2HPO_4) added as microwave absorber	Quartz flask inside a MW cavity; power input 875 W and 20 min reaction time at 450–550°C Both the condensable vapors and the condensates adhering to the interior wall of the flask collected and weighted as liquid (bio-oil)	KAc, Al_2O_3, $MgCl_2$, H_3BO_3, Na_2HPO_4 all increase bio-oil yield. Increase in $MgCl_2$ dosage (by mixing 0/2/4/8 g $MgCl_2$ with 100 g biomass) yields more bio-oil from corn stover and aspen: both produce ca. 42% bio-oil when 8 g $MgCl_2$ is used; charcoal yield also increases for aspen but does not change much for corn stover. $MgCl_2$ also effective in improving product selectivity	Wan et al. 2009

Table 5. contd....

Table 5. contd.

Feedstock and pretreatment and use of additives/catalysts/microwave absorbers	Reactor and process conditions	Pyrolysis products	References
Aspen pellets: 5 × 20 mm (D × L); 240 kg/m³ bulk density; 5.9% moisture; 17.9 MJ/kg; C/H/N/S/O = 45.4/4.8/0.5/0/49.3 (wt.%, DAF basis) No pretreatment	50–250 g samples placed in a quartz flask inside a microwave reactor: constant microwave power input of 700 W, 20 min irradiation. The evolved volatiles pass through a catalyst column (under controlled temperature 300–600°C) before being condensed on water-cooling columns and collected	Solid acids are effective catalysts to decompose pyrolysis vapors, while solid alkaline and other catalysts do not affect the composition of the liquid product. Increasing the temperature of the catalyst bed and the ratio of catalysts to biomass adversely affected the liquid yield	Zhang et al. 2010
Chlorella sp. (a wide-type algae strain): naturally-dried; 13.7% moisture, 68.4% volatiles, 10.1% fixed carbon, 7.8% ash; C/H/N/O = 49.7/6.98/10.92/24.6 (wt.%, dry basis) Pretreatment: Char (produced from pyrolysis of biomass) used as a microwave absorber	30 g algae (blended with 6 g solid char) placed in a 500 ml quartz flask in a microwave cavity, 500 ml/min N_2 as carrier gas, 20 min reaction time, reaction temperature monitored using an infrared optical pyrometer, final temperature measured by inserting a thermocouple immediately at the end of reaction. The final temperature measured: ca. 462°C at 500 W, 569°C at 750 W, 600°C at 1000 W, and 627°C at 1250 W	Water yield is ca. 21–22%, independent of input power. Oil increases from 26% at 500 W to a max. yield of 28.6% at 750 W and then decreases gradually (to 18% at 1250 W). Gas yield increases over the entire range of power (from 24% at 500 W to 35% at 1250 W). Char yield decreases from 28% at 500 W to 25% at 750 W and then remains constant. Bio-oils composed of aliphatic hydrocarbons, aromatic hydrocarbons, phenols, long chain fatty acids, nitrogenated compounds, among which aliphatic and aromatic hydrocarbons account for 22.18 area%	Du et al. 2011
DDGS (distillers dried grain with solubles): 36.74% neutral detergent fiber (composed of cellulose, lignin and hemicellulose), 29.93% crude protein, 11.07% starch, 10.5% fat, and 12.82% crude ash (all on dry basis). Ca. 6% moisture in DDGS No pretreatment	100 g DDGS placed in a 500 mL quartz flask inside a microwave oven with a rated power of 1000 W. Heating rate of 100 g DDGS at various power inputs: 105, 130, 255, 500, and 570°C/min at 600, 700, 800, 900, and 1000 W, respectively. N_2 as the reactor purge gas; heavier volatiles condensed and collected as bio-oils, char left in the quartz flask. Effects of reaction temperatures (516–684°C), reaction time (the time under the desired temperature, 4.6–21.4 min), and power input (600–1000 W) are studied	Power input affects DDGS biofuel yields. At same reaction time and lower temperature, higher power inputs result in increased yields of bio-oil and syngas. At same reaction time and higher temperature, higher power inputs reduce bio-oil yield and increase syngas yield. Bio-oils are phase-separated to two levels at room temperature: the lower level as aqueous phase (2–5 MJ/kg); the upper level as oil phase (20–28 MJ/kg). C6–14 chemical compounds are up to 95% of DDGS bio-oils	Lei et al. 2011

Material / Pretreatment	Conditions	Results	Reference
Douglas fir: 7 mm × 15 mm (D × L) Activated carbon used as microwave absorber and different carbon-to fir ratios (0, 2:1, 3:1 and 4:1) tested and compared	Different additives-biomass ratios (0, 2, 3, and 4), reaction temperatures (316, 350, 400, 450 and 484°C, measured by an infrared optical pyrometer), and reaction times under the desired temperature (4, 8, 12 and 15 min)	Liquid yields range from 6.8 wt.% (12 min under 350°C, additive/biomass = 4/1) to 48.1 wt.% of biomass (12 min under 350°C, additive/biomass = 2/1), with the majority in 25–35%. Reaction temperatures and additive-biomass ratios greatly affect liquid yields, and also affect liquid compositions (e.g., phenol, phenolics, furfural and its derivatives). The reaction time has less important effect on product distribution	Bu et al. 2011
Sewage sludge from urban wastewater treatment plant; ca. 80% moisture in sludge sample; 75.5% volatiles and 24.5% ash (on dry basis). Element mass fraction (%, DAF basis): C/H/N/S/O = 39.4/5.7/4.75/1.18/24.4 Pretreatment: Graphite (1 × 1 mm) used as microwave absorber	100 g sludge (homogenously blended with 25 g graphite) placed in a quartz tube in a multimode microwave cavity; sample temperature measured by an infrared optical pyrometer; 100 cm³/min N₂ as carrier gas; microwave power 200–1200 W tested, 400 W (max. temperature 490°C) leading to max. oil yield; 10 min reaction time Pyrolysis in electric furnace at 500°C (for 30 min) was also performed	MW at 400 W yields 49.79% bio-oil (35 MJ/kg, 929 kg/m³, viscosity 45cSt at room temperature, flash point 45°C, 29.5wt% monoaromatics), 10.84% gases, 39.37% solids. 800 W (800°C) results in a reduced oil yield (11.71%) due to cracking into gases. Electrical pyrolysis yields 37% oils (30 MJ/kg, with higher oxygen content), 23% gases, 40% solids	Tian et al. 2011
OPF (oil palm fiber): 300–600 μm (< 1 mm in width) OPS (oil palm shell): 1–100 μm (0.5–4 mm in width) On as received basis, OPF has 10% moisture and OPS has 8% moisture. On dry basis, both contain ~75% volatile matters, 20% fixed carbon, and 5% ash Pretreatment: char from conventional pyrolysis of OPS used as microwave absorber, 100–300 μm in size	Fluidized bed quartz glass reactor (0.1 m i.d. × 0.15 m height), with perforated steel distributor plate of 1 mm holes. Purge gas ~10 L/min N₂ (ca. 7 s of vapor residence time); multimode microwave of 450 W, irradiation time 25 min Temperatures in bed and just above the bed surface both measured via thermocouples. In pyrolysis of OPF (1:0.5 blended with char): maximum in-bed temperature 1460°C, very high heating rate (~0.1 s to reach 1000°C). In pyrolysis of OPS (also 1:0.5 blended with char): maximum in-bed temperature ~237°C, low heating rate (18°C/min)	The 1:0.5 biomass-char blend results in the highest oil yield and lowest char yield among various blend ratios (e.g., 1:0.25, 1:1) for both OPF and OPS. For OPF, pyrolysis products include 22 wt.% bio-oil (incl. water), 29 wt.% gas, 49 wt.% char. For OPS, the products include 25 wt.% bio-oil (incl. water), 30 wt.% gas, 45 wt.% char (for large OPS particles, the particle center temperature is believed to be much higher than the measured value). Bio-oils: pH 2.7–3.0, viscosity ~0.015 cm²/s at room temperatures	Salema and Ani 2011

Table 5. contd....

Table 5. contd.

Feedstock and pretreatment and use of additives/catalysts/microwave absorbers	Reactor and process conditions	Pyrolysis products	References
Wheat straw: 7.4% moisture, 69.3% volatiles, 17.4% fixed carbon, 5.9% ash (air-dried) Corn straw: 7.5%, 70.2%, 17.5%, 4.7%, respectively Both straws are baled after air-drying: bale size 1 × 0.6 × 0.6 m³, bulk density 80 kg/m³ No pretreatment	Microwave-assisted pyrolysis of 1 × 0.6 × 0.6 m³ straw bale (20 sets of magnetron, total power 18 kW), 1 L/min N_2 as carrier gas. Reaction time of 85 min when 0.334 kW/kg-straw microwave power used; ca. 46 min when 0.668 kW/kg straw used. Final temperature ca. 600°C	Pyrolysis products: close to 1:1:1 for gas, liquid and char (due to medium pyrolysis temperature and slow heating rate), in which solids are the slightly highest The pure pyrolysis gas from both kinds of straw bales: ~18 vol.% CO, 36 vol.% H_2, 22 vol.% CH_4, 19 vol.% CO_2; LHV 15 MJ/Nm³ Electricity consumption: 0.58–0.65 kWh/kg straw	Zhao et al. 2011
Wheat straw: 7.4% moisture, 69.3% volatiles, 17.4% fixed carbon, 5.9% ash (air-dried); C/H/O/N/S = 41.41/5.86/45.22/1.44/0.17 (wt.%, DAF). Shredded and sieved to particles with sizes < 0.09 mm Pyrolysis residue used as microwave absorber	MW pyrolysis of shredded wheat straw. 5–30 g shredded straw mixed at 2:1 in mass ratio with additives, placed in a quartz reaction tube (18 cm length, 8 cm outer diameter) in a modified 3000 W domestic microwave oven cavity. N_2 (3 L/min) used as purge gas. In case of need, hot air (up to 400°C) can flush the microwave cavity to prevent the condensation of liquids on walls of the reaction tube and pipelines. Low heating rate (< 60°C/min even using SiC as the sample in the reactor); 10 min reaction time under target pyrolysis temperatures (400, 500, 600°C)	Yield of gas/liquid/solid (wt.%): 18/25/57 at 400°C; 21/31/48 at 500°C; 23/31/46 at 600°C Gas product: heating value 11–12 MJ/Nm³; $CO/CO_2/H_2/CH_4$ (vol.%): 35/34/22/8 at 400°C; 33/33/25/8 at 500°C; 27/24/44/6 at 600°C Solid product: C/H/O/N/S = 53/2.5/36.5/1/1 wt.%; temperatures have no big effect on the compositions. Increase in temperatures greatly increases specific surface area and pore volume, and decreases average pore size	Zhao et al. 2012
Douglas fir (7 mm in diameter and 15 mm in length) Activated carbon as catalyst Zn powder was used for the first time as a catalyst with formic acid as a hydrogen donor and ethanol as reaction medium	The ratio of activated carbon (GAC 830 PLUS) to biomass (Douglas fir) between 1.32 and 4.86 with a fixed total loading of 120 g. Microwave power of about 700 W with a constant heating rate of 60 K min⁻¹. Reaction temperatures between 589 K–757 K.	Activated carbon - Bio-oils with high concentrations of phenol (38.9%) and phenolics (66.9%) Zn powder as catalyst and formic acid/ethanol as reaction medium - high concentration of esters (42.2% in the upgraded bio-oil) Long chain fatty acid esters (85.15%), including 11.04% of hexadecanoic acid, methyl ester, 2.37% of hexadecenoic acid, ethyl ester, 9.74% of 8-octadecenoic acid, methyl ester, 2.33% of octadecanoic acid, methyl ester and 2.18% of ethyl oleate	Bu et al. 2012

Microalga *Scenedesmus almeriensis* and its extraction residue was carried out at 400 and 800°C.	4 g of sample, single mode MW oven at 400 and 800°C. Reactor was purged with He for 30 min at 100 mL STP min^{-1} flow rate and then set to 20 mL STP min^{-1} for the pyrolysis experiments	MW pyrolysis produced higher gas yields but also to greater syngas and H$_2$ (50 vol%) production whereas the conventional process produced high bio-oil yields. Bio-oil yield from electric furnace was around 48 wt.% at 800°C. MW pyrolysis of the residue at 800°C produced a maximum syngas concentration (94 vol.%), whereas the MIP of microalga at 400°C produced the gas fraction with the highest H$_2$/CO ratio (2.3)	Beneroso et al. 2013
Microcrystalline cellulose	A 35 mL CEM Discover SP MW unit with a maximum pressure of 300 psi and power output of 300 W (150 → 270°C). A variety of analytical techniques (e.g., HPLC, 13C NMR, FTIR, CHN analysis, hydrogen–deuterium exchange) used for the analysis of the obtained liquid and solid products	The highest selectivity toward glucose was found to be ~75% while the highest glucose yield obtained was 21%	Fan et al. 2013
Pyrolysis of oil palm shell (OPS) biomass with activated carbon as absorber	MW power of 800 W at 2.45 GHz frequency with mechanical stirring	The highest bio-oil yield of 28 wt.% was obtained at 25% MW absorber and 50 rpm stirrer speed. Bio-char showed highest calorific value of the 29.5 MJ/kg at 50% MW absorber and 100 rpm stirrer speed. Bio-oil was rich in phenol with highest detected as 85 area% in the GC–MS	Abubakar et al. 2013
Black liquor degradation from a pulping company with formic acid. Concerning the product yield and distribution of phenolic compounds against reaction temperature (110–180°C) and reaction time (5–90 min)	Reaction temperature: 110–180°C and reaction time: 5–90 min 1 g lignin sample (150–200 mesh) was firstly mixed with 12 mL formic acid in the digestion tank maintained at 4 MPa and <250°C. MW power: 600 W, set reaction temperature and reaction time), and then the digestion tank was placed in microwave reactor and started to be heated	The liquid product consisting of bio-oil achieved a maximum yield of 64.08% at 160°C and 30 min	Dong et al. 2014
Olive tree pruning to isolate lignin catalyzed by different supported metal nanoparticles on mesoporous AlSBA-15 including -nickel (2, 5 and 10 wt.%), palladium (2 wt.%), platinum (2 wt.%) and ruthenium (2 wt.%). The liquid fraction (where lignin was dissolved) was separated from the solid fraction by filtration	The treatment for lignin extraction consisted on the digestion of the feedstock in a mixture of ethanol–water (70 wt.%) at 200°C for 90 min in a pressure reactor. Dissolved lignin was isolated by precipitation with two acidified portions of an aqueous solution (pH around 2). The suspension was centrifuged at 4000 rpm for 20 min to recover lignin which was then dried at 50°C	The bio-oil obtained was primarily composed of monomers, dimers and trimers. Biochar yield was >35% due to oligomerization reactions. The catalyst at 10 wt.% nickel achieved the highest lignin depolymerization degree, with a maximum yield of 30% bio-oil after a short time of reaction (typically 30 min of MW irradiation)	Toledano et al. 2013

Table 5. contd....

Table 5. contd.

Feedstock and pretreatment and use of additives/catalysts/microwave absorbers	Reactor and process conditions	Pyrolysis products	References
Organosolv olive tree pruning lignin possessed the following composition: acid insoluble lignin 71.90% ± 0.79, acid soluble lignin 1.63% ± 0.08, total sugars 2.94% ± 0.14 (glucose 1.75% ± 0.12, xylose 1.10% ± 0.03 and arabinose 0.09% ± 0.01) and ash content 0.39% ± 0.01			
Lignin sulphonic acid sodium salt degradation with NaF as catalyst	1 g of lignin and 10 mg NaF dissolved in 10 ml distilled water under MW with reaction temperature between 100–150°C and average power 85–90 W	Maximum conversion (~100%) of lignin to monolignols was observed after 90 min. Effect of NaF was also observed when catalytic amount of NaF (10 mg) was used in each reaction. The presence of NaF reduced reaction time from 90 min to 30 min	Shaveta et al. 2014
Pulp mill sludge to char with alkali metal hydroxide is an effective microwave absorber	1200 W MW oven at 2.45 GHz. The cavity of the microwave oven had a size of 30 × 30 × 18 cm. 5% of KOH to dried sludge enabled a complete pyrolysis in a few minutes, compared to hours in conventional pyrolysis. Microwave-assisted pyrolysis, however, retained 20% less carbon from the sludge, likely caused by the faster release of volatiles	The secondary sludge yielded more char than the primary sludge. MW pyrolysis and activation was feasible due to the dual roles of alkali metal hydroxides as activating agent and microwave absorber. The combined one step (pyrolysis and activation) process generated an activated carbon with a specific surface area of 660 m^2 g^{-1} from the sludge in three minutes	Namazi et al. 2015
Co-pyrolysis of mixtures of cellulose, paraffin oil, kitchen waste and garden waste to mimic municipal solid wastes (MSW) Ten different microwave absorbing materials (or susceptors) such as aluminium, activated carbon, garnet, iron, silica beads, cement, SiC, TiO$_2$, fly ash and graphite	Pyrolysis at 600°C, and feed to susceptor ratio and composition of the model MSW mixture on (i) overall bio-oil, gas and char yields, (ii) heating value of the bio-oil, and (iii) composition of the bio-oil and gases, were evaluated	The bio-oil contained oxygenated compounds (furans, phenolics, cyclo-oxygenates), aliphatic and aromatic hydrocarbons (mono and polycyclics). Highest bio-oil yield of 53 wt% at an equal composition mixture at 1:1 wt/wt of MSW : graphite. This corresponded to nearly 95% energy recovery and 85% deoxygenation in bio-oil. High selectivities of monoaromatics such as benzene, toluene, xylene and styrene, and C$_8$–C$_{20}$ aliphatic hydrocarbons, and low selectivity of polycyclic aromatics were obtained with a majority of the susceptor–MSW combinations. Methane, ethylene, propylene, isobutylene and hydrogen were the major gaseous products	Suriapparao et al. 2015

MW assisted liquefaction of wheat straw alkali lignin (WAL) to obtain monophenolic compounds as the precursor of bio-fuel	1.0 g WAL, 20.0 mL ethylene glycol, 5–15 wt% H_2SO_4 (or 10 wt% $Al_2(SiO_4)_3$, Al_2O_3 or FeS) and 0–10 wt% of phenol were added into the microwave reactor (ETHOS ONE, Milestone Inc., Italy) after ultrasonic dispersion pretreatment with an ultrasonicator at room temperature for 20 min. Liquefaction reaction at 100–180°C for 10–60 min and 300 W	Total yield of monophenolic compounds is significantly improved under MW irradiation, reaching 15.77% under a relatively mild liquefaction condition of 10 wt% H_2SO_4 as the catalyst, 10 wt% phenol as the hydrogen-donor reagent at 120°C for 40 min. MW irradiation promotes the cleavage of C–C bonds with an extra 29% of C_{aryl}–$C\alpha$ bond cleavage, and increases the yield of monophenolic compounds from 0.92% to 13.61% under the same conditions as conventional process	Ouyang et al. 2015
Solid oil palm shell (OPS) waste biomass with uniformly distributed coconut activated carbon (CAC) microwave absorber	The effects of CAC loading (wt%), microwave power (W) and N_2 flow rate (LPM) were investigated on heating profile, bio-oil yield and its composition	Bio-oil was rich in phenol and 1,1-dimethyl hydrazine. The phenol content was 32.24–58.09% GC–MS area. The bio-oil also contain 1,1-dimethyl hydrazine of 10.54–21.20% GC–MS area. The presence of phenol and 1,1-dimethyl hydrazine implies that the microwave pyrolysis of OPS with carbon absorber has the potential to produce valuable fuel products	Mushtaq et al. 2015
Algal biomass and peanut shell Carbon as microwave absorber N_2 gas was supplied at 10 liter per minute for 15 min	MW power of 800 W operated at 2.54 GHz. Temperatures as high as 1170 and 1015°C for peanut shell and Chlorella vulgaris and up to 15 minutes. Activation energy values for MW pyrolysis were 221.96 and 214.27 kJ/mol for peanut shell and C. vulgaris, respectively	Bio-oil yields reached to 27.7 wt.% and 11.0 wt.% during pyrolysis of C. vulgaris and peanut shell, respectively. Microalgae pyrolysis contained more nitrogen-containing species	Wang et al. 2015
Larch woodchips with water mass fraction of 26.24% sieved through 4 and 12 mm meshes	6 kW microwave generator (2.45 GHz) with A TM_{01n} applicator with power densities of 108 W/m^3. N_2 (2 L/min) was used as purge/sweep gas. Experiments with 5 to 20 g sample 1–6 kW for 12–216 s, specific energy range 150 to 4000 kWh/t	The yields of levoglucosan and phenolic compounds were much higher under MW when compared with conventional fast pyrolysis. The quality of the liquid product is much higher C-H-N (Wt%) compositions were 40.8, 7.8, and 0.5 for MW and 33.4, 8.3, < 0.5 with water 27 and 30% for conventional process, respectively	Robinson et al. 2015

Microwave process related characteristics

Microwave output power control is important for uniform heating and process optimization. Reaction time and power output can be manipulated to achieve the final pyrolysis temperature. The microwave reactor performance, especially, for multi-mode configurations should be dealt with carefully due to variations in the results and to confirm reliability and reproducibility. Mixing can have a significant effect in distribution of microwave heat distribution because non-uniform mixing of absorber with material can result in hot spots and uneven heating within the material.

Inert gas flow and its configuration affect the microwave pyrolysis process significantly. The type of gas, and flow rate are very important. High gas flow rate reduces vapor–gas interaction with absorber, which can affect the product yield and quality, and vice versa. On the other hand, high flow rate increases heat carrying capacity of reaction bed while low flow rate increases the chances of volatile sticking on the reactor wall. Heat carrying capacity depends on type of inert gas used, optimization of inert gas flow rate and system design to control process temperature.

Microwave Pyrolysis Reactor Configurations

In a pyrolysis process, biomass is thermally degraded under an inert environment resulting in liquid, solid and gaseous products depending on the process condition. The reactor design and inert gas circulation scheme are very important in optimizing the pyrolysis process. Figure 5 shows the common process schemes for microwave mediated pyrolysis (Abubakar et al. 2013). Conventionally, quartz glass reactor

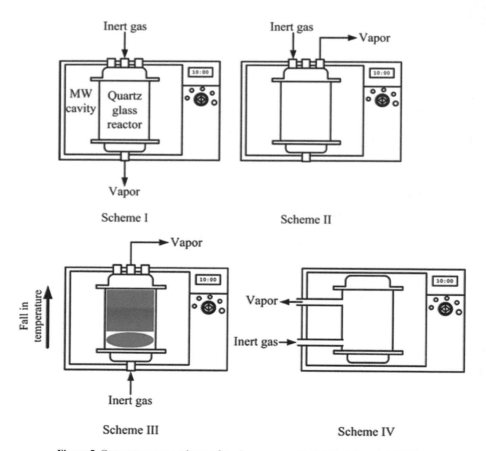

Figure 5. Common process schemes for microwave pyrolysis (Abubakar et al. 2013).

was utilized to pyrolyze the biomass under microwave irradiation. Different techniques were used by which the bio-oil could be collected during MW pyrolysis (see Figure 5). Table 6 summarizes various studies involving microwave pyrolysis of biomass in different process schemes. A variety of feedstock including coffee hulls, wood chips, rice straw, sawdust, microalgae, wheat straw, paper mill sludge, rice husk, corn stalk bale, oil palm biomass, pinewood sawdust, and sewage sludge were studied.

Scheme I shows a continuous flow scheme where the inert gas is continuously purged through the reactor entraining the products. This scheme has some advantages. Firstly, there is no bio-oil deposition on the wall of the reactor due to continuous purging of inert gas from the top of the reactor and immediate exit of vapor from the bottom of the reactor. Secondly, the vapor is instantaneously condensed as soon as it is released from the biomass. The need of glass connector to link the reactor outlet and condenser can be avoided and the time consuming process of cleaning the equipment, particularly the reactor after each run can be circumvented. Importantly, there is no loss of bio-oil in the process.

Schemes II, III, and IV are the commonly used process configurations. The inert gas (nitrogen, helium or CO_2) is introduced either from the bottom or from the top or from the side of the reactor. There are several drawbacks with each scheme. The vapors can get entrained from the top of the reactor. As seen in Scheme III, the material is usually at high temperature compared to other regions of the microwave reactor. This causes deposition of bio-oil on the wall of the glass reactor and other equipment. The bio-oil yield gets affected due to leftover of heavy fraction of bio-oil in the reactor. The problem of bio-oil deposition during microwave pyrolysis is an important concern (Zhao et al. 2012). Hot air can be passed to avoid any vapor condensation in the experimental apparatus. However, additional step of drying might increase energy consumption of the process.

Bio-oils from Pyrolysis

Bio-oils are usually dark brown, free-flowing liquids that have a distinctive odor. During bio-oil production, a large number of reactions occur, including hydrolysis, dehydration, isomerization, dehydrogenation, aromatization, retro-condensation and coking. The exact composition of the bio-oil is dependent on various factors are listed below (Huber et al. 2006).

- The feedstock (including dirt and moisture content)
- Organic nitrogen or protein content of the feedstock
- Heat transfer rate and final char temperature during pyrolysis
- Extent of vapor dilution in the reactor
- Time and temperature of vapors in the reactor
- Time and temperature of vapors in heated lines from the reactor to the quench zone
- If the vapors are able to pass through the accumulated char during filtration
- Efficiency of the char removal system
- Efficiency of the condensation equipment to recover the volatile components from the non-condensable gas stream
- If the condensates have been filtered to remove suspended char fines
- Water content of the feedstock
- Extent of contamination of the bio-oil during storage by leaching of containers
- Exposure of air during storage
- Length of storage time
- Storage temperature

The operating conditions, bio-oil and gas product characteristics for microwave mediated pyrolysis process from various biomass feedstock are shown in Table 7. The bio-oil has a wide composition including carboxylic acids, phenols, guaiacols, cresols, condensed aromatics, furfural, cresol, eugenol, arabinoses, xylitol, levoglucosan, ketones, aldehydes, carboxylic acids, nitrogenous compounds, alcohols, esters, ethers, etc. These depend on the type of microwave process and the process configuration and the type of microwave absorber used in the process.

Table 6. Microwave systems used to carry out biomass pyrolysis (Abubakar et al. 2013).

Configuration	Feedstock, absorber and process details	Reference
Scheme I	Wood Sample was hung in the MW oven type A. In type B MW oven, the sample was placed on the rotating table Nitrogen was introduced from the top of the reactor and bio-oil was collected at the bottom. Aspirator was used to extract the vapors out of the MW chamber. Bio-oil remaining on the interior wall of the reactor was washed down into a collection bottle or tank with methanol	Miura et al. 2004
Scheme II	Coffee hulls Char A quartz reactor was used, which was placed in the center of a waveguide and radiated with single mode MW Helium gas was introduced from the top and the vapor was also entrained from top	Dominguez et al. 2007
	Rice straw and sawdust Ionic liquids A three-necked round-bottom flask was used Nitrogen gas was introduced from the top and the vapor was also entrained from top	Jun et al. 2010
	Microalgae Nitrogen gas was introduced from the top and the vapor was also entrained from top	Du et al. 2011
	Waste engine oil Nitrogen gas was introduced from the top and the vapor was also entrained from top	Lam et al. 2012
	Wheat straw Nitrogen gas was used as inert gas. The circulation air heated by an electric heating device was used to prevent the condensation of liquid-phase products on the walls of the quartz reactor and pipelines	Zhao et al. 2012
	Paper mill sludge Carbon dioxide and nitrogen was used as inert gas. They were introduced from the top and the vapor was also entrained from top	Jiang and Ma 2011
	Microalgae Nitrogen gas was introduced from the top and the vapor was also entrained from top	Hu et al. 2012
Scheme III	Rice husk and sugarcane Nitrogen gas was introduced from the bottom and vapor was purged from the top of the reactor	Wang et al. 2012
	Corn stalk bale Nitrogen gas was introduced from the bottom and vapor was purged from the top of the reactor. Electrical heating tube was used before the condenser to avoid deposition of bio-oil	Zhao et al. 2010
	Oil palm biomass Nitrogen gas was introduced from the bottom and vapor was purged from the top of the reactor. Problem of bio-oil deposition on the walls of the reactors and equipment	Salema and Ani 2011, Salema and Ani 2012a,b
	Oil palm empty fruit bunches Nitrogen gas was introduced from the bottom and vapor was purged from the top of the reactor. Bio-oil was expected to condense on the reactor	Omar et al. 2010

Table 6. cntd....

Table 6. cntd.

Configuration	Feedstock, absorber and process details	Reference
Scheme IV	Sewage sludge	Tian et al. 2011
	Nitrogen gas was introduced from the side of the MW chamber and the vapor was entrained from side	
Other configurations	Pine wood sawdust	Chen et al. 2008
	Inorganic additives in liquid form	
	Beaker-shaped quartz reactor was placed in the MW cavity	
	The reactor and tube system were purged by Nitrogen	
	Rice straw	Huang et al. 2008
	Single mode MW was used. A quartz reaction tube and sample holder were used	
	Nitrogen was purged into the system	
	Wheat straw	Budarin et al. 2009
	The MW reactor was fitted with a vacuum module to remove the vapor	

Effect of microwave power

The output power of microwave irradiation has a significant effect on the reaction temperature. The ability of the microwave irradiation to increase the temperature of the biomass depends on various parameters, as discussed before. The correlation between the microwave power output and the temperature rise is usually linear (Huang et al. 2013) and in some cases non-linear especially when the biomass has had sufficient exposure and interaction with microwave irradiation. The heating rate of the biomass also depends on the microwave power input and the maximum temperature of the biomass is not as sensitive but the heating rate is dependent on the microwave power. Higher power output allows for reaching higher biomass temperatures quickly.

Relationship between microwave power and the energy yield

Optimum microwave power output should be determined to minimize energy losses and product wastes. The mass yield of the process can be calculated using equation [1]. The mass yield means the ratio of solid residues mass to biomass sample mass:

$$Mass\ yield = \frac{Mass\ of\ solid\ residues}{Mass\ of\ biomass\ sample} \times 100\% \qquad [1]$$

Figure 6 shows the HHV, the mass yield of solid residues and the total energy values of the solid residues at different microwave power levels for microwave pyrolysis of rice straw by Huang et al. (2008). It can be noted from Figure 6 that the mass yield was lower at higher microwave power level, which represents that more rice straw was converted into liquid and gaseous products. The highest HHV of solid residues occurred at the middle microwave power level (300 W). This should be owing to that part of the fixed carbon content which was removed when the microwave power level was too high (Huang et al. 2008).

The solid residues of the pyrolysis process usually contain a higher energy value. The energy densification ratio can be calculated using equation [2] using the high heating values of the biomass sample and biochar product.

$$Energy\ densification\ ratio = \frac{HHV\ of\ solid\ residues}{HHV\ of\ biomass\ sample} \times 100\% \qquad [2]$$

The energy yield can be determined using equation [3]:

$$Energy\ yield = Mass\ yield \times Energy\ densification\ ratio \qquad [3]$$

Table 7. Summary of bio-oil production studies from microwave mediated pyrolysis process (Mushtaq et al. 2014).

Feedstock	Operating conditions	Bio-oil yield (wt%)	Bio-oil composition (% area of GC–MS)	Gas yield (wt%)	Gas composition (vol% of the total gas)	Reference
With carbonaceous microwave absorbers						
Wheat straw pellet (cylindrical shape) 1 × 1 cm^2	Sweeping gas helium, 1000–2000 W, 1–5 min	27.90	Phenols, guaiacols, cresols, condensed aromatics	46.10	H_2 = 1.3, CO = 24.4, CO_2 = 14.5, CH_4 = 2.6, C_2H_2 = 1.2, C_2H_4 = 1.5, C_2H_6 = 0.6, Others = 53.9	Krieger-Brockett 1994
Larch samples of diameter 60, 80, 100 and 300 mm, weight 80, 190, 370 and 12,000 g, respectively	Sweeping gas N_2, variable power 100–1500 W, 3–18 min	26.4a	Carboxylic acids, furfural, cresol, guaiacol, eugenol, arabinoses, xylitol, levoglucosan, etc.	na	na	Miura et al. 2004
Rice straw	Sweeping gas N_2 50 mL/min, reaction time 30 min, 50–500 W, max. temperature 105–563°C	max. 22.56b	Alkanes = 43.68, polars = 25.84, polycyclic aromatic hydrocarbons = 8.94, others = 21.54	max. 49.37c	H_2 = 55, CO = 13, CO_2 = 17, CH_4 = 10, Others = 5	Huang et al. 2008
Rice straw	Sweeping gas N_2 50 mL/min, 200–500 W, 260–500°C	na	na	na	(At 300 W and max. 400°C) H_2 = 50.67, CO_2 = 22.56, CO = 10.09, CH_4 = 7.42, Others = 9.26	Huang et al. 2010
Wheat straw pellets	Vacuumed, heating rate with additives 17°C/ min, 1000 Wa (max. 1200 W), 130a–180°Cf, time 130–165 min, pre-treatment of feedstock with additives (H_2SO_4, HCl and NH_3)	max. 22.1d	1,4:3,6-Dianhydro-α-D-glucopyranose = 1.27e–6.69 g, levoglucosan = 27.87e–46.44 h, levoglucosenone = 1.51d–23.63 g, phenols = 15.85 h–20 e, desaspidinol = 1.3e–2.74 h	max. 20 h	na	Budarin et al. 2009
Corn stover, air dried, particle size 0.5–4 mm	Sweeping gas N_2, 700 W, reaction temperatures 515–685°C, reaction time 4–22 min	max. 36.98i	Phenols = 28–40, aliphatic hydrocarbons = 11–24, aromatic hydrocarbons, furan derivatives, some acids, etc.	Max. 42.36j	na	Lei et al. 2009
Rice straw	Sweeping gas N_2, 50 mL/min, reaction time 30 min, 50–500 W	na	–	40	H_2 = 50.67, CO = 16.09, CO_2 = 22.56, CH_4 = 7.4, Others = 3.26	Huang et al. 2010
Wheat straw and corn straw bales 1 0.6 0.6 m^3	Sweeping gas N_2, reaction time 85 min, 18,000 W, max. temperature 600°C	Approx. 33.30k	na	Approx. 33.30k	H_2 = 36, CO = 18, CO_2 = 19, CH_4 = 22, Others = 5	Zhao et al. 2011

Without carbonaceous microwave absorbers

Cylindrical coffee hulls pellets (3 mm dia. 2 cm length)	Fixed amount of bio-char, N_2 at 60 ml/min, reaction time 15 min, 130–420 W, 500–1000°C, single mode microwave	7.90a–9.19b	68.72c	$H_2 = 40.06c$, $CO = 32.75c$, $CH_4 = 6.74c$, $CO_2 = 17.73c$, $C_2H_4 = 2.15c$, $C_2H_6 = 0.56c$	Dominguez et al. 2007
Oil palm fiber, oil palm shells (OPF 300–600 mm and OPS 0.001–0.002 m)	OPF and OPS to bio-char ratio (1:0.25, 1:0.50 and 1:0.75), N_2 at 10 mL/min, reaction time 25 min, fluidized MW heating, 450 W	max. 22d, max. 25e	Ketones, aldehydes, carboxylic acids, nitrogenous compounds, alcohols, esters, ethers, etc.	max. 29d, max. 30e na		Salema and Ani 2011
Oil palm shells (OPS-850 mm)	OPS to CAC ratio (1:0.25, 1:0.50 and 1:0.75), N_2 at 5 mL/min, reaction time 25 min, overhead stirred fluidized MW heating, 300 W and 450 W, fixed stirrer speed 200 rpm	max. 17f	(At 450 W, 50% CAC) Phenol = 72.1, o-cresol = 3, p-cresol = 2.4, 2-methoxy-4-methyl-phenol = 5, 2-Methoxy-phenol = 9, 4-Ethyl-2-methoxy-phenol = 4.4, 2,6-Dimethoxy-phenol = 3.9, Others = not detected	max. 47g	na	Salema and Ani 2012a
Empty fruit bunch pellets (0.4–4 cm long having diameter 0.7 cm)	EFB pellets to CAC ratio (1:0.25, 1:0.50 and 1:0.75), N_2 at 4 mL/min, reaction time 25 min fluidized MW heating, 300 W and 450 W	max. 22g	(At 450 W, 75% CAC) o-cresol/46.1, p-cresol/45.3, 2-methoxyphenol/26, 2-methoxy-4-methyl-phenol/9.8, 4-ethyl-2-methoxyphenol/10.5, 2,6-dimethoxy-phenol/14.3, 1-(2,6-dihydroxy-4-methoxyphenyl)-ethanone/4.1	max. 31g	na	Salema and Ani 2012b
Oil palm shell (1.4 mm)	OPS to CAC ratio (1:0.25, 1:0.50 and 1:0.75), N_2 at 7 ml/min, reaction time 30 min, overhead stirred at 50, 100 and 150 rpm, 450 W	max. 27.5h	(At 450 W, 75% CAC, 100 rpm) Phenol = 84.7, 2-methoxy-phenol = 6.7, 2-methoxy-4-methyl-phenol = 2.1	max. 47i	na	Abubakar et al. 2013
Pelletized douglas fir 0.7 cm in diameter and 1.5 cm in length	Douglas fir to activate carbon ratio (1.32:1–4.86:1), 700 W, temperature 350–484°C, time 1.27–15 min	6.8j–48.1k	Phenolic compounds = 66.9 including 38.9% area of phenol	13.1l –66.2m	na	Bu et al. 2011

Table 7. contd....

Table 7. contd.

Feedstock	Operating conditions	Bio-oil yield (wt%)	Bio-oil composition (% area of GC–MS)	Gas yield (wt%)	Gas composition (vol% of the total gas)	Reference
Without carbonaceous microwave absorbers						
Wheat straw of size 0–0.090 mm	Wheat straw to pyrolysis residue 2:1, N$_2$ at 5–30 mL/min, reaction time 10 min, 900 W, max. temperature 1000°C using 200°C hot air	25–31n	na	18–22n	H$_2$ = 22.1–43.7n, CO = 34.7n, CO$_2$ = 23.6–33.8n, CH$_4$ = 7.9n	Zhao et al. 2012
Douglas fir sawdust pellet (0.7 cm diameter and 1.5 cm length)	Four acid washed activated carbons (coconut shell based, wood based, bituminous coal based and lignite based), DF sawdust pellets to four AC ratios 1:3, N$_2$ flow rate na, 700 W, 8 min	26.5p, 28.97r, 31q, 33.21, 45.2s	Phenols = 2.54s–74.77r, guaiacols = 1.33p–48.78s, furans = 3.9s –14.24q, ketones/aldehydes = 2.02r–16.85q	11.8s, 44.6q, 44.8l, 47.53r, 52.67p	na	Bu et al. 2013

na—not available; [a]100 mmdia, 0.351 kg, 7 min; [b]20/40 meshsize, 407°C, 300 W; [c]20/40 meshsize, 407°C, 300 W; [c]with HCl (3% v/v); [e]without additive; [f]with additive; [g]with H$_2$SO$_4$ (10% w/w); [h]with NH$_3$ (3% v/v); [i]1 mm particle size, 650°C, 18 min; [j]3 mm particle size, 650°C, 8 min; [j]based on 1:1:1 of liquid:gases:solids.

With carbonaceous microwave absorbers, na—not available: [a]500°C, 130 W, 15 min; [b]800°C, 270 W, 15 min; [c]1000°C, 420 W, 15 min; [d]OPF to bio-char ratio 1:0.5, 450 W, 25 min; [e]OPS to bio-char ratio 1:0.5, 450 W, 25 min; [f]OPS to CAC ratio 1:0.25; [h]same as [g]with 50 rpm; [i]same as [g]with 100 rpm; [j]350°C, 12 min, Douglas fir to activated carbon ratio 4:1; [k]350°C, 12 min, Douglas fir to activated carbon ratio 2:1; [l]400°C, 15 min, no activated carbon; [m]350°C, 12 min, Douglas fir to activated carbon ratio 4:1; [n]400–600°C; [o]coconut shell based; [p]wood based; [q]bituminous based; [r]lignite based; [s]no activated carbon.

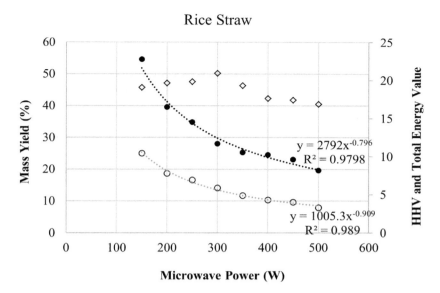

Figure 6. The mass yield and higher heating value (HHV) of solid residues and caloric values of solid residues under different microwave power (Huang et al. 2008).

Effect of temperature

The products and their composition depend on the temperature of the pyrolysis process. The influence of the pyrolysis method (conventional vs. microwave heating) and temperature have significant effect on the product yields and on the characteristics of the pyrolysis products. In general, microwave mediated process yields a larger fraction of gas when compared with other conventional pyrolysis even at low temperatures (Dominguez et al. 2007). The energy accumulated in gas increases with pyrolysis temperature and the corresponding oil and char fractions decrease.

Microwave pyrolysis can be conducted at much lower temperatures than the conventional process. Microwaves were found to efficiently deliver comparable evolution of bio-gas in the system when compared with conventional pyrolysis at significantly reduced temperatures (120–180°C vs. 250–400°C) from a variety of feedstock (Shuttleworth et al. 2012). The gas obtained from microwave mediated pyrolysis was found to contain CO_2, CH_4 and CO as major components as well as other related chemicals (e.g., acids, aldehydes, and alkanes) which were obtained in different proportions depending on the selected feedstock. Table 8 shows the char, gas and oil yields for different feedstock evaluated at low temperatures under microwave mediated pyrolysis. Budarin et al. (2015) reported similar results with a total mass recovery of 51% and carbon recovery of 71.4% from wheat straw biomass.

Table 8. Mass balance for microwave pyrolysis experiments (Shuttleworth et al. 2012).

	Oil yield (%)	Char yield (%)	Gas evolution ratio (mL/g)
Wheat straw	26	35	32
Reed canary grass	15.0	41.5	37.8
Softwood pelletized	20.8	43.3	35.4
Bracken pelletized	10.3	43.9	38.0
Waste office paper	18.0	41.0	35.2
Barley dust	20.4	38.1	41.0
Macroalgae	11.0	54.8	305

Effect of feedstock on microwave pyrolysis

Gas, oil and solid char yields depend on the type of feedstock. Microwave pyrolysis reportedly produces higher gas yields from lignocellulosic biomass. In some cases, oil yields are better for microwave mediated pyrolysis process. Table 9 shows the comparison of various products and their distribution among the conventional pyrolysis and microwave mediated pyrolysis from different biomass feedstock (Luque et al. 2012).

Table 9. Comparison of various products and their distribution among the conventional pyrolysis and microwave mediated pyrolysis (Luque et al. 2012).

Product and Yield	Feedstock	Microwave pyrolysis	Conventional pyrolysis
Gas yield (wt.%)	Sewage sludge[a]	33–45	22–35
	Coffee hulls[b]	60–70	55–65
	Glycerol[c]	55–85	47–82
Production of syngas [H_2 + CO] (L/g feedstock)	Sewage sludge[a]	0.50–0.56	0.25–0.40
	Coffee hulls[b]	0.41–0.62	0.17–0.36
	Glycerol[c]	0.34–0.93	0.20–0.87
	Wheat straw	54% total gas volume (37% H_2)	< 40% total gas volume
	Corn straw	54% total gas volume (35% H_2)	< 40% total gas volume
Production of CO_2 (L/g feedstock)	Sewage sludge[a]	0.02–0.15	0.04–0.18
	Coffee hulls[b]	0.20–0.34	0.45–0.65
	Glycerol[c]	0.00–0.04	0.00–0.15
Oil yields (wt.%)	Waste oil[d]	85	46–80
	Sewage sludge A[e]	10.3	3.1
	Sewage sludge B[f]	2.2–4.0	0.9
Petroleum fractions in waste oil pyrolysis (%)	Waste automotive oil[g]	Gasoline (C4-C12):69	Gasoline (C4-C12):40
		Kerosene (C11-C15):16	Kerosene (C11-C15):18
		Diesel (C15-C19):15	Diesel (C15-C19):13
		Heavy oil (>C19):4	Heavy oil (>C19):34

[a]T = 1000°C, moisture content: 0–81 wt%, two different feedstocks. [b]T =500–1000°C. [c]T = 400–900°C. [d]T = 550°C. [e]T = 1000°C. [f]T = 1000°C, two different microwave devices and microwave absorbers. [g]MW-P vs. C-P using an electric oven.

Effect of microwaves on the biomass feedstock

The physical effects of microwaves are quite distinct from the conventional heating method. This mainly depends on how the microwaves interact with the material and material composition. Numerous observations were reported in literature. An example of microwave effect on oil palm shell (OPS) pyrolysis is shown in Figure 7 (Salema and Ani 2011). The surface image analysis of the biomass samples treated with microwave heating and conventional heating showed a clear distinction in the surface structure of the char product. Large and deep cracks were found in conventional pyrolyzed OPS char while pores were observed in microwave pyrolyzed char with absence of any cracks. Microwave creates volumetric heating compared to conventional heating. Unlike the conventional heating where the heat is transferred from the outer surface of the material to the inner part, the microwave heat is generated in entire volume of surface and more particular from the core of the materials towards the surface of the materials. Hence, in

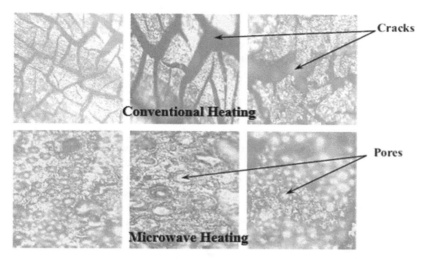

Figure 7. Surface image analysis of OPS chars after conventional heating and microwave heating (Salema and Ani 2011).

conventional pyrolysis of OPS the outer surface is at higher temperature than the inner core. By the time the inner surface gets the heat, the outer surface undergoes overheating and creates the cracks or fissures. This diminishes the quality and makes the char more fragile.

Considerations in Microwave Mediated Pyrolysis

Microwave power and reaction temperature control

Microwave power is usually fixed in a reaction scheme and the treatment time is varied to control the microwave heating process. This method does not provide clear information about the actual power absorbed by biomass (Waheed ul Hasan and Ani 2014). The specific energy consumption defined as the product of power and treatment time per unit product or biomass cannot be used as a manipulated variable for scale-up of the microwave reactor to pilot or industrial scale pyrolysis. Often, laboratory scale procedures are limited by the equipment design and operating feasibility, for example, in a domestic microwave oven. For precise temperature control, microwave power serves as the default manipulating variable coupled with a temperature sensor which would provide signals to the microwave power control in a feedback control loop (Tong et al. 1993). A less precise and simple method for controlling the temperature in an open loop control system is to provide a circulating water jacket around the quartz glass reactor. The temperature of the solids inside the reactor then can be controlled by manipulating the water flow rate in the cooling jacket (Pagnotta et al. 1992).

Dielectric properties of biomass and reaction mixture

The dielectric properties change with the change in microwave frequency, temperature, and orientation (direction) of the biomass and the reaction contents such as acid and alkali substances and salts (Keshwani and Cheng 2010). Therefore, the evaluation of change in dielectric properties with all these parameters and their monitoring during the phase of reaction is important for understanding the energy dissipation, predicting the rates of heat generation, yield of the reaction, and subsequently the design of the reactor. High moisture content in biomass increases microwave induced heat generation, increasing the efficiency of the process. On the other hand, it also increases vaporization leading to energy loss. Therefore, biomass should be dried to an optimal moisture level before introducing into a MW reactor, and moisture contents should be reported in the suitability studies of various kinds of biomass for pyrolysis process.

Thermal runaway and voltage breakdown

The power dissipation density P can be calculated by the internal electric field stress by $P = 2\pi f \varepsilon_o \varepsilon'' E_i^2$, where f is MW frequency, ε_o is the permittivity of free space, and ε'' is the loss factor. Voltage breakdown occurs if any of the internal or external electric field stress reaches a critical value which results in ionization of the gases present in the reactor. Ionized gases form an arc, establishing a low resistance electric path within the reactor which results in high power loss. As a result the biomass is burnt or denatured, and the equipment may undergo a meltdown crisis. Moreover, breakdown voltage is likely to happen at low pressure or at high temperature, leading to low density or pressure of the gas inside the reactor (Metaxas and Meredith 1983, Meredith 1998). Therefore, care must be taken while using a low loss tangent biomass (requiring high external field stress) and experimenting at elevated temperatures (causing lower density in the reactor) as the conditions are conducive to arching. Thermal runaway occurs when the rate of heat generation in the material exceeds its rate of heat dissipation which results in localized heating and subsequently voltage breakdown. These factors become very important for high throughput pilot scale reactors and are the primary safety considerations during a scale-up operation.

Reactor configuration and mode of operation

Feedback temperature control using microwave power as a manipulating variable is necessary for the precise temperature control in the reactor. Fiber optic probes or infrared sensors are not suitable for the measurement of temperature inside the solids reactor, and thermocouples are the only available option, though the result may be inaccurate for the case of solids. Agitation and use of microwave absorption materials within the reactor increases temperature uniformity and improves the accuracy of thermocouple results. The considerations of dielectric properties like loss factor and loss tangent are essential in scale-up of the microwave reactor for estimation of the rate of heat generation and temperature profile within the reactor. Voltage breakdown and thermal runaway conditions are the primary safety concerns in scale-up of microwave reactor for the production of activated carbon.

Pyrolysis Economics

Economics of bio-oil production through conventional processes were studied by many researchers. The price per unit of bio-oil decreases with increase in production capacity. The bio-oil costs are significantly influenced by the feedstock costs similar to other biofuel production. Feedstock costs vary among different types of biomass. The capital costs typically increase with the plant capacity and the sophistication of the chosen process. Table 10 shows a summary of bio-oil production costs for various plant capacities.

The following factors should be considered for economic analysis (Wright et al. 2010a):

- Plant Size, Location, and Construction (typical size between 5–2000 tons/day)
- Biomass collection area and transportation costs (23% of feedstock costs with linear scaling for smaller sized areas)
- Products from the plant (pyrolysis gas, bio-oil, and charcoal)
- Process technology and design
- Production time (~350 days/year or equivalent capacity factor of 96%)
- Construction time (< 24 months) with startup period of 25% of the construction time (6 months)
- Moisture content in the feedstock is 15% (wet basis)
- Variation of feedstock compositions
- Feedstock costs
- Material balance (basis: ultimate and proximate analyses data)
- Energy Balance
- Biomass handling (pretreatment and preparation)

Table 10. Bio-oil Production Cost/Selling Price (Ringer et al. 2006).

Plant Size (tonne/d)	Feed Cost ($/dry tonne)	Feed Cost ($/GJ)[1]	Bio-Oil Cost ($/kg)	Bio-Oil Cost ($/gal)	Bio-Oil Cost ($/GJ)	Total Capital Investment	Source
2.4	$22	$1.1	$0.38	$1.73	$21.2	$97,000	Islam and Ani 2000
24	$22	$1.1	$0.18	$0.82	$10.1	$389,000	Islam and Ani 2000
100	$36	$1.8	$0.26	$1.21	$14.5	$6.6 M[2]	Mullaney 2002
200	$36	$1.8	$0.21	$0.99	$11.7	$8.8 M	Mullaney 2002
400	$36	$1.8	$0.19	$0.89	$10.6	$14 M	Mullaney 2002
1,000	$46.5	$2.3	$0.09	$0.41	$5.0	(Not given)	Cottam and Bridgwater 1994
1,000	$44	$2.2	$0.11	$0.5	$6.1	$46 M	Gregoire and Bain 1994
250	$44	$2.2	$0.11	$0.5	$6.1	$14 M	Gregoire 1992
1,000	$20–$42.5	$1.0–$2.13	$0.13–$0.54	$0.59–$2.46	$7.3–$30.0	$44–143 M	Solantausta et al. 1992
250	$11	$0.55	$0.10	$0.46	$5.6	$14 M	Arthur Power and Assoc. 1991
1,000	$44	$2.20	$0.09	$0.41	$5.0	$37 M	Arthur Power and Assoc. 1991

[1]Assumes 20 MJ/kg HHV for wood, 17.9 MJ/kg HHV of bio-oil and 4.55 kg/gal density of bio-oil (Mullaney 2002, EPRI 1997); [2]Million

- Chemical Costs (Costs for acids and other chemicals)
- Operating Costs
- Annual maintenance materials cost (~2% of the total installed equipment cost)
- Personnel costs (total salaries and coverage items such as safety, general engineering, general plant maintenance, payroll overhead (including benefits), plant security, janitorial and similar services, phone, light, heat, and plant communications)
- Financial packages, geographical location, interest rates, return on investment periods

Microwave Pyrolysis Economics

Pyrolysis process can be economically sized for small-scale applications to meet the feedstock supply capacities. Ruan et al. (2008) developed a small-scale pyrolysis process (6 dry tons/day) for on-farm application as on-farm "mobile" units or at slightly larger scales as central hubs that source biomass from surrounding farms. They also presented an economic analysis for this scenario. A farm with 1,000 acres of corn crop and annual yields of 3,000 tons of corn stover, with 50% collected as feedstock was considered. A small scale pyrolysis facility with 0.75 ton/h capacity at $200,000, similar to the cost of a combine harvester used for the corn crop, receives this feedstock. It had a 10 year lifetime with 8 hours/day operation time and processing 6 tons/day. Facility operation time was 250 days per year. Pyrolysis of the feedstock yielded 50% bio-oil, 20% char, and 30% pyrolytic gas, the compositions and properties of these products being similar to those produced by other commercial pyrolytic processes. The bio-oil price was estimated at $1/gallon. The pyrolytic gas was used to generate electricity and heat, which could provide all the electricity needed for the pyrolysis process and additional electricity was purchased to power peripheral equipment, such as the stover chopping system. The char price was estimated at $50/ton. Table 11 shows an estimate of income and costs. The feedstock cost is significantly lower than that for large ethanol or pyrolysis

Table 11. Cost-benefit analysis of microwave mediated bio-oil production (Ruan et al. 2008).

Items	Quantity	Value/unit	Amount
Revenue			
Sale of Bio-oil (Dry tons)	168,000	$1.00	$168,000
Sale of Char as fertilizer and liming (dry tons)	300	$50	$15,000
	Total sales		$183,000
Costs			
Feedstock (dry tons)	1500	$32.91	$49,365
Processing labor (hours)	1000	$12	$12,000
Pyrolysis machinery depreciation			$30,000
Electricity purchased			$8,640
Consumables			$9,150
Maintenance			$10,000
Other expenses			$18,300
Transportation of bio-oil to market (ton)	750	$4.00	$3,000
	Total costs		$140,455
	Net Income		$42,545

Calculations based on data from Perlack and Turhollow 2003, Petrolia 2006.

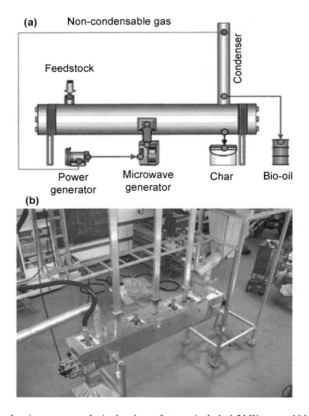

Figure 8. A commercial scale microwave pyrolysis chamber—features include 4.5 kW power, 10 kg/h throughput, horizontal cylindrical reaction chamber, three microwave inlets, pressure relief vent, dry fraction and vapor outlets, inert gas line, temperature measurement and process automation (Ruan et al. 2008).

plants, mainly due to savings in transportation. The net income based on the conservative estimates was suggested to be attractive to many farmers seeking an enterprise that utilizes their biomass and employs their labor. Compared with current large-scale biomass energy systems, this small scale system is more technologically feasible, economically viable, and sustainable.

Environmental life cycle assessment (LCA) studies were conducted for fast pyrolysis of woody biomass used in large scale (2000 dry tons/day) pyrolysis to produce transportation fuels (Hsu 2012) and at small scale (400 dry tons/day) to produce electricity (Fan et al. 2011). These studies assume that biochar is used for process energy. Gaunt and Lehmann (2008) compared biochar for electricity generation (via slow pyrolysis) or for land amendment when produced in agricultural residue- and energy crop-fed slow pyrolysis systems, and concluded that greenhouse gas avoidance was greater for the land amendment case. Avoided emissions were between 2 and 5 times greater when biochar was applied to agricultural land (2–19 Mg CO_2 ha^{-1} y^{-1}) than used solely for fossil energy offsets.

Pourhashem et al. (2013) evaluated the life cycle greenhouse gas (GHG), energy, and cost tradeoffs for farm-scale bio-oil production via fast pyrolysis of corn stover feedstock and subsequent utilization for power generation. GHG emissions of 217 and 84 g CO_{2eq} per kWh of bio-oil electricity for coal cofiring and land amendment, respectively were reported. The electricity produced from burning pyrolysis oil and biochar with variable operating costs of $93/MWh and $18/MWh, respectively, were competitive with the fuel oil and coal electricity markets. At a higher scale (e.g., 2000 dry metric tons per day), the estimated cost of upgrading pyrolysis bio-oil falls within the range of $2.11 to $8.32 per gallon of gasoline equivalent, depending upon the hydrogen production pathway (Wright et al. 2010b). The estimated cost of producing pyrolysis bio-oil is $0.72/gallon bio-oil ($0.65/gal bio-oil to $0.85/gal bio-oil, low and high, respectively), which in turn is competitive with fuel oil in peak electricity markets at $93/MWh in the near term. These studies suggest that local pyrolysis plants have the potential to offer a low cost, low GHG emission, and low nonrenewable energy feedstock for power generation.

Concluding Remarks

Microwave heating has been demonstrated as a new tool for high and low temperature pyrolysis reaction and process development. The reaction can be completed more efficiently and rapidly in comparison with other conventional heating due to effective heat transfer profiles. Microwave mediated pyrolysis technology and research and development is more advanced than other microwave applications in biofuel production.

Several advantages associated with microwave heating seem to be even more complementing for the pyrolysis process. However, issues related to voltage breakdown and thermal runaway should be considered. Understanding the fundamental mechanisms of microwave mediated pyrolysis process is very important to successfully design the microwave reactor. Small-scale and pilot-scale studies are still warranted prior to large-scale process development. The techno-economics and environmental aspects related to the microwave mediated pyrolysis technology seem to be much better and offer numerous benefits for sustainable development. Where economics are not favorable, integrated heat and electricity production, reuse of process heat, valuable byproduct recovery and land applications should be considered to improve the overall economic feasibility and environmental impacts.

References

Abubakar, Z., Salema, A.A. and Ani, F.N. 2013. A new technique to pyrolyse biomass in a microwave system: effect of stirrer speed. *Bioresource Technology*. 128: pp.578–585.

Allan, G.G., Krieger, B.B. and Work, D.W. 1980. Dielectric loss microwave degradation of polymers: cellulose. *Journal of Applied Polymer Science*. 25(9): pp.1839–1859.

Arthur, J. Power and Associates, Inc. Feasibility Study: One thousand tons per day feedstock wood to crude pyrolysis oils plant 542,000 pounds per year using fast pyrolysis of biomass process. Golden, CO: Prepared for Solar Energy Research Institute, 1991.

Bain, R.L. 2004. An introduction to biomass thermochemical conversion. In *DOE/NASLUGC Biomass and Solar Energy Workshop, Golden*.

Basu, P. 2010. *Biomass Gasification and Pyrolysis: Practical Design and Theory*. Academic press.

Beneroso, D., Bermúdez, J.M., Arenillas, A. and Menéndez, J.A. 2013. Microwave pyrolysis of microalgae for high syngas production. *Bioresource Technology*. 144: pp.240–246.

Bridgwater, A.V. 2003. Renewable fuels and chemicals by thermal processing of biomass. *Chemical Engineering Journal*. 91(2): pp.87–102.

Bu, Q., Lei, H., Ren, S., Wang, L., Holladay, J., Zhang, Q., Tang, J. and Ruan, R. 2011. Phenol and phenolics from lignocellulosic biomass by catalytic microwave pyrolysis. *Bioresource Technology*. 102(13): pp.7004–7007.

Bu, Q., Lei, H., Ren, S., Wang, L., Zhang, Q., Tang, J. and Ruan, R. 2012. Production of phenols and biofuels by catalytic microwave pyrolysis of lignocellulosic biomass. *Bioresource Technology*. 108: pp.274–279.

Bu, Q., Lei, H., Wang, L., Wei, Y., Zhu, L., Liu, Y., Liang, J. and Tang, J. 2013. Renewable phenols production by catalytic microwave pyrolysis of Douglas fir sawdust pellets with activated carbon catalysts. *Bioresource Technology*. 142: pp.546–552.

Bu, Q., Lei, H., Wang, L., Yadavalli, G., Wei, Y., Zhang, X., Zhu, L. and Liu, Y. 2015. Biofuel production from catalytic microwave pyrolysis of Douglas fir pellets over ferrum-modified activated carbon catalyst. *Journal of Analytical and Applied Pyrolysis*. 112: pp.74–79.

Budarin, V.L., Clark, J.H., Lanigan, B.A., Shuttleworth, P., Breeden, S.W., Wilson, A.J., Macquarrie, D.J., Milkowski, K., Jones, J., Bridgeman, T. and Ross, A. 2009. The preparation of high-grade bio-oils through the controlled, low temperature microwave activation of wheat straw. *Bioresource Technology*. 100(23): pp.6064–6068.

Budarin, V.L., Shuttleworth, P.S., Farmer, T.J., Gronnow, M.J., Pfaltzgraff, L., Macquarrie, D.J. and Clark, J.H. 2015. The potential of microwave technology for the recovery, synthesis and manufacturing of chemicals from bio-wastes. *Catalysis Today*. 239: pp.80–89.

Campanone, L.A. and Zaritzky, N.E. 2005. Mathematical analysis of microwave heating process. *Journal of Food Engineering*. 69(3): pp.359–368.

Chen, M.Q., Wang, J., Zhang, M.X., Chen, M.G., Zhu, X.F., Min, F.F. and Tan, Z.C. 2008. Catalytic effects of eight inorganic additives on pyrolysis of pine wood sawdust by microwave heating. *Journal of Analytical and Applied Pyrolysis*. 82(1): pp.145–150.

Ciacci, T., Galgano, A. and Di Blasi, C. 2010. Numerical simulation of the electromagnetic field and the heat and mass transfer processes during microwave-induced pyrolysis of a wood block. *Chemical Engineering Science*. 65(14): pp.4117–4133.

Cottam, M.L. and Bridgwater, A.V. 1994. Techno-economic modelling of biomass flash pyrolysis and upgrading systems. *Biomass and Bioenergy*. 7(1): pp.267–273.

Cuccurullo, G., Berardi, P.G., Carfagna, R. and Pierro, V. 2002. IR temperature measurements in microwave heating. *Infrared Physics & Technology*. 43(3): pp.145–150.

Demirbaş, A., 2002. Biodiesel from vegetable oils via transesterification in supercritical methanol. *Energy Conversion and Management*. 43(17): pp.2349–2356.

Domínguez, A., Menéndez, J.A., Inguanzo, M. and Pis, J.J. 2006. Production of bio-fuels by high temperature pyrolysis of sewage sludge using conventional and microwave heating. *Bioresource Technology*. 97(10): pp.1185–1193.

Dominguez, A., Menéndez, J.A., Fernandez, Y., Pis, J.J., Nabais, J.V., Carrott, P.J.M. and Carrott, M.R. 2007. Conventional and microwave induced pyrolysis of coffee hulls for the production of a hydrogen rich fuel gas. *Journal of Analytical and Applied Pyrolysis*. 79(1): pp.128–135.

Dong, C., Feng, C., Liu, Q., Shen, D. and Xiao, R. 2014. Mechanism on microwave-assisted acidic solvolysis of black-liquor lignin. *Bioresource Technology*. 162: pp.136–141.

Du, Z., Li, Y., Wang, X., Wan, Y., Chen, Q., Wang, C., Lin, X., Liu, Y., Chen, P. and Ruan, R. 2011. Microwave-assisted pyrolysis of microalgae for biofuel production. *Bioresource Technology*. 102(7): pp.4890–4896.

Electric Power Research Institute/U.S. DOE. 1997. Renewable Energy Technology Characterizations. EPRI TR-109496. Available from National Renewable Energy Laboratory, Golden, CO.

Fan, J., Kalnes, T.N., Alward, M., Klinger, J., Sadehvandi, A. and Shonnard, D.R. 2011. Life cycle assessment of electricity generation using fast pyrolysis bio-oil. *Renewable Energy*. 36(2): pp.632–641.

Fan, J., De Bruyn, M., Budarin, V.L., Gronnow, M.J., Shuttleworth, P.S., Breeden, S., Macquarrie, D.J. and Clark, J.H. 2013. Direct microwave-assisted hydrothermal depolymerization of cellulose. *Journal of the American Chemical Society*. 135(32): pp.11728–11731.

Farag, S. 2013. *Production of Chemicals by Microwave Thermal Treatment of Lignin* (Doctoral dissertation, ÉcolePolytechnique de Montréal).

Fernández, Y. and Menéndez, J.A. 2011. Influence of feed characteristics on the microwave-assisted pyrolysis used to produce syngas from biomass wastes. *Journal of Analytical and Applied Pyrolysis*. 91(2): pp.316–322.

Gaunt, J.L. and Lehmann, J. 2008. Energy balance and emissions associated with biochar sequestration and pyrolysis bioenergy production. *Environmental Science & Technology*. 42(11): pp.4152–4158.

Gregoire, Catherine E. 1992. *Technoeconomic Analysis of the Production of Biocrude from Wood*. NREL/TP-430-5435. Golden, CO: National Renewable Energy Laboratory.

Gregoire, C.E. and Bain, R.L. 1994. Technoeconomic analysis of the production of biocrude from wood. *Biomass and Bioenergy*. 7(1-6): pp.275–283.

Gunasekaran, S. and Yang, H.W. 2007. Effect of experimental parameters on temperature distribution during continuous and pulsed microwave heating. *Journal of Food Engineering*. 78(4): pp.1452–1456.

Guo, X., Zheng, Y. and Zhou, B. 2008. Influence of absorption medium on microwave pyrolysis of fir sawdust. In *Bioinformatics and Biomedical Engineering, 2008. ICBBE 2008. The 2nd International Conference on* pp.798–800. IEEE.

Hsu, D.D. 2012. Life cycle assessment of gasoline and diesel produced via fast pyrolysis and hydroprocessing. *Biomass and Bioenergy.* 45: pp.41–47.

Hu, Z., Ma, X. and Chen, C. 2012. A study on experimental characteristic of microwave-assisted pyrolysis of microalgae. *Bioresource Technology.* 107: pp.487–493.

Hu, Z., Ma, X. and Chen, C. 2012. A study on experimental characteristic of microwave-assisted pyrolysis of microalgae. *Bioresource Technology.* 107: pp.487–493.

Huang, Y.F., Kuan, W.H., Lo, S.L. and Lin, C.F. 2008. Total recovery of resources and energy from rice straw using microwave-induced pyrolysis. *Bioresource Technology.* 99(17): pp.8252–8258.

Huang, Y.F., Kuan, W.H., Lo, S.L. and Lin, C.F. 2010. Hydrogen-rich fuel gas from rice straw via microwave-induced pyrolysis. *Bioresource Technology.* 101(6): pp.1968–1973.

Huang, Y.F., Chiueh, P.T., Kuan, W.H. and Lo, S.L. 2013. Microwave pyrolysis of rice straw: Products, mechanism, and kinetics. *Bioresource Technology.* 142: pp.620–624.

Huber, G.W., Iborra, S. and Corma, A. 2006. Synthesis of transportation fuels from biomass: chemistry, catalysts, and engineering. *Chemical Reviews.* 106(9): pp.4044–4098.

Islam, M.N. and Ani, F.N. 2000. Techno-economics of rice husk pyrolysis, conversion with catalytic treatment to produce liquid fuel. *Bioresource Technology.* 73(1): pp.67–75.

Jiang, J. and Ma, X. 2011. Experimental research of microwave pyrolysis about paper mill sludge. *Applied Thermal Engineering.* 31(17): pp.3897–3903.

Jun, D.U., Ping, L.I.U., Liu, Z.H., Sun, D.G. and Tao, C.Y. 2010. Fast pyrolysis of biomass for bio-oil with ionic liquid and microwave irradiation. *Journal of Fuel Chemistry and Technology.* 38(5): pp.554–559.

Keshwani, D.R. and Cheng, J.J. 2010. Modeling changes in biomass composition during microwave-based alkali pretreatment of switchgrass. *Biotechnology and Bioengineering.* 105(1): pp.88–97.

Krieger-Brockett, B. 1994. Microwave pyrolysis of biomass. *Res. Chem. Intermed.* 20: 39–49.

Kržan, A. and Žagar, E. 2009. Microwave driven wood liquefaction with glycols. *Bioresource Technology.* 100(12): pp.3143–3146.

LaClaire, C.E., Barrett, C.J. and Hall, K. 2004. Technical, environmental and economic feasibility of bio-oil in new hampshire's north country. *Durham, NH: University of New Hampshire.*

Laird, D.A., Brown, R.C., Amonette, J.E. and Lehmann, J. 2009. Review of the pyrolysis platform for coproducing bio-oil and biochar. *Biofuels, Bioproducts and Biorefining.* 3(5): pp.547–562.

Lam, S.S., Russell, A.D., Lee, C.L. and Chase, H.A. 2012. Microwave-heated pyrolysis of waste automotive engine oil: Influence of operation parameters on the yield, composition, and fuel properties of pyrolysis oil. *Fuel.* 92(1): pp.327–339.

Lei, H., Ren, S. and Julson, J. 2009. The effects of reaction temperature and time and particle size of corn stover on microwave pyrolysis. *Energy & Fuels.* 23(6): pp.3254–3261.

Lei, H., Ren, S., Wang, L., Bu, Q., Julson, J., Holladay, J. and Ruan, R. 2011. Microwave pyrolysis of distillers dried grain with solubles (DDGS) for biofuel production. *Bioresource Technology.* 102(10): pp.6208–6213.

Luque, R., Menéndez, J.A., Arenillas, A. and Cot, J. 2012. Microwave-assisted pyrolysis of biomass feedstocks: the way forward?. *Energy & Environmental Science.* 5(2): pp.5481–5488.

Luque, R., Menéndez, J.A., Arenillas, A. and Cot, J. 2012. Microwave-assisted pyrolysis of biomass feedstocks: the way forward? *Energy & Environmental Science.* 5(2): pp.5481–5488.

Menéndez, J.A., Domínguez, A., Fernández, Y. and Pis, J.J. 2007. Evidence of self-gasification during the microwave-induced pyrolysis of coffee hulls. *Energy & Fuels.* 21(1): pp.373–378.

Meredith, R.J. 1998. *Engineers' Handbook of Industrial Microwave Heating* (No. 25). IET.

Metaxas, A.A. and Meredith, R.J. 1983. *Industrial microwave heating* (No. 4). IET.

Miura, M., Kaga, H., Sakurai, A., Kakuchi, T. and Takahashi, K. 2004. Rapid pyrolysis of wood block by microwave heating. *Journal of Analytical and Applied Pyrolysis.* 71(1): pp.187–199.

Miura, M., Kaga, H., Yoshida, T. and Ando, K. 2001. Microwave pyrolysis of cellulosic materials for the production of anhydrosugars. *Journal of Wood Science.* 47(6): pp.502–506.

Mohan, D., Pittman, C.U. and Steele, P.H. 2006. Pyrolysis of wood/biomass for bio-oil: a critical review. *Energy & Fuels.* 20(3): pp.848–889.

Mullaney, H. 2002. Technical, Environmental and Economic Feasibility of Bio-Oil in New Hampshire's North Country. Durham, NH: University of New Hampshire.

Mushtaq, F., Abdullah, T.A.T., Mat, R. and Ani, F.N. 2015. Optimization and characterization of bio-oil produced by microwave assisted pyrolysis of oil palm shell waste biomass with microwave absorber. *Bioresource Technology.* 190: pp.442–450.

Mushtaq, F., Mat, R. and Ani, F.N. 2014. A review on microwave assisted pyrolysis of coal and biomass for fuel production. *Renewable and Sustainable Energy Reviews.* 39: pp.555–574.

Namazi, A.B., Allen, D.G. and Jia, C.Q. 2015. Microwave-assisted pyrolysis and activation of pulp mill sludge. *Biomass and Bioenergy.* 73: pp.217–224.

Omar, R., Idris, A., Yunus, R. and Khalid, K. 2010, November. Microwave absorber addition in microwave pyrolysis of oil palm empty fruit bunch. In *Third International Symposium on Energy from Biomass and Waste, Venice, Italy.*

Omar, R., Idris, A., Yunus, R., Khalid, K. and Isma, M.A. 2011. Characterization of empty fruit bunch for microwave-assisted pyrolysis. *Fuel*. 90(4): pp.1536–1544.

Ouyang, X., Huang, X., Zhu, Y. and Qiu, X. 2015. Ethanol-enhanced liquefaction of lignin with formic acid as an *in situ* hydrogen donor. *Energy & Fuels*. 29(9): pp.5835–5840.

Pagnotta, M., Nolan, A. and Kim, L. 1992. A simple modification of a domestic microwave oven for improved temperature control. *Journal of Chemical Education* 69, pp.599.

Pandit, R.B. and Prasad, S. 2003. Finite element analysis of microwave heating of potato—transient temperature profiles. *Journal of Food Engineering*. 60(2): pp.193–202.

Perlack, R.D. and Turhollow, A.F. 2003. Feedstock cost analysis of corn stover residues for further processing. *Energy*. 28(14): pp.1395–1403.

Petrolia, D.R. 2006. Economics of Harvesting and Transporting Corn Stover for Conversion to Fuel Ethanol [electronic resource].

Pourhashem, G., Spatari, S., Boateng, A.A., McAloon, A.J. and Mullen, C.A. 2013. Life cycle environmental and economic tradeoffs of using fast pyrolysis products for power generation. *Energy & Fuels*. 27(5): pp.2578–2587.

Rajvanshi, A.K. 1986. Biomass gasification. *Alternative Energy in Agriculture*. 2(4): pp.82–102.

Rattanadecho, P. 2006. The simulation of microwave heating of wood using a rectangular wave guide: Influence of frequency and sample size. *Chemical Engineering Science*. 61(14): pp.4798–4811.

Raveendran, K., Ganesh, A. and Khilar, K.C. 1995. Influence of mineral matter on biomass pyrolysis characteristics. *Fuel*. 74(12): pp.1812–1822.

Richel, A. and Jacquet, N. 2015. Microwave-assisted thermochemical and primary hydrolytic conversions of lignocellulosic resources: a review. *Biomass Conversion and Biorefinery*. 5(1): pp.115–124.

Ringer, M., Putsche, V. and Scahill, J. 2006. Large-Scale Pyrolysis Oil. *Assessment*.

Robinson, J.P., Kingman, S.W., Barranco, R., Snape, C.E. and Al-Sayegh, H. 2009. Microwave pyrolysis of wood pellets. *Industrial & Engineering Chemistry Research*. 49(2): pp.459–463.

Robinson, J., Dodds, C., Stavrinides, A., Kingman, S., Katrib, J., Wu, Z., Medrano, J. and Overend, R. 2015. Microwave pyrolysis of biomass: control of process parameters for high pyrolysis oil yields and enhanced oil quality. *Energy & Fuels*. 29(3): pp.1701–1709.

Ruan, R.R., Chen, P., Hemmingsen, R., Morey, V. and Tiffany, D. 2008. Size matters: small distributed biomass energy production systems for economic viability. *International Journal of Agricultural and Biological Engineering*. 1(1): pp.64–68.

Salema, A.A. and Ani, F.N. 2011. Microwave induced pyrolysis of oil palm biomass. *Bioresource Technology*. 102(3): pp.3388–3395.

Salema, A.A. and Ani, F.N. 2012a. Microwave-assisted pyrolysis of oil palm shell biomass using an overhead stirrer. *Journal of Analytical and Applied Pyrolysis*. 96: pp.162–172.

Salema, A.A. and Ani, F.N. 2012b. Pyrolysis of oil palm empty fruit bunch biomass pellets using multimode microwave irradiation. *Bioresource Technology*. 125: pp.102–107.

Sanderson, K. 2009. From plant to power. pp.710–7111.

Sarotti, A.M., Spanevello, R.A. and Suárez, A.G. 2007. An efficient microwave-assisted green transformation of cellulose into levoglucosenone. Advantages of the use of an experimental design approach. *Green Chem*. 9(10): pp.1137–1140.

Shaveta, H.K. and Singh, P. 2014. Microwave assisted degradation of lignin to Monolignols. *Pharmaceutica Analytica Acta*. 5: pp.308.

Shuttleworth, P., Budarin, V., Gronnow, M., Clark, J.H. and Luque, R. 2012. Low temperature microwave-assisted vs. conventional pyrolysis of various biomass feedstocks. *Journal of Natural Gas Chemistry*. 21(3): pp.270–274.

Solantausta, Y., Beckman, D., Bridgwater, A.V., Diebold, J.P. and Elliott, D.C. 1992. Assessment of liquefaction and pyrolysis systems. *Biomass and Bioenergy*. 2(1): pp.279–297.

Suriapparao, D.V. and Vinu, R. 2015. Bio-oil production via catalytic microwave pyrolysis of model municipal solid waste component mixtures. *RSC Advances*. 5(71): pp.57619–57631.

Tian, Y., Zuo, W., Ren, Z. and Chen, D. 2011. Estimation of a novel method to produce bio-oil from sewage sludge by microwave pyrolysis with the consideration of efficiency and safety. *Bioresource Technology*. 102(2): pp.2053–2061.

Toledano, A., Serrano, L., Balu, A.M., Luque, R., Pineda, A. and Labidi, J. 2013. Fractionation of organosolv lignin from olive tree clippings and its valorization to simple phenolic compounds. *ChemSusChem*. 6(3): pp.529–536.

Tong, C.H., Lentz, R.R. and Lund, D.B. 1993. A microwave oven with variable continuous power and a feedback temperature controller. *Biotechnology Progress*. 9(5): pp.488–496.

Torgovnikov, G.I. 1993. Dielectric Properties of Wood-Based Materials. *Dielectric Properties of Wood and Wood-Based Materials*. pp.135–159. Springer Berlin Heidelberg.

Waheed ul Hasan, S. and Ani, F.N., 2014. Review of limiting issues in industrialization and scale-up of microwave-assisted activated carbon production. *Industrial & Engineering Chemistry Research*. 53(31): pp.12185–12191.

Wan, Y., Chen, P., Zhang, B., Yang, C., Liu, Y., Lin, X. and Ruan, R. 2009. Microwave-assisted pyrolysis of biomass: Catalysts to improve product selectivity. *Journal of Analytical and Applied Pyrolysis*. 86(1): pp.161–167.

Wang, M.J., Huang, Y.F., Chiueh, P.T., Kuan, W.H. and Lo, S.L. 2012. Microwave-induced torrefaction of rice husk and sugarcane residues. *Energy*. 37(1): pp.177–184.

Wang, L., Lei, H., Lee, J., Chen, S., Tang, J. and Ahring, B. 2013. Aromatic hydrocarbons production from packed-bed catalysis coupled with microwave pyrolysis of Douglas fir sawdust pellets. *Rsc Advances*. 3(34): pp.14609–14615.

Wang, N., Tahmasebi, A., Yu, J., Xu, J., Huang, F. and Mamaeva, A. 2015. A Comparative study of microwave-induced pyrolysis of lignocellulosic and algal biomass. *Bioresource Technology*. 190: pp.89–96.

Wright, M.W., Satrio, J.A., Brown, R.C., Daugaard, D.E., Hsu, D. 2010a. Techno-Economic Analysis of Biomass Fast Pyrolysis to Transportation Fuel; NREL/TP-6A20-46586; National Renewable Energy Laboratory: Golden, CO, November. p.62.

Wright, M.M., Daugaard, D.E., Satrio, J.A. and Brown, R.C. 2010b. Techno-economic analysis of biomass fast pyrolysis to transportation fuels. *Fuel*. 89: pp.S2–S10.

Yang, H.W. and Gunasekaran, S. 2001. Temperature profiles in a cylindrical model food during pulsed microwave heating. *Journal of Food Science*. 66(7): pp.998–1004.

Yin, C. 2012. Microwave-assisted pyrolysis of biomass for liquid biofuels production. *Bioresource Technology*. 120: pp.273–284.

Yu, F., Deng, S., Chen, P., Liu, Y., Wan, Y., Olson, A., Kittelson, D. and Ruan, R. 2007. Physical and chemical properties of bio-oils from microwave pyrolysis of corn stover. *Applied Biochemistry and Biotechnology*. 137(1-12): pp.957–970.

Zhang, Z. and Zhao, Z.K. 2010. Microwave-assisted conversion of lignocellulosic biomass into furans in ionic liquid. *Bioresource Technology*. 101(3): pp.1111–1114.

Zhang, B., Yang, C., Moen, J., Le, Z., Hennessy, K., Wan, Y., Liu, Y., Lei, H., Chen, P. and Ruan, R. 2010. Catalytic conversion of microwave-assisted pyrolysis vapors. *Energy Sources, Part A: Recovery, Utilization, and Environmental Effects*. 32(18): pp.1756–1762.

Zhao, X., Zhang, J., Song, Z., Liu, H., Li, L. and Ma, C. 2011. Microwave pyrolysis of straw bale and energy balance analysis. *Journal of Analytical and Applied Pyrolysis*. 92(1): pp.43–49.

Zhao, X., Wang, M., Liu, H., Li, L., Ma, C. and Song, Z. 2012. A microwave reactor for characterization of pyrolyzed biomass. *Bioresource Technology*. 104: pp.673–678.

Zhu, L., Lei, H., Wang, L., Yadavalli, G., Zhang, X., Wei, Y., Liu, Y., Yan, D., Chen, S. and Ahring, B. 2015. Biochar of corn stover: Microwave-assisted pyrolysis condition induced changes in surface functional groups and characteristics. *Journal of Analytical and Applied Pyrolysis*. 115: pp.149–156.

Zinoviev, S., Müller-Langer, F., Das, P., Bertero, N., Fornasiero, P., Kaltschmitt, M., Centi, G. and Miertus, S. 2010. Next-generation biofuels: survey of emerging technologies and sustainability issues. *ChemSusChem*. 3(10): pp.1106–1133.

8

Heat and Mass Transfer and Reaction Kinetics Under Microwaves

Microwaves are a group of electromagnetic waves within a frequency band of 300 MHz to 300 GHz. It can be represented by equation [1] where the frequency f is linked by the velocity of light c to a corresponding wavelength λ.

$$c = \lambda \times f \tag{1}$$

Wavelength of microwaves in vacuum is in the range between 1 m and 1 mm. The "microwave" name rather points to their wavelength within the matter, where their wavelength can be in the micrometer range. In practice, for microwave heating applications not all the microwave spectrum is used, in effect there are some discrete frequency bands, which have been set aside from telecommunication applications for industrial, scientific and medical (so called ISM) applications. The most important and most used ISM microwave frequency bands are 915 ± 25 MHz and 2450 ± 50 MHz, where a certain limited radiation level has to be tolerated by other applications (like communication devices).

The dissipated power inside a microwave cavity can be represented as energy generated inside a heated material. For nonmagnetic materials, it can be represented in the form of Equation [2] (Grant and Halstead 1998, Farag et al. 2012).

$$P = \sigma E_{rms}^2 = 2\pi \varepsilon_0' \varepsilon_{eff}'' E_{rms}^2 = 2\pi f \varepsilon' \tan \delta \, E_{rms}^2 \tag{2}$$

where P is the absorbed power per unit volume [W/m³], ε_{eff}'' is the effective dielectric loss factor ($\varepsilon_0 \varepsilon_{eff}'' = \varepsilon' \tan \delta$) [–], ω is the angular frequency ($\omega = 2\pi f$) [s⁻¹], and E_{rms} is the root mean square of the electric field [V/m].

In the case of magnetic materials, equation [2] should be replaced by equation [3] (Ramasamy and Moghtaderi 2010).

$$P = \omega \varepsilon_0' \varepsilon_{eff}'' E_{rms}^2 + \omega \mu_0' \mu_{eff}' H_{rms}^2 \tag{3}$$

Where μ_{eff} is the effective magnetic loss factor [–], and H_{rms} is the magnetic field [A/m].

Penetration Depth and Power Penetration Depth

Electric fields and generated power decay exponentially inside the material. The penetration depth (D) is defined as the depth where the magnitude of the electric field drops by a factor $1/e$ with respect to the surface value. In a similar manner, the power penetration depth (D_p) is the distance where the power density is reduced by a factor $1/e$ of the surface. Equation [4] describes the relationship between D and D_p, where α is called the attenuation factor and can be represented generally by equation [5]. For high loss ($\varepsilon''_{eff} \gg \varepsilon'$) and low loss ($\varepsilon''_{eff} \ll \varepsilon'$) mediums, equations [6] and [7] apply, respectively (Campanone and Zaritzky 2005).

$$D_p = \frac{D}{2} = \frac{1}{2\alpha} \quad [4]$$

$$\alpha = \omega \sqrt{\frac{(\mu_0 \mu'_{eff} \varepsilon_0 \varepsilon'')[(1+(\varepsilon_0'/\varepsilon_0')^2)^{1/2} - 1]}{2}} \quad [5]$$

$$\alpha = \frac{\omega}{2} \sqrt{\frac{(\mu_0 \mu'_{eff} \varepsilon_0 \varepsilon'')}{\varepsilon''}} \varepsilon'' \quad [6]$$

$$\alpha = \frac{\pi \varepsilon''}{\lambda_0 \sqrt{\varepsilon''}} \quad [7]$$

Microwaves and Thermal Energy

Microwave energy absorbed by a substance is converted into heat as mentioned earlier in previous chapters. It is converted through dipolar momentum, ionic conduction and interfacial polarization which is a combination of the first two mechanisms. The electromagnetic field which is represented by Maxwell's equations is dissipated as heat source to result in a temperature increase and is represented by a heat balance equation (Figure 1). The temperature field or distribution estimation requires complex permittivity which again varies with the temperature field. Therefore an iterative process is required to estimate the variables and the heat transfer phenomenon in microwave mediated operations.

Power absorbed by the reaction medium is reflected by the temperature increase which can be expressed as:

$$P = \frac{mC_p \Delta T}{t} \quad [8]$$

where P is the microwave power supplied to the reaction medium, ΔT is the temperature gradient, t is the reaction time, and Cp is the heat capacity of the water.

Theoretically, the rate at which electromagnetic energy is converted to thermal energy in a load, which is heated by microwave power at a frequency of F, is given by the equation:

$$\frac{dT}{dt} = \frac{k\varepsilon'' f E_{rms}^2}{\rho C_p} \quad [9]$$

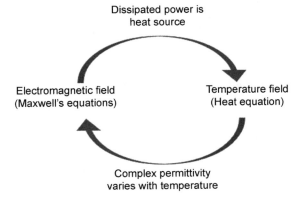

Figure 1. Coupling of Maxwell's and heat equations in microwave heating (Vaz et al. 2014).

where k is a constant, E_{rms} is the electric field intensity and ρ is the density of the reaction medium. The electric field intensity E_{rms} is dependent on ε'.

Explanations for Microwave Enhancement in Reaction Kinetics

Microwaves increase reaction temperatures which in turn promote higher reaction kinetics. There are many explanations in literature about the rate enhancements caused by microwave mediated reactions (Conner and Tompsett 2008). These are: microwave energy heats the reaction mixture more rapidly, increasing the net rate early in the heating process; microwave energy leads to a more uniform heating of the reactor; microwaves change the association between species within the reacting mixture; microwave heating superheats the reaction mixture; hot spots are created within the reacting mixture which is in contrast to the uniform heating; and microwave energy enhances the dissolution of reacting species. Some reactions can be enhanced as a result of the absorption of microwave energy by reactants, intermediate-species, and/or products and especially at interfaces of reaction mixtures in different phases. Therefore, it is also possible that microwave adsorption changes the temperatures or the energies of the intermediate states. Microwave irradiation effect, in general, can be associated with four major effects:

1) Higher temperature differences;
2) Lower activation energy;
3) A combination of the two at same energy levels; and
4) Enhancement at interfaces of reaction mixtures.

Microwave-induced temperature differences

Superheating or hot spots are the two words that have often been used to explain the reaction rate enhancement due to microwave exposure. "Superheating" is assumed to pertain to the total reaction system covering the whole reactor in the process. Hot spots are cited when it is believed that the system is not uniform but varies in position. A mechanism based on "excess dipole energy" was also proposed wherein the localized energy (temperature) of the dipole groups is higher compared to the nonpolar bonds within these systems.

Microwave effect on activation energy

Activation energy of different reactions (especially in conventional and microwave mediated processes varies greatly) depends on the differences in dielectric properties of the reactions, intermediates, and products, allowing for selective heating. Activation energies for microwave mediated reactions are generally lower than the reaction conducted under conventional heating. Reduction in activation energy has been reported under microwave irradiation for numerous reactions involving organic and inorganic reactions with a few exceptions such as nucleation and growth. Some specific ions and intermediate products that are susceptible to microwave irradiation have the ability to increase the pre-exponential factor thereby compensating the activation energy.

Changes in activation energy or effective temperature

The changes in temperature or in activation energies due to microwave exposure were most often estimated from differences in reaction rates as interpreted with an Arrhenius expression. In each case, the other variable (ΔT or E_a) was assumed to be constant. The relation between an effective temperature and the equivalent change in energy levels for the interpretation of changes in reaction rates can be expressed in the Arrhenius equation. The reaction rate constants can be represented in terms of the Gibbs free energy or the activation energy between the reacting and transition states, respectively. Pre-exponential factor A accounts for transition probabilities and the transition entropy in the case of the Arrhenius equation.

$Rate\ constant = A^{-Ea/RT}$ [10]

The ratio of reaction rates for different activation (or Gibbs transition) energies at the same temperature can be expressed assuming that the pre-exponential factor, a, is the same when the transition Gibbs free energy is employed.

$$\frac{Rate_1}{Rate_2} = e^{-[(\Delta G_1^* - \Delta G_2^*)]/RT} = \frac{A_1}{A_2} e^{-[(Ea_1 - Ea_2)]/RT} \qquad [11]$$

If the energy profile is the same but the temperatures change, $T_1 \rightarrow T_2$, then the rate ratio could be expressed with constant pre-exponential factors.

$$\frac{Rate_1}{Rate_2} = e^{-[\Delta G^*/R][1/T_1 - 1/T_2]} = e^{-E_a/R[1/T_1 - 1/T_2]} \qquad [12]$$

If $\Delta G^* = \Delta G_1^*$ and $T_1 = T$, then

$$\frac{\Delta G_2^*}{\Delta G_1^*} = \frac{T_1}{T_2} ....or... \frac{E_{a2}}{E_{a1}} = \frac{T_1}{T_2} \qquad [13]$$

Thus, a change in the effective temperature can be analyzed as if there were a change in the relative energies either of the transition states or of the relative energies of intermediates. Prior analyses of the influence of microwave exposure have elucidated the proposed differences in temperature due to exposure to microwave radiation (Conner and Tompsett 2008). These analyses show that an equivalent kinetic or equilibrium analysis can be conducted by analyzing activation energy or free energy differences between different states. These changes in free energies are examined in these analyses. This would mean a 100 K temperature change from room temperature is equivalent to a 25% change in activation energy.

Enhancement at interfaces

Interfacial reactions are particularly susceptible to microwave influences. Maxwell-Wagner interfacial microwave polarizations are a well documented phenomena that occur for heterogeneous systems. Simple systems with known geometries and differences in dielectric properties can be analyzed to demonstrate that the interfaces between materials with differing dielectric properties provide loci for microwave interactions. Adsorption of a high dielectric adsorbent on a surface provides two significant new interfaces: between the vapor phase and the sorbed phase, and between the sorbed phase and the solid surface. Species sorbed (or formed) on a surface are more susceptible to microwave interactions.

Enhancement of diffusion coefficient

Microwaves have shown to increase the diffusion rate of reactants in rather diffusion limited reactions under conventional heating. The diffusion rate enhancement may be attributed to the superior temperatures in the medium. The dependence of diffusion on temperature in liquids and solids is often fitted by an Arrhenius equation (Antonio and Deam 2007).

$$D_C = A e^{\frac{\Delta E}{RT}} \qquad [14]$$

where T is the temperature, R is the gas constant (8.3144 Jmol^{-1} K^{-1}) and D$_e$ is the diffusion coefficient.

The pre-exponential, A, is considered the "free diffusion" coefficient that would arise if no local potential well existed to delay the random walk. ΔE is the activation energy required for a molecule to jump out of the local potential well and undertake a random step to the next site. An Arrhenius plot of the logarithm of diffusion coefficient versus inverse temperature will help determine the activation energy under microwave heating.

Microwaves in Drying Operations

Drying of biomass is a first step in pretreatment process for biofuel production. Removing moisture from the biomass can increase the yields, thereby increasing energy efficiency. Microwave irradiation can be used conveniently to release moisture from biomass by heating. It is critical to understand the heat (microwave power) and mass transport (water vapor and air) within the biomass to optimize the heating process. Figure 2 shows the modes of energy, water and vapor transport in a porous medium such as biomass. Datta and Rakesh (2013) use a scanning electron microscope image to illustrate various types of liquid, vapor and energy transport which are relevant to biomass as a porous medium. Liquid transport can be pressure driven or capillary driven whereas the vapor transport is both pressure and diffusion drive. In both cases, evaporation phenomenon results in loss of mass. Energy transport is a result of conduction, convection from mass transport, absorption of microwave irradiation and evaporation of moisture content.

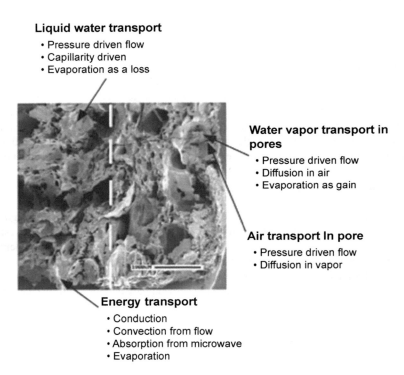

Figure 2. Modes of transport of energy, water, and water vapor in food as a porous medium, illustrated using a scanning electron microscope (SEM) image of bread.

Heat and Mass Transfer in Microwave Field

Heating process under microwave irradiation is quite different from conventional method. Figure 3 shows a representation of qualitative temperature and moisture profiles for conventional, infrared, and microwave heating processes. The x-axis represents the distance from the surface to the center of the biomass while the y-axis represents the temperature or mass transport profiles under different heating processes. It can be noted that microwave heating produced increasing thermal gradient towards the center of the biomass whereas the other types of heating produced inverse thermal gradient. The same applies to the mass transport which is expressed as moisture content. Microwave irradiation allows the moisture to be expelled from the center of the medium to the outer surface.

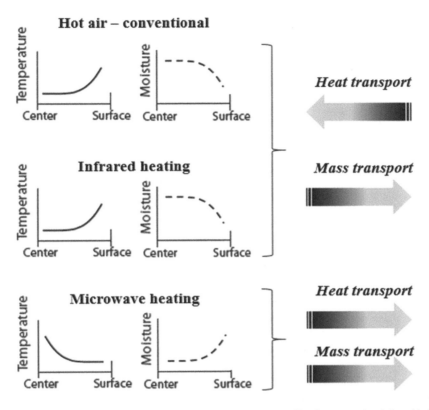

Figure 3. Simplified qualitative representations of temperature and moisture profiles for conventional (hot air), infrared and microwave heating.

It can be noted from Figure 3 that higher temperatures exist in the center of the biomass in microwave heating process. This means that heat transfer occurs from the center to the surface of the medium. It is also shown that the moisture content in the center of the medium is low because the moisture is transported to the outer layers of the medium. In both cases, the heat and mass transfer occur from the center to the outer surface in the same direction. These trends for conventional and infrared heating are quite opposite. As Figure 3 demonstrates, these occur in opposite directions for reasons mentioned earlier.

Illustration

An iterative process is necessary for determining the temperature distribution under microwave heating mode. The solution involves equations of Maxwell and energy which need to be solved in iterations to account for the temperature dependence of the electromagnetic and thermodynamic properties of the dielectric fluid. The iterative process is shown in Figure 4 which includes five steps (Peterson and Paul 2014).

At the start, the amount of heat generated in the system is set to zero and the starting temperature is set to the ambient value measured at the time of the experiments. The first iterative step involves updating all properties where the electromagnetic and thermodynamic properties are found as a function of the bulk mean temperature of the solution within the applicator. Next, the electromagnetic (EM) field is solved within the cavity which involves solving Maxwell's equations for a two dimensional rectangular field in the XZ plane (orthogonal to the flow direction of the applicator). As shown in Figure 4 this XZ plane is at the center line of the system in the Y direction. Then the absorbed power is determined as the difference between the original microwave field without attenuation and the unabsorbed microwave energy which passes through the sample determined using the Beer-Lambert law. This is followed by

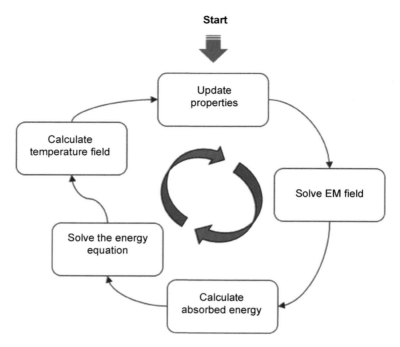

Figure 4. An iterative process for estimating temperature distribution in microwave field (Peterson and Paul 2014).

solving the energy equation, with absorbed microwave energy accounted for as a source term, yielding a change in temperature. Finally, a new estimated temperature profile is calculated which is the sum of the incoming sample temperature and the change in temperature from microwave heating at each locale within the process volume. The difference between the new estimated temperature and the temperature of the previous iteration is compared to a convergence criterion to determine if the solution has converged on a solution. If the difference is less than the convergence criteria, then the estimated temperature profile is reported as the final temperature profile. Otherwise the numerical process starts again using the newly approximated temperature profile to update the dielectric fluid properties. The initial five modeling steps and the subsequent iterative steps are shown in Table 1.

Table 1. Model iteration steps (Peterson and Paul 2014).

Event	Initial	Iterations
Solve energy equation	(1) $q = 0$	(6) Determine the amount of absorbed energy by subtracting the unabsorbed field from the original EM field
Calculate the temperature field	(2) $T = T_{in}$	(7) Solve the energy equation to determine the temperature change; Add temperature change to previous temperature; Apply convergence criteria
Update properties	(3) Determine values at T_{in}	(8) Assuming no convergence, find the new property values at the new temperature
Solve EM field	(4) Solve for the ideal EM field	(9) Solve for the new ideal field
Calculate the absorbed power	(5) Solve Beer-Lambert law for unabsorbed energy	(10) Solve Beer-Lambert law with the updated properties

Heat and mass transfer in liquid-solid suspensions

Heat transfer in two-phase systems containing solid-liquid phases are common in oil extraction phases. The efficiency of extraction process is often hindered by heat and mass transfer limitations. For example, microalgae oil extraction system would necessarily consist of two phases, algae cells and extracting liquid.

The heat (energy) transfer equation for an algae cell-solvent suspension can be written following the model equations provided by Dincov et al. (2004).

$$\rho_{a-l}\frac{\partial H_{a-l}}{\partial t} + \nabla\left(\rho_{a-l}u_l\frac{C_{pl}}{C_{pa-l}}H_{a-l} - k_{a-l}\nabla T_{a-l}\right) = Q + S_{a-l}^{int} \quad [15]$$

where thermal conductivity, k can be averaged for the algae cell-liquid suspension as

$$k_{a-l} = (1-\phi)k_a + \phi M k_l \quad [16]$$

Specific heat C_p and density ρ are averaged in a similar way.

The heating function is included in the solid–liquid heat transfer equation. There will be an interphase transfer between the two-phases, represented by the interface (algae cell-liquid) source S_i^{int} defined as $S_i^{int} = h_{ij}A_s(H_i^{int} - H_i)$ where h_{ij} is a bulk-to interface heat transfer coefficient, A_s is the total interface area and H_{int} are interface enthalpies.

Similarly, a mass transfer equation can be written using continuity equation as:

$$\frac{\partial(\rho_a V_a)}{\partial t} + \nabla \bullet (\rho_a V_a u_a) + \dot{m} = 0; \quad [17]$$

$$\frac{\partial(\rho_l V_l)}{\partial t} + \nabla \bullet (\rho_l V_l u_l) + \dot{m} = 0 \quad [18]$$

V_a and V_l are algae and liquid volume fractions, respectively and u is the momentum transfer rate. The interface mass transfer rate, $\dot{m}$, is determined from the balance of heat through the interface between the (algae-liquid) phases (Ishii et al. 1975):

$$\dot{m} = \frac{h_1^m(H_1^{int} - H_1) + h_2^m(H_2^{int} - H_2)}{(H_1^{int} - H_2^{int})} \quad [19]$$

where h_1^m and h_2^m are bulk-to interface mass transfer coefficients.

Solving both heat and mass transfer equations requires an iterative and simultaneous calculation procedure for determining the reaction temperature as shown in Figure 5.

Two main problems that hinder microwave application are: (1) thermal runaway due to the reflection and absorption of microwave irradiation by the reactants which change nonlinearly with time during the reaction; and (2) hot spot formation in the reaction due to inhomogeneous heating under microwave heating. Therefore, to overcome these difficulties, thermal analysis of microwave assisted biodiesel production process based on the multi-physics calculation should be studied. Effective permittivity can be used to describe the reflection and absorption of microwave irradiation in the reaction mixture during the reaction. For a liquid chemical reaction, the effective permittivity can be determined by the reaction mixture components and temperature. Therefore, obtaining the effective permittivity determined by the reaction mixture's temperature and concentration of one component is preliminary and necessary for multi-physics calculation. The calculation requires such parameters as effective permittivity to be determined for different materials on an iterative basis. Finite-difference-time-domain (FDTD) method and Phoenics programs were used to conduct this task (Dinčov et al. 2004).

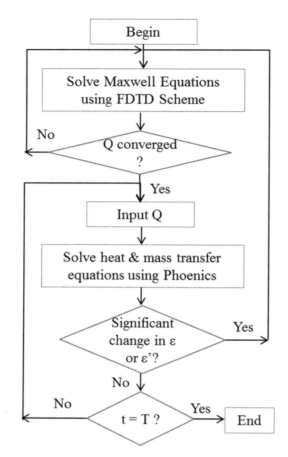

Figure 5. An algorithm for temperature distribution in microwave heating.

Heat and Mass Transfer in Oil Extraction

Oil extraction can be conveniently represented by the first order reactions where the concentration of the product is directly proportional to the initial concentration of the reactant. However, many studies have reported a better correlation with other reaction orders such as pseudo-first order and second order reactions. A few examples will be discussed here.

Oil extraction from *Scenedesmus obliquus* microalgae biomass using a continuous microwave system design, optimization, and quality characterization was studied by Balasubramanian et al. (2011). A single mode focused cavity continuous microwave system (1.2 kW, 2.45 GHz) was used for the extraction experiments. Algae-water suspension (1:1 w/w) was heated to 80 and 95°C, and subjected to extraction for up to 30 min. Maximum oil yield was achieved at 95°C and 30 min. The microwave system extracted 76–77% of total recoverable oil at 20–30 min and 95°C, compared to only 43–47% for water bath control. Extraction time and temperature had significant influence (p < 0.0001) on extraction yield. Figure 6 shows a comparison of the effects of conventional heating and microwave heating for oil extraction from the algal biomass. It can be noted that microwave interaction with microalgae biomass is more severe than the conventional water bath heating as shown by the deformation of the cell structure and the disruption of the microalgae biomass.

Figure 6. Effect of conventional and microwave heat treatment on algal cells compared with the untreated cells. (a) Untreated thawed, 5000X; (b) Water bath, 95°C, 5 min at 5000X; (c) Microwave, 95°C, 5 min at 5000X; and (d) Microwave, 95°C, 5 min at 8000X (Balasubramanian et al. 2011).

Analysis of the temperature profiles by curve fitting using high R-squared values indicated that the heating patterns within the holding tubes follow a first order heating curve obeying the equation (Marlin 2000):

$$y = y_o + A(1 - e^{-\beta t}) \qquad [20]$$

where y = outlet or target process temperature, °C, y_o = initial sample temperature, °C, A = temperature difference between the initial and target temperature, ΔT, °C, β = time constant dependent on the rate of temperature increase, s^{-1}, t = time, s.

A similar correlation was used to represent the effect of different process parameters including extraction time on the extracted oil yields expressed as percent of oil per microalgae biomass (dry). Figure 7 clearly shows the benefits of microwave extraction over conventional water bath heating in terms of higher extraction yields. The parameters fitting the trends are significantly higher for microwave heating at both extraction temperatures 80°C and 95°C, respectively. In the case of microwave assisted extraction, the profile of the extraction curve (in terms of oil yield) does not appear to attain a steady state value at 30 min. The highest oil yield for control was 19.08 ± 1.84% per dry mass of algae obtained at 95°C for an extraction time of 30 min. Under similar conditions, microwave assisted solvent extraction resulted in a significantly higher oil recovery per dry mass of algae (31.38 ± 2.06%). This study validated for the first time the efficiency of a continuous microwave system for extraction of lipids from algae. Higher oil yields, faster extraction rates and superior oil quality demonstrate this system's feasibility for oil extraction from a variety of feedstock.

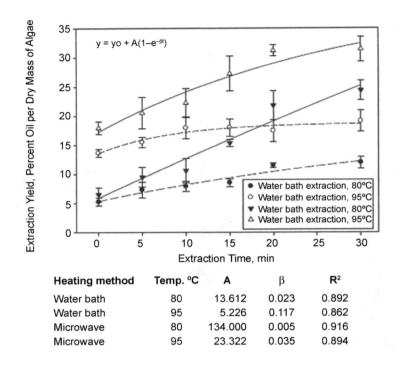

Figure 7. Extraction yields obtained at two process temperatures and various extraction times while using water bath and microwave as heating medium (Balasubramanian et al. 2011).

In another study, a continuous microwave-assisted extraction method (CMAE) was developed at laboratory- and pilot-scale for rapid extraction of oil from soybean flour and rice bran using ethanol as a solvent (Terigar et al. 2011). The two systems were optimized and their extraction capabilities compared to those of conventional extraction systems. Prediction equations for extraction yield with respect to time showed to follow a first-order response as shown in Figure 8a and Figure 8b for both laboratory-scale CMAE and conventional extractions and is represented by:

$$C_{(t)} = C_f(1 - e^{-kt}) \quad [21]$$

where $C_{(t)}$ = mean relative amount of oil extracted, %; C_f = final oil yield extracted, %; k = extraction rate constant; t = time, min.

In predicting the amount of oil extracted vs. temperature, a simple linear regression fitted well all the data (Figure 8c and Figure 8d).

$$y = y_o + a \times t \quad [22]$$

where y = mean relative amount of oil extracted, %; y_o = initial relative amount of oil extracted at the start of experiment, %; a = temperature constant dependent on the rate of oil extraction; T = temperature, °C.

The laboratory-scale microwave system was optimized for extraction performance using various extraction times and temperatures; with a maximum extraction efficiency at 73°C and 21 min. Based on the results from the laboratory-scale extraction process, a pilot-scale CMAE process was designed and used to extract soybean and rice bran oil at two flow rate conditions of 0.6 and 1.0 L/min, at an extraction temperature of 73°C, and up to 1 h extraction time. Oil yields obtained from both CMAE (laboratory- and pilot-scale) systems were better than the yields obtained by conventional extraction. The pilot-scale system extracted more than 90% of recoverable oil in both feedstocks with an extraction time of 8 min. The quality of the extracted oils indicated that they met biodiesel feedstock standard specifications. These

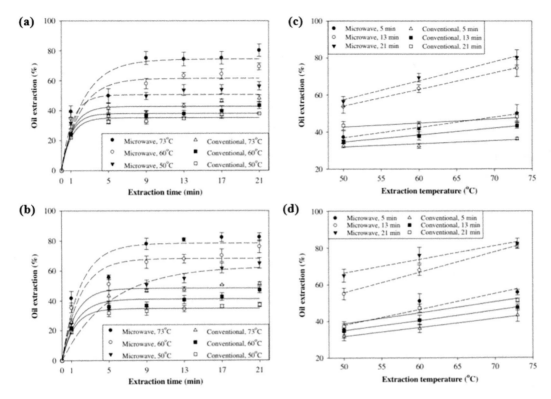

Figure 8. Regression analysis fitted curves for laboratory-scale CMAE and conventional methods vs. time for: (a) soybean oil and (b) rice bran oil; selected regression analysis fitted curves for laboratory-scale CMAE and conventional methods vs. temperature for: (c) soybean oil and (d) rice bran oil (Terigar et al. 2011).

results indicate that a microwave-based extraction process can be a commercially viable alternative for traditional extraction methods. Future research will focus on a detailed process economics and energy requirements for establishing a microwave-assisted oil extraction process on an industry scale set-up (Terigar et al. 2011).

Microalgae Oil Extraction Under Microwave Irradiation

Cheng et al. (2013) calculated the microalgae cell disruption efficiency based on a modified mathematical model of an olive oil extraction process developed by Amarni and Kadi (2010). The microalgae were divided into broken and unbroken cells. The extraction was conducted based on two diffusion processes: lipid release of broken cells through unhindered diffusion and lipid extraction from unbroken cells through hindered diffusion. The lipid yield q_t (% of dry biomass) at any time can be expressed by the following equation:

$$q_t = q_e^{d1} \times (1 - e^{-K_{d1} \times t}) + q_e^{d2} \times (1 - e^{-K_{d2} \times t}) \qquad [23]$$

where q_e^{d1} and q_e^{d2} (% dry biomass) represent the lipid yield at equilibrium for the hindered and unhindered diffusion, respectively. K_{d1} and K_{d2} (1/min) represent the mass transfer coefficient for the hindered and unhindered diffusion, respectively. The extraction time is represented by t (min).

Therefore, the total lipid extraction rate R can be obtained as follows:

$$R = \frac{dq_t}{dt} = K_{d1} \times q_e^{d1} \times e^{(-K_{d1} \times t)} + K_{d2} \times q_e^{d2} \times e^{(-K_{d2} \times t)} \qquad [24]$$

$$R = \left(\frac{dq_t}{dt}\right)_{t=0} = K_{d1} \times q_e^{d1} + K_{d2} \times q_e^{d2} \quad [25]$$

The values of the model parameters were calculated using the fitting experimental data of the program Origin 8.0. All measurements were conducted in triplicate. These values of lipid yields q^{d1}_e and q^{d2}_e and the mass transfer coefficients K_{d1} and K_{d2} were calculated as follows: q^{d1}_e = 4.2741 (% dry biomass), q^{d2}_e = 14.6891 (% dry biomass), K_{d1} = 0.1280 (L/min), K_{d2} = 19.4371 (L/min), correlation coefficient R² = 0.9968.

Considering that bio-oil extraction rate and yield vary with different chloroform and methanol amounts, the corresponding mass transfer coefficients can be calculated using this formula. The basic assumption is that all of the algae cells contain the same lipid amount. In addition, all of the lipids will be extracted at a period of time equal to infinity. Thus, the percentage of disrupted cells can be determined by the following formula:

$$\text{Percentage of disrupted cells (\%)} = \frac{\rho_e^{d2}}{\rho_e^{d1} + \rho_e^{d2}} \quad [26]$$

$$\rho_e^d = \rho_e^{d1} + \rho_e^{d2} \quad [27]$$

30 min was required for cell disruption under microwave irradiation at 60°C without sulfuric acid and extraction agents (chloroform and methanol), where transesterification reaction rarely occurs. The disrupted cells accounted for 77.5% of the total cells. With sulfuric acid and extraction agents, 2 min was sufficient for cell disruption under microwave irradiation at 60°C. The fatty acid methyl ester yield under microwave irradiation was 94 wt.% (Cheng et al. 2013).

Moisture can be heated approximately 100 times faster than organic carbon using microwave irradiation (Cheng et al. 2008). The large amount of water in the cells heated by the 2.45 GHz microwave electromagnetic field generated a micro-region of high temperature and pressure, as well as a micro-explosion of steam, which resulted in cell wall disruption and lipid release. The initial rate of total lipid extraction R was 286.06%/min), which resulted from the rapid heating of intracellular water under microwave irradiation. It was explained that localized high temperature and pressures produced under microwave irradiation forced the lipid release from algae cells. After 1 min, the total lipid extraction rate was significantly lower (0.48%/min), which indicates that the residual lipids were released mainly through unhindered diffusion. After 10 min, the total lipid extraction rate significantly decreased (0.01%/min), which indicates that the residual lipids were released mainly through hindered diffusion.

An interesting approach was proposed by Chan et al. (2013), where extraction is parameterized as a function of the absorbed energy density with independent parameters for the microwave exposition and the extraction kinetics (Figure 9). This method can be adjusted for models using microwave extraction by means of steam (Navarrete et al. 2010). An optimization method based on absorbed power density (APD) and absorbed energy density (AED) can be developed based on the extraction mechanism of microwave mediated extraction using sequential single factor experiments. This method optimizes MAE operating parameters based on the extraction mechanisms as demonstrated in Figure 10 (Chan et al. 2013). In general, there are three extraction mechanisms and each of them is affected by a group of operating parameters. The first mechanism is associated with the penetration of solvent into the plant matrices. In the second, the polar solvent in the plant cells is heated up by microwave and gradually the built up internal pressure ruptures the cells. The final mechanism involves the elution of active compounds from the ruptured cells and dissolution of the compounds into the solvent. Hypothetically, the rupturing of plant cells in mechanism 2 is rate limiting as it requires heating energy to proceed. The mechanisms 1 and 3 are relatively fast and the operating parameters associated are extraction solvent, solvent to feed (S/F) ratio (constant volume) and particle size of sample. These parameters do not exhibit interactive effects with each other thus they can be investigated individually and can be specified prior to the optimization of the mechanism 2. The rate limiting mechanism of MAE (rupturing of plant cells) is crucial as it determines both the rate of extraction and the yields of the extraction significantly.

The operating parameters that affect the mechanism are microwave power and extraction time. They exhibit interactive effect and are usually optimized together with other interactive parameters such as S/F ratio at constant sample mass (Chen et al. 2010, Yang and Zhai 2010) and also extraction temperature when

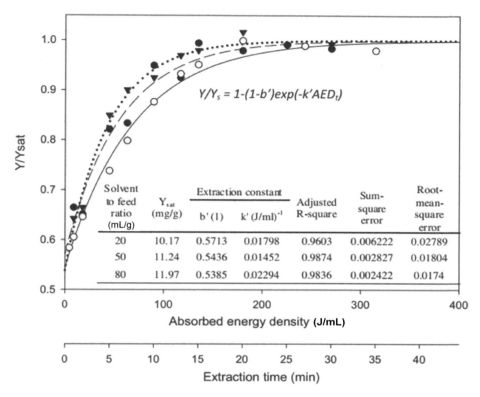

Figure 9. Curve fitting of extraction profiles of MAE with the proposed model in Eq. (6) for solvent to feed ratio (d 20 mL/g, s 50 mL/g and 80 mL/g) under applied microwave power of 100 W and solvent loading of 100 mL (APD of 0.15 W/mL). The constant b' and k' were determined with 95% confidence bounds.

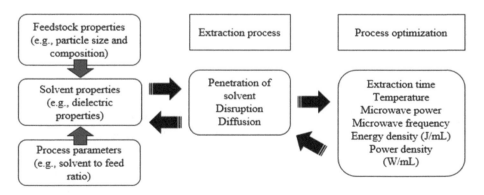

Figure 10. Strategy for optimizing microwave mediated extraction based on its extraction mechanisms.

involved with thermal sensitive compounds (Wang et al. 2009). It was reported that APD and AED could serve as an appropriate alternative in place of the microwave power and extraction time in the study of MAE (Chan et al. 2013). The intensive optimum extraction conditions offer great operational flexibility as they describe the intrinsic criteria for optimum extraction regardless of the size of extraction. The intensive parameter such as S/F ratio is closely related to the concentration gradient effect and which indirectly affects the diffusion and dissolution of the active compounds.

Absorbed power density and absorbed energy density parameters indicate the amount of heating power and energy required to heat up the extraction solvent in order to rupture the plant cells to achieve equilibrium or optimum extraction yield. In view of the above, the intensive optimum conditions of MAE can be easily modified to fit certain optimization objective without having the need to investigate the interactions between the parameters concerned. Apart from varying the S/F ratio, different combinations of APD and AED result in different microwave mediated extraction performances. There are nine performance regimes of MAE which can be classified under the effects of APD and AED as illustrated in Figure 11. MAE conducted below optimum AED values give incomplete extraction due to inadequate heating time; above the optimum AED, the extraction is risked of thermal degradation due to prolonged extraction and hence may give poor equilibrium extraction yields. On one hand, MAE which is conducted below optimum APD would give poor extraction yield as the heating power is insufficient to rupture all the plant cells, on the other hand high APD beyond the optimum value would expose the extraction to high temperature causing undesirable effect on thermal sensitive compounds (Chan et al. 2014).

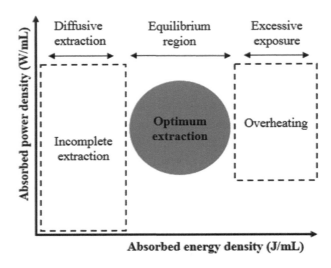

Figure 11. Performance regimes of MAE based on APD and AED.

Specific Energy Consumption

The energy consumption of an extraction process depends on various parameters such as feedstock quality and composition and compound of interest, and the extraction conditions like temperature, pressure, solvent and catalyst. More importantly, specific energy consumption depends on the size of the extraction sample. A wide range of specific energy consumption values were reported in literature for numerous feedstock extraction processes. Specific energy consumption values are, in general, reported to be very high in small scale studies and it decreases with the scale-up of the process. This is especially true for microwave mediated processes. Therefore, it is very important to study the optimum energy required for microwave mediated extraction to minimize the energy consumption.

Bermúdez et al. (2015) reported normalized specific energy consumption values for six different reactions. The specific energy consumption of six different microwave mediated processes and equipment was studied and it was found that the scale of the process had a significant effect (see Figure 12). Increasing the amount of sample employed from 5 to 100 g leads to a reduction in the specific energy consumption of 90–95%. When the amount of sample is 200 g or higher, the specific energy consumption remains practically constant. This means that to assess the real energy efficiency of a microwave-driven process

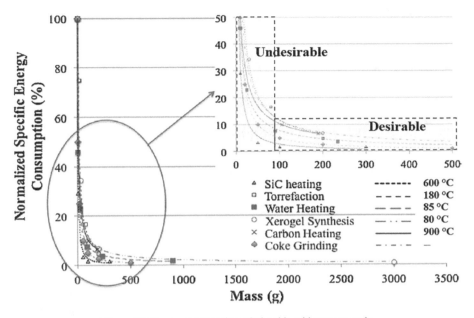

Figure 12. Energy consumption relationship with process scale.

a minimum mass of about 200 g needs to be used. The energy results can then be easily extrapolated to larger scales. Otherwise, a correlation should be used to avoid overestimated energy values and inaccurate energy efficiencies (Bermúdez et al. 2015).

Reaction Kinetics

It is important to study reaction kinetics to optimize the reactor design and process parameters. Conventional heating process is very different from microwave heating. Much higher reaction rate constants are usually reported for microwave heating when compared to conventional heating. A first order reaction rate is commonly used in kinetics studies. Few studies have evaluated pseudo order, and second order reaction rates. The kinetics of a reaction depends on the process conditions such as temperature, chemical composition (reactants), catalyst and operating pressure. Kinetics can be studied for extraction, esterification and transesterification reactions in biodiesel production. Whereas in bioethanol production, the degradation and solubilization of feedstock can be studied to determine the dimensions of the pretreatment reactors.

Reaction Kinetics in Biodiesel Production

The kinetics of the homogeneously catalyzed alcoholysis reaction has been most frequently studied, and different kinetic models were used (Table 2). Some complex models are based on the stepwise reversible alcoholysis reaction and sometimes include side saponification reactions (Berchmans et al. 2013). However, there are simple models that assume the irreversible overall reaction. Bikou et al. (1999) studied the effect of water content in ethanol on the cottonseed oil ethanolysis reaction kinetics. The dependence of both the reaction rate and equilibrium constants are well correlated. For the pongamia oil methanolysis, Karmee et al. (2006) observed that the fastest reaction was the forward reaction of the first step (methanolysis of TAG to diacylglycerols, DAG), while the second step (reaction of DAG to monoacylglycerols MAG) was the slowest step. The most complex kinetics model, which includes saponification reactions of all glycerides and esters, was developed for the methanolysis of a jatropha oil–waste food oil mixture. The comparison of the second-order kinetics based on the stepwise methanolysis reaction, and the new proposed kinetic model showed that the latter had smaller deviation from experimental data.

Table 2. Kinetic models used in conventional transesterification studies (Banković-Ilić et al. 2012).

Feedstock - Pongamia Alcohol - Methanol Temperature - 60°C Reference - Karmee et al. 2006 Kinetic equations: $$-\frac{d[TAG]}{dt} = k_1[TAG][MeOH] - k_{-1}[DAG][FAME]$$ $$-\frac{d[DAG]}{dt} = k_2[DAG][MeOH] - k_{-2}[MAG][FAME] - k_1[TAG][MeOH] + k_{-1}[DAG][FAME]$$ $$-\frac{d[MAG]}{dt} = k_2[MAG][MeOH] - k_{-3}[MAG][FAME] - k_{-3}[GL][FAME] + k_2[DAG][MeOH]$$ $$+ k_{-2}[MAG][FAME]$$
Feedstock - Cottonseed[a] Alcohol - Ethanol Temperature - 78.2 ± 2°C Reference - Bikou et al. 1999 Kinetic equations: $$-\frac{d[TAG]}{dt} = k_1[TAG][EtOH]^3 - k_{-1}[DAG][FAEE]$$ $$-\frac{d[DAG]}{dt} = k_2[DAG][EtOH]^3 - k_{-2}[MAG][FAEE] - k_1[TAG][EtOH]^3 + k_{-1}[DAG][FAEE]$$ $$-\frac{d[MAG]}{dt} = k_3[MAG][EtOH]^3 - k_{-3}[GL][FAEE] - k_2[DAG][EtOH]^3 + k_{-2}[MAG][FAEE]$$
Feedstock - Jatropha Mahua Alcohol - Methanol Temperature - 28 and 45°C Reference - Kumar et al. 2011 One step reaction (overall reaction) $$-\frac{d[TAG]}{dt} = k[TAG]^2[MeOH]$$
Feedstock - Cottonseed[b] Alcohol – Methanol Temperature - 60°C Reference - Georgogianni et al. 2008 One irreversible reaction (overall reaction) $$-\frac{d[TAG]}{dt} = k[TAG]$$
Feedstock - Castor Alcohol – Ethanol Temperature - 30–70°C Reference - Da Silva et al. 2009 One irreversible reaction (overall reaction) $$-\frac{d[TAG]}{dt} = k[TAG]$$ $$-\frac{d[TAG]}{dt} = [OH](-k_1[TAG][MeOH] + k_{-1}[DAG][FAME] - k_5[TAG])$$ $$-\frac{d[TAG]}{dt} = [OH](k_1[TAG][MeOH] + k_{-1}[DAG][FAME] - k_2[DAG][MeOH] + k_{-2}[MAG][FAME]$$ $$+ k_5[TAG] - k_6[DAG])$$ $$-\frac{d[MAG]}{dt} = [OH](k_2[DAG][MeOH] + k_{-2}[MAG][FAME] - k_3[MAG][MeOH] + k_{-3}[GL][FAME]$$ $$+ k_6[DAG] - k_7[MAG])$$

Table 2. contd....

Table 2. contd....

Feedstock -Jatropha–waste food oil mixture Alcohol - Methanol Temperature -50°C Reference - Berchmans et al. 2010 Three consecutive reversible (methanolysis) + five saponification reactions $$-\frac{d[GL]}{dt} = [OH](-k_3[MAG][MeOH] + k_{-3}[GL][FAME] - k_7[MAG])$$ $$-\frac{d[FAME]}{dt} = -\frac{d[MeOH]}{dt} = [OH](k_1[TAG][MeOH] - k_{-1}[DAG][FAME] + k_2[DAG][MeOH])$$ $$- k_{-2}[MAG][FAME] + k_3[MAG][MeOH] - k_{-3}[GL][FAME] - k_4[FAME])$$ $$-\frac{d[OH]}{dt} = -\frac{d[A]}{dt} = [OH](-k_4[FAME] - k_5[TAG] - k_6[DAG] - k_7[MAG] - k_8[FFA])$$ $$-\frac{d[W]}{dt} = -\frac{d[FFA]}{dt} = -k_8[FFA][OH]$$
Feedstock -Pre-esterified jatropha oil Alcohol - Methanol Temperature -32–51°C Reference - Lu et al. 2009 One irreversible reaction (overall reaction) $$-\frac{d[TAG]}{dt} = k[TAG]^2$$

[a]In the presence of water. [b]Using both mechanical stirring and low frequency ultrasound. [c][TAG], [DAG], [MAG], [FAME], [FAEE], [MeOH], [EtOH], [Gl], [FFA], [OH], [A], [W] – concentrations of TAG, DAG, MAG, FAME, FAEE, methanol, ethanol, glycerol, FFA, OH−, soap and water, respectively, t – time, k_1, k_2 and k_3 – forward alcoholysis reaction rate constant, k_{-1}, k_{-2} and k_{-3} – reverse alcoholysis reaction rate constant, k – overall reaction rate constant, k_4, k_5, k_6, k_7, k_8 – saponification reactions rate constants.

Kinetics of Microwave Enhanced Biodiesel Production

Similar to conventional transesterification reaction, microwave mediated transesterification depends on the oil feedstock, catalyst employed and the alcohol and other reaction parameters. Several examples can be found in literature to discuss the effects of these factors on reaction kinetics and thermodynamic parameters.

Effect of feedstock on transesterification kinetics

The reaction kinetics vary among feedstock depending on the length of the carbon chain and polarity of the feedstock. Terigar et al. (2010) studied the transesterification kinetics of rice bran oil and soybean oil using ethanol in a continuous flow microwave reactor. Transesterification reaction was assumed to follow a first order reaction as shown below.

$$C_t = C_f(1 - e^{-kt}) \qquad [28]$$

$$k = A \times e^{-E_a/RT} \qquad [29]$$

Where $C_{(t)}$ = concentration of product at time t (%); C_f = final concentration assumed to be 100% (complete conversion); k = reaction rate constant (s^{-1}); t = time (s); A = pre-exponential/frequency factor, describing the molecular motion of the reactants (s^{-1}); E_a = activation energy (J/mol); R = gas constant (8.3144 J/mol K); and T = temperature (K).

Activation energy E_a and pre-exponential factor A were determined by using Arrhenius equation [29] analytically for the two temperatures and reaction rate constants at each temperature investigated. The reaction rate constant was calculated from equation [29] as presented in Table 3. A higher value of pre-

Table 3. Comparison of kinetic data for soybean and rice bran oils at two different temperatures.

Parameter	Soybean oil		Rice-bran oil	
	50°C	73°C	50°C	73°C
k (s^{-1})	0.0567	0.0747	0.0684	0.0800
Ea (J/mol)	11,147		6,334	
A (s^{-1})	3.5932		0.7228	

exponential factor, A indicates a higher reaction rate, as its value is largely associated with the frequency of the vibrating molecules in the reaction mixture. Microwaves increase molecular vibrations due to the oscillatory electric field, resulting in an increased value of A and subsequently the reaction rate. The activation energy of soybean oil is higher than that of the rice bran oil at 50°C because rice bran oil had a higher conversion rate. At a higher temperature (73°C) the conversions were almost the same, with the activation energy playing less of a role as nearly complete conversion was observed for both the oils.

Conversion rates ranged from 96.7 to 99.3% for biodiesel derived from soybean oil and from 98.4 to 99.3% for rice bran oil. Microwave mediated biodiesel production achieved higher yields with lower alcohol to oil ratio (5:1 molecular ratio), and shorter reaction times when compared with conventional methods. Reaction rate constants ranged from 0.0567 to 0.0800 s^{-1}, whereas activation energies were 11,147 and 6,334 J/mol for soybean and rice bran ethyl esters, respectively. Pre-exponential factors were 3.593 and 0.723 s^{-1} for soybean and rice bran oil ethyl esters, respectively. Soybean has slightly lower conversion rate than rice bran oil at lower temperatures, but this difference becomes negligible at higher temperatures.

It was explained that rice bran oil has slightly higher reaction rates than soybean oil due to its higher ratio of oleic acid (with one double bond) to linoleic acid (with two double bonds). The presence of the single double bond bends the molecule only once in a single direction, whereas the double bonds in the linoleic acid (predominant in the soybean oil) bend it twice, reducing the molecule's gyration radius and subsequently the chance of reaction. As such, the rice bran oil interacts better with the microwave field. Regardless of these small differences in reaction rates, at all parameters investigated and for both oils, the reaction rate was observed to increase at the very beginning of the reaction (first minute). The tremendous increase in the reaction rate in the first minute represents the major advantage of microwaves over the conventional heating methods, as when using conventional heating, a very slow reaction rate is usually observed at the beginning for a significantly longer time (15 min) followed by a sudden increase in the rate of reaction.

Effect of alcohol on transesterification kinetics

Among alcohols used for transesterification, methanol is the most commonly used one, due to its low price and chemical and physical advantages. It is the most polar and shortest chain alcohol (Zhang et al. 2003). The solubility of a catalyst such as NaOH or KOH in methanol is faster when compared to other alcohols and it can easily react with triglycerides leading to faster reaction rates (Sanli and Canakci 2008). By comparison, the formation of ethyl esters is more difficult than of methyl esters due to the different reactivities of alcohols with catalysts in order to produce alkoxide ion (the active moiety in transesterification reaction) (Meher et al. 2006, Sanli and Canakci 2008). As the reaction essentially proceeds by formation of oil–alcohol emulsions, in the case of methanolysis, these emulsions quickly break to form glycerol and methyl esters (FAME). In the case of ethanolysis, on the other hand, these emulsions are more stable and interfere with esters separation (Meher et al. 2006).

Kanitkar et al. (2011) studied the differences in kinetics of transesterification of soybean and rice bran oils using two different alcohols namely methanol and ethanol. The transesterification reactions were carried out at different temperatures of 60, 70, and 80°C for 5, 10, 15 and 20 min. The heating time was 5 min with a chamber cooling time of 15 min. These additional times were accounted for in determining reaction kinetic parameters. A first-order reaction was assumed to determine the rate constants using Arrhenius relationship. Tables 4a and 4b show the reaction kinetic parameters for the transesterification reactions for two different oils and two different alcohols with and without cooling times between the samples.

Table 4a. Kinetic and thermodynamic data for transesterification of soybean oil under microwaves (Kanitkar et al. 2011).

Parameters	Soybean oil					
Alcohol	Methanol			Ethanol		
	With 15 min cooling time					
T (°C)	60	70	80	60	70	80
k × 10⁻³ (s⁻¹)	2.256	2.312	2.375	2.049	2.127	2.312
E_a (J/mole)	2521.54			5862.297		
A (s⁻¹)	0.005602			0.016877		
	Without 15 min cooling time					
T (°C)	60	70	80	60	70	80
k × 10⁻³ (s⁻¹)	5.844	5.878	6.107	5.239	5.481	5.976
E_a (J/mole)	2132.608			6410.409		
A (s⁻¹)	0.012553			0.052622		

Table 4b. Kinetic and thermodynamic data for transesterification of Rice Bran oil under microwaves.

Parameters	Rice Bran oil					
Alcohol	Methanol			Ethanol		
	With 15 min cooling time					
T (°C)	60	70	80	60	70	80
k × 10⁻³ (s⁻¹)	2.184	2.358	2.368	1.832	1.931	2.041
E_a (J/mole)	3994.204			5274.178		
A (s⁻¹)	0.00934			0.012288		
	Without 15 min cooling time					
T (°C)	60	70	80	60	70	80
k × 10⁻³ (s⁻¹)	5.608	5.967	6.101	4.629	4.957	5.221
E_a (J/mole)	4137.188			5886.555		
A (s⁻¹)	0.025124			0.038851		

The activation energy values for ethanol were higher than those for methanol for both soybean and rice bran oil (methyl ester conversion yields were also significantly higher than the ethyl esters ones). Methanol being more polar and shortest chain length alcohol, it is proposed that it interacts better with the microwaves as it has a lower gyration radius and less molecular inertia at this particular frequency. This allows it to rotate faster and induce higher reaction rates than ethanol. For soybean oil, the activation energy for methyl ester was found to be less than half of that needed for ethyl ester formation. For rice bran oil, however, the reduction in activation energy for methyl ester vs. ethyl ester formation was not nearly as pronounced. It was suggested that this difference is mainly due to the difference in the fatty acid composition of the two oils and specific interaction of the triglyceride molecules with the different polarity alcohols used in this study.

Effect of catalyst on transesterification kinetics

The reaction kinetics of transesterification under microwave irradiation could be very different from conventional heating. Reactants respond to microwave heating depending on their polarity, dipole momentum and microwave absorption capacity. Therefore, each of the reactants in a microwave enhanced reaction may have a different reaction rate. The reaction rates with respect to individual reactants such as oil and methanol are not reported in many studies. This is important because the kinetics of transesterification reaction depends on both the reactants and their individual interaction with microwaves. The following mathematical model can be used to account for the reaction kinetics of individual reactants and the overall reaction order under microwave irradiation.

Chemistry of the Reaction

The overall triglyceride transesterification reaction is reversible and excess amount of alcohol is commonly used to shift the equilibrium towards the formation of esters. Generalized transesterification reaction is given by Eq. [30], where A is the triglyceride, B is the methanol, C is the biodiesel (FAME) and D is glycerol.

Mathematical analysis

The effect of microwave irradiation on the triglycerides, methanol and catalyst can be different. To account for these differences, individual reaction rate orders with respect to the reactants in the reactant mixture (oil and methanol) can be changed in the overall reaction kinetic equation (Patil et al. 2010). The development of the reaction kinetics is as follows (Singh and Fernando 2007):

$$A + 3B \leftrightarrow C + D \qquad [30]$$

The general rate equation for the Eq. [27] is:

$$-\frac{dC_A}{dt} = kC_A^\alpha C_B^\beta \qquad [31]$$

where C_A/t is the consumption of reactant A per unit time, k is a rate constant, C_A is the concentration of A after time t, C_B is the concentration of B after time t, α is the order of reactant A, and β is the order of reactant B. In addition:

$$C_A = C_{A0}(1 - X) \qquad [32]$$

$$C_B = C_{A0}(\theta_B - 3X) \qquad [33]$$

$$\theta_B = \frac{C_{B0}}{C_{A0}} \qquad [34]$$

Where C_{A0} and C_{B0} are the initial concentrations of A and B, X is the conversion of triglycerides, and θ_B is the ratio of C_{B0} to C_{A0}. Equation [28] can be rewritten as:

$$\frac{dX}{dt} = kC_{A0}^{\alpha+\beta-1}(1-X)^\alpha(\theta_B - 3X)^\beta \qquad [35]$$

Eight different cases can be considered in order to obtain the reaction order. These cases are ($\alpha = 0, \beta = 0$), ($\alpha = 1, \beta = 0$), ($\alpha = 0, \beta = 1$), ($\alpha = 1, \beta = 1$), ($\alpha = 2, \beta = 0$), ($\alpha = 0, \beta = 2$), ($\alpha = 2, \beta = 1$), ($\alpha = 1, \beta = 2$). For each case, definite integrals of Eq. [28] can be calculated from a conversion of X = 0 to a conversion of X = X in the time span of t = 0 to t = t. The calculated equation for each case will then be transferred into a linear equation passing through origin (y = mx). The transferred equations for each of the case are as follows:

a) Case 1: ($\alpha = 0, \beta = 0$)

$$C_{A0} X = kt \qquad [36]$$

b) Case 2: ($\alpha = 1, \beta = 0$)

$$\ln\left[\frac{1}{1-X}\right] = kt \qquad [37]$$

c) Case 3: ($\alpha = 0, \beta = 1$)

$$-\frac{1}{3}\left[\ln\frac{\theta_B - 3X}{\theta_B}\right] = kt \qquad [38]$$

d) Case 4: ($\alpha = 1, \beta = 1$)

$$\frac{1}{(\theta_B - 3)}\ln\left[\frac{(\theta_B - 3X)}{(1-X)\theta_B}\right] = kC_{A0}t \qquad [39]$$

e) Case 5: ($\alpha = 2, \beta = 0$)

$$\frac{1}{(1-X)} = kC_{A0}t \qquad [40]$$

f) Case 6: ($\alpha = 0, \beta = 2$)

$$\frac{1}{(\theta_B - 3X)\theta_B} = kC_{A0}t \qquad [41]$$

g) Case 7: ($\alpha = 2, \beta = 1$)

$$\frac{1}{(\theta_B - 3)}\left\{\frac{X}{(1-X)} - \frac{3}{(\theta_B - 3)}\ln\left[\frac{(\theta_B - 3X)}{(1-X)\theta_B}\right]\right\} = kC^2_{A0}t \qquad [42]$$

h) Case 8: ($\alpha = 1, \beta = 2$)

$$\frac{1}{(3-\theta_B)}\left\{\frac{3X}{(\theta_B - 3X)\theta_B} - \frac{1}{(3-\theta_B)}\ln\left[\frac{(1-X)\theta_B}{(\theta_B - 3X)}\right]\right\} = kC^2_{A0}t \qquad [43]$$

For Eqs. [33–40], if it is assumed that the left-side component is an ordinate (y variable) and t (for Eqs. [33–35]), $C_{A0}t$ (for Eqs. [36–38]) and C_{A0}^2t (for Eqs. [39] and [40]) are abscissas (x variables), respectively, the equations are in the form of $y = mx$ (a straight line passing through origin). For all eight cases, the y variable was plotted against the corresponding x variable and the coefficient of determination was estimated. In all cases for Eqs. [33–40], the slope of the straight line is the rate constant, k, for the reaction. The highest correlation coefficient, R^2, for each case was observed and the case that gave the highest correlation coefficient was used to determine the reaction order and reaction rate constants.

Determination of activation energy

The Arrhenius equation gives a relationship between the specific reaction rate constant (k), absolute temperature (T) and the energy of activation (E_a) as:

$$k = A * \exp\left[-\frac{E_a}{RT}\right] \qquad [44]$$

Where A is the frequency factor and R is universal gas constant (Jmol^{-1}K^{-1}). This equation can be rewritten as:

$$\ln(k) = -\frac{E_a}{RT} + \ln(A) \qquad [45]$$

A plot of ln (k) vs. 1/T (the Arrhenius plot) gives slope equal to ($-E_a/R$) from which activation energy can be determined.

Kinetics of the transesterification of *Camelina Sativa* oil was studied using the theoretical equations developed and the optimum reaction conditions obtained (Patil et al. 2011). The effect of the reaction time with the best combination of other parameters is evaluated using the eight cases discussed in the kinetics section. The FAME's yields obtained for different heterogeneous catalysts with respect to the reaction time (as shown in Eq. [32] are considered for the analysis. Different cases were plotted assuming different reaction orders for reaction compounds namely methanol and triglycerides. The plots were modified to fit the data in the form $y = mx$, to determine the reaction rate constant which is the slope of the straight line passing through the origin. Accordingly, the reaction rate constants and reaction orders for individual reaction compounds and overall reaction orders were determined from the best-fit plot with highest R^2 value. The values of R^2 for different catalysts are shown in Table 5 for conventional method of heating. The reaction orders with respect to methanol and triglycerides, overall reaction order and the rate constants are shown in Table 6. From Tables 5 & 6, it is interesting to note that both BaO and SrO fit the case 7 which is second order with respect to triglycerides and first order with respect to methanol. The reaction rate constants were 0.0526 and 0.0493 g^2 mol^{-2} min^{-1}, respectively. For CaO, the reaction rate constant observed was the minimum which is about two orders of magnitude (87 times) smaller than that for BaO with overall first order reaction.

The reaction order and the rate constant for microwave-assisted transesterification are shown in Table 6. The reaction order for BaO with respect to triglycerides and methanol are similar to conventional heating method except that the reaction rate constant is two orders of magnitude (98.7 times) higher for microwave-assisted heating method. This is reflected by the short reaction times required for microwave heating and efficient heating of the reaction compounds. Similarly, for SrO, the reaction rate constant for microwave-assisted heating is one order of magnitude higher than conventional heating method. The above study can be very instrumental in choosing the appropriate amounts of reaction compounds, catalyst concentrations and the method of heating. The reaction kinetics allow for better understanding of the reactor design and efficient utilization of energy with higher FAME's yields.

The transesterification reaction of the *Camelina Sativa* oil is dependent on the type of catalyst employed and individual reaction orders of triglycerides and methanol. The heterogeneous metal oxide catalysts used in this study had varying selectivity toward the transesterification reaction depending upon their acid/base-site strength and surface area. Among the catalysts tested, BaO produced superior results compared to SrO catalyst. The reaction rate constant was improved by two orders of magnitude in microwave-assisted transesterification reaction. This study reveals that the rate of the transesterification

Table 5. The values of "R^2" for all eight cases of each catalyst for microwave heating.

Catalyst	Case 1	Case 2	Case 3	Case 4	Case 5	Case 6	Case 7	Case 8
BaO	0.8783	0.9095	0.8319	0.8935	0.9078	0.7773	**0.9161**	0.9039
SrO	0.7818	0.8249	0.7209	0.7941	0.8286	0.6476	**0.8291**	0.8203

Table 6. Reaction order for each of the reactants, overall order and rate constants.

Catalyst	Method	Order w.r.t. TG	Order w.r.t. MeOH	Overall order	Rate constant	Unit
BaO	CH	2	1	3	0.0526	g^2 mol^{-2} min^{-1}
SrO		2	1	3	0.0493	g^2 mol^{-2} min^{-1}
BaO	MW	2	1	3	5.195	g^2 mol^{-2} min^{-1}
SrO		2	0	2	1.584	g mol^{-1} min^{-1}

CH: Conventional heating; MW: Microwave; TG: Triglyceride; MeOH: Methanol

Estimation of Thermodynamic Parameters

To determine the effects of microwave heating, the activation energy and the pre-exponential factor can be calculated by using the Arrhenius equation which correlates these two parameters with the rate constant. Further, the estimation of thermodynamic reaction parameters can be calculated using the Eyring equation derived from the transition state theory as shown below (Mazo et al. 2012).

$$k = \alpha\left(\frac{k_B T}{h}\right) \times e^{-\Delta G/RT} = \alpha\left(\frac{k_B T}{h}\right) \times e^{\Delta S/R} \times e^{-\Delta H/RT} \quad [46]$$

$$E_a = \Delta H^\circ + RT \quad [47]$$

$$\Delta G^\circ = \Delta H^\circ - T\Delta S^\circ \quad [48]$$

where k_B (1.38066E−23 J/K), h (6.626068E−34 J s), and α are the Boltzmann's constant, Planck's constant, and the transmission coefficient, respectively, and ΔG°, ΔS° and ΔH° are the standard-state free energy of activation, standard-state entropy of activation, and standard-state enthalpy of activation of transition state at the absolute temperature T, respectively. Assuming that α is equal to 1, i.e., there is no equilibrium between the transition state and the reactants, it is possible to calculate the thermodynamic properties in the transition state.

Mazo et al. (2012) estimated reaction thermodynamic parameters for self-esterification of maleated castor oil without a catalyst under conventional and microwave heating methods. A first order kinetic model fitted the experimental data very well with respect to hydroxyl group concentrations and with respect to acid group. Figure 13 shows the plotting of Arrhenius equation (a) and Eyring equation (b) for conventional and microwave heating. ΔS° was calculated from the slope of the line and ΔH° was calculated from the intercept. ΔG° was calculated by using Eq. [48].

Table 7 shows the kinetic and thermodynamic properties for the reaction. The decrease (ca. 10%) in the activation energy and the increase (ca. 182%) in the pre-exponential factor confirm the presence of non-thermal effects when using microwave heating. It follows from Figure 13a that to obtain any specific value of the rate constant a lower temperature is required with microwaves than with conventional heating. This indicates that thermal effects of microwaves are also present. The latter can be either rapid dielectric heating or specific thermal effects.

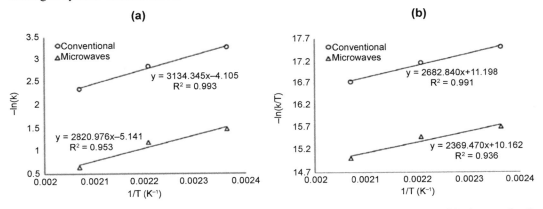

Figure 13. (a) Plot of Arrhenius equation for conventional and microwave heating; and (b) plot of Eyring equation for conventional and microwave heating (Mazo et al. 2012).

Table 7. Kinetic and thermodynamic parameters for the self-esterification reaction, using conventional and microwave heating.

Parameter	Temperature (°C)	Conventional heating	Microwave heating
k (L/mol h)	150	0.04	0.23
	180	0.06	0.30
	210	0.09	0.53
E_a (kJ/mol)		26.06	23.45
A (L/mol h)		60.64	170.91
$\Delta H^{\ddagger}$ (kJ/mol)		22.31	19.70
$\Delta S^{\ddagger}$ (kJ/mol K)		−0.29	−0.28
$\Delta G^{\ddagger}$ (kJ/mol)		145.29	139.04

Kinetics of Pretreatment in Bioethanol Production

Sugar hydrolysis is the key step in fermentation process which leads to bioethanol production. Understanding the reaction kinetics of hydrolysis process can be instrumental in optimizing the energy requirements. Adnadjevic and Jovanovic (2012) studied the isothermal kinetics of sucrose hydrolysis using acidic ion-exchange resin type IR-120 H under conventional and microwave heating methods in the temperature range from 30 to 60°C (303–333K) for both heating methods. A first order kinetics model was found suitable to represent the kinetics of sucrose hydrolysis under both conventional and microwave heating modes. A comparison of normalized conversion curves and the first order reaction order relation are shown in Figure 14.

The values of the activation energy (E_a) and pre-exponential factor ($\ln A$) for sucrose hydrolysis were found to be lower under microwave heating than under conventional heating (see Table 8). Application

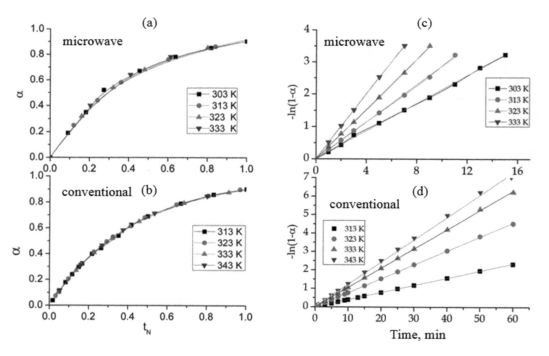

Figure 14. (a) Normalized experimental conversion curves and (b) [−ln(1−α)] vs. reaction time for conventional heating and microwave heating sucrose hydrolysis (adapted from Adnadjevic and Jovanovic 2012).

Table 8. Thermodynamic parameters for the microwave mediated and conventional sucrose hydrolysis: influence of temperature on the kinetic parameters.

T, K	Conventional heating		Microwave heating	
	(dα/dt)max, min^{-1}	Kinetics parameters	(dα/dt)max, min^{-1}	Kinetics parameters
303		$E_{a,s}$ = 33.3 ± 0.5 kJ/mol ln[Af(α max/min^{-1})] = 9.6 ± 0.2	0.19	$E_{a,s}$ = 20.8 ± 0.5 kJ/mol ln A[Af(max/min^{-1})] = 6.6 ± 0.2
313	0.038		0.25	
323	0.073		0.32	
333	0.099		0.4	
343	0.117			

of the differential isoconversional method showed that sucrose hydrolysis was kinetically an elementary reaction. It was found that the increased rate of hydrolysis observed under microwave heating was not a consequence of overheating. A new explanation of the established effects of microwave heating based on a model of selective energy transfer during the chemical reaction is suggested. The established decreases in the activation energy and in the pre-exponential factor under microwave heating in comparison to conventional heating is explained by an increase in the energy of the ground vibrational level of the –OH out-of-plane deformation in the sucrose molecule and with a decrease in the anharmonicity factor, which is caused by the selective resonant transfer of energy from the catalyst to the –OH oscillators in the sucrose molecules (Adnadjevic and Jovanovic 2012).

The relationship between the activation energy and conversion factor is shown in Figure 15. It can be noted that the activation energy is independent of the degree of sucrose hydrolysis for both heating modes and that the values of E_a for microwave heating were lower overall degrees of sucrose hydrolysis than those for the conventional heating hydrolysis.

Because the independence of E_a on the degree of sucrose hydrolysis is characteristic for an elementary (single-stage) processes, it was concluded that the investigated process of acid catalyzed sucrose hydrolysis on an ion-exchange resin presents an elementary chemical reaction with a unique kinetic model and mechanism and that the microwave field does not lead to changes in the mechanism of the process.

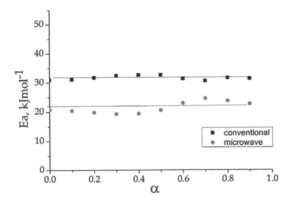

Figure 15. The dependence of $E_{a,α}$ on the degree of sucrose hydrolysis (adapted from Adnadjevic and Jovanovic 2012).

Kinetics of HMF and Furfural Production

The hydrolysis of lignocellulosic materials is a complex process provided that furfural (pentose decomposition product), HMF (hexose decomposition product) and phenolic compounds (decomposition products of lignin) are produced (da Silva Lacerda et al. 2015). It is necessary to study simplified models to determine the reaction kinetics and verify the optimum conditions for separate production of furfural and HMF. The hydrolysis of lignocellulosic biomass involves a series of irreversible reactions from the raw material (cellulose chains), to oligosaccharide formation, to HMF/furfural formation and finally to levulinic acid and humic acid. Therefore, the hydrolysis process can be described by two pseudo first order, homogeneous consecutive irreversible reactions, which can be determined from the concentrations of furfural and HMF as a function of time (min) and temperature (°C) according to Saeman's model (Saeman 1945). In this model, the concentrations of HMF (H) and furfural (F) versus time (t) are expressed as shown in Eqs. [49] and [50]:

$$H = H_o e^{k_2 t} + H_o \frac{k_1}{k_2 - k_1} (e^{-k_1 t} - e^{-k_2 t})$$ [49]

$$F = F_o e^{k_2 t} + F_o \frac{k_1}{k_2 - k_1} (e^{-k_1 t} - e^{-k_2 t})$$ [50]

where H is the HMF concentration (g/L), H_o is the HMF initial concentration (determined by regression, g/L), F is the furfural concentration (g/L), F_o is the initial furfural concentration (determined by regression, g/L), k_1 is the reaction rate for the conversion of glucose into HMF/furfural (min^{-1}) and k_2 is the rate of the decomposition reaction of HMF/furfural into 2,5-dimethylfuran (min^{-1}).

Kinetics of Anaerobic Digestion Pretreatment

Pretreatment of biomass with microwave irradiation can produce beneficial results. It may result in higher biogas production due to increased digestibility and increased sugar release from biomass. Very few studies reported the kinetics of the microwave pretreatment effect on biogas production. Similar to other processes, it is often convenient to evaluate the kinetics of biogas production assuming a first order reaction. Jackowiak et al. (2011) used first order model to describe the effect of microwave pretreatment on methane production from switchgrass and to define a kinetic constant which was presented as shown in Equation [51].

$$Vol_t = Vol_\infty \times (1 - e^{-kt})$$ [51]

where $Vol_{(t)}$ is the cumulative volume of methane as function of time t, $Vol_{(\infty)}$ is maximum volume of methane and k is the kinetic constant. The values of kinetic constants for the different temperatures of microwave pretreatment are shown in Table 9. At 90 and 105°C no significant difference was observed. The enhancement of the kinetic constant appeared at 120°C with an improvement of 21% compared to untreated switchgrass and continue to increase at 135, 150 and 180°C with an improvement of 39%, 44% and 68%, respectively.

To evaluate the reduction in digestion time due to microwave pretreatment on methane production, number of days to reach 80% of the methane volume obtained from untreated switchgrass was evaluated. Data shown in Table 9 reveal that a treatment at 90°C reduced 19% of digestion time and at 105°C there

Table 9. Values of maximum volume of methane produced and kinetic constant.

Temperature (°C)	Maximum volume of methane (LCH$_4$/kgTVS)	Kinetic constant (day^{-1})	80% of maximum volume of native sample (equal to 237 LCH$_4$/kgTVS) (day)
Native sample	296 ± 17	0.080 ± 0.001	16
90	260 ± 32	0.095 ± 0.022	19
105	291 ± 2	0.081 ± 0.010	16
120	296 ± 29	0.097 ± 0.003	13.5
135	275 ± 16	0.111 ± 0.001	14.5
150	320 ± 5	0.115 ± 0.001	11.5
180	284 ± 12	0.134 ± 0.013	12

was no significant reduction. Therefore, pretreatment below 105°C decreases or does not change the biodegradability of switchgrass even if there was an improvement of matter solubilization of 9%. At 150°C the highest enhancement was observed in reduction of 28% digestion time which represents a reduction of 4.5 days compared to untreated switchgrass biomass and was almost equivalent at 180°C. The difference between the treatment at 150 and 180°C is in significant to conclude if 150°C is the optimal temperature and if a higher temperature reduces microwave pretreatment effectiveness, or if temperatures above 150°C allow better methane production. Jin et al. found in their study about microwave pretreatment on dairy manure, an optimal temperature for methane production at 147°C (Jin et al. 2009). The fact that the microwave pretreatment induced an enhancement of the kinetic constant value without gain of the ultimate volume of methane, could be explained by a breakdown of biomass matrix without transformation of weakly degradable material into easily degradable material. Indeed, if the structure of the lignocellulosic biomass is affected, the enzymes will access easily to their targets (Jackowiak et al. 2011).

The mechanisms of anaerobic digestibility of cattail improved by microwave irradiation pretreatment were explored using process optimization techniques (Hu et al. 2012). The optimum conditions were substrate loading of 17.0 g volatile solids/L, microwave intensity of 500 W, and irradiation time of 14.0 min. The following Monod relationship as shown in Equation [52] was used in this study:

$$V = V_{max} \frac{C_s}{K_s + C_s} \qquad [52]$$

Where V is the degradation rate of substrate, V_{max}, C_s and K_s are maximum degradation rate, substrate loading and half-saturation constant, respectively. The Ks and V_{max} were determined by the measurement of *CODpro* formed during the anaerobic digestion with various concentrations of cattail pretreated by microwave irradiation and conventional heating. The kinetic parameters were reported as 357.1, 150.4 for V_{max} (mg COD h^{-1} L^{-1}) and K_s(gVS/L), respectively for microwave pretreatment and 270.3 and 152.2 for conventional pretreatment, respectively. Methane production rate increased by 32% and product yield by 19% after microwave irradiation pretreatment at 100°C when compared with the conventional (water bath) heating pretreatment under the same conditions.

Kinetics of Microwave Pyrolysis

There is little consensus in the literature of pyrolysis of lignocellulosic biomass on the kinetic model and activation energies for the parallel reaction scheme. Widely varying kinetic data have been reported in recent years (see Figure 16). Although the inputs of the different and difficult-to-separate components in biomass overlap, the activation energies of the pseudo-components in the parallel reaction scheme usually resemble the activation energies of the original components (Grønli et al. 2002). A general consensus exists on the kinetic model for pure cellulose based on a first order reaction with high activation energy: 228 (191–253) kJ/mol (Anca-Couce et al. 2014). The literature usually reports a lower activation energy value for the hemicellulose pseudo-component than for cellulose but it is still high (150–200 kJ/mol). The value for the lignin pseudo-component is usually quite low (< 100 kJ/mol), albeit higher values are also reported.

Wang et al. (2015) studied the kinetics modeling of lignocellulosic biomass and microalgae biomass pyrolysis under microwave irradiation. Profiles of total weight loss and yields of bio-oil, gas, and char from MW pyrolysis of peanut shell at 700 W using Fe_3O_4 as MW absorber with respect to the reaction time are shown in Figure 17. Temperatures as high as 1170 and 1015°C were achieved for peanut shell and *Chlorella vulgaris*. The activation energy for MW pyrolysis was calculated by Coats–Redfern method and the values were 221.96 and 214.27 kJ/mol for peanut shell and *C. vulgaris*, respectively. Bio-oil yields reached 27.7 wt.% and 11.0 wt.% during pyrolysis of *C. vulgaris* and peanut shell, respectively. Bio-oil from lignocellulosic biomass pyrolysis contained more phenolic compounds while that from microalgae pyrolysis contained more nitrogen-containing species. The kinetics scheme developed to investigate the thermal decomposition of biomass is described as follows (Wang et al. 2015):

Biomass → Char + Volatiles

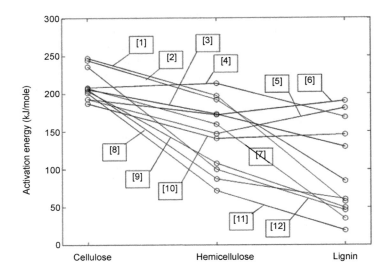

Figure 16. Activation energies reported in the literature for the biomass pseudo-components (Anca-Couce et al. 2014). Data obtained from the following studies.1: Manya et al. 2003, 2: Gomez et al. 2004, 3: Koufopanos et al. 1989, 4: Volker and Rieckmann 2001, 5: Branca et al. 2005, 6: Caballero et al. 1997, 7: Teng and Wei 1998, 8: Barneto et al. 2010, 9: Varhegyi et al. 1997, 10: Mészáros et al. 2004, 11: Grønli et al. 2002, 12: Orfao et al. 1999).

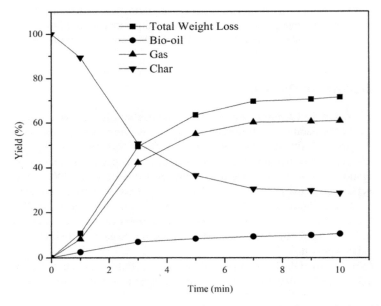

Figure 17. Total weight loss and yields of bio-oil, gas, and char from MW pyrolysis of peanut shell at 700 W using Fe_3O_4 as MW absorber (Wang et al. 2015).

The reaction rate is a function of remaining material in the reactor and follows an Arrhenius law dependent on temperature. The thermal decomposition reaction of biomass is a typically solid state decomposition reaction. The conversion of biomass can be written as:

$$\alpha = \frac{m_o - m_t}{m_o - m_\infty}$$ [53]

where m_o is the initial mass of the sample, m_t the mass of sample at time t, and m_∞ is the final indecomposable mass of the sample after the reaction. The time rate of degradation or conversion, $d\alpha/dt$, is a function of a temperature-dependent rate constant, k, and is given as:

$$\frac{d\alpha}{dt} = k \times g(\alpha) \tag{54}$$

The reaction rate constant, k, has been described by the Arrhenius expression as:

$$k = A \exp\left(-\frac{E}{RT}\right) \tag{55}$$

where A is the pre-exponential factor, E is the activation energy, R is the gas constant, and T is temperature in Kelvin. The combination Eqs. [54] and [55] gives:

$$\frac{d\alpha}{dt} = A \exp\left(-\frac{E}{RT}\right) f(\alpha) \tag{56}$$

With heating rate, $\beta = dT/dt$, the rearrangement of Eq. [56] gives:

$$\frac{d\alpha}{dt} = \frac{A}{\beta} \exp\left(-\frac{E}{RT}\right) f(\alpha) \tag{57}$$

The integrated form of Eq. [56] is generally expressed as:

$$G(\alpha) = \int_0^\alpha \frac{d\alpha}{f(\alpha)} = \frac{A}{\beta} \int_{T_0}^{T} \exp\left(-\frac{E}{RT}\right) dt \tag{58}$$

where $G(\alpha)$ is the integrated form of the conversion-dependent function $f(\alpha)$. Based on these equations, the Coats–Redfern method was used to calculate the kinetic parameters and to determine the reaction order and mechanism. Coats–Redfern equation can be expressed as follows (Tian et al. 2014):

$$\ln\frac{g(\alpha)}{T^2} = \ln\frac{AR}{\beta E} - \frac{E}{RT} \tag{59}$$

A plot of $\ln[g(\alpha)/T^2]$ against reciprocal of temperature gives a straight line with a slope equal to $-E/R$. Table 10 shows different expressions of $G(\alpha)$ used in this study for the estimation of reaction mechanisms for MW pyrolysis of biomass.

The Coats–Redfern method was used to determine the kinetic parameters and the mechanism of MW pyrolysis of biomass. The reaction mechanism and order are predefined in Coats–Redfern method (Table 10). The activation energies and correlation coefficients for different reaction mechanisms were calculated for peanut shell and *C. vulgaris* pyrolysis which are shown in Table 10. The most appropriate reaction mechanism was selected based on the adjusted coefficient of determination (R^2). From Table 10, it can be noted that the 1.2 order chemical reaction mechanism produced the best approximation of the temperature range where the reaction took place and therefore was chosen as the most appropriate mechanism to describe the MW pyrolysis of biomass. The activation energies obtained for MW pyrolysis of peanut shell and *C. vulgaris* were 221.96 and 214.27 kJ/mol, respectively. The values of activation energy from MW pyrolysis of the two biomass samples were lower compared to those in conventional fixed-bed pyrolysis. The values of activation energy for fixed-bed pyrolysis of peanut shell and *C. vulgaris* calculated by using thermogravimetric analysis and Flynn–Wall–Ozawa (FWO) method were 253.9 and 230.79 kJ/mol, respectively (Yuan et al. 2015).

The catalytic effects on the microwave pyrolysis of sugarcane bagasse were studied through the reaction performance, product distribution, and kinetic analysis (Kuan et al. 2013). The effect of addition of cheap and commonly available metal-oxides as catalysts such as NiO, CuO, CaO, and MgO on the microwave pyrolysis of sugarcane bagasse was studied. Reaction results such as mass reduction ratio and reaction rate

Table 10. Mathematical models for the most common solid-state reaction mechanisms. Activation energies and correlation coefficients for different mechanisms calculated by Coats–Redfern method for peanut shell and *C. vulgaris* biomass pyrolysis (Wang et al. 2015).

Mechanism	Function name	Differential form f(a)	Integral form g(a)	E (kJ/mol) Peanut shell	E (kJ/mol) *C. vulgaris*	Adjusted R² Peanut shell	Adjusted R² *C. vulgaris*	A Peanut shell	A *C. vulgaris*
F0.8	Chemical reaction n = 0.8	$1/0.8(1-\alpha)[-\ln(1-\alpha)]^{0.2}$	$1-\ln(1-\alpha)^{0.2}$	161.74	160.74	0.79301	0.62436	18.68	17.29
F0.9	Chemical reaction n = 0.9	$1/0.9(1-\alpha)[-\ln(1-\alpha)]^{0.1}$	$1-\ln(1-\alpha)^{0.1}$	176.79	174.13	0.81089	0.62594	20.19	18.88
F1	Chemical reaction n = 1	$1-\alpha$	$-\ln(1-\alpha)$	191.85	187.51	0.82425	0.62463	21.72	20.48
F1.1	Chemical reaction n = 1.1	$1/1.1(1-\alpha)[-\ln(1-\alpha)]^{-0.1}$	$1-\ln(1-\alpha)^{-0.1}$	206.91	200.89	0.83431	0.62154	23.24	22.08
F1.2	Chemical reaction n = 1.2	$1/1.2(1-\alpha)[-\ln(1-\alpha)]^{-0.2}$	$1-\ln(1-\alpha)^{-0.2}$	221.96	214.27	0.84192	0.62737	24.76	23.69
D1 1-D diffusion	Parabola law	$1/(2\alpha)$	α^2	113.79	55.77	0.82383	0.14663	12.06	5.81
D2 2-D diffusion	Valensi equation	$-[\ln(1-\alpha)]^{-1}$	$(1-\alpha)\ln(1-\alpha) + \alpha$	171.47	117.09	0.86739	0.3652	17.54	13.86
D3 3-D diffusion	Jander equation	$3/2[1-(1-\alpha)^{1/3}]^{-1}(1-\alpha)^{2/3}$	$[1-(1-\alpha)^{1/3}]^2$	260.40	224.59	0.87279	0.52809	25.67	27.84
R2 phase boundary contracting reaction	Contracting surface	$2(1-\alpha)^{1/2}$	$1-(1-\alpha)^{1/2}$	106.38	85.39	0.85225	0.57416	11.95	9.17
R3 phase boundary contracting reaction	Contracting volume	$3(1-\alpha)^{2/3}$	$1-(1-\alpha)^{1/3}$	134.87	119.43	0.83915	0.60033	14.53	13.65
A1.5 random nucleation and nuclei growth	Avrami–Erofeev equation	$3/2(1-\alpha)[-\ln(1-\alpha)]^{1/3}$	$[-\ln(1-\alpha)]^{2/3}$	141.66	142.90	0.75911	0.61482	16.60	15.19
A2 random nucleation and nuclei growth	Avrami–Erofeev equation	$2(1-\alpha)[-\ln(1-\alpha)]^{1/2}$	$[-\ln(1-\alpha)]^{1/2}$	116.57	120.59	0.69145	0.58161	14.00	12.59
A3 random nucleation and nuclei growth	Avrami–Erofeev equation	$3(1-\alpha)[-\ln(1-\alpha)]^{2/3}$	$[-\ln(1-\alpha)]^{1/3}$	91.47	98.29	0.57733	0.50557	11.33	10.08
A4 random nucleation and nuclei growth	Avrami–Erofeev equation	$4(1-\alpha)[-\ln(1-\alpha)]^{3/4}$	$[-\ln(1-\alpha)]^{1/4}$	78.92	87.14	0.49477	0.44349	9.94	8.89

increased with addition of metal oxide catalysts although the maximum temperature decreased. Adding either NiO or CaO slightly increased the production of H_2, while adding either CuO or MgO slightly decreased the H_2 production. The addition of either CaO or MgO enhanced the gaseous production, and either NiO or CuO addition enhanced the liquid production. There could be several secondary reactions such as self-gasification and interactions among the gases originally produced during the pyrolysis stage to alter the composition of gaseous product and the final three-phase product distribution. The catalyst addition slightly increased the activation energy but greatly increased the pre-exponential factor as shown in Table 11. Compared with the X-ray diffraction (XRD) peaks of original metal oxides, the peaks of metal oxides after microwave heating did not shift. Therefore, it can be concluded that the microwave irradiation does not affect the properties of the metal oxide catalysts. The particle sizes of NiO, CuO, MgO, and CaO were 0.5–6, 1–4, 3–6 mm, and < 20 mesh, respectively. It was noted that the particle size did not have significant effect due to addition of small quantities.

Table 11. Effect of metal oxides on microwave pyrolysis of sugarcane bagasse.

Catalyst	Activation energy, E_a (kJ/mol)	Pre-exponential factor, A (1/s)	Reaction rate, k (1/s)	Coefficient of determination, R^2
None				0.99
NiO 10%	20.24	0.247	4.05×10^{-3}	0.99
CuO 10%	20.85	0.289	4.05×10^{-3}	0.99
CaO 10%	21.39	0.337	4.11×10^{-3}	0.99
MgO 10%	19.95	0.225	3.88×10^{-3}	0.99

Microwave Mediated Gasification

Microwave mediated CO_2 gasification of oil palm shell (OPS) char to produce CO through the Boudouard reaction (C + CO_2 ↔ 2CO) was investigated (Lahijani et al. 2014). The influence of char particle size, temperature and gas flow rate on CO_2 conversion and CO evolution was considered. About 99% of CO_2 conversion was achieved during 60 min microwave mediated gasification of iron-catalyzed char. When similar gasification experiments were performed in conventional electric furnace, the superior performance of microwave over thermal driven reaction was elucidated.

Char gasification was assumed to follow a first-order reaction. Table 12 shows the kinetic parameters, CO evolution rate and char reaction rate under microwave heating with and without catalyst and conventional thermal heating. The activation energies of 36.0, 74.2 and 247.2 kJ/mol were obtained for catalytic and non-catalytic microwave and thermal heating, respectively. The activation energy was considerably lower than that of conventional electric furnace heating. It can also be inferred that use of catalyst had a significant role in reducing the activation energy, where the activation energy for microwave gasification of catalyzed char reduced to half of non-catalyzed char. The rates of char reaction and CO evolution increased with

Table 12. Kinetic parameters, CO evolution rate and char reaction rate under microwave and thermal heating (Lahijani et al. 2014).

	Microwave heating		Catalytic microwave		Conventional heating	
Arrhenius parameters	E_a = 74.22 kJ/mol $A = 3.91 \times 10^{-1}$ s^{-1}		E_a = 36.04 kJ/mol $A = 5.79 \times 10^{-3}$ s^{-1}		E_a = 247.24 kJ/mol $A = 8.39 \times 10^{6}$ s^{-1}	
T (°C)	CO production rate (mmol/min)	Char reaction rate (mmol/min)	CO production rate (mmol/min)	Char reaction rate (mmol/min)	CO production rate (mmol/min)	Char reaction rate (mmol/min)
750	2.23	1.46	4.67	2.64	0.17	0.15
800	3.30	1.72	5.51	3.06	0.57	0.39
850	4.97	2.67	6.54	3.26	1.97	1.11
900	6.52	3.31	7.55	3.93	4.83	2.61

temperature between 750°C and 900°C. The highest rate was observed for catalytic microwave reaction followed by non-catalytic microwave and thermal heating. The large difference between the rates of microwave induced and thermal reaction suggests that the rate constants must have fundamentally different dependencies on temperature which are manifested in the Arrhenius parameters.

The rate of microwave driven reaction at low temperatures was significantly higher than that of thermal heating, where the microwave induced reaction was 13.7 times faster than thermal driven reaction at 750°C. This confirms the beneficial effect of microwave heating at lower temperature. The difference in rates followed a decreasing trend and reached to 1.3 at 900°C. By equating the Arrhenius equations for the microwave and thermal reactions, the temperature of 926°C was found to be optimum at which an equal rate of CO production could be expected through the two heating routes.

Concluding Remarks

This chapter discussed the reaction kinetics of various biofuel and chemical production processes related to biodiesel, bioethanol, biogas (methane), bio-oil, HMF and furfural and biochar production. The effects of various process parameters on the reaction kinetics were elucidated with the aid of various kinetic models. Most of the studies applied the first-order reaction kinetics for easy and convenient estimation of reaction kinetic parameters. From the literature review and discussions made in this chapter, it can be concluded that the reaction kinetics of a biofuel production process depends on the feedstock characteristics, process parameters and the method of heating and the scale of reaction. There are also several variations among the kinetics data reported in the literature for various processes, especially, in pyrolysis which can be due to the aforementioned differences in different studies. It could be beneficial to design the experimental studies to represent standard procedures which will help produce consistent and reliable reaction kinetic data suitable for large-scale process design and demonstration for various biofuel and chemical production processes.

References

Adnadjevic, B.K. and Jovanovic, J.D. 2012. A comparative kinetics study on the isothermal heterogeneous acid-catalyzed hydrolysis of sucrose under conventional and microwave heating. *Journal of Molecular Catalysis A: Chemical*. 356: pp.70–77.
Amarni, F. and Kadi, H. 2010. Kinetics study of microwave-assisted solvent extraction of oil from olive cake using hexane: comparison with the conventional extraction. *Innovative Food Science & Emerging Technologies*. 11(2): 322–327.
Anca-Couce, A., Berger, A. and Zobel, N. 2014. How to determine consistent biomass pyrolysis kinetics in a parallel reaction scheme. *Fuel*. 123: pp.230–240.
Antonio, C. and Deam, R.T. 2007. Can "microwave effects" be explained by enhanced diffusion?. *Physical Chemistry Chemical Physics*. 9(23): pp.2976–2982.
Balasubramanian, S., Allen, J.D., Kanitkar, A. and Boldor, D. 2011. Oil extraction from Scenedesmus obliquus using a continuous microwave system–design, optimization, and quality characterization. *Bioresource Technology*. 102(3): pp.3396–3403.
Banković-Ilić, I.B., Stamenković, O.S. and Veljković, V.B. 2012. Biodiesel production from non-edible plant oils. *Renewable and Sustainable Energy Reviews*. 16(6): pp.3621–3647.
Barneto, A.G., Carmona, J.A., Alfonso, J.E.M. and Serrano, R.S. 2010. Simulation of the thermogravimetry analysis of three non-wood pulps. *Bioresource Technology*. 101(9): pp.3220–3229.
Barton, G. 1996. Process control: Designing processes and control systems for dynamic performance: by Thomas E. Marlin (McGraw-Hill, New York, 1995, ISBN 0-07-040491-7, pp.954+ xxii).
Berchmans, H.J., Morishita, K. and Takarada, T. 2013. Kinetic study of hydroxide-catalyzed methanolysis of Jatropha curcas–waste food oil mixture for biodiesel production. *Fuel*. 104: pp.46–52.
Bermúdez, J.M., Beneroso, D., Rey-Raap, N., Arenillas, A. and Menéndez, J.A. 2015. Energy consumption estimation in the scaling-up of microwave heating processes. *Chemical Engineering and Processing: Process Intensification*. 95: pp.1–8.
Bikou, E., Louloudi, A. and Papayannakos, N. 1999. The effect of water on the transesterification kinetics of cotton seed oil with ethanol. *Chemical Engineering & Technology*. 22(1): pp.70–75.
Branca, C., Albano, A. and Di Blasi, C. 2005. Critical evaluation of global mechanisms of wood devolatilization. *Thermochimica Acta*. 429(2): pp.133–141.
Caballero, J.A., Conesa, J.A., Font, R. and Marcilla, A. 1997. Pyrolysis kinetics of almond shells and olive stones considering their organic fractions. *Journal of Analytical and Applied Pyrolysis*. 42(2): pp.159–175.
Campanone, L.A. and Zaritzky, N.E. 2005. Mathematical analysis of microwave heating process. *Journal of Food Engineering*. 69(3): pp.359–368.

Chan, C.H., Yusoff, R. and Ngoh, G.C. 2013. Modeling and prediction of extraction profile for microwave-assisted extraction based on absorbed microwave energy. *Food Chemistry.* 140(1): pp.147–153.

Chan, C.H., Yusoff, R. and Ngoh, G.C. 2014. Optimization of microwave-assisted extraction based on absorbed microwave power and energy. *Chemical Engineering Science.* 111: pp.41–47.

Chen, Y., Gu, X., Huang, S.Q., Li, J., Wang, X. and Tang, J. 2010. Optimization of ultrasonic/microwave assisted extraction (UMAE) of polysaccharides from Inonotus obliquus and evaluation of its anti-tumor activities. *International Journal of Biological Macromolecules.* 46(4): pp.429–435.

Cheng, J., Yu, T., Li, T., Zhou, J. and Cen, K. 2013. Using wet microalgae for direct biodiesel production via microwave irradiation. *Bioresource Technology.* 131: pp.531–535.

Conner, W.C. and Tompsett, G.A. 2008. How could and do microwaves influence chemistry at interfaces? *The Journal of Physical Chemistry B.* 112(7): pp.2110–2118.

Cui, Y. and Yuan, W. 2013. Thermodynamic modeling of algal cell–solid substrate interactions. *Applied Energy.* 112: pp.485–492.

da Silva Lacerda, V., López-Sotelo, J.B., Correa-Guimarães, A., Hernández-Navarro, S., Sánchez-Bascones, M., Navas-Gracia, L.M., Martín-Ramos, P., Pérez-Lebeña, E. and Martín-Gil, J. 2015. A kinetic study on microwave-assisted conversion of cellulose and lignocellulosic waste into hydroxymethylfurfural/furfural. *Bioresource Technology.* 180: pp.88–96.

Datta, A.K. and Rakesh, V. 2013. Principles of microwave combination heating. *Comprehensive Reviews in Food Science and Food Safety.* 12(1): pp.24–39.

de Lima da Silva, N., Benedito Batistella, C., Maciel Filho, R. and Maciel, M.R.W. 2009. Biodiesel production from castor oil: optimization of alkaline ethanolysis. *Energy & Fuels.* 23(11): pp.5636–5642.

Dinčov, D.D., Parrott, K.A. and Pericleous, K.A. 2004. Heat and mass transfer in two-phase porous materials under intensive microwave heating. *Journal of Food Engineering.* 65(3): pp.403–412.

Farag, S., Sobhy, A., Akyel, C., Doucet, J. and Chaouki, J. 2012. Temperature profile prediction within selected materials heated by microwaves at 2.45 GHz. *Applied Thermal Engineering.* 36: pp.360–369.

Flórez, N., Conde, E. and Domínguez, H. 2015. Microwave assisted water extraction of plant compounds. *Journal of Chemical Technology and Biotechnology.* 90(4): pp.590–607.

Georgogianni, K.G., Kontominas, M.G., Pomonis, P.J., Avlonitis, D. and Gergis, V. 2008. Alkaline conventional and *in situ* transesterification of cottonseed oil for the production of biodiesel. *Energy & Fuels.* 22(3): pp.2110–2115.

Gomez, C.J., Manya, J.J., Velo, E. and Puigjaner, L. 2004. Further applications of a revisited summative model for kinetics of biomass pyrolysis. *Industrial & Engineering Chemistry Research.* 43(4): pp.901–906.

Grant, E. and Halstead, B.J. 1998. Dielectric parameters relevant to microwave dielectric heating. *Chemical Society Reviews.* 27(3): pp.213–224.

Grønli, M.G., Varhegyi, G. and Di Blasi, C. 2002. Thermogravimetric analysis and devolatilization kinetics of wood. *Industrial & Engineering Chemistry Research.* 41(17): pp.4201–4208.

Hu, Z.H., Yue, Z.B., Yu, H.Q., Liu, S.Y., Harada, H. and Li, Y.Y. 2012. Mechanisms of microwave irradiation pretreatment for enhancing anaerobic digestion of cattail by rumen microorganisms. *Applied Energy.* 93: pp.229–236.

Jackowiak, D., Frigon, J.C., Ribeiro, T., Pauss, A. and Guiot, S. 2011. Enhancing solubilisation and methane production kinetic of switchgrass by microwave pretreatment. *Bioresource Technology.* 102(3): pp.3535–3540.

Jin, Y., Hu, Z. and Wen, Z. 2009. Enhancing anaerobic digestibility and phosphorus recovery of dairy manure through microwave-based thermochemical pretreatment. *Water Research.* 43(14): pp.3493–3502.

Kanitkar, A., Balasubramanian, S., Lima, M. and Boldor, D. 2011. A critical comparison of methyl and ethyl esters production from soybean and rice bran oil in the presence of microwaves. *Bioresource Technology.* 102(17): pp.7896–7902.

Karmee, S.K., Chandna, D., Ravi, R. and Chadha, A. 2006. Kinetics of base-catalyzed transesterification of triglycerides from Pongamia oil. *Journal of the American Oil Chemists' Society.* 83(10): pp.873–877.

Koufopanos, C.A., Lucchesi, A. and Maschio, G. 1989. Kinetic modelling of the pyrolysis of biomass and biomass components. *The Canadian Journal of Chemical Engineering.* 67(1): pp.75–84.

Kuan, W.H., Huang, Y.F., Chang, C.C. and Lo, S.L. 2013. Catalytic pyrolysis of sugarcane bagasse by using microwave heating. *Bioresource Technology.* 146: pp.324–329.

Kumar, G.R., Ravi, R. and Chadha, A. 2011. Kinetic studies of base-catalyzed transesterification reactions of non-edible oils to prepare biodiesel: the effect of co-solvent and temperature. *Energy & Fuels.* 25(7): pp.2826–2832.

Lahijani, P., Zainal, Z.A., Mohamed, A.R. and Mohammadi, M. 2014. Microwave-enhanced CO_2 gasification of oil palm shell char. *Bioresource Technology.* 158: pp.193–200.

Lu, H., Liu, Y., Zhou, H., Yang, Y., Chen, M. and Liang, B. 2009. Production of biodiesel from Jatropha curcas L. oil. *Computers & Chemical Engineering.* 33(5): pp.1091–1096.

Marlin, T.E. 2000. Process Control, Designing Processes and Control Systems for Dynamic Performance. McGraw-Hill Inc.: New York, New York.

Manya, J.J., Velo, E. and Puigjaner, L. 2003. Kinetics of biomass pyrolysis: a reformulated three-parallel-reactions model. *Industrial & Engineering Chemistry Research.* 42(3): pp.434–441.

Mazo, P., Rios, L., Estenoz, D. and Sponton, M. 2012. Self-esterification of partially maleated castor oil using conventional and microwave heating. *Chemical Engineering Journal.* 185: pp.347–351.

Meher, L.C., Dharmagadda, V.S. and Naik, S.N. 2006. Optimization of alkali-catalyzed transesterification of Pongamia pinnata oil for production of biodiesel. *Bioresource Technology.* 97(12): pp.1392–1397.

Mészáros, E., Várhegyi, G., Jakab, E. and Marosvölgyi, B. 2004. Thermogravimetric and reaction kinetic analysis of biomass samples from an energy plantation. *Energy & Fuels*. 18(2): pp.497–507.

Navarrete, A., Mato, R.B. and Cocero, M.J. 2012. A predictive approach in modeling and simulation of heat and mass transfer during microwave heating. Application to SFME of essential oil of Lavandin Super. *Chemical Engineering Science*. 68(1): pp.192–201.

Orfao, J.J.M., Antunes, F.J.A. and Figueiredo, J.L. 1999. Pyrolysis kinetics of lignocellulosic materials—three independent reactions model. *Fuel*. 78(3): pp.349–358.

Patil, P., Gude, V.G., Pinappu, S. and Deng, S. 2011. Transesterification kinetics of Camelina sativa oil on metal oxide catalysts under conventional and microwave heating conditions. *Chemical Engineering Journal*. 168(3): pp.1296–1300.

Peterson, D. and Paul, B.K. 2014. Evaluating applicator designs for heating nanoparticle flow chemistries using single-mode microwave energy. *Journal of Microwave Power and Electromagnetic Energy*. 48(2): pp.113–130.

Ramasamy, S. and Moghtaderi, B. 2010. Dielectric properties of typical Australian wood-based biomass materials at microwave frequency. *Energy & Fuels*. 24(8): pp.4534–4548.

Saeman, J.F. 1945. Kinetics of wood saccharification-hydrolysis of cellulose and decomposition of sugars in dilute acid at high temperature. *Industrial & Engineering Chemistry*. 37(1): pp.43–52.

Sanli, H. and Canakci, M. 2008. Effects of different alcohol and catalyst usage on biodiesel production from different vegetable oils. *Energy & Fuels*. 22(4): pp.2713–2719.

Singh, A.K. and Fernando, S.D. 2007. Reaction kinetics of soybean oil transesterification using heterogeneous metal oxide catalysts. *Chemical Engineering & Technology*. 30(12): pp.1716–1720.

Teng, H. and Wei, Y.C. 1998. Thermogravimetric studies on the kinetics of rice hull pyrolysis and the influence of water treatment. *Industrial & Engineering Chemistry Research*. 37(10): pp.3806–3811.

Terigar, B.G., Balasubramanian, S., Lima, M. and Boldor, D. 2010. Transesterification of soybean and rice bran oil with ethanol in a continuous-flow microwave-assisted system: yields, quality, and reaction kinetics. *Energy & Fuels*. 24(12): pp.6609–6615.

Terigar, B.G., Balasubramanian, S., Sabliov, C.M., Lima, M. and Boldor, D. 2011. Soybean and rice bran oil extraction in a continuous microwave system: from laboratory-to pilot-scale. *Journal of Food Engineering*. 104(2): pp.208–217.

Tian, L., Tahmasebi, A. and Yu, J. 2014. An experimental study on thermal decomposition behavior of magnesite. *Journal of Thermal Analysis and Calorimetry*. 118(3): pp.1577–1584.

Varhegyi, G., Antal, M.J., Jakab, E. and Szabó, P. 1997. Kinetic modeling of biomass pyrolysis. *Journal of Analytical and Applied Pyrolysis*. 42(1): pp.73–87.

Vaz, R.H., Pereira, J.M., Ervilha, A.R. and Pereira, J.C. 2014. Simulation and uncertainty quantification in high temperature microwave heating. *Applied Thermal Engineering*. 70(1): pp.1025–1039.

Volker, S. and Rieckmann, T. 2001. Thermochemical biomass conversion. Blackwell Science. pp.1076–1090.

Wang, J., Zhang, J., Wang, X., Zhao, B., Wu, Y. and Yao, J. 2009. A comparison study on microwave-assisted extraction of Artemisia sphaerocephala polysaccharides with conventional method: Molecule structure and antioxidant activities evaluation. *International Journal of Biological Macromolecules*. 45(5): pp.483–492.

Wang, J., Zhang, J., Zhao, B., Wang, X., Wu, Y. and Yao, J. 2010. A comparison study on microwave-assisted extraction of Potentilla anserina L. polysaccharides with conventional method: Molecule weight and antioxidant activities evaluation. *Carbohydrate Polymers*. 80(1): pp.84–93.

Wang, N., Tahmasebi, A., Yu, J., Xu, J., Huang, F. and Mamaeva, A. 2015. A Comparative study of microwave-induced pyrolysis of lignocellulosic and algal biomass. *Bioresource Technology*. 190: pp.89–96.

Yang, Z. and Zhai, W. 2010. Optimization of microwave-assisted extraction of anthocyanins from purple corn (Zea mays L.) cob and identification with HPLC–MS. *Innovative Food Science & Emerging Technologies*. 11(3): pp.470–476.

Yuan, T., Tahmasebi, A. and Yu, J. 2015. Comparative study on pyrolysis of lignocellulosic and algal biomass using a thermogravimetric and a fixed-bed reactor. *Bioresource Technology*. 175: pp.333–341.

Zhang, Y., Dube, M.A., McLean, D. and Kates, M. 2003. Biodiesel production from waste cooking oil: 1. Process design and technological assessment. *Bioresource Technology*. 89(1): pp.1–16.

9

Microwaves and Ionic Liquids in Biofuel Production

Introduction to Ionic Liquids

Ionic liquids have attracted the broad scientific community (chemists, biologists, and other industrial research scientists and engineers) due to their potential greenness and promising sustainable characteristics over the recent years. The history of ionic liquids dates back to 1914 or even before and they were originally called as "molten salts" (Walden 1914). The earliest use of ionic liquids was as a propellant in warfare especially ethyl ammonium nitrate (Suresh and Sandhu 2011). Ionic liquids are considered to be ionic salts substances having a melting point up to 100°C. Ionic liquids have several advantages over other solvents (Sheldon 2001, Suresh and Sandhu 2011). A comparison of various differences between the organic solvents and ionic liquids is shown in Table 1. Ionic liquids provide superior functionality and flexibility over the organic solvents.

Low or no volatility—ionic liquids are different from conventional solvents. They do not evaporate and are considered as non-volatile with no or negligible vapor pressure. Their loss due to evaporation is at least not measurable.

Stability—ionic liquids have high chemical and thermal stability. They are stable over a long temperature range which enables them suitable for a variety of organic and inorganic chemical synthesis and separation processes. Their stability allows for storage without decomposition for a long time. For example, emimBF$_4$ is reportedly stable up to 300°C and emim-(CF$_3$SO$_2$)$_2$N up to 400°C.

Solubility—ionic liquids are considered universal solvents. They can dissolve a large range of organic compounds. The solubility of gases, e.g., H$_2$, CO and O$_2$, is generally good which makes them attractive solvents for catalytic hydrogenations, carbonylations, hydroformylations, and aerobic oxidations. They are immiscible with some organic solvents, e.g., alkanes, and, hence, can be used in two-phase systems. Similarly, lipophilic ionic liquids can be used in aqueous biphasic systems.

Process flexibility—ionic liquids can dissolve gases such as H$_2$, CO, O$_2$, and CO$_2$. They can be used even under supercritical CO$_2$; and separation and purification processes.

Ionic reactivity—the solubility of ionic liquids depends on the type of cations and anions of which these are composed and this ionic character makes them suitable for microwave chemistry and accelerates the rate of reaction even under MW irradiations.

Table 1. Comparison between organic solvents and ionic liquids (Sowmiah et al. 2009).

Property	Organic Solvents	Ionic Liquids
Number of solvents	> 1,000	> 1,000,000
Applicability	Single function	Multi-function
Vapor pressure	Obeys Clausius-Clapeyron	Equation negligible under normal atmospheric conditions
Flammability	Usually flammable	Usually non-flammable
Chirality	Rare	Common and tunable
Catalytic ability	Rare	Common and tunable
Tunability	Limited solvents available	Unlimited range "designer solvents"
Polarity	Conventional concepts apply	Questionable
Solvation	Weakly solvating	Strongly solvating
Cost	Normally inexpensive	2 to 10 times the cost of organic solvents
Recyclability	Green imperative	Economic imperative
Density (g/cm^3)	0.6–1.7	0.8–3.3
Viscosity (cP)	0.2–100	22–40,000
Refractive index	1.3–1.6	1.5–2.2

Safe chemistry—They do not participate in co-ordination with metal complexes, macrocycles like enzymes, etc.; they are appropriate for extensive use in the control of stereoselectivity.

Viscosity—The viscosity of ionic liquids derived from imidazoles can be manipulated by variations in branching.

Low toxicity—Ionic liquids have minimum toxicity to humans and environment.

Application of Ionic Liquids

Ionic liquids (ILs) are employed in a broad area of applications (Suresh and Sandhu 2011) such as solvent extraction, physico-chemical processes, as media for nucleophilic substitution reactions, electrodeposition of metals and semiconductors in ILs, chemical analysis, dye-sensitized solar cells, electrodeposition and extraction, nuclear-based separations, oil shale processing, separation of petrochemical relevance, synthesis of functional nanoparticles and other inorganic nanostructures, solvents for electrochemistry, as solvents for polymerization processes, chemical and biochemical transformations, materials chemistry, and biocatalysts. Other applications of ionic liquids are shown in Figure 1.

Ionic liquids are organic salts with melting points around or below the ambient temperature. ILs are composed of organic cations and either organic or inorganic anions. The most common ILs are divided into four groups according to their cations, which includes the following: quaternary ammonium ILs, N-alkylpyridinium ILs, N-alkyl-isoquinolinium ILs, and 1-alkyl-3-methylimidazolium ILs as shown in Figure 2.

Classification of Ionic Liquids

Ionic liquids consist of anions and cations which determines their application and role in a chemical reaction or process operation. In general, ionic liquids can be classified as anionic or cationic. Examples of some ionic liquids used in organic and inorganic synthesis are described below (Ho et al. 2013, Suresh and Sandhu 2011).

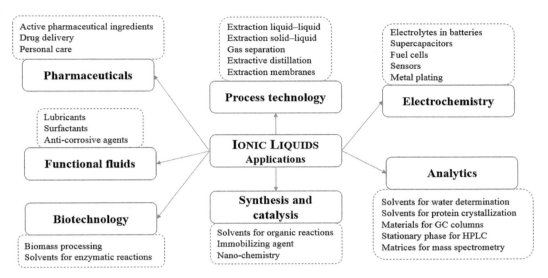

Figure 1. Potential applications of ionic liquids in various industries.

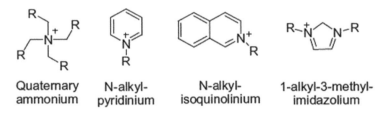

Figure 2. Four groups of most commonly used ionic liquids.

Ionic Liquids-Anions

A) Borate:
- Tetracyanoborate TCB: $[B(CN)_4]$
- Tetrafluoroborate TFB: $[BF_4]$
- Oxalatoborate BOB: $[B(C_2O_4)_2]$

B) Dicyanamide DCN: $[N(CN)_2]$

C) Halide: Br^-, Cl^-, F^-, I^-
- Bromide: Br^-
- Chloride: Cl^-
- Fluoride: F^-
- Iodide: I^-

D) Bis(trifluoromethylsulfonyl)imide NTF: $[N(SO_2CF_3)_2]$

E) Nonaflate NON: $[C_4H_9SO_3]$

F) Phosphate
- Alkylphosphate
- Fluoroalkylphosphate FAP: $[(C_2F_5)_3PF_3]$
- Hexafluorophosphate HFP: $[PF_6]$

G) Sulfate HSO4: [HSO$_4$], and Alkylsulfate: [RSO$_4$]
H) Sulfonate
- Methanesulfonate MSO: [CH$_3$SO$_3$]
- Tosylate TOS: [CH$_3$C$_6$H$_4$SO$_3$]
- Trifluoromethanesulfonate OTF: [CF$_3$SO$_3$]

I) Thiocyanate SCN: [SCN]
J) Tricyanomethide TCC: [C(CN)$_3$]

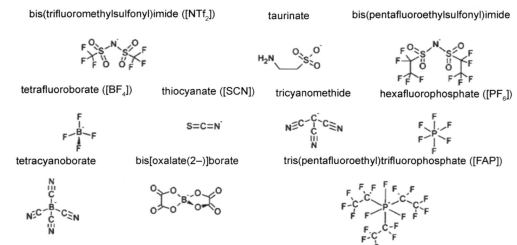

Ionic Liquids-Cations

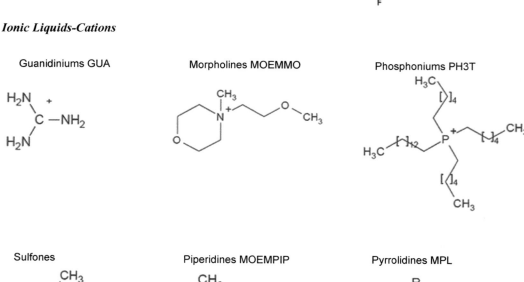

Preparation of Ionic Liquids

Ionic liquids are available in two main forms, namely, simple salts (made of a single anion and cation) and binary ionic liquids (salts where equilibrium is involved). For example, [EtNH$_3$][NO$_3$] is a simple salt whereas mixtures of aluminum(III) chloride and 1,3-dialkylimidazolium chlorides (a binary ionic liquid system) contain several different ionic species, and their melting point and properties depend upon the mole fractions of aluminum(III) chloride and 1,3-dialkylimidazolium chloride present. The synthesis of ionic liquids can be described in two steps (Figure 3): (1) formation of the desired cation and (2) anion exchange. The desired cation can be synthesized either by the protonation of the amine by an acid or through quaternization reactions of amine with a haloalkane and heating of the mixture. Anion exchange reactions can be carried out by treatment of halide salts with Lewis acids to form Lewis acid-based ionic liquids or by anion metathesis (Ratti 2014). The hydrophilicity/lipophilicity of an ionic liquid can be modified by a suitable choice of anion, e.g., bmimBF$_4$ is completely miscible with water while the PF$_6$ salt is largely immiscible with water. The lipophilicity of dialkylimidazolium salts, or other ionic liquids, can also be increased by increasing the chain length of the alkyl groups (Sheldon 2001).

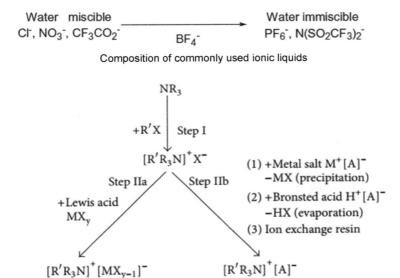

Figure 3. Synthesis path for the preparation of ionic liquids.

Example 1

The preparation of choline hydroxide is shown in Scheme 1. Equimolar amount of choline chloride and KOH is dissolved in methanol separately and charged into a round-bottom flask by stirring at ambient temperature. Then the reaction mixture is stirred at 60°C for 24 h followed by cooling down to room temperature (Reddy et al. 2014).

Example 2

The preparation of choline imidazole is shown in Scheme 2. Equimolar amount of choline chloride and sodium imidazole is dispersed in diethyl ether at ambient temperature. Then the reaction mixture is refluxed at 60°C with continuous stirring followed by cooling down to room temperature.

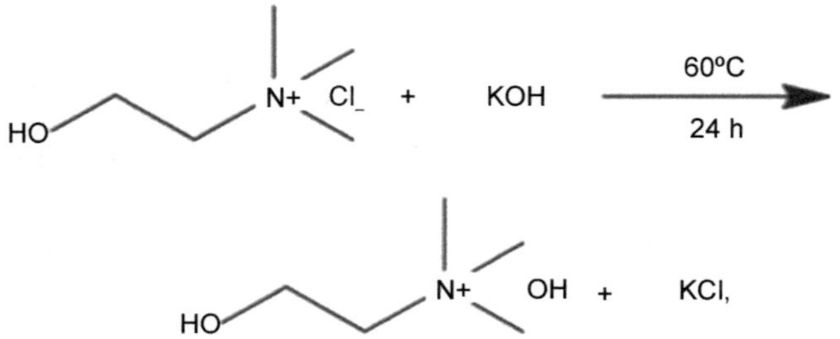

Scheme 1. Preparation of choline hydroxide using potassium hydroxide.

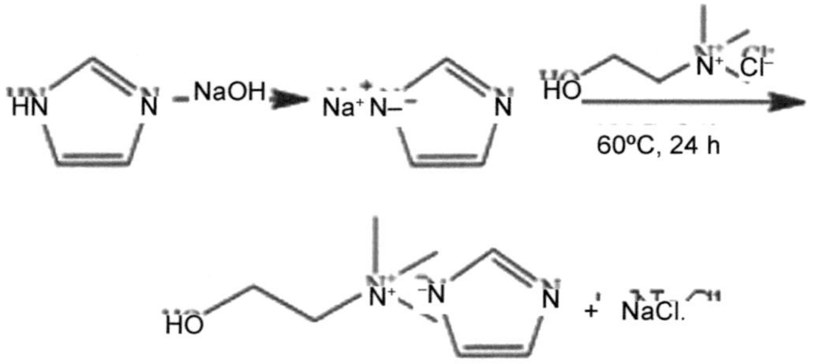

Scheme 2. Preparation of choline imidazole using sodium imidazole (adapted from Reddy et al. 2014).

Ionic Liquids and Microwaves

Many characteristics of ionic liquids render them suitable for applications in microwave assisted organic and inorganic synthesis. The main properties of ionic liquids are discussed below (Greaves and Drummond 2008, Suresh and Sandhu 2011, Ho et al. 2013):

Extremely low vapor pressure—Different from the classical organic solvents, ionic liquids are known to have a negligible vapor pressure below their decomposition temperature. This is the main reason why ILs are considered environmental friendly solvents.

Thermal Stability—Thermal stability of ionic liquids is limited by the strength of their heteroatom-carbon and their heteroatom-hydrogen bonds, respectively. The nature of the ionic liquids, containing organic cations, restricts upper stability temperatures. Pyrolysis generally occurs between 350 and 450°C, if no other lower temperature decomposition pathways are accessible. In most cases, decomposition occurs with complete mass loss and volatilization of the component fragments.

Solubility—Many ionic liquids possess the ability to dissolve a wide range of inorganic and organic compounds. This is important for dissolving disparate combinations of reagents into the same phase. Ionic liquids are considered polar solvents, but can be non-coordinating (mainly depending on the ionic liquid anions). The polarity of many ionic liquids is intermediate between water and chlorinated organic solvents and varies, depending on the nature of the ionic liquid components. Several processes would be impossible with conventional solvents because of their limited liquid range or miscibility. Even greater potential is in the use of ionic liquids for chemical synthesis because the charged nature of these solvents can influence the synthesis itself.

Electrochemical Stability—Ionic liquids often have wide electrochemical potential windows, and reasonably good electrical conductivity. The electrochemical window of an ionic liquid is influenced by the stability of the cation against electrochemical reduction-processes and the stability of the anion against oxidation-processes.

Catalytic properties—The catalytic properties in organic and inorganic synthesis have been widely described including biocatalytic transformations using a range of different enzymes, mainly in biphasic aqueous systems using hydrophobic dialkylimidazolium ionic liquids.

Polarity and hydrophilicity or lipophilicity—These properties can be readily adjusted by a suitable choice of cation/anion and ionic liquids have been referred to as 'designer solvents'. They are often composed of weakly coordinating anions, e.g., BF_4^- and PF_6^- and, hence, have the potential to be highly polar yet non-coordinating solvents. They can be expected, therefore, to have a strong rate-enhancing effect on reactions involving cationic intermediates. Ionic liquids containing chloroaluminate ions are strong Lewis, Franklin and Brønsted acids. Protons present in emimAlCl$_4$ have been shown to be superacidic with Hammett acidities up to 218. Such highly acidic ionic liquids are easily handled and offer potential as non-volatile replacements for hazardous acids such as HF in several acid-catalyzed reactions.

Ionic liquids have been utilized in a wide range of reactions resulting in production of various valuable chemicals and fuels. Table 2 shows the substances that can be produced from various processes utilizing ionic liquids (Zhao et al. 2005).

Table 2. Examples of ionic liquids and microwaves in organic synthesis (Zhao et al. 2005, Martínez-Palou 2007).

	Substances	Ionic Liquid	Reference
Phenolic compounds	Phthalic acid, aniline, 4-hydroxybenzoic acid, benzoic acid, *p*-toluic acid, benzene, chlorobenzene, 1,2,4-trichlorobenzene, 1,4-dichlorobenzene, 4,4´-dichlorobiphenyl	[BMIM][PF$_6$]	Huddleston et al. 1998
	Phenol, tyrosol, *p*-hydroxybenzoic acid	[C$_n$MIM][BF$_4$] (*n* = 1, 3, 6, 8, 10) [C$_n$MIM][PF$_6$] (*n* = 6, 10)	Vidal et al. 2004
	Chlorophenols Amino acids	[C$_4$MIM][PF$_6$], [EMIM][Beti]	Bekou et al. 2003
Amino acids	Tryptophan, glycine, alanine, leucine, lysine, arginine	[BMIM][PF$_6$]	Smirnova et al. 2004
Carbohydrates	Xylose, fructose, glucose, sucrose	[C$_n$MIM][X] (*n* = 4, 6, 8, 10; X = Cl$^-$, PF$_6^-$, BF$_4^-$)	Swatloski et al. 2002
	Glucose, sucrose, lactose, cyclodextrin	[BMIM][dca] (carbohydrate solubility is around 200 g L^{-1})	Liu et al. 2005
	Cellulose	[C$_n$MIM]X (*n* = 4, 6, 8)	Swatloski et al. 2002
Organic acids	Lactic acid, acetic acid, glycolic acid, propionic acid, pyruvic acid, butyric acid	[C$_n$MIM][PF$_6$] (*n* = 4, 6, 8)	Matsumoto et al. 2004
Biofuels	Butyl alcohol (from fermentation broth)	[BMIM][PF$_6$], [C$_8$MIM][PF$_6$]	Fadeev and Meagher 2001
Antibiotic	Erythromycin-A	[BMIM][PF$_6$]	Cull et al. 2000
Hydrocarbons	Olefins (such as ethylene, propylene, and butanes) from paraffins C4–8 diolefin (such as butadiene) from C1–18 paraffins	[C$_n$MIM][X], [HPy][X] (*n* = 4, 6; X = BF$_4^-$, PF$_6^-$) [BMIM][BF$_4$]	Munson et al. 2002, 2003 Smith et al. 2004

Ionic Liquids and Microwaves in Biofuel Production

Biofuel production involves several stages of feedstock processing which include pretreatment, extraction, conversion and separation of the process reaction mixtures and solvents. Among the biofuels discussed in earlier chapters of this book, biodiesel, bioethanol and biogas have been identified as the priority renewable fuels due to their availability, and their demands for various applications.

Biodiesel production essentially requires five steps: feedstock preparation and pretreatment, oil extraction, conversion of oils into biodiesel, separation and purification. Ionic liquids can be used as solvents in oil extraction and as catalysts in transesterification reaction. Lignocellulosic feedstock can be processed using ionic liquids to enhance hydrolysis and sugar recovery suitable for bioethanol and biogas production. Ionic liquids can be used in monophasic or biphasic or triphasic systems (Sheldon 2001). In monophasic systems, the substrate and catalyst are dissolved in the ionic liquid which serves as both a catalyst and a solvent. In biphasic systems, the catalyst is included in ionic liquid phase while the substrate and product remain in another phase. In a triphasic system, catalyst remains in the ionic liquid, the substrate and product in organic phase and the salts formed are absorbed aqueous phase. Examples of ionic liquids are as follow:

- Monophasic system—dialkylimidazolium chloroaluminates as Friedel–Crafts catalysts
- Biphasic system—a sulfonated phosphine ligand
- Triphasic system—Heck reactions

Ionic Liquids in Biodiesel Production

Ionic liquids can be used as catalysts and solvents in biodiesel reactions. To develop sustainable processes, a combination of environmentally friendly catalysts and solvents should be considered. Due to their unique and greener properties, ionic liquids were identified as a promising option to meet the requirement.

Ionic liquids as catalysts

Ionic liquids are non-volatile except at low pressures and high temperatures (Earle et al. 2006), and their properties can be altered to suit a particular process need. They are often referred to as "designer" solvents. An example of design is to alter the structure of ionic liquid such that its phase separates from the product of a reaction, making product isolation easier (Carmichael et al. 1999). Another approach is to make either the cation or anion or both of the ionic liquids acidic or basic (either Lewis or Brønsted). Specific examples of ionic liquids which exhibit Brønsted acidity due to functionalized cations include sulfonic acid derivatives of the imidazolium cation, e.g., 4-(3-methylimidazolium) butanesulfonic acid (Cole et al. 2002). This last approach could be used to enable the ionic liquid to catalyze esterification, transesterification, and specifically biodiesel forming reactions. The esterification reaction is an equilibrium process driven towards right by using an excess of methanol with water as the only by-product. Reaction schemes of vegetable oil esterification and transesterification reactions using ionic liquids are shown in Figure 4. The

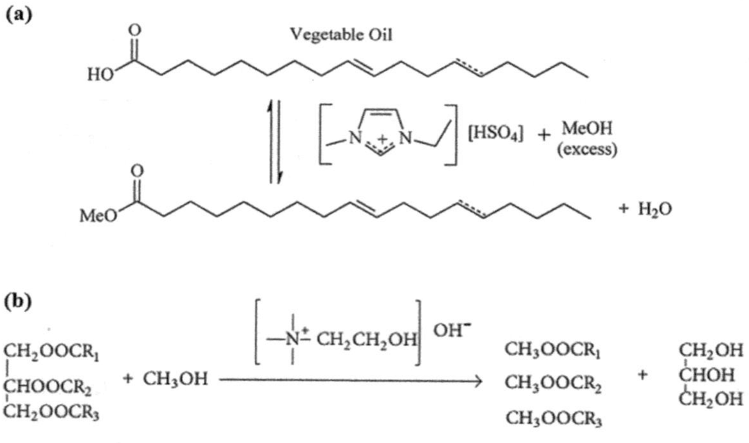

Figure 4. (a) Esterification of vegetable oil (Priolene 6927) with [emim][HSO$_4$] (Earle et al. 2009); (b) Transesterification using choline hydroxide ionic liquid (Fan et al. 2012, Ratti 2014).

advantage of this method is that the ionic liquid, water, and methanol mixture obtained at the end of the reaction is immiscible with the FAME product and forms a separate phase, allowing gravity separation of the biodiesel. Another advantage of this reaction is that the reaction occurs at room temperature and hence no energy input is required in this step.

Ionic liquids as solvent

Ionic liquids are non-volatile except at low pressures and high temperatures (Earle et al. 2006), and their properties can be altered to suit a particular need. Ionic liquids are mostly used in combination with enzymes in biodiesel production. The enzymatic transesterification method has many advantages over chemical methods which include mild reaction conditions, flexibility in selection of enzymes for various types of substrates, reusability of enzymes, reduced water requirement in substrates lowering the waste treatment.

However, the enzyme catalysts can be inactivated or inhibited by the alcohol in the transesterification reaction. In addition, other impurities in the oil, feedstock and glycerol produced in the reaction can reduce their reusability and catalytic effect (Muhammad et al. 2015). The enzymatic transesterification of vegetable oils in ionic liquids has been studied recently by many researchers (Li et al. 2009b, Yang et al. 2010, De Los Rios et al. 2011). Enzymes combined with imidazolium type ionic liquids produced more active and stable catalysts as the ionic liquids were able to protect the lipase from deactivation induced by methanol (De Diego et al. 2005).

Ionic liquids containing shorter chain 1,3-dialkylimidazolium cation (e.g., [BMIM][PF$_6$] or [BMIM][Tf$_2$N] or [BMIM][BF$_4$]) were used in a biphasic system which requires a certain amount of water. In this combination, a maximum biodiesel yield of 96.3% was reported from soybean oil (Gamba et al. 2008). Hydrophobic ionic liquids usually do not dissolve triglycerides and the lipase, resulting in a multi-phase reaction. Yang et al. (2010) studied the addition of a salt hydrate directly to the non-aqueous system and found to be unsuitable to meet the water requirement of the lipase-catalyzed reaction in [BMIM][PF$_6$]. The salt hydrate affected both the water buffering and the specific ion effect on the enzyme activities but the latter was found to be more dominant in the ionic liquid system. When the water is added to the originally biphasic system, it causes a formation of a third phase. Normally, after transesterification, the water phase will help in capturing the by-products and the acyl acceptor, thereby facilitating the recovery and reuse of the ILs. However, it was shown in a study that the lipase activity demonstrated a decline in its efficiency after the sixth cycle of reused (Ruzich and Bassi 2010).

Ha et al. (2007) screened 23 ionic liquids for methanolysis of soybean oil catalyzed by immobilized *Candida antarctica* lipase (Novozym1 435), and identified the hydrophilic ionic liquid (i.e., [EMIM][OTf]) as the best solvent for achieving the highest yield (80%) of fatty acid methyl esters based on 12 h reaction time. On the other hand, Sunitha et al. (2007) obtained 98–99% yields of fatty acid methyl esters within 10 h of methanolysis of sunflower oil in hydrophobic [BMIM][PF$_6$] and [EMIM][PF$_6$] when catalyzed by Novozym 435. Zhao et al. (2013) introduced choline-based deep eutectic solvents (such as choline chloride/glycerol at 1:2 molar ratio) as inexpensive, non-toxic, biodegradable and lipase-compatible solvents for the enzymatic preparation of biodiesel from soybean oil. Through the evaluation of different eutectic solvents and different lipases and a process parametric study (i.e., methanol concentration, Novozym 435 loading and reaction time), they achieved up to 88% triglyceride conversions in 24 h.

Reaction mechanisms of ionic liquids

A possible reaction mechanism of a cationic ionic liquid is shown in Scheme 2. Similar to the conventional biodiesel synthesis, cationic and anionic ionic liquids play their roles as regular acid or base catalysts. The acidic ionic liquid (cationic) behaves similar to the conventional acid catalyst in biodiesel production and it is suitable for feedstock that contain high free fatty acid content (FFA). In solid acid catalyst, the esterification occurs when a carbonyl carbon group of a triglyceride/FFA is protonated. Then a nucleophilic carbonium is attacked by an alcohol to form a tetrahydral intermediate, which further forms the fatty acid methyl ester. In the transesterification, a proton is attached to the α-carbonyl group of a triglyceride. And an alcohol attacks a carbocation to construct a tetrahydral intermediate that later breaks down to form FAME via proton mitigation.

The basic ionic liquid catalyst is similar to the conventional basic catalyst, in which OH⁻ takes H⁺ atom and converts methanol to CH_3O^-(methoxide), and then binds with the imidazolium cation of ionic liquid. A carbonyl group of triglyceride is attacked by CH_3O^-, to form a tetrahydral intermediate and re-arranges proton to produce fatty acid methyl esters or biodiesel. Several basic ionic liquid catalysts used in transesterification have high catalytic activities and produce yields up to 95%. Both acidic IL and basic ionic liquid can be recovered and reused up to 7 times without any significant reduction in the catalytic activity. Acidic ionic liquids seem to show high catalytic activity at high temperatures, whereas basic ionic liquid can be operated at milder temperatures. However, the Brønsted acidic ionic liquid can be operated at mild temperature when it has higher number of acid groups (e.g., 4 acid group).

A comparison of existing methods for the transesterification of vegetable oils is shown in Table 3. Process reaction parameters, yields, catalysts costs and recovery and environmental impacts were compared. It can be noted that transesterification processes involving enzyme catalysts with and without ionic liquids require longer reaction times when compared with other catalysts. However, they offer such benefits as easy separation and environmentally friendly production.

Fauzi and Amin (2013) optimized an esterification process using 1-butyl-3-methyl-imidazolium hydrogen sulfate (BMIMHSO$_4$) as a catalyst. The optimum conditions for biodiesel production were reported as a reaction temperature of 87°C, methanol:oil ratio of 9:1, catalyst loading of 6%, with a reaction time of 5.2 h. The obtained yield of methyl oleate and oleic acid conversion reached to 81.8% and 80.4%, respectively. In addition, acidic ionic liquid namely BMIMHSO$_4$, butylimidazolium hydrogen sulfate (BMIMHSO$_4$), methylimidazolium hydrogen sulfate (MIMHSO$_4$) were used as catalysts in esterification of waste cooking oil, and then followed by transesterification using an alkaline catalyst. When using BMIMHSO$_4$, 95.7% of FAMEs with longer alkyl chain were produced at optimal condition of methanol:oil ratio 15:1, 160°C, 1 h, and agitation speed at 600 rpm (Scheme 3).

Preparation of biodiesel from soybean oils catalyzed by five acidic ionic liquids with three cationic functional groups was investigated (Fan et al. 2012). The improvement of the catalytic activities was affected by various functional groups including pyridine group, Nmethylimidazole group, triethylamine group. Among them [C$_4$SO$_3$Hpy]HSO$_4$ with pyridine group showed better catalytic activity with the biodiesel yield of 94.5%, and still yielded more than 90% after six successive uses. The possible mechanism was also discussed by two reaction paths in detail.

The proposed reaction mechanism of ionic liquid catalyst [C$_4$SO$_3$Hpy]HSO$_4$ is illustrated in Scheme 4, based on the experimental studies by Fan et al. (2012). It is well known that the H⁺ of Brønsted acid acts as the active center on biodiesel synthesis through transesterification of methanol. In this case, the ionic liquid catalyst has two H⁺ of Brønsted acid. It is suggested that both H⁺ of Brønsted acid groups as active centers could interact with isolated electron pair at O atom in carbonyl group, which would promote two reaction paths. As shown in mechanism 1, the carbonyl group of the triglyceride is first attacked by H⁺ of cationic group of [C$_4$SO$_3$Hpy]HSO$_4$ catalyst. During this process, an intermediate with the carbocation will be produced. Then the intermediate product reacts rapidly with the nucleophilic reagent (CH$_3$OH) to produce methyl ester and diglyceride. Finally, the active species are reformed after H⁺ release. As a result, [C$_4$SO$_3$Hpy]HSO$_4$ catalyst shows good activity and stability in biodiesel synthesis. In mechanism 2, H⁺ of anionic group of ionic liquid catalyst follows the same pathway of H⁺ of cationic group. The presence of both H⁺ of cationic group and anionic group not only provides two active species but also promotes the activity and stability for the catalysts.

Biodiesel feedstock containing high FFA content should be processed through esterification and transesterification reactions to avoid soap formation. Ghiaci et al. (2011) used Brønsted acidic ionic liquid, 1-benzyl-1Hbenzimidazole and found that high number of Brønsted acidic sites provide higher yields in the transesterification reaction up to 95%. Another Brønsted acidic ionic liquid, triethylammonium hydrogen sulfate (Et$_3$NHSO$_4$), was also used as the catalyst for biodiesel production from crude palm oil (Man et al. 2013). Et$_3$NHSO$_4$ was able to convert 82.1% of FFA under a high temperature of 170°C for 3 h with 15:1 ratio of methanol:oil, followed by transesterification reaction, generating 96.9% biodiesel yield where 1.0% KOH was used as catalyst (under 60°C for 50 min with agitation speed of 600 rpm).

Table 3. Comparison of methods commonly used for the transesterification of vegetable oil with the use of ILs.

Catalyst	Feedstock	Reaction time (h), temperature (°C)	Yield (%)	Catalyst cost	Environmental impact	Catalyst recovery	Remarks	References
Ionic liquids (ILs)	Cotton seed, Jatropha	5–9, 100–170	92	Expensive	Low	Easy	Corrosive, new varieties requirement	Li et al. 2010c, Wu et al. 2006
ILs + acid/alkaline	Soybean	1.5–24, 72	93–98	Expensive	Moderate	Difficult	Acid/alkaline loss from ILs	Lian et al. 2009, Lapis et al. 2008
ILs + lipase	Soybean, corn	3–30, 25–60	80–96	Very Expensive	Low	Moderate	Lipase loss, increase of solvent recovery cost	Ha et al. 2007, Gamba et al. 2008, Yang et al. 2010, Zhang et al. 2011
Acid (Homogeneous)	Soybean	1–50, 65–120	>95	Cheap	High	Difficult	Corrosive, low catalytic activity	Leung et al. 2010, Helwani et al. 2009, Vyas et al. 2010, Shahla et al. 2010
Alkali (Homogeneous)	Soybean, Karanja, Pongamia pinnata, Rapeseed, Sunflower	0.5–10, 60–95	>96	Cheap	High	Difficult	Low free fatty acid and anhydrous requirement	Leung et al. 2010, Helwani et al. 2009, Vyas et al. 2010, Shahla et al. 2010, Robels-Medina et al. 2009
Heterogeneous acid/alkali	Soybean, Karanja, Palm, Jatropha, Babassu Rapeseed, Sunflower	0.5–10, 65–210	>95	Expensive	Moderate	Easy	Diffusion limitations	Leung et al. 2010, Helwani et al. 2009, Vyas et al. 2010, Shahla et al. 2010
Lipase	Soybean, Rapeseed, Cotton seed, Palm, Jatropha, Sunflower, Tallow	4–90, 25–70	>80	Expensive	Low	Difficult	Enzyme denaturation	Leung et al. 2010, Helwani et al. 2009, Vyas et al. 2010, Shahla et al. 2010, Bajaj et al. 2010, Al-Zuhair 2007
Immobilized lipase	Soybean, Sunflower, Palm, Jatropha, Karanja	8–60, 20–60	>90	Expensive	Low	Easy	Enzyme denaturation, Diffusion limitations	Leung et al. 2010, Helwani et al. 2009, Vyas et al. 2010, Bajaj et al. 2010, Tan et al. 2010

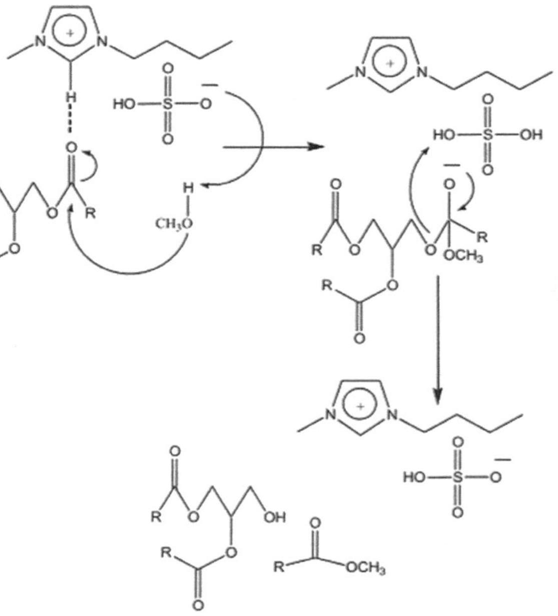

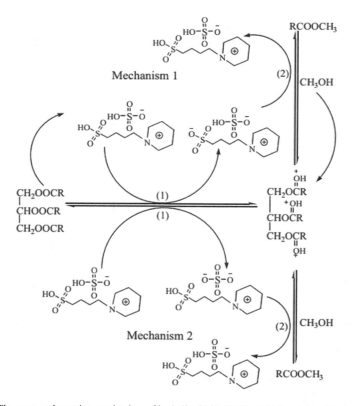

Scheme 3. Mechanism of the reaction between triglycerides and methanol catalyzed by ionic liquid BMIMHSO$_4$.

Scheme 4. The proposed reaction mechanism of ionic liquid [C$_4$SO$_3$Hpy]HSO$_4$ catalyst (Fan et al. 2012).

Although two-step biodiesel production has high efficiency, it leads to complications in the industrial-scale process. A one-step process was developed to reduce this complication by copolymerization between a novel Brønsted acidic polymeric IL, 3-vinyl-1-(4-sulfonic acid) butyl imidazolium hydrogen sulfate ([$SO_3H(CH_2)_3$ VIm]-[HSO_4]), and divinylbenzene (DVB) (Liang 2013). The polymeric ionic liquid can be used as a catalyst both in esterification and transesterification simultaneously and produce higher yield (99%) than acidic IL (less than 96%). The interaction between water and active site of the catalyst reduces the catalyst efficiency. However, the polymeric ionic liquid has large hydrophobic surface that prevents this interaction. In addition, high stability of polymeric ionic liquid enables recycling up to 6 times without significant change in biodiesel yield.

Effect of Functional Groups

Ionic liquids can be designed to include task specific groups which can provide superior benefits in any process. Five acidic ionic liquids including various functional groups were investigated as the possible catalysts for synthesis of biodiesel from soybean oil (Guo et al. 2013). As shown in Table 4, five acidic ionic liquids with various functional groups had shown different catalytic performance to a certain extent. When the number of carbon atom was same, the catalytic activity of ionic liquid was closely related with the functional group containing nitrogen element. The catalytic activities of ionic liquids with different functional groups were in the following order: pyridine > N-methy-limidazole > triethylamine. When the functional group was same, the catalytic activities of ionic liquids increased with the increase of the number of carbon atom. The [C_4SO_3Hpy] HSO_4 catalyst had shown favorable catalytic properties, and the yield of biodiesel was about 94.5%.

Synthesis of the Novel IL

Ionic liquids present some limitations (see Table 3) such as high cost, high viscosity, catalyst lost, and low purity of product. Synthesizing novel ionic liquids will help improve their benefits. A novel ionic liquid was synthesized from Choline Chloride·$xZnCl_2$, a deep eutectic solvent (DES), in lieu of using imidazolium to reduce the costs and simplify the preparation method (Long et al. 2010). However, biodiesel yield was only 54.5% (methanol to oil ratio of 16:1 with 10% catalyst for 72 hours). In another study, ammonium based DES was prepared by mixing N,N-diethylenethanol ammonium chloride (DEAC) with p-toluenesulfonic acid monohydrate (PTSA) to form DEAC-DES catalyst, for the pretreatment of low grade crude palm oil prior to the transesterification. The FFA in biodiesel was reduced to less than 1% which indicates the possibility of high FAME production during transesterification (Hama and Kondo 2013).

Hayyan et al. (2014) applied a similar procedure for esterification of crude palm oil with high FFA content using choline chloride-DES (ChCl-DES) which reduced the FFA content from 9% to less than 1% and transesterification of this treated crude palm oil resulted in 96% FAME yield. DES is considered inexpensive, biodegradable and non-toxic. In this reaction, DES mixes with methanol, which allows phase formation with the FAME. Saponification can be reduced due to FAME standing alone in a single phase limiting its interaction with the homogeneous base catalyst. The base catalyst dissolves in DES/methanol mixture which causes more interactions and increases the FAME yields up to 98%. Glycerol produced in the reaction can be deposited by DES. However, excess concentration of DES is undesirable due to the competition between glycerol and DES to interact with methanol. Excess volumes of DES to base catalyst leads to reaction inhibition which will decrease FAME production (Mendow et al. 2011).

Ionic Liquids and Microwaves in Biodiesel Production

Dipolar polarization and ionic conduction are two principles that are applied in the microwave method. Dipolar polarization occurs when dipoles of molecules align with the electrical field once the microwave introduced into the system. The oscillation of dissolved charged particles that are induced by the microwave is called as ionic conduction (Mazubert et al. 2014). Microwave induces the molecular agitation of atoms

Table 4. Effect of functional groups, and oil to ionic liquids ratios on biodiesel production (Fan et al. 2012, Guo et al. 2013, Muhammad et al. 2015).

Catalysts	Oil:ionic liquid (w/w)	Esterification (%)	Transesterification Biodiesel Yield (%)
[C$_3$SO$_3$HEt$_3$N]HSO$_4$*	5 wt.%. of oil		60.1
[C$_3$SO$_3$Hmim]HSO$_4$	5 wt.%. of oil		66.8
[C$_3$SO$_3$Hpy]HSO$_4$	5 wt.%. of oil		87.7
[C$_4$SO$_3$Hmim]HSO$_4$	5 wt.%. of oil		71.3
[C$_4$SO$_3$Hpy]HSO$_4$	5 wt.%. of oil		94.5
H$_2$SO$_4$	5 wt.%. of oil		97.3
BMIm PF$_6$[a,b]	1:2		97 ± 0.8
	1:1		98 ± 0.6
	4:3		95 ± 1
	2:1		38 ± 1.5
	4:1		-
EMIm PF$_6$	1:2		98 ± 0.8
	1:1		98 ± 1.6
	4:3		96 ± 1.4
	2:1		46 ± 3
	4:1		-
HMIm BF$_4$	1:2		10 ± 2.4
BMIm BF$_4$	1:2		-
[BMIm][TS]		35[x]	63.7[y]
[BMIm][TS]-LiCl		41	-
[BMIm][TS]-FeCl$_3$		90	85
[BMIm][TS]-AlCl$_3$		91	64
[BMIm][TS]-ZnCl$_2$		46	88
[BMIm][TS]-CuCl$_2$		86	52
[BMIm][TS]-CoCl$_2$		42	80
[BMIm][TS]-MnCl$_2$		45	87

*Reaction conditions: reaction temperature = 120°C, reaction time = 8 h, methanol:soybean ratio of 8:1, m(catalyst)/m(reactants) = 5 wt.%.

a) Reaction conditions: sunflower oil (1 g, average equivalence 1.3 mM), methanol(0.4 ml, 10.4 mM), *Candida antarctica* (10%, 0.1 g) at 60°C for 4 h. Values are reported as the mean of three independent reactions SD. b) [BMIm][BF$_4$], 1-butyl-3-methyl imidazolium tetrafluoroborate; [BMIm][PF$_6$], 1-butyl-3-methyl imidazolium hexafluorophosphate; [EMIm][PF$_6$], 1-ethyl-3-methyl imidazolium hexafluoro-phosphate; [HMIm][BF$_4$], 3-methyl imidazolium tetrafluoroborate.

x: Esterification rate (esters, %) (oleic acid) at 80°C a) Reaction conditions (in flask): oleic acid 42.3 g, IL 3 mmol [x(MCln) = 0.7], methanol/oleic-acid molar ratio 2/1, and reaction time 5 h.

y: (crude Jatropha oil) at 200°C, b) Reaction conditions: Jatropha oil 45 g (acid value of 13.8 mg KOH/g), IL 9 mmol [x(MCln) = 0.7], methanol/Jatropha-oil molar ratio 12/1, initial N$_2$ pressure 2.8 MPa, reaction pressure 4.8 MPa and reaction time 5 h.

through dipole rotation or ion migration that promotes collision and molecular attrition (Teixeira et al. 2014). Collision and attrition of molecular movement generates heat, which speeds up chemical reaction by increasing the kinetic energy, and resulting in short reaction time. The polar molecules of vegetable oil, methanol and ionic liquid can absorb the microwave energy. In addition, the ionic and polar components in the system help distribute heat rapidly via conduction and convection (Kanitkar et al. 2011). In this case, the FAME yield in the microwave heating system is higher than the conventional method when both processes were performed with ionic liquids.

A schematic representation of the transesterification process for yellow horn oil biodiesel production is shown in Figure 5 (Zhang et al. 2016) in which an enzyme catalyst (Novozym 435—*Candida antarctica* lipase B immobilized on polyacrylic resin) was used in combination with a deep eutectic solvent under microwave irradiation. Several combinations of DESs were evaluated. The heterogeneous reaction mixture turns into a pseudo-homogeneous system after stirring due to the good compatibility of DESs with substrates and enzymes. The system is then subjected to microwave irradiation to complete transesterification of oil using methanol and enzyme catalyst. The biodiesel produced is insoluble in DESs, eliminating possible side-reactions. The byproduct glycerol is mostly captured by DESs during the reaction, which shifts the reaction equilibrium to the product side so as to increase the biodiesel yield. After the reaction is completed, the mixture is centrifuged to form a single phase layer of biodiesel at upper level, a middle layer containing enzymes, and a third bottom layer with a mixture of DESs, glycerol, methanol, residual oil and other byproducts. Thus, the apparent layering can lead to easy and efficient separation of products and enzymes from reaction mixture.

The mechanism of the transesterification reaction catalyzed by Novozym 435 is shown in Figure 6. In the first stage of the reaction, the triglyceride which acted as acyl-donor is connected with enzyme and acylated enzyme is formed. Then in the second stage, the acylated enzyme is combined with methanol which acted as acyl-acceptor, resulting in FAME production and the enzyme regeneration occurring in the next round of reaction. The diglyceride generated in first stage continued to react with enzyme, starting another catalytic cycle.

Several studies demonstrated the use of ionic liquids in biodiesel production. Lin et al. (2013) conducted the transesterification of waste cooking oil using ionic liquid as a catalyst with microwave heating. The 4-allyl-4-methylmorpholin-4-iumbromine ([MorMeA][Br]) was combined with NaOH. They found that

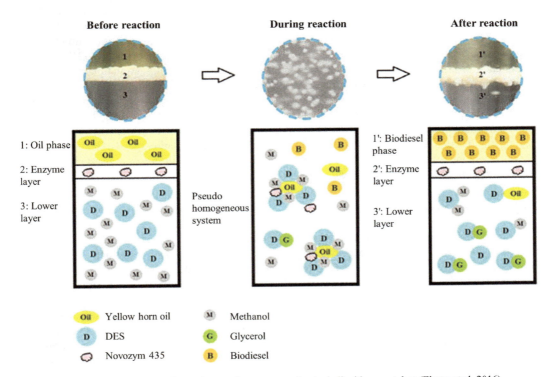

Figure 5. Biodiesel production and separation process using ionic liquid as a catalyst (Zhang et al. 2016).

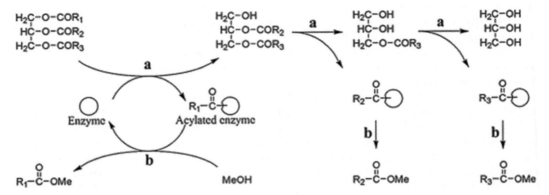

Figure 6. Proposed mechanism of enzymatic transesterification reaction (Zhang et al. 2016).

increasing the amount of ionic liquid with constant amount of NaOH enhanced the biodiesel yield up to 98.1% under microwave heating for 7 min at 70°C and a methanol:oil ratio of 9:1. However, higher methanol:oil ratio, and longer reaction time together reduced the biodiesel yield due to excess glycerol production. Heteropolyacid-based (HPA) ionic liquid derived from choline-chloride was also evaluated as heterogeneous catalyst in esterification process under microwave irradiation (Duan et al. 2013). The ionic liquid catalyst combined good properties of homogeneous and heterogeneous catalysts, and achieved up to 97% conversion under microwave heating. Another ionic liquid catalyst, 2,3-dimethylimidazolium hydrogen sulfate (MMBIMHSO$_4$), also produced comparable FAME yields with shorter retention time under the microwave heating when compared with conventional heating (Kotadia and Soni 2013).

Microwave heating efficiency was evaluated in a system of enzymatic transesterification by Novozyme 435 with several different types of ionic liquids. An optimum yield of transesterification was obtained when using 1-ethyl-3-methy limidazolium hexafluorophosphate (EMIMPF$_6$) as the reaction medium with a [EMIM][PF$_6$]/oil volume ratio of 2:1 at a temperature of 60°C. After 6 h of reaction with 6% (of oil weight) Novozyme 435 under microwave irradiation at 480 W, the FAME yield reached 92%. However, the agitation of mixture was necessary due to high viscosity of ionic liquids. Agitation speed between 50–200 rpm enhanced the rate of reaction as interfacial area was increased significantly (Yu et al. 2011).

Ullah et al. (2015) performed a two-step esterification and transesterification of waste cooking oil. The ionic liquid butyl-methyl imidazolium hydrogensulfate (BMIMHSO$_4$) was found to be very effective with a maximum biodiesel yield of 96%. The esterification step at 5 wt.% BMIMHSO$_4$, methanol:oil of 15:1, 60 min reaction time at a reaction temperature of 160°C, and an agitation speed of 600 rpm, reduced the waste cooking oil acid value to lower than 1.0 mg KOH/g. The second step of transesterification was catalyzed by 1.0 wt.% KOH at 60°C, and 60 min of reaction time.

Ionic Liquid Recovery

Ionic liquids are more expensive than other conventional acids and base catalysts including homogeneous and heterogeneous catalysts and other conventional solvents. It is important to develop ionic liquids suitable for easy separation so that they can be recycled in the process. The design of ionic liquids should also consider eliminating any potential ionic liquid waste. Vacuum distillation and solvent extraction are the methods available to separate the ionic liquids from a reaction mixture. In biodiesel production, excess methanol can be separated by stripping with warm air in lieu of expensive conventional distillation process. Vacuum distillation is another preferred method to separate ionic liquids from glycerol. Because the boiling points of ionic liquids are much higher than glycerol. Solvent extraction can be used to separate mixtures of organic solvents (*n*-butanol, ethyl ethanoate, or dichloromethane) and water. The ionic liquid would preferentially dissolve in the organic solvents, especially if it is a hydrophobic ionic liquid and glycerol would prefer the aqueous phase; and finally, the best approach is to design an ionic liquid in such a way that it is not soluble in either biodiesel or glycerol, so that the triphasic system is obtained at the end of the reaction. More work is required in this direction.

Ionic liquids enable simple recovery due to generation of biphasic or triphasic systems once the biodiesel synthesis is completed. Several types of ILs were studied to stabilize the lipase enzyme from methanol deactivation in transesterification (De Diego et al. 2011). After the transesterification, triphasic system was formed, which consists of 3 different layers of FAME at the upper layer, glycerol and unconverted methanol in the middle layer, and IL and enzyme in the bottom layer. A decantation-vacuum separation-filtration treatment process was employed by Soni et al. (2014) with MMBIMHSO$_4$ as a catalyst in biodiesel production. Effective recovery of ionic liquid can be achieved based on its low melting point property. FAME, methanol, glycerol, and ionic liquid mixture was subjected to the aforementioned separation process. FAME was collected by simple decantation, while methanol was separated under vacuum, which left the glycerol and ionic liquid in the system. The ionic liquid was recovered from glycerol by filtration.

Ionic liquids are classified as molten salt, and the recovery as a solvent could be possible by the crystallization process (Hayyan et al. 2010). Furthermore, the non-volatile and thermal stability property of ionic liquid enables the thermal-based separation, such as distillation. Super critical CO_2 ($scCO_2$) can also be used to separate ionic liquids from organic compounds as where organic material is highly soluble in $scCO_2$, while ionic liquid is insoluble in $scCO_2$ and settles down to the bottom of the reactor (Mai et al. 2014). The adsorption separation of ionic liquid using activated carbon is applicable according to the hydrophilic or hydrophobic content of ionic liquid. If ionic liquid is a highly polar type, the activated carbon should contain high polarity on its surface to occupy the hydrogen bonding between adsorbate and adsorbent (Lemus et al. 2012). Recycling the ionic liquids as catalysts or solvent is another option to improve the economics. The recyclability of ionic liquid was found up to 6 times without significant reduction of the yield, and no ionic liquid degradation was observed (Lozano et al. 2015). Other studies also found that ionic liquid catalyst can be reused for 7 times of operation with high yield of biodiesel. However, it can be noticed that at the optimal condition of ionic liquid recycle, the decline in biodiesel yield was observed (Ghiaci et al. 2011). It was explained that the decreasing yield was affected by the remaining amount of glycerol in ionic liquid, which decreased the ionic liquid activity by the reduction of active sites (Lin et al. 2013).

Ionic Liquids and Microwaves in Bioethanol Production

Pretreatment of lignocellulosic feedstock is the primary step in bioethanol production. Pretreatment processes should increase the surface area accessible to water (largely excluded due to the packing of cellulose fibrils) and provide more accessible enzyme binding sites to substantially increase glucose and xylose yields (Bradshaw et al. 2007). Therefore a pretreatment process should consider (a) conversion of the initial lignocellulosic material into an easily hydrolysable format and (b) elimination of any degradation products (Chen et al. 2007a) or inhibitory compounds that will negatively impact the downstream processing of that treated biomass into biofuels (Berson et al. 2006). Ionic liquids show great promise to meet all of these requirements.

There are several possible basic strategies for ionic liquid pretreatment of lignocellulosic biomass: (1) cellulose dissolution in processed cellulose materials such as pulp cellulose in an ionic liquid, (2) lignin (and hemicellulose) removal by dissolving the lignocellulosic biomass in an ionic liquid while leaving cellulose behind, and (3) dissolving all the biomass components and then selectively recovering the needed component(s). An effective ionic liquid for cellulose dissolution in biomass pretreatment should satisfy the following three criteria:

- Its anion is a good hydrogen bond acceptor,
- Its cation should be a moderate hydrogen bond donator, and
- Its cation should not be too bulky.

From the perspective of process engineering, a good ionic liquid for cellulose dissolution should have:

- High cellulose solubility,
- Low melting point,
- Low viscosity,
- Good stability, and
- Low toxicity (Sun et al. 2011).

Ionic Liquids and Microwaves in Cellulose Dissolution

Cellulose is a polysaccharide consisting of a linear chain of several hundred to over ten thousand b-(1-4)-linked glucose repeating units, and there are numerous intermolecular and intramolecular hydrogen bonds in cellulose (Figure 7), which result in the difficulty in dissolving cellulose in water and common organic solvents, much less separating it from natural lignocellulosic biomass. A few aqueous and non-aqueous cellulose solvents were previously developed for the dissolution of cellulose, but all of these solvents suffer drawbacks, such as high cost, volatility, toxicity, generation of poisonous gas, difficulty in solvent recovery, or insufficient solvation power for cellulose (Olivier-Bourbigou et al. 2010, Wang et al. 2012).

The mechanism of cellulose dissolution in ionic liquids involves the oxygen and hydrogen atoms of cellulose hydroxyl groups, which form electron donor-electron acceptor complexes that interact with ionic liquids (Olivier-Bourbigou et al. 2010, Wang et al. 2012). Upon interaction between cellulose's hydroxyl groups and ionic liquids, hydrogen bonds are broken, leading to opening of the hydrogen bonds between molecular chains of the cellulose, resulting in cellulose dissolution. Solubilized cellulose can be rapidly precipitated with anti-solvents such as ethanol, methanol, acetone, or water. The recovered cellulose was found to have the same degree of polymerization and polydispersity as the initial cellulose, but significantly different macro- and micro-structures, especially decreased crystallinity and increased porosity.

Some of the common solvents used for cellulose dissolution are shown in Figure 8. Experimental data suggest that with the same cation, the decreasing order for cellulose solubility for different anions is: $(CH_3CH_2O)_2PO_2^- \approx OAc^- > SHCH_2COO^- > HCOO^- > Cl^- > Br^- \approx SCN^-$ (Swatloski et al. 2002). The

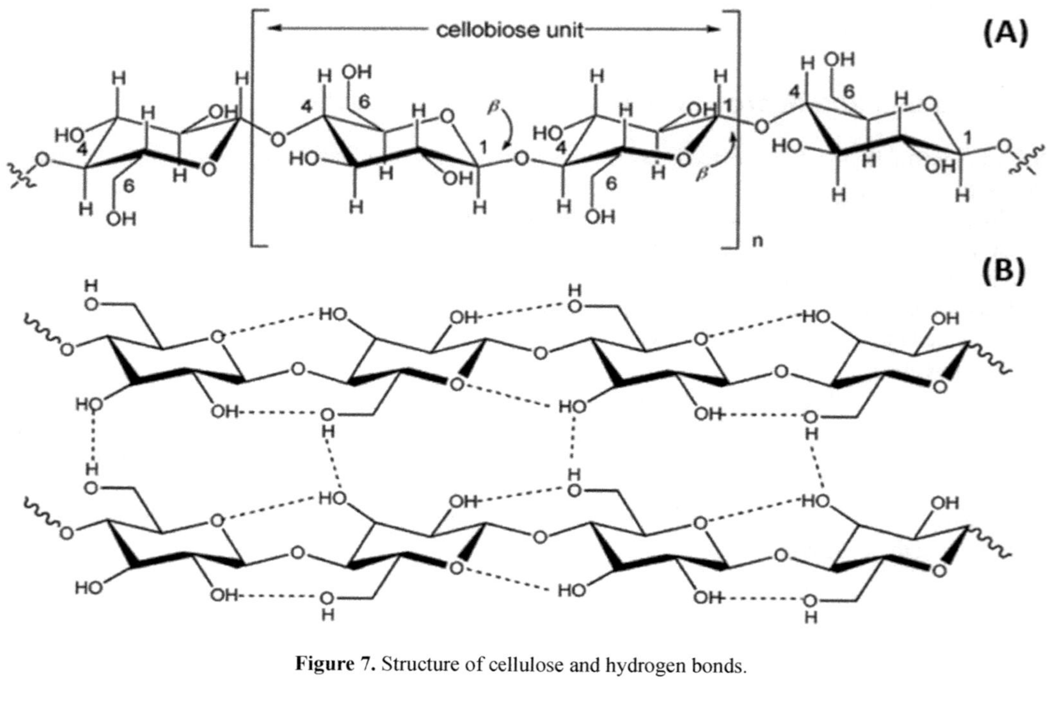

Figure 7. Structure of cellulose and hydrogen bonds.

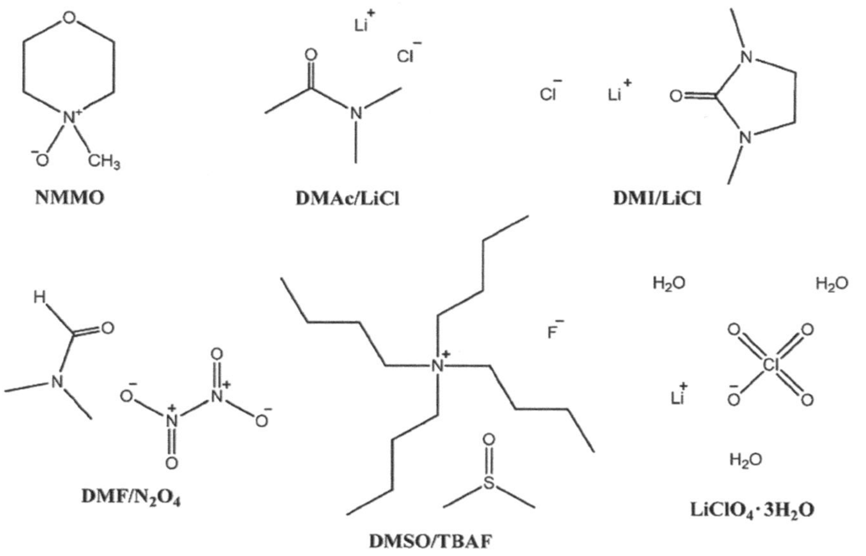

Figure 8. Structure of traditional solvents for cellulose dissolution (N-methylmorpholine oxide (NMMO), N,N-dimethylacetamide/lithium chloride (DMAc/LiCl), 1,3-dimethyl-2-imidazolidinone/lithium chloride (DMI)/LiCl,N,N-dimethylformamide/nitrous tetroxide (DMF/N$_2$O$_4$), dimethyl sulfoxide (DMSO)/tetrabutylammonium fluoride (TBAF), some molten salt hydrates, such as LiClO$_4$•3H$_2$O, LiSCN•2H$_2$O, and some aqueous solutions of metal complexes.

most effective cations for cellulose dissolution are found to be those based on the methylimidazolium and methylpyridinium cores, and contain allyl-, methyl-, ethyl-, or butyl- side chains (Wang et al. 2012). Some ionic liquids can be used to remove lignin from biomass while leaving behind cellulose. For example, [C$_2$mim]OAc removes lignin from triticale (a hybrid of wheat and rye) straw efficiently at 70–150°C in 1.5 h with some hemicellulose removal, but little of cellulose (Fu et al. 2010).

Compared with other volatile organic cellulose solvents, ionic liquids possess the advantages of low toxicity, low hydrophobicity, low viscosity, thermal stability, broad selection of anion and cation combinations, enhanced electrochemical stability, high reaction rates, low volatility with potentially minimal environmental impact, and non-flammable properties (Zheng et al. 2014). A comparison of ionic liquids with other methods used for lignocellulosic biomass pretreatment is shown in Table 5. Recovery of chemicals and solvents from ionic liquids based pretreatment products is easier than other processes. Energy costs are lower when compared with most other processes but the chemical costs are higher than other methods. Solubilities of various lignocellulosic feedstock are presented in Table 6.

The combination of ionic liquids, as novel solvents, and microwaves, as an efficient energy source, has been studied intensively for numerous organic syntheses, including cross-coupling reactions and olefin metathesis. When compared with conventional heating, microwave irradiation produces higher yields and significant reduction of reaction times, from several hours to several minutes (Leadbeater et al. 2004). Cellulose dissolution in ionic liquids with the aid of microwave heating has been reported frequently as well (Swatloski et al. 2002, Vitz et al. 2009, Sun et al. 2009). As shown in Table 7, microwave heating increases the cellulose dissolution when compared with ultrasonication and conventional heating methods using same ionic liquid. It should also be noted that some ionic liquids may not dissolve cellulose under microwave irradiation.

Ha et al. (2011) found that microwave heating not only increased cotton cellulose solubility in ionic liquids, but also reduced the degree of polymerization in the regenerated cellulose obtained after the pretreatment compared with pretreatment at 110°C without microwave. They suggested that internal heating by microwave irradiation was more effective for polar solvents such as ionic liquids. Casas et al. (2012) used microwave irradiation as a thermal source to reduce dissolution time required for *Pinus radiata* and *Eucalyptus globulus* woods in the following ionic liquids: [C$_2$mim]OAc, [C$_4$mim]OAc, [C$_2$mim]Cl,

Table 5. Comparison of methods commonly used for the pretreatment of lignocellulose with the use of ILs.

Method	Operation temperature (°C)	Inhibitor production	Recovery of chemicals	Equipment cost	Chemical cost	Energy cost	Remarks	Reference
Ionic liquids (ILs)	80–130	Low	Easy	Low	Very high	Low	Commercial recovery methods and equipment are not currently available	Pu et al. 2007; Tan et al. 2009; Liu and Chen 2006; Li et al. 2009a; Li et al. 2010a; Li et al. 2010b; Nguyen et al. 2010; Shill et al. 2011
Mechanical disruption	Room Temperature	—	—	Moderate	—	Very high	Pretreatment for other methods	Alvira et al. 2010; Banerjee et al. 2010; Mora-Pale et al. 2011
Dilute acid	120–140	High	Difficult	High	High	Moderate	Moderate formation of degradation products	Pedersen and Meyer 2010; Alvira et al. 2010; Banerjee et al. 2010; Mora-Pale et al. 2011
Alkaline (lime)	54–160	Moderate	Difficult	Low	High	Moderate	Long reaction time (3–7 days)	Pedersen and Meyer 2010; Alvira et al. 2010; Banerjee et al. 2010; Mora-Pale et al. 2011
Organosolv	100–250	Low	Difficult	High	High	Moderate	Biologic inhibition of residual solvent	Pedersen and Meyer 2010; Alvira et al. 2010; Banerjee et al. 2010; Mora-Pale et al. 2011
Ammonia fiber explosion	50–180	Low	Easy	Difficult	Very high	High	Low digestibility for high-lignin containing materials	Pedersen and Meyer 2010; Alvira et al. 2010; Banerjee et al. 2010; Mora-Pale et al. 2011
Steam explosion	160–230	High	—	Very high	Low	Very high	Hemicellulose is partly degraded	Pedersen and Meyer 2010; Alvira et al. 2010; Banerjee et al. 2010; Mora-Pale et al. 2011

Oxidative (ozone)	RT	Low	Moderate	Low	Very high	Low	Large consumption of ozone	Pedersen and Meyer 2010; Alvira et al. 2010; Banerjee et al. 2010; Mora-Pale et al. 2011
Wet oxidation	175–195	Low	–	Very high	Low	Moderate	High pressure and cost of oxygen	Pedersen and Meyer 2010; Alvira et al. 2010; Banerjee et al. 2010
Biological degradation	28–40	Low	–	Low	–	Low	Low hydrolysis rate	Pedersen and Meyer 2010; Alvira et al. 2010; Mora-Pale et al. 2011

Table 6. Solubility of lignocellulose in ionic liquids (Liu et al. 2012).

Ionic liquids	Reaction conditions	Lignocellulose	Solubility	References
[AMIM]Cl	90°C, 12 h	a-Cellulose, Spruce, Common beech, Chestnut	Soluble	Zavrel et al. 2009
[AMIM]Cl	80°C, 8 h	Ball-milled southern pine powder	8 wt.%	Kilpeläinen et al. 2007
[AMIM]Cl	110°C, 8 h	Norway spruce sawdust	8 wt.%	Kilpeläinen et al. 2007
[AMIM]Cl	80°C, 24 h	Norway spruce sawdust	5 wt.%	Kilpeläinen et al. 2007
[AMIM]Cl	110°C, 8 h	Southern pine TMP	2 wt.%	Kilpeläinen et al. 2007
[AMIM]Cl	90°C, 24 h	Indulin AT (kraft lignin)	30 wt.%	Lee et al. 2009
[BMIM]Cl	90°C, 12 h	a-Cellulose	Soluble	Zavrel et al. 2009
[BMIM]Cl	90°C, 12 h	Spruce, Common beech, Chestnut	Partially soluble	Zavrel et al. 2009
[BMIM]Cl	110°C, 8 h	Norway spruce sawdust	8 wt.%	Kilpeläinen et al. 2007
[BMIM]Cl	130°C, 8 h	Southern pine TMP	5 wt.%	Kilpeläinen et al. 2007
[BMIM]Cl	110°C, 16 h	Southern yellow pine	52.6 wt.%	Vanoye et al. 2009
[BMIM]Cl	130°C, 15 h	Wood chips	Partially soluble	Kilpeläinen et al. 2007
[BMIM]Cl	90°C, 24 h	Indulin AT (kraft lignin)	10 wt.%	Lee et al. 2009
[EMIM]Cl	90°C, 12 h	a-Cellulose	Soluble	Kilpeläinen et al. 2007
[EMIM]Cl	90°C, 12 h	Spruce	Partially soluble	Kilpeläinen et al. 2007
[EMIM]Cl	90°C, 12 h	Common beech, Chestnut	Partially soluble	Zavrel et al. 2009
[EMIM]OAc	90°C, 12 h	a-Cellulose, Spruce, Common beech, Chestnut	Soluble	Zavrel et al. 2009
[EMIM]OAc	90°C, 24 h	Indulin AT (kraft lignin)	30 wt.%	Lee et al. 2009
[EMIM]OAc	110°C, 16 h	Southern yellow pine	5 wt.%	Li et al. 2009b
[EMIM]OAc	110°C, 16 h	Red oak	5 wt.%	Sun et al. 2009
[Bemim]Cl	130°C, 8 h	Indulin AT (kraft lignin)	5 wt.%	Kilpeläinen et al. 2007
[Bemim]Cl	130°C, 8 h	Norway spruce TMP	5 wt.%	Kilpeläinen et al. 2007
[Bemim]Cl	90°C, 24 h	Southern pine TMP	10 wt.%	Lee et al. 2009

Table 7. A comparison of heating methods on cellulose dissolution (Swatloski et al. 2002).

Ionic Liquid	Method	Solubility (wt. %)
[C$_4$mim]Cl	heat (100°C)	10%
	(70°C)	3%
[C$_4$mim]Cl	heat (80°C) + sonication	5%
[C$_4$mim]Cl	microwave heating	25%, clear
	3–5 s pulses	viscous solution
[C$_4$mim]Br	microwave	5–7%
[C$_4$mim]SCN	microwave	5–7%
[C$_4$mim][BF$_4$]	microwave	insoluble
[C$_4$mim][PF$_6$]	microwave	insoluble
[C$_6$mim]Cl	heat (100°C)	5%
[C$_8$mim]Cl	heat (100°C)	slightly soluble

[C$_4$mim]Cl, and [Amim]Cl. Despite its advantages, microwave heating also has significant drawbacks. Apart from increased equipment cost and scale-up difficulties, another drawback for microwave heating is that uneven heating may cause pyrolysis of cellulose due to high local temperatures (Feng and Chen 2008).

Confocal fluorescence images of switchgrass dissolution in ethyl methyl imidazolium acetate ionic liquid under microwave irradiation are shown in Figure 9 (Singh et al. 2009). Pretreatment with EmimAc (ethyl methyl imidazolium acetate) caused swelling of the cell wall by a factor of 8× within 10 min of

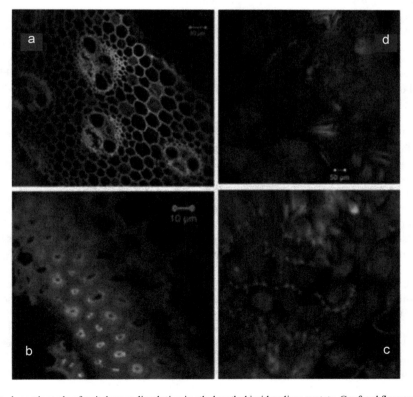

Figure 9. *In situ* dynamic study of switchgrass dissolution in ethyl methyl imidazolium acetate. Confocal fluorescence images of switchgrass stem section before pretreatment (a), and after 20 min (b) and 50 min of pretreatment. (c) Complete breakdown of organized plant cell wall structure (d) is observed after 2 h (Singh et al. 2009).

pretreatment. This swelling may be a result of breaking the inter- and intramolecular hydrogen bonding responsible for the rigid and highly compact crystalline cellulose polymer structure within biomass. The highly polar ionic liquid EmimAc may disrupt the hydrogen bonding interactions in biomass due to interactions between the ionic liquid anion and hydroxyl group hydrogen atoms within the biomass. Remsing et al. (2006) proposed a mechanism, in which hydrogen bonding between the carbohydrate hydroxyl protons and the ionic liquid anions are the critical factors in the ionic liquid solubilization process.

As shown in Figure 9a, EmimAc is able to completely solubilize both the cellulose and lignin in switchgrass. The lignin-rich sclerenchyma and middle lamella (Figure 9a) in plant tissues disintegrate immediately and simultaneously as primary and secondary cell walls start separating from the middle lamella (Figure 9b). Within 30–50 min of pretreatment at 120°C (Figure 9c), intact stem sections formed a very viscous solution, and complete solubilization was accomplished after 2.5–3 h (Figure 9d).

Ionic Liquids in Biogas Production

Ionic liquids can be used for pretreatment of lignocellulosic biomass in biogas production. They are also attractive because large amounts of cellulose can be dissolved at mild conditions (90–130°C and ambient pressure) with low energy inputs, and it is feasible to recover nearly 100% of the used ionic liquids to its initial purity and leave minimum residues for the downstream anaerobic digestion (AD) process. Various ionic liquids, including N-methylmorpholine-N-oxide monohydrate (NMMO), 1-n-butyl-3 methylimidazolium chloride (BMIMCl), 1-allyl-3-methylimidazolium chloride, 3-methyl-Nbytylpyridinium chloride (MBPCl), and benzyldimethyl (tetradecyl) ammonium chloride, have been studied for pretreatment of lignocellulosic biomass to improve enzymatic digestibility. Ionic liquid pretreatment achieved significant enhancement of sugar yield from enzymatic hydrolysis (Dadi et al. 2006, Liu and Chen 2006). Therefore, it is expected to have similarly positive effect on the improvement of biogas production from lignocellulosic biomass in the AD process.

Among the previously tested ionic liquids, NMMO has been the most frequently applied ionic liquid for biomass pretreatment for biogas production. A NMMO water solution with a concentration of 83–87% (w/w) has been defined as the "dissolution mode", in which cellulose fibers can be completely disintegrated and dissolved. When NMMO concentration decreases to 76–82%, the cellulose fibers exhibit ballooning with partial dissolution inside the balloons. In a NMMO-water mixture containing 70–75% NMMO, only the swelling and partial ballooning were observed. Further dilution of the NMMO solution had no effect on cellulose fibers (Cuissinat and Navard 2006).

Akhand and Méndez Blancas (2012) optimized NMMO treatment on wheat straw and found the highest methane yield was 470 L/kg VS (47% higher than that of untreated straw), which was achieved after 7 h treatment with 85% NMMO at 90°C. Decreased crystallinity and increased porosity of treated straw were observed.

Teghammar et al. (2012) conducted NMMO treatment on softwood spruce, rice straw, and triticale straw (a hybrid of rye and wheat) and reported that methane yields of tested materials increased by 400e 1200% and digestion time was also shortened to 1–1.5 weeks. Longer treatment time usually produced better results, except for rice straw, which showed a decrease of methane yield by 3.6-fold as the treatment time increased from 1–3 h to 15 h. Traditional ethanol fermentation inhibitors such as furans and phenolic compounds were not detected.

Jeihanipour et al. (2010a) carried out NMMO pretreatment on high-crystalline cellulose (defatted and bleached cotton linter) and obtained the best results of biogas production from the cellulose treated with 73% NMMO (swelling mode) and 79% NMMO (ballooning mode) at 90°C for 1 h. The NMMO treatment significantly increased the methane production rate and yield, with the 99% theoretical methane yield achieved in 15 days of digestion, and the methane yield increased by 16% compared with untreated material.

Purwandari et al. (2013) indicated that the best improvement in biogas production was achieved by a dissolution mode pretreatment of oil palm empty fruit bunch, using conditions of 85% NMMO, 3 h, and 120°C that resulted in 48% enhancement of methane yield. Using NMMO to treat straw separated from manure (120°C for 15 h), Aslanzadeh et al. (2011) achieved 53% and 51% more methane yield from pretreated straw separated from cattle and horse manure, respectively, compared with untreated samples. The lignin and carbohydrate contents decreased by 10% and increased by 13%, respectively. The crystallinity of straw decreased with increased treatment time. The biogas yield from anaerobic digestion of waste textiles was doubled by pretreatment with 85% NMMO at 120°C for 3 h (Jeihanipour et al. 2013).

Other research on NMMO pretreatment of cotton linter, waste textiles, and spruce wood also achieved positive results on the improvement of biogas production and/or reactor rate (Jeihanipour et al. 2010b). One study reported on the use of both BMIMCl and 1-ethyl-3-methlyimidazolium acetate for biogas production. Ionic liquids and co-solvents were used to treat water hyacinth and biogas yield increased by 97.6% when compared with untreated water hyacinth (Gao et al. 2013). Meanwhile, cellulose content increased by 27.9% and 49.2% of the lignin content was removed.

Unlike physical and biological pretreatment, chemical pretreatment may leave chemical residues which may influence the downstream anaerobic digestion (AD) process. Some chemical residues, such as alkali residues, which can increase alkalinity (buffering capacity) of the digester and help to stabilize the pH, may be beneficial to the AD process, especially at high organic loading rates. On the contrary, acid residues may reduce the pH of the digester and negatively affect the AD processes that follow. Na^+ ions were reported to inhibit the AD process for feedstocks pretreated with NaOH. In oxidative pretreatment, oxidants such as ozone almost resulted in zero residues, while excessive peroxides such as H_2O_2 remained in the feedstock and might have caused inhibition to methanogens and other microorganisms. For ionic liquid pretreatment, it is possible to recover nearly 100% of the used ionic liquid, so chemical residues might not be a problem. Currently, few publications have been found on the effect of chemical pretreatment residues on the AD process. However, compared to ethanol fermenting microorganisms, AD microbes are commonly used for treating wastes with high levels of inhibitors so that they may adapt to the residues better and faster. However, future investigations are still needed to quantitatively evaluate these inhibitory effects on AD. Pretreatment methods which generate minimum inhibitory residues, byproducts, and waste streams, such as biological pretreatment, should be more intensively studied.

Ionic Liquids in Preparation of HMF, Furfural and Levulinic Acid

Ionic liquids were also used in HMF, furfural, and levulinic acid production under microwave heating. A comparison of 5-HMF yields using ionic liquids for conventional and microwave heating methods is shown in Table 8. It can be noted that similar to other processes, microwave heating reduced the reaction time significantly from hours to minutes with higher yields in comparison with conventional heating. It can also be concluded the microwave heating yields are higher than the conventional heating for the same process. A few studies can be discussed here. Zhang and co-workers have reported a number of studies on the use of microwave irradiation for HMF production. In one case, they used microwave irradiation to produce

Table 8. Conventional vs. microwave heating and 5-HMF yields in ionic liquid systems (Qi et al. 2010).

Solvent	Catalyst, amount[a]	T [°C]	Time	5-HMF yield [%]	Heating method[b]	Reference
[EMIM][Cl]	$CrCl_2$, 6 mol%	100	3 h	68	Oil bath	Zhao et al. 2007
[BMIM][Cl]	NHC-$CrCl_2$, 9 mol%	100	6 h	81	Oil bath	Yong et al. 2008
[EMIM][BF_4]	$SnCl_4$, 10 mol%	100	3 h	60	Oil bath	Hu et al. 2009
[BMIM][Cl]	$CrCl_3$, 9 mol%	Unknown[c]	1 min	91	MW	Li et al. 2009c
		100	1 h	17	Oil bath	Li et al. 2009c
[BMIM][Cl]	$CrCl_3$, 10 mol%	100	1 h	67	MW	Qi et al. 2010
		120	5 min	45	Oil bath	
		120	5 min	67	MW	
		140	0.5 min	71	MW	
		140	0.5 min	48	Oil bath	

[a] Based on glucose. [b] MW: Microwave heating. [c] The sample was treated under MW irradiation at 400 W for 1 min, and the reaction temperature was unknown.

HMF from cellulose and glucose in the ionic liquid [BMIM]Cl using CrCl$_3$ catalyst (Li et al. 2011). HMF yields of around 60% from cellulose and 90% from glucose were obtained, which was much higher than the corresponding yields obtained by conventional oil-bath heating, even when a much longer time was used for the latter. In another work, they studied the effectiveness of different metal salts in conversion of glucose to HMF in the ionic liquid [BMIM]Cl and found that ZrCl$_4$ was the most effective catalyst (Liu et al. 2013). A ZnCl$_4$-[BMIM]Cl system was used to directly convert cellulose to HMF which produced a maximum yield of 50%. This study also showed that the microwave irradiation power had an optimal value, beyond which there was a decrease in yields caused by degradation of the HMF. The ZnCl$_4$-[BMIM]Cl system could be reused multiple times without any significant decrease in HMF yield. The same team also studied the [BMIM]Cl-CrCl$_3$ system for converting lignocellulosic biomass like corn stalk, rice straw and pine wood directly to HMF and furfural, obtaining yields of up to 52% and 31% respectively (Zhang and Zhao 2010). Similar to their previous work, the yields obtained under microwave irradiation were much higher than those obtained using oil bath heating. A possible mechanism of HMF production from lignocellulosic biomass is shown in Figure 10.

De et al. (2011) also investigated the use of microwave irradiation for HMF production from various feedstock. The efficacy of AlCl$_3$ in aqueous or aqueous biphasic solvents, was evaluated for substrates such as fructose, glucose, sucrose, starch and inulin and water, water/MIBK or DMSO as the solvent. The HMF yields obtained by microwave heating were as high as 71%, while for conventional heating, the maximum yield was only about 36%. DMSO resulted in a higher HMF yield than aqueous media under microwave irradiation, which was attributed to the higher microwave absorbing ability of DMSO, the occurrence of less side reactions than in aqueous media, and the catalytic ability of DMSO. In case of glucose and sucrose, the maximum HMF yields obtained were 52.4% and 42.5%, respectively, while for biopolymers like starch and inulin, the yields were lower, at around 30% and 39%, respectively.

Datta et al. (2012) used a dimethylacetamide (DMA)-lithium chloride (LiCl) system to obtain HMF from cellulose and sugarcane bagasse under microwave irradiation (Datta et al. 2012). A number of metal chloride catalysts were evaluated and it was discovered that Zr(O)Cl$_2$ was the most active. The effect of addition of the ionic liquid [BMIM]Cl was tested and it was found that the HMF yield from cellulose increased from around 30% to around 75%. [BMIM]Cl is believed to accelerate cellulose hydrolysis by increasing the concentration of Cl$^-$ ions in the reaction medium. For sugarcane bagasse, a Zr(O)Cl$_2$/CrCl$_3$-[BMIM]Cl system produced an HMF yield of up to 42%. Möller et al. (2012), investigated the hydrothermal conversion of glucose and fructose under microwave irradiation. It was seen that the decomposition of glucose requires higher temperatures than fructose, with the decomposition rate constant at 220°C being eight times higher for fructose than for glucose. A summary of studies presenting various feedstock and ionic liquids for HMF preparation under microwave heating method is shown in Table 9.

Ionic Liquid Costs

Ionic liquids are about 20–30 times more expensive than the conventional catalysts. For example, the most commonly used ionic liquids, 1-n-butyl-3-methylimidazolium tetrafluoroborate ([BMI][BF$_4$]) and 1-n-butyl-3-methylimidazolium hexafluorophosphate ([BMI][PF$_6$]) are available at US$ 493 and US$ 590 for 250 g, respectively from suppliers like Sigma-Aldrich and Acros Organics. However, the most popular alkaline catalyst used for transesterification of vegetable oils, sodium hydroxide, costs US$ 20 for 250 grams from the same suppliers. Synthesizing the ionic liquids might reduce the costs to US$ 150 and US$ 205 respectively which is still 10 times more expensive than the conventional catalysts (Andreani and Rocha 2012).

Some ionic liquids can be relatively inexpensive if composed from low cost raw chemicals for anions and cations (Andreani and Rocha 2012). For example, choline based salts are relatively inexpensive, even from the industrial point of view. Choline chloride costs about US$ 25 (250 grams), but can be bought in 5 kg bottles for US$ 273 from traditional suppliers. Holbrey et al. (2002) prepared an easy and efficient low cost ionic liquid containing methyl and ethyl sulfate anions through the reaction of 1-methylimidazolium (US$ 65 per 250 g) with dimethyl sulfate (US$ 20 per 250 mL) and diethylsulfate (US$ 30 per 250 mL), all prices obtained on the Sigma-Aldrich website. Furthermore, ionic liquids are commonly used in smaller

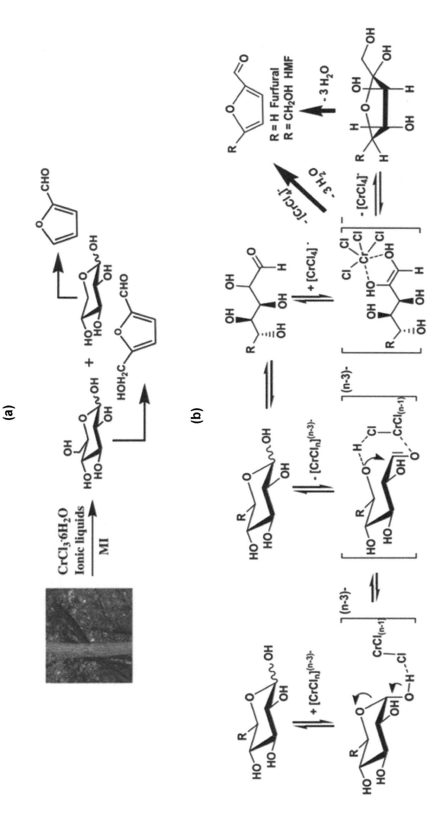

Figure 10. (a) Schematic illustration of HMF and furfural production from lignocellulosic biomass; (b) Putative mechanism of $CrCl_3$ promoted conversion of glucose and D-xylose into HMF and furfural.

Table 9. Use of ionic liquid systems to produce HMF from different feedstocks.

Substrate & conc. (wt%)	Solvent	Catalyst	MW power (W)	Primary final products	Temp (°C)	Reaction/ residence time	Conversion (%)	Yield (wt%)	Reference
Cellulose (2)	[BMIM]Cl	CrCl$_3$/LiCl (1:1 molar ratio to glucose units)	N/A	HMF	160	10 min	N/A	49	Wang et al. 2011
Cellulose (4)	[BMIM]Cl	ZrCl$_4$ (10 mol%)	400	HMF	N/A	3 min 30 s	N/A	40	Liu et al. 2013
Cellulose (5)	[BMIM]Cl	CrCl$_3$·6H$_2$O (0.5 wt%)	400	HMF	-	2 min 30 s	N/A	48	Zhang and Zhao 2010
Cellulose (5)	[BMIM]Cl	CrCl$_3$·6H$_2$O (10 wt% of cellulose)	400	HMF	N/A	2 min	N/A	48	Li et al. 2009a
Cellulose (8)	[C$_3$SO$_3$Hmim]HSO$_4$-Water (1:2 w/w)	-	800	LA	160	30 min	N/A	32	Ren et al. 2013
Corn stalk-hexose content 36.1%- (5)	[BMIM]Cl	CrCl$_3$·6H$_2$O (0.5 wt%)	400	HMF	-	3 min	N/A	32	Zhang and Zhao 2010
Foxtail straw (5)	DMA-LiCl [DMA]$^+$	[CH$_3$SO$_3$]$^-$ (0.5 wt%)	300	HMF	120	2 min	N/A	58	Alam et al. 2012
Fructose (10)	Acetone-DMSO (7:3 w/w)	DOWEX 50WX8-100 strong acidic ion-exchange resin (4 wt%)	700	HMF	150	20 min	99	58	Qi et al. 2008b
Fructose (10)	Water	Phosphate buffer system (pH ~ 2.1)	N/A	HMF	150	30 min	94	45	Lu et al. 2012
Fructose (2)	Water	TiO$_2$ (0.4 wt%)	700	HMF	200	10 min	90.4	29	Qi et al. 2008a
Fructose (2)	Water	H$_2$SO$_4$ (1 wt%)	700	HMF	200	5 min	97.3	33	Qi et al. 2008a
Fructose (2)	Acetone-DMSO (70:30 w/w)	Dowex 50wx8-100 ion-exchange resin (2 wt%)		HMF	150	15 min	95.1	51	Qi et al. 2008a
Fructose (2)	Acetone-DMSO (7:3 w/w)	Sulphated zirconia (20 wt% of substrate)	700	HMF	180	5 min	91.3	46	Qi et al. 2009
Fructose (3)	DMSO	ILIS-SO$_3$H	200	HMF	100	4 min	100	49	Bao et al. 2008
Fructose (3)	DMSO	SiO$_2$-SO$_3$H	200	HMF	100	4 min	95	44	Bao et al. 2008
Fructose (3)	DMSO	SiO$_2$-SO$_3$H	200	HMF	100	4 min	92.1	42	Bao et al. 2008
Fructose (3)	DMSO	ILIS-SO$_2$Cl	200	HMF	100	4 min	100	47	Bao et al. 2008
Fructose (4)	[BMIM]Cl	ZrCl$_4$ (10 mol%)	400	HMF	N/A	2 min	98.9	65	Liu et al. 2013
Fructose (5)	DMSO	AlCl$_3$ (50 mol%)	300	HMF	140	5 min	N/A	49	De et al. 2011

Substrate	Solvent	Catalyst	Power (W)	Product	Temp (°C)	Time	Yield	Yield 2	Reference
Fructose (5)	[BMIM]Cl-MIBK (1:1 v/v)	Zr(O)Cl$_2$ (10 mol%)	300	HMF	120	5 min	N/A	59	Saha et al. 2013
Fructose (8)	Isopropanol	NH$_4$Cl (1 wt%)	600	HMF	120	10 min	92	26	Liu et al. 2012
Fructose (8)	Ethanol	NH$_4$Cl (1 wt%)	600	HMF	120	10 min	99	39	Liu et al. 2012
Glucose (1)	Water-THF (1:3 v/v)*	AlCl$_3$·6H$_2$O (40 mol% of substrate); HCl (40 mol% of substrate)	N/A	HMF	160	15 min	100	43	Yang et al. 2013
Glucose (10)	Water	Phosphate buffer system (pH ~ 2.1)	N/A	HMF	150	1 h 30 min	33	13	Lu et al. 2012
Glucose (10)	Water	Phosphate buffer system (pH ~ 2.1); Sodium borate (0.875 mol/mol glucose)	N/A	HMF	150	1 h 30 min	70.4	29	Lu et al. 2012
Glucose (2)	Water	TiO$_2$ (0.4 wt%)	700	HMF	200	5 min	63.8	13	Qi et al. 2008
Glucose (4)	[BMIM]Cl	ZrCl$_4$ (10 mol%)	400	HMF	N/A	3 min 30 s	72.3	33	Liu et al. 2013
Glucose (5)	DMSO	AlCl$_3$ (50 mol%)	300	HMF	140	5 min	N/A	37	De et al. 2011
Glucose (9)	[BMIM]Cl	CrCl$_3$ (3.6 wt%)	400	HMF	N/A	1 min	N/A	64	Li et al. 2009a
Glucose (9)	[BMIM]Cl	H$_2$SO$_4$ (10 wt%)	400	HMF	N/A	1 min	N/A	34	Li et al. 2009b
Inulin (5)	DMSO	AlCl$_3$ (50 mol%)	300	HMF	140	5 min	N/A	30	De et al. 2011
Pine wood-hexose content 54%- (5)	[BMIM]Cl	CrCl$_3$·6H$_2$O (0.5 wt%)	400	HMF	-	3 min	N/A	36	Zhang and Zhao 2010
Rice straw-hexose content 37.5%- (5)	[BMIM]Cl	CrCl$_3$·6H$_2$O (0.5 wt%)	400	HMF	-	3 min	N/A	33	Zhang and Zhao 2010
Starch (5)	DMSO	AlCl$_3$ (50 mol%)	300	HMF	140	5 min	N/A	24	De et al. 2011
Starch (5)	[BMIM]Cl-MIBK (1:1 v/v)	[Zr(O)Cl$_2$ (10 mol%)	300	HMF	120	5 min	N/A	34	Saha et al. 2013
Sucrose (5)	DMSO	AlCl$_3$ (50 mol%)	300	HMF	140	5 min	N/A	33	De et al. 2011
Sugarcane bagasse (4)	[BMIM]Cl	Zr(O)Cl$_2$ + CrCl$_3$ (20 mol% of cellulose)	300	HMF	120	5 min	N/A	33	Datta et al. 2012
Wheat straw cellulose 38.5% (2)	[BMIM]Cl	CrCl$_3$/LiCl (1:1 M ratio to glucose units)	N/A	HMF	160	15 min	N/A	48	Wang et al. 2011

quantities than traditional catalysts and can be reused in most applications. Ionic liquids would represent the same cost per cycle, or can be even cheaper than conventional catalysts If they can be reused for 10 to 20 times. Therefore, the two factors that will lead to a decision on the feasibility of large scale ionic liquid systems are the ability to reuse the catalysts without activity decrease and attainment of efficient product separation without contamination by the ionic liquid (Plechkova and Seddon 2008).

Concluding Remarks

Ionic liquids provide numerous opportunities for safe and efficient chemical synthesis if the present obstacles can be resolved. The major obstacle for their application is their high costs which increases the product costs. The costs of ionic liquids are 5 to 20 times higher than the conventional organic solvents. The high costs of ionic liquids are due to the raw materials and energy consumption during their preparation. Use of cheaper anions and cations and appropriate scale of production of ionic liquids may bring the economics to be more attractive. Ionic liquids preparation should also include microwave heating for energy-efficient processing. Recyclability of ionic liquids should be considered as well when comparing with conventional organic solvents. Ionic liquids can be used for 10–20 cycles.

Ionic liquids pose several environmental hazards. Their preparation involves non-renewable raw materials derived from petroleum feedstock and other harmful volatile organic solvents. Although ionic liquids can be degraded by up to 99% by oxidation and thermal decomposition, they have been suggested to cause toxicity towards aquatic life and terrestrial organisms due to their relatively poor biodegradability and environmental persistence. The environmental impact of ionic liquids is greatly affected by their structure and therefore careful attention should be given in their manufacturing. Ionic liquids are currently feasible in preparation of high value products such as specialty chemicals and pharmaceutical products and electronics. To make these catalyst/solvent agents suitable for biofuel production, the above mentioned obstacles should be overcome. When these issues are resolved, they can serve as excellent alternatives for conventional organic solvents in all respects in all applications.

A techno-economic analysis modelling of ionic liquid deconstruction by Klein-Marcuschamer et al. (2011) suggested a few sensitive parameters for ionic liquid based process development as ionic liquid characteristics > biomass related factors > recycling and reuse. Further investigations will help to determine whether the identified challenges of ionic liquid processing of lignocellulose are inherent disadvantages or can be improved upon by optimization or technical innovation. Future research efforts should focus on development of low cost ionic liquids and co-solvents (ideally ≤ \$2.50 per kg); reduction of pretreatment time, investigate biomass feedstock related factors (such as biomass type, geographical location and time of year, increased moisture tolerance; reduced energy input during drying), derivation of toxicological data and knowledge about basic physico-chemical characteristics; improving yield and high selectivity fractionation; reducing ionic liquid losses; optimized recycling, inhibitor generation issues; end of use ionic liquid recovery, reduce regeneration requirements; reduced impact of residual ionic liquid on downstream processing (toxicity on fermenting organisms and inhibition of enzymes); improve 'Greenness' of ionic liquid and co-solvent (health and environmental impact).

References

Akhand, M.M. and Méndez Blancas, A. 2012. Optimization of NMMO pre-treatment of straw for enhanced biogas production. MS Thesis, University of Boras.

Alam, M.I., De, S., Dutta, S. and Saha, B. 2012. Solid-acid and ionic-liquid catalyzed one-pot transformation of biorenewable substrates into a platform chemical and a promising biofuel. *RSC Advances*. 2(17): pp.6890–6896.

Alvira, P., Tomás-Pejó, E., Ballesteros, M. and Negro, M.J. 2010. Pretreatment technologies for an efficient bioethanol production process based on enzymatic hydrolysis: a review. *Bioresource Technology*. 101(13): pp.4851–4861.

Al-Zuhair, S. 2007. Production of biodiesel: possibilities and challenges. *Biofuels, Bioproducts and Biorefining*. 1(1): pp.57–66.

Andreani, L. and Rocha, J.D. 2012. Use of ionic liquids in biodiesel production: a review. *Brazilian Journal of Chemical Engineering*. 29(1): pp.1–13.

Aslanzadeh, S., Taherzadeh, M.J. and Horváth, I.S. 2011. Pretreatment of straw fraction of manure for improved biogas production. *BioResources*. 6(4): pp.5193–5205.

Bajaj, A., Lohan, P., Jha, P.N. and Mehrotra, R. 2010. Biodiesel production through lipase catalyzed transesterification: an overview. *Journal of Molecular Catalysis B: Enzymatic*. 62(1): pp.9–14.

Banerjee, S., Mudliar, S., Sen, R., Giri, B., Satpute, D., Chakrabarti, T. and Pandey, R.A. 2010. Commercializing lignocellulosic bioethanol: technology bottlenecks and possible remedies. *Biofuels, Bioproducts and Biorefining*. 4(1): pp.77–93.

Bao, Q., Qiao, K., Tomida, D. and Yokoyama, C. 2008. Preparation of 5-hydroymethylfurfural by dehydration of fructose in the presence of acidic ionic liquid. *Catalysis Communications*. 9(6): pp.1383–1388.

Bekou, E., Dionysiou, D.D., Qian, R.Y. and Botsaris, G.D. 2003. Extraction of Chlorophenols from Water using Room Temperature Ionic Liquids. 544–560.

Berson, R.E., Young, J.S. and Hanley, T.R. 2006. Reintroduced solids increase inhibitor levels in a pretreated corn stover hydrolysate. *Twenty-Seventh Symposium on Biotechnology for Fuels and Chemicals*. pp.612–620. Humana Press.

Binder, J.B. and Raines, R.T. 2009. Simple chemical transformation of lignocellulosic biomass into furans for fuels and chemicals. *Journal of the American Chemical Society*. 131(5): pp.1979–1985.

Bradshaw, T.C., Alizadeh, H., Teymouri, F., Balan, V. and Dale, B.E. 2007. Ammonia fiber expansion pretreatment and enzymatic hydrolysis on two different growth stages of reed canarygrass. *Applied Biochemistry and Biotechnology*. pp.395–405. Humana Press.

Carmichael, A.J., Earle, M.J., Holbrey, J.D., McCormac, P.B. and Seddon, K.R. 1999. The Heck reaction in ionic liquids: a multiphasic catalyst system. *Organic Letters*. 1(7): pp.997–1000.

Chatel, G. and Rogers, R.D. 2013. Review: Oxidation of lignin using ionic liquids- an innovative strategy to produce renewable chemicals. *ACS Sustainable Chemistry & Engineering*. 2(3): pp.322–339.

Chen, S.F., Mowery, R.A., Chambliss, C.K. and van Walsum, G.P. 2007. Pseudo reaction kinetics of organic degradation products in dilute-acid-catalyzed corn stover pretreatment hydrolysates. *Biotechnology and Bioengineering*. 98(6): pp.1135–1145.

Cole, A.C., Jensen, J.L., Ntai, I., Tran, K.L.T., Weaver, K.J., Forbes, D.C. and Davis, J.H. 2002. Novel Brønsted acidic ionic liquids and their use as dual solvent-catalysts. *Journal of the American Chemical Society*. 124(21): pp.5962–5963.

Cuissinat, C. and Navard, P. 2006. Swelling and Dissolution of Cellulose Part 1: Free Floating Cotton and Wood Fibres in N-Methylmorpholine-N-oxide–Water Mixtures. In *Macromolecular Symposia*. 244(1): pp.1–18. WILEY-VCH Verlag.

Cull, S.G., Holbrey, J.D., Vargas-Mora, V., Seddon, K.R. and Lye, G.J.–2000. Room-temperature ionic liquids as replacements for organic solvents in multiphase bioprocess operations. *Biotechnology and Bioengineering*. 69(2): pp.227–233.

Dadi, A.P., Varanasi, S. and Schall, C.A. 2006. Enhancement of cellulose saccharification kinetics using an ionic liquid pretreatment step. *Biotechnology and Bioengineering*. 95(5): pp.904–910.

De Diego, T., Lozano, P., Gmouh, S., Vaultier, M. and Iborra, J.L. 2005. Understanding structure-stability relationships of Candida antartica lipase B in ionic liquids. *Biomacromolecules*. 6(3): pp.1457–1464.

De Diego, T., Manjón, A., Lozano, P. and Iborra, J.L. 2011. A recyclable enzymatic biodiesel production process in ionic liquids. *Bioresource Technology*. 102(10): pp.6336–6339.

De los Ríos, A.P., Fernández, F.H., Gómez, D., Rubio, M. and Víllora, G. 2011. Biocatalytic transesterification of sunflower and waste cooking oils in ionic liquid media. *Process Biochemistry*. 46(7): pp.1475–1480.

De, S., Dutta, S. and Saha, B. 2011. Microwave assisted conversion of carbohydrates and biopolymers to 5-hydroxymethylfurfural with aluminium chloride catalyst in water. *Green Chemistry*. 13(10): pp.2859–2868.

Duan, X., Sun, G., Sun, Z., Li, J., Wang, S., Wang, X., Li, S. and Jiang, Z. 2013. A heteropolyacid-based ionic liquid as a thermoregulated and environmentally friendly catalyst in esterification reaction under microwave assistance. *Catalysis Communications*. 42: pp.125–128.

Dutta, S., De, S., Alam, M.I., Abu-Omar, M.M. and Saha, B. 2012. Direct conversion of cellulose and lignocellulosic biomass into chemicals and biofuel with metal chloride catalysts. *Journal of Catalysis*. 288: pp.8–15.

Earle, M.J., Esperança, J.M., Gilea, M.A., Lopes, J.N.C., Rebelo, L.P., Magee, J.W., Seddon, K.R. and Widegren, J.A. 2006. The distillation and volatility of ionic liquids. *Nature*. 439(7078): pp.831–834.

Earle, M.J., Plechkova, N.V. and Seddon, K.R. 2009. Green synthesis of biodiesel using ionic liquids. *Pure and Applied Chemistry*. 81(11): pp.2045–2057.

Fadeev, A.G. and Meagher, M.M. 2001. Opportunities for ionic liquids in recovery of biofuels. *Chemical Communications*. (3): pp.295–296.

Fan, M.M., Zhou, J.J., Han, Q.J. and Zhang, P.B. 2012. Effect of various functional groups on biodiesel synthesis from soybean oils by acidic ionic liquids. *Chinese Chemical Letters*. 23(10): pp.1107–1110.

Feng, L. and Chen, Z.L. 2008. Research progress on dissolution and functional modification of cellulose in ionic liquids. *Journal of Molecular Liquids*. 142(1): pp.1–5.

Fu, C., Kuang, Y., Huang, Z., Wang, X., Du, N., Chen, J. and Zhou, H. 2010. Electrochemical co-reduction synthesis of graphene/Au nanocomposites in ionic liquid and their electrochemical activity. *Chemical Physics Letters*. 499(4): pp.250–253.

Gamba, M., Lapis, A.A. and Dupont, J. 2008. Supported ionic liquid enzymatic catalysis for the production of biodiesel. *Advanced Synthesis & Catalysis*. 350(1): pp.160–164.

Gao, J., Chen, L., Yan, Z. and Wang, L. 2013. Effect of ionic liquid pretreatment on the composition, structure and biogas production of water hyacinth (Eichhornia crassipes). *Bioresource Technology*. 132: pp.361–364.

Gericke, M., Fardim, P. and Heinze, T. 2012. Ionic liquids—promising but challenging solvents for homogeneous derivatization of cellulose. *Molecules*. 17(6): pp.7458–7502.

Ghiaci, M., Aghabarari, B., Habibollahi, S. and Gil, A. 2011. Highly efficient Brønsted acidic ionic liquid-based catalysts for biodiesel synthesis from vegetable oils. *Bioresource Technology*. 102(2): pp.1200–1204.

Greaves, T.L. and Drummond, C.J. 2008. Protic ionic liquids: properties and applications. *Chemical Reviews*. 108(1): pp.206–237.
Gu, T. 2013. *Green Biomass Pretreatment for Biofuels Production*. Springer.
Guo, F., Fang, Z., Tian, X.F., Long, Y.D. and Jiang, L.Q. 2013. One-step production of biodiesel from Jatropha oil with high-acid value in ionic liquids [Bioresour. Technol. 102 (11)(2011)]. *Bioresource Technology*. 140: pp.447–450.
Ha, S.H., Lan, M.N., Lee, S.H., Hwang, S.M. and Koo, Y.M. 2007. Lipase-catalyzed biodiesel production from soybean oil in ionic liquids. *Enzyme and Microbial Technology*. 41(4): pp.480–483.
Ha, S.H., Mai, N.L., An, G. and Koo, Y.M. 2011. Microwave-assisted pretreatment of cellulose in ionic liquid for accelerated enzymatic hydrolysis. *Bioresource Technology*. 102(2): pp.1214–1219.
Hama, S. and Kondo, A. 2013. Enzymatic biodiesel production: an overview of potential feedstocks and process development. *Bioresource Technology*. 135: pp.386–395.
Hayyan, M., Mjalli, F.S., Hashim, M.A. and AlNashef, I.M. 2010. A novel technique for separating glycerine from palm oil-based biodiesel using ionic liquids. *Fuel Processing Technology*. 91(1): pp.116–120.
Hayyan, A., Hashim, M.A., Hayyan, M., Mjalli, F.S. and AlNashef, I.M. 2014. A new processing route for cleaner production of biodiesel fuel using a choline chloride based deep eutectic solvent. *Journal of Cleaner Production*. 65: pp.246–251.
Helwani, Z., Othman, M.R., Aziz, N., Fernando, W.J.N. and Kim, J. 2009. Technologies for production of biodiesel focusing on green catalytic techniques: a review. *Fuel Processing Technology*. 90(12): pp.1502–1514.
Ho, T.D., Zhang, C., Hantao, L.W. and Anderson, J.L. 2013. Ionic liquids in analytical chemistry: fundamentals, advances, and perspectives. *Analytical Chemistry*. 86(1): pp.262–285.
Holbrey, J.D., Reichert, W.M., Swatloski, R.P., Broker, G.A., Pitner, W.R., Seddon, K.R. and Rogers, R.D. 2002. Efficient, halide free synthesis of new, low cost ionic liquids: 1, 3-dialkylimidazolium salts containing methyl-and ethyl-sulfate anions. *Green Chemistry*. 4(5): pp.407–413.
Hu, S., Zhang, Z., Song, J., Zhou, Y. and Han, B. 2009. Efficient conversion of glucose into 5-hydroxymethylfurfural catalyzed by a common Lewis acid SnCl 4 in an ionic liquid. *Green Chemistry*. 11(11): pp.1746–1749.
Huddleston, J.G., Willauer, H.D., Swatloski, R.P., Visser, A.E. and Rogers, R.D. 1998. Room temperature ionic liquids as novel media for 'clean'liquid–liquid extraction. *Chemical Communications*. (16): pp.1765–1766.
Jeihanipour, A., Aslanzadeh, S., Rajendran, K., Balasubramanian, G. and Taherzadeh, M.J. 2013. High-rate biogas production from waste textiles using a two-stage process. *Renewable Energy*. 52: pp.128–135.
Jeihanipour, A., Karimi, K. and Taherzadeh, M.J. 2010a. Enhancement of ethanol and biogas production from high-crystalline cellulose by different modes of NMO pretreatment. *Biotechnology and Bioengineering*. 105(3): pp.469–476.
Jeihanipour, A., Karimi, K., Niklasson, C. and Taherzadeh, M.J. 2010b. A novel process for ethanol or biogas production from cellulose in blended-fibers waste textiles. *Waste Management*. 30(12): pp.2504–2509.
Kanitkar, A., Balasubramanian, S., Lima, M. and Boldor, D. 2011. A critical comparison of methyl and ethyl esters production from soybean and rice bran oil in the presence of microwaves. *Bioresource Technology*. 102(17): pp.7896–7902.
Kilpeläinen, I., Xie, H., King, A., Granstrom, M., Heikkinen, S. and Argyropoulos, D.S. 2007. Dissolution of wood in ionic liquids. *Journal of Agricultural and Food Chemistry*. 55(22): pp.9142–9148.
Klein-Marcuschamer, D., Simmons, B.A. and Blanch, H.W. 2011. Techno-economic analysis of a lignocellulosic ethanol biorefinery with ionic liquid pre-treatment. *Biofuels, Bioproducts and Biorefining*. 5(5): pp.562–569.
Kotadia, D.A. and Soni, S.S. 2013. Symmetrical and unsymmetrical Brønsted acidic ionic liquids for the effective conversion of fructose to 5-hydroxymethyl furfural. *Catalysis Science & Technology*. 3(2): pp.469–474.
Lapis, A.A., de Oliveira, L.F., Neto, B.A. and Dupont, J. 2008. Ionic Liquid Supported Acid/Base-Catalyzed Production of Biodiesel. *ChemSusChem*. 1(8-9): pp.759–762.
Leadbeater, N.E., Torenius, H.M. and Tye, H. 2004. Microwave-promoted organic synthesis using ionic liquids: a mini review. *Combinatorial Chemistry & High Throughput Screening*. 7(5): pp.511–528.
Lee, S.H., Doherty, T.V., Linhardt, R.J. and Dordick, J.S. 2009. Ionic liquid-mediated selective extraction of lignin from wood leading to enhanced enzymatic cellulose hydrolysis. *Biotechnology and Bioengineering*. 102(5): pp.1368–1376.
Lemus, J., Palomar, J., Heras, F., Gilarranz, M.A. and Rodriguez, J.J. 2012. Developing criteria for the recovery of ionic liquids from aqueous phase by adsorption with activated carbon. *Separation and Purification Technology*. 97: pp.11–19.
Leung, D.Y., Wu, X. and Leung, M.K.H. 2010. A review on biodiesel production using catalyzed transesterification. *Applied Energy*. 87(4): pp.1083–1095.
Li, C., Zhang, Z. and Zhao, Z.K. 2009a. Direct conversion of glucose and cellulose to 5-hydroxymethylfurfural in ionic liquid under microwave irradiation. *Tetrahedron Letters*. 50(38): pp.5403–5405.
Li, N.W., Zong, M.H. and Wu, H. 2009b. Highly efficient transformation of waste oil to biodiesel by immobilized lipase from Penicillium expansum. *Process Biochemistry*. 44(6): pp.685–688.
Li, Q., He, Y.C., Xian, M., Jun, G., Xu, X., Yang, J.M. and Li, L.Z. 2009c. Improving enzymatic hydrolysis of wheat straw using ionic liquid 1-ethyl-3-methyl imidazolium diethyl phosphate pretreatment. *Bioresource Technology*. 100(14): pp.3570–3575.
Li, B., Asikkala, J., Filpponen, I. and Argyropoulos, D.S. 2010a. Factors affecting wood dissolution and regeneration of ionic liquids. *Industrial & Engineering Chemistry Research*. 49(5): pp.2477–2484.
Li, C., Knierim, B., Manisseri, C., Arora, R., Scheller, H.V., Auer, M., Vogel, K.P., Simmons, B.A. and Singh, S. 2010b. Comparison of dilute acid and ionic liquid pretreatment of switchgrass: biomass recalcitrance, delignification and enzymatic saccharification. *Bioresource Technology*. 101(13): pp.4900–4906.
Li, K.X., Chen, L., Yan, Z.C. and Wang, H.L. 2010c. Application of pyridinium ionic liquid as a recyclable catalyst for acid-catalyzed transesterification of Jatropha oil. *Catalysis Letters*. 139(3-4): pp.151–156.
Li, T., Wang, S.J., Yu, C.S., Ma, Y.C., Li, K.L. and Lin, L.W. 2011. Direct conversion of methane to methanol over nano-[Au/SiO 2] in [Bmim] Cl ionic liquid. *Applied Catalysis A: General*. 398(1): pp.150–154.

Liang, X. 2013. Synthesis of biodiesel from waste oil under mild conditions using novel acidic ionic liquid immobilization on poly divinylbenzene. *Energy.* 63: pp.103–108.

Lin, Y.C., Yang, P.M., Chen, S.C. and Lin, J.F. 2013. Improving biodiesel yields from waste cooking oil using ionic liquids as catalysts with a microwave heating system. *Fuel Processing Technology.* 115: pp.57–62.

Liu, Q., Janssen, M.H., van Rantwijk, F. and Sheldon, R.A. 2005. Room-temperature ionic liquids that dissolve carbohydrates in high concentrations. *Green Chemistry.* 7(1): pp.39–42.

Liu, L. and Chen, H. 2006. Enzymatic hydrolysis of cellulose materials treated with ionic liquid [BMIM] Cl. *Chinese Science Bulletin.* 51(20): pp.2432–2436.

Liu, C.Z., Wang, F., Stiles, A.R. and Guo, C. 2012. Ionic liquids for biofuel production: opportunities and challenges. *Applied Energy.* 92: pp.406–414.

Liu, B., Zhang, Z. and Zhao, Z.K. 2013. Microwave-assisted catalytic conversion of cellulose into 5-hydroxymethylfurfural in ionic liquids. *Chemical Engineering Journal.* 215: pp.517–521.

Lozano, P., Bernal, J.M., García-Verdugo, E., Vaultier, M. and Luis, S.V. 2015. Biocatalysis in ionic liquids. pp.31–66. *In:* Environmentally friendly syntheses using ionic liquids. Taylor & Francis Group. USA.

Lu, J., Yan, Y., Zhang, Y. and Tang, Y. 2012. Microwave-assisted highly efficient transformation of ketose/aldose to 5-hydroxymethylfurfural (5-HMF) in a simple phosphate buffer system. *RSC Advances.* 2(20): pp.7652–7655.

Mai, N.L., Ahn, K. and Koo, Y.M. 2014. Methods for recovery of ionic liquids—A review. *Process Biochemistry.* 49(5): pp.872–881.

Man, Z., Elsheikh, Y.A., Bustam, M.A., Yusup, S., Mutalib, M.A. and Muhammad, N. 2013. A Brønsted ammonium ionic liquid-KOH two-stage catalyst for biodiesel synthesis from crude palm oil. *Industrial Crops and Products.* 41: pp.144–149.

Martínez-Palou, R. 2007. Ionic liquid and microwave-assisted organic synthesis: a "green" and synergic couple. *Journal of the Mexican Chemical Society.* 51(4): 252–264.

Matsumoto, M., Mochiduki, K., Fukunishi, K. and Kondo, K. 2004. Extraction of organic acids using imidazolium-based ionic liquids and their toxicity to Lactobacillus rhamnosus. *Separation and Purification Technology.* 40(1): pp.97–101.

Mazubert, A., Taylor, C., Aubin, J. and Poux, M. 2014. Key role of temperature monitoring in interpretation of microwave effect on transesterification and esterification reactions for biodiesel production. *Bioresource Technology.* 161: pp.270–279.

Mendow, G., Veizaga, N.S. and Querini, C.A., 2011. Ethyl ester production by homogeneous alkaline transesterification: influence of the catalyst. *Bioresource Technology.* 102(11): pp.6385–6391.

Mendow, G., Veizaga, N.S., Sánchez, B.S. and Querini, C.A. 2012. Biodiesel production by two-stage transesterification with ethanol by washing with neutral water and water saturated with carbon dioxide. *Bioresource Technology.* 118: pp.598–602.

Möller, M., Harnisch, F. and Schröder, U. 2012. Microwave-assisted hydrothermal degradation of fructose and glucose in subcritical water. *Biomass and Bioenergy.* 39: pp.389–398.

Mora-Pale, M., Meli, L., Doherty, T.V., Linhardt, R.J. and Dordick, J.S. 2011. Room temperature ionic liquids as emerging solvents for the pretreatment of lignocellulosic biomass. *Biotechnology and Bioengineering.* 108(6): pp.1229–1245.

Muhammad, N., Elsheikh, Y.A., Mutalib, M.I.A., Bazmi, A.A., Khan, R.A., Khan, H., Rafiq, S. and Man, Z. 2015. An overview of the role of ionic liquids in biodiesel reactions. *Journal of Industrial and Engineering Chemistry.* 21: pp.1–10.

Mukherjee, A., Dumont, M.J. and Raghavan, V. 2015. Review: Sustainable production of hydroxymethylfurfural and levulinic acid: Challenges and opportunities. *Biomass and Bioenergy.* 72: pp.143–183.

Munson, C.L., Boudreau, L.C., Driver, M.S. and Schinski, W.L., Chevron USA Inc. 2002. Separation of olefins from paraffins using ionic liquid solutions. U.S. Patent 6,339,182.

Munson, C.L., Boudreau, L.C., Driver, M.S. and Schinski, W.L., Chevron USA Inc. 2003. Separation of olefins from paraffins using ionic liquid solutions. U.S. Patent 6,623,659.

Nguyen, T.A.D., Kim, K.R., Han, S.J., Cho, H.Y., Kim, J.W., Park, S.M., Park, J.C. and Sim, S.J. 2010. Pretreatment of rice straw with ammonia and ionic liquid for lignocellulose conversion to fermentable sugars. *Bioresource Technology.* 101(19): pp.7432–7438.

Olivier-Bourbigou, H., Magna, L. and Morvan, D. 2010. Ionic liquids and catalysis: Recent progress from knowledge to applications. *Applied Catalysis A: General.* 373(1): pp.1–56.

Pedersen, M. and Meyer, A.S. 2010. Lignocellulose pretreatment severity–relating pH to biomatrix opening. *New Biotechnology.* 27(6): pp.739–750.

Plechkova, N.V. and Seddon, K.R. 2008. Applications of ionic liquids in the chemical industry. *Chemical Society Reviews.* 37(1): pp.123–150.

Pu, Y., Jiang, N. and Ragauskas, A.J. 2007. Ionic liquid as a green solvent for lignin. *Journal of Wood Chemistry and Technology.* 27(1): pp.23–33.

Purwandari, F.A., Sanjaya, A.P., Millati, R., Cahyanto, M.N., Horváth, I.S., Niklasson, C. and Taherzadeh, M.J. 2013. Pretreatment of oil palm empty fruit bunch (OPEFB) by N-methylmorpholine-N-oxide (NMMO) for biogas production: Structural changes and digestion improvement. *Bioresource Technology.* 128: pp.461–466.

Qi, X., Watanabe, M., Aida, T.M. and Smith, R.L. 2008a. Catalytic conversion of fructose and glucose into 5-hydroxymethylfurfural in hot compressed water by microwave heating. *Catalysis Communications.* 9(13): pp.2244–2249.

Qi, X., Watanabe, M., Aida, T.M. and Smith, Jr., R.L. 2008b. Selective conversion of D-fructose to 5-hydroxymethylfurfural by ion-exchange resin in acetone/dimethyl sulfoxide solvent mixtures. *Industrial & Engineering Chemistry Research.* 47(23): pp.9234–9239.

Qi, M., Wu, G., Li, Q. and Luo, Y. 2008c. γ-Radiation effect on ionic liquid [bmim][BF 4]. *Radiation Physics and Chemistry*. 77(7): pp.877–883.

Qi, X., Watanabe, M., Aida, T.M. and Smith, R.L. 2009. Efficient Catalytic Conversion of Fructose into 5-Hydroxymethylfurfural in Ionic Liquids at Room Temperature. *ChemSusChem*. 2(10): pp.944–946.

Qi, X., Watanabe, M., Aida, T.M. and Smith, R.L. 2010. Fast transformation of glucose and Di-/polysaccharides into 5-hydroxymethylfurfural by microwave heating in an ionic liquid/catalyst system. *ChemSusChem*. 3(9): pp.1071–1077.

Ramadhan, A.R., Pornwongthong, P., Rattanaporn, K. and Sriariyanun, M. 2015. Review of ionic liquid as a catalyst for biodiesel prodution. *J. Sci. Technol. MSU*. 34(4): pp.404-412.

Ratti, R., 2014. Ionic Liquids: Synthesis and Applications in Catalysis. *Advances in Chemistry*, 2014. Article ID 729842, 16 pages http://dx.doi.org/10.1155/2014/729842.

Reddy, E.R., Sharma, M., Chaudhary, J.P., Bosamiya, H. and Meena, R. 2014. One-pot synthesis of biodiesel from high fatty acid Jatropha curcas oil using bio-based basic ionic liquid as a catalyst. *Current Science*. 106(10): p.1394.

Remsing, R.C., Swatloski, R.P., Rogers, R.D. and Moyna, G. 2006. Mechanism of cellulose dissolution in the ionic liquid 1-n-butyl-3-methylimidazolium chloride: a 13 C and 35/37 Cl NMR relaxation study on model systems. *Chemical Communications*. (12): pp.1271–1273.

Ren, H., Zhou, Y. and Liu, L. 2013. Selective conversion of cellulose to levulinic acid via microwave-assisted synthesis in ionic liquids. *Bioresource Technology*. 129: pp.616–619.

Robles-Medina, A., González-Moreno, P.A., Esteban-Cerdán, L. and Molina-Grima, E. 2009. Biocatalysis: towards ever greener biodiesel production. *Biotechnology Advances*. 27(4): pp.398–408.

Ruzich, N.I. and Bassi, A.S. 2010. Investigation of enzymatic biodiesel production using ionic liquid as a co-solvent. *The Canadian Journal of Chemical Engineering*. 88(2): pp.277–282.

Saha, B., De, S. and Fan, M. 2013. Zr (O) Cl 2 catalyst for selective conversion of biorenewable carbohydrates and biopolymers to biofuel precursor 5-hydroxymethylfurfural in aqueous medium. *Fuel*. 111: pp.598–605.

Shahla, S., Cheng, N.G. and Yusoff, R. 2010. An overview on transesterification of natural oils and fats. *Biotechnology and Bioprocess Engineering*. 15(6): pp.891–904.

Sheldon, R. 2001. Catalytic reactions in ionic liquids. *Chemical Communications*. 23: pp.2399–2407.

Shill, K., Padmanabhan, S., Xin, Q., Prausnitz, J.M., Clark, D.S. and Blanch, H.W. 2011. Ionic liquid pretreatment of cellulosic biomass: enzymatic hydrolysis and ionic liquid recycle. *Biotechnology and Bioengineering*. 108(3): pp.511–520.

Singh, S., Simmons, B.A. and Vogel, K.P. 2009. Visualization of biomass solubilization and cellulose regeneration during ionic liquid pretreatment of switchgrass. *Biotechnology and Bioengineering*. 104(1): pp.68–75.

Smirnova, S.V., Torocheshnikova, I.I., Formanovsky, A.A. and Pletnev, I.V. 2004. Solvent extraction of amino acids into a room temperature ionic liquid with dicyclohexano-18-crown-6. *Analytical and Bioanalytical Chemistry*. 378(5): pp.1369–1375.

Smith, K., Liu, S. and El-Hiti, G.A. 2004. Use of ionic liquids as solvents for epoxidation reactions catalysed by a chiral Katsuki-type salen complex: enhanced reactivity and recovery of catalyst. *Catalysis Letters*. 98(2-3): pp.95–101.

Soni, S.S., Kotadia, D.A., Patel, V.K. and Bhatt, H. 2014. A synergistic effect of microwave/ultrasound and symmetrical acidic ionic liquids on transesterification of vegetable oils with high free fatty acid. *Biomass Conversion and Biorefinery*. 4(4): pp.301–309.

Sowmiah, S., Srinivasadesikan, V., Tseng, M.C. and Chu, Y.H. 2009. On the chemical stabilities of ionic liquids. *Molecules*. 14(9): pp.3780–3813.

Sun, N., Rahman, M., Qin, Y., Maxim, M.L., Rodríguez, H. and Rogers, R.D. 2009. Complete dissolution and partial delignification of wood in the ionic liquid 1-ethyl-3-methylimidazolium acetate. *Green Chemistry*. 11(5): pp.646–655.

Sun, N., Rodríguez, H., Rahman, M. and Rogers, R.D. 2011. Where are ionic liquid strategies most suited in the pursuit of chemicals and energy from lignocellulosic biomass? *Chemical Communications*. 47(5): pp.1405–1421.

Sunitha, S., Kanjilal, S., Reddy, P.S. and Prasad, R.B.N. 2007. Ionic liquids as a reaction medium for lipase-catalyzed methanolysis of sunflower oil. *Biotechnology Letters*. 29(12): pp.1881–1885.

Suresh and Sandhu, J.S. 2011. Recent advances in ionic liquids: green unconventional solvents of this century: part I. *Green Chemistry Letters and Reviews*. 4(4): pp.289–310.

Swatloski, R., Spear, S., Holbrey, J. and Rogers, R. 2002. Dissolution of cellulose with ionic liquids. *JACS*. 124: pp.4974–4975.

Swatloski, R.P., Spear, S.K., Holbrey, J.D. and Rogers, R.D. 2002. Dissolution of cellose with ionic liquids. *Journal of the American Chemical Society*. 124(18): pp.4974–4975.

Tan, S.S., MacFarlane, D.R., Upfal, J., Edye, L.A., Doherty, W.O., Patti, A.F., Pringle, J.M. and Scott, J.L. 2009. Extraction of lignin from lignocellulose at atmospheric pressure using alkylbenzenesulfonate ionic liquid. *Green Chemistry*. 11(3): pp.339–345.

Tan, T., Lu, J., Nie, K., Deng, L. and Wang, F. 2010. Biodiesel production with immobilized lipase: a review. *Biotechnology Advances*. 28(5): pp.628–634.

Tao, L.O.N.G., Yuefeng, D.E.N.G., Shucai, G.A.N. and Ji, C.H.E.N. 2010. Application of choline chloride xZnCl 2 ionic liquids for preparation of biodiesel. *Chinese Journal of Chemical Engineering*. 18(2): pp.322–327.

Teghammar, A., Karimi, K., Horváth, I.S. and Taherzadeh, M.J. 2012. Enhanced biogas production from rice straw, triticale straw and softwood spruce by NMMO pretreatment. *Biomass and Bioenergy*. 36: pp.116–120.

Teixeira, C.B., Junior, J.V.M. and Macedo, G.A. 2014. Biocatalysis combined with physical technologies for development of a green biodiesel process. *Renewable and Sustainable Energy Reviews*. 33: pp.333–343.

Ullah, Z., Bustam, M.A. and Man, Z. 2015. Biodiesel production from waste cooking oil by acidic ionic liquid as a catalyst. *Renewable Energy*. 77: 521–526.

Vidal, F., Plesse, C., Teyssié, D. and Chevrot, C. 2004. Long-life air working conducting semi-IPN/ionic liquid based actuator. *Synthetic Metals*. 142(1): pp.287–291.

Vyas, A.P., Verma, J.L. and Subrahmanyam, N. 2010. A review on FAME production processes. *Fuel*. 89(1): pp.1–9.

Walden, P. 1914. Molecular weights and electrical conductivity of several fused salts. *Bull. Acad. Imper. Sci. (St. Petersburg)*. 8: pp.405–422.

Wang, H., Gurau, G. and Rogers, R.D. 2012. Ionic liquid processing of cellulose. *Chemical Society Reviews*. 41(4): 1519–1537.

Wang, P., Yu, H., Zhan, S. and Wang, S. 2011. Catalytic hydrolysis of lignocellulosic biomass into 5-hydroxymethylfurfural in ionic liquid. *Bioresource Technology*. 102(5): pp.4179–4183.

Wu, Q., Chen, H., Han, M.H., Wang, J.F., Wang, D.Z. and Jin, Y. 2006. Transesterification of cottonseed oil to biodiesel catalyzed by highly active ionic liquids. *Chinese Journal of Catalysis*. 27(4): pp.294–296.

Yang, Z., Zhang, K.P., Huang, Y. and Wang, Z. 2010. Both hydrolytic and transesterification activities of Penicillium expansum lipase are significantly enhanced in ionic liquid [BMIm][PF 6]. *Journal of Molecular Catalysis B: Enzymatic*. 63(1): pp.23–30.

Yang, Y., Hu, C. and Abu-Omar, M.M. 2013. The effect of hydrochloric acid on the conversion of glucose to 5-hydroxymethylfurfural in AlCl 3–H 2 O/THF biphasic medium. *Journal of Molecular Catalysis A: Chemical*. 376: pp.98–102.

Yong, G., Zhang, Y. and Ying, J.Y. 2008. Efficient catalytic system for the selective production of 5-hydroxymethylfurfural from glucose and fructose. *Angewandte Chemie*. 120(48): pp.9485–9488.

Yu, D., Wang, C., Yin, Y., Zhang, A., Gao, G. and Fang, X. 2011. A synergistic effect of microwave irradiation and ionic liquids on enzyme-catalyzed biodiesel production. *Green Chemistry*. 13(7): pp.1869–1875.

Zavrel, M., Bross, D., Funke, M., Büchs, J. and Spiess, A.C. 2009. High-throughput screening for ionic liquids dissolving (ligno-) cellulose. *Bioresource Technology*. 100(9): pp.2580–2587.

Zhang, Z. and Zhao, Z.K. 2010. Microwave-assisted conversion of lignocellulosic biomass into furans in ionic liquid. *Bioresource Technology*. 101(3): pp.1111–1114.

Zhang, Y., Xia, X., Duan, M., Han, Y., Liu, J., Luo, M., Zhao, C., Zu, Y. and Fu, Y. 2016. Green deep eutectic solvent assisted enzymatic preparation of biodiesel from yellow horn seed oil with microwave irradiation. *Journal of Molecular Catalysis B: Enzymatic*. 123: pp.35–40.

Zhao, H., Xia, S. and Ma, P. 2005. Use of ionic liquids as 'green' solvents for extractions. *Journal of Chemical Technology and Biotechnology*. 80(10): 1089–1096.

Zhao, H., Holladay, J.E., Brown, H. and Zhang, Z.C. 2007. Metal chlorides in ionic liquid solvents convert sugars to 5-hydroxymethylfurfural. *Science*. 316(5831): pp.1597–1600.

Zhao, X., Cheng, K. and Liu, D. 2009. Organosolv pretreatment of lignocellulosic biomass for enzymatic hydrolysis. *Applied Microbiology and Biotechnology*. 82(5): pp.815–827.

Zhao, H., Zhang, C. and Crittle, T.D. 2013. Choline-based deep eutectic solvents for enzymatic preparation of biodiesel from soybean oil. *Journal of Molecular Catalysis B: Enzymatic*. 85: pp.243–247.

Zhang, Z., Wang, Q., Xie, H., Liu, W. and Zhao, Z.K. 2011. Catalytic Conversion of Carbohydrates into 5-Hydroxymethylfurfural by Germanium (IV) Chloride in Ionic Liquids. *ChemSusChem*. 4(1): pp.131–138.

Zheng, Y., Zhao, J., Xu, F. and Li, Y. 2014. Pretreatment of lignocellulosic biomass for enhanced biogas production. *Progress in Energy and Combustion Science*. 42: pp.35–53.

Zhu, S., Wu, Y., Chen, Q., Yu, Z., Wang, C., Jin, S., Ding, Y. and Wu, G. 2006. Dissolution of cellulose with ionic liquids and its application: a mini-review. *Green Chem*. 8(4): pp.325–327.

Zhu, S. 2008. Use of ionic liquids for the efficient utilization of lignocellulosic materials. *Journal of Chemical Technology and Biotechnology*. 83(6): pp.777–779.

10

Microwave Hybridization for Advanced Biorefinery

Conventional energy sources are limited and non-renewable and their consumption contributes to greenhouse gas emissions. Often, fuel recovery alone from the biomass feedstock is not the most economical option. Recovery of multiple high value products through biorefinery concepts is an attractive route for sustainable development. Advanced biorefineries should be developed to meet ever growing energy demands associated with population growth and economic development. An advanced biorefinery should use renewable and sustainable (both in quality and quantity) feedstock that gives rise to higher energy gains with minimum non-renewable energy and resource consumption.

Development of advanced biorefineries is currently encircled by two major issues. The first issue is to ensure adequate biofuel feedstock supplies while the second issue is to develop resource-efficient technologies for the feedstock conversion to maximize energy, economic and environmental benefits. While microalgae, microbial derived oils, agricultural biomass and other energy crops show great potential for meeting current energy demands in a sustainable manner, process intensification and associated synergism can improve the resource utilization efficiency. Among the many process intensification methods, this chapter provides a perspective on the essential role of microwaves and ultrasound and their synergy in biofuel production. Individual, sequential, and simultaneous application of microwaves and ultrasound irradiations can be utilized for process intensification of various biofuels production and selective recovery of high value bioproducts. This chapter focuses on recent developments in microwave and ultrasound mediated process intensification in biofuel synthesis and associated issues in their synergism followed by a discussion on current challenges and future prospective is presented.

Introduction

Escalating environmental pollution associated with fossil fuel consumption has created an urge for nations around the world to investigate into renewable and sustainable energy and fuel supplies such as biofuels. The stimulus for research in biofuel synthesis comes from their additional benefits of high energy density, high capacity factor (resource availability) and ease in process utilization. However, current biofuel industry is encircled by two major issues. First, ensuring adequate biofuel (biomass-derived) feedstock supplies that can make significant contributions toward the total global energy load (~10 TW) generated by the fossil fuel sources (Rittmann 2008) and the second is to address the first without adverse environmental impacts by efficient resource utilization, and produce biofuels without causing energy-food-environment trilemma. Various types of feedstock were utilized for biofuel production which include vegetable plants, oil seed crops (edible such as peanut, soybean, and corn) seeds known as first generation feedstock; animal fats, non-edible oils (jatropha, karanja and other tropical oil seeds), waste oils, lignocellulosic feedstock, grass, crop residues and waste biomass called second generation feedstock; and more recently algae, cyanobacteria and microbial (from wastewater sludge) oils called third generation feedstock (Gude et al. 2011). Considering the escalating demands for the transportation and other fuels for various industrial uses, algae seems to be the most promising and reliable feedstock since it has the potential to sustain the biofuel production at current consumption and meet the process economics by delivering valuable bioproducts. Apart from ensuring adequate feedstock supplies for biofuel production, another major hurdle lies in their conversion processes (Gude et al. 2011). Conventional and ambient pressure or high pressure and high temperature processing methods are not chemical- or energy-efficient or even cost-effective. In this context, process intensification has gained increasing interest in conventional and emerging chemical industries. Process intensification also became an essential endeavor in conventional petroleum and other oil refining industries to improve the energy and material utilization efficiencies (Energy 2007). Process intensification and the synergism promoted by its effects can lead to development of resource-efficient technologies (Huang et al. 2007). This article describes the benefits of process intensification and its synergism for development of resource-efficient technologies. Microwaves and ultrasound have been discussed as two potential novel and unique process intensification methods for developing resource-efficient advanced biorefineries.

Resource-Efficient Technologies for Biorefineries

Development of novel technologies should focus on reducing the energy and material (resource) utilization inefficiencies and increasing economic and environmental benefits. Process intensification and associated synergistic effects may help develop resource-efficient technologies for biorefineries. Process intensification refers to development of novel equipment and/or methods that produce significantly higher yields or superior benefits in comparison with the existing equipment and/or methods in practice. Several techniques or technologies that contribute to process intensification are shown in Figure 1 along with techniques and associated benefits. These benefits can be realized in the form of dramatic reduction in processing times, significant improvements in product quality or quantity and decreasing the equipment size, reducing the complexity of production schemes, improving the energy efficiency, minimizing the waste production, and finally resulting in cheaper and sustainable technologies (Stankiewicz et al. 2000, 2006). The process intensification developments in equipment may focus on developing novel reactor design with intense mixing to promote heat- and mass-transfer while the developments in methods may focus on integrating the reaction-separation processes (minimizing process steps), use of alternative energy sources, and new

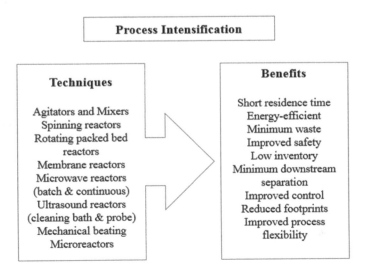

Figure 1. Benefits of process intensification (adapted from Kumar and Nigam 2012).

process control techniques. For example, in the context of biodiesel production, process intensification efforts refer to increasing mass and heat transfer rates among the reaction products whether in extraction and/or transesterification and/or separation and/or purification stages.

Process intensification makes use of process conditions that are far from conventional practices. This involves the use of high temperatures, high pressures, high concentrations (solvent-free), new chemical transformations, explosive conditions, and process simplification and integration to boost synthetic chemistry on both the laboratory and production scale. Such harsh reaction conditions can be safely reached in small scale and continuous flow reactors due to their excellent transport intensification properties. This chapter will provide several examples for different chemical and process intensification techniques and biorefinery concepts (Stouten et al. 2013).

Synergism by process intensification

Process intensification by combining two individual process tools or mechanisms may lead to synergism (magnified impact) (Huang et al. 2007). Synergism can be defined as a phenomenon resulting from the effect of a combination of technologies, tools, or reagents that exceeds the sum of their individual effects (Martinez-Guerra and Gude 2014a).

Synergistic effect by process intensification should be based on the following major criteria (Van Gerven and Stankiewicz 2009):

i) maximize the effectiveness of intramolecular and intermolecular interactions by creating dynamic conditions to promote kinetic regimes with higher conversion and efficiency;
ii) ensure uniform gradient-less mixing and heating;
iii) optimize driving forces and maximizing specific surface areas to improve the heat and mass transfer, and
iv) maximize the synergistic effects from conventional or partial processes.

The most relevant issues addressed by process intensification are structure (in molecular reactions, catalysis), energy (thermodynamic domain in which energy is imparted to the chemistry, hydrodynamic and transport processes), synergy (functional domain in multi-functional tools developed) and time (temporal domain in which timing of events, application of dynamics and process control) (Stankiewicz 2006). However, it is important to identify suitable process configurations when combining two conventional process effects to promote process intensification thereby synergy among them. It is often realized as a process related issue (Huang et al. 2007). All process combinations may not result in process intensification. Even if they provide a synergism, several additional issues may arise from the novel processes with regard to process control and optimization.

Metrics for process intensification

Process intensification is aimed to provide benefits that are related to mass intensity, energy intensity, and waste intensity, extraction index and productivity or footprint (Brunetti et al. 2015). These are defined in equations [1–5] below. The values for mass intensity, energy intensity, and waste intensity should be as low as possible whereas the values for productivity or footprint and extraction index should be as high as possible.

$$\text{Mass intensity, } MI = \frac{\text{Total intel mass}}{\text{Reference product mass}} \qquad [1]$$

$$\text{Energy intensity, } EI = \frac{\text{Total energy consumption}}{\text{Reference product mass}} \qquad [2]$$

$$\text{Waste intensity, } WI = \frac{\text{Total waste mass}}{\text{Reference product mass}} \qquad [3]$$

$$\text{Productivity/Footprint, } PF = \frac{\text{Reference product flowrate}}{\text{Footprint}} \quad [4]$$

$$\text{Extraction Index, } ExI = \frac{\text{Reference product recovered}}{\text{Reference product available in feed}} \quad [5]$$

Microwave Hybridization

Microwaves can be conveniently integrated with various chemical, electrical, mechanical, physical and thermal technologies that complement the effects as shown in Figure 2. Process intensification can be achieved through various complementing technologies such infrared heating, steam heating, hot air circulation, ohmic heating and mechanical and physical mixing such as high shear mixers and ultrasound. Figure 2 depicts the process parameters that need attention in combination heating. Because each technique is different in nature, process design for combination heating could be challenging but it may result in superior benefits.

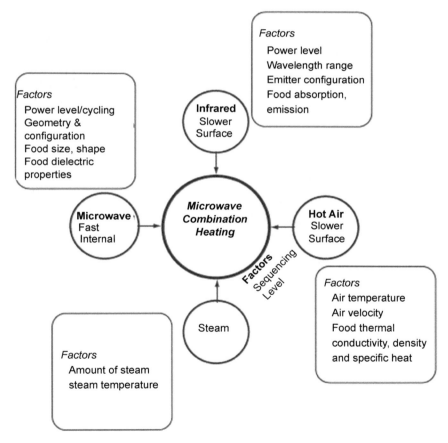

Figure 2. Microwave combination heating—advantages and influencing factors (adapted from Datta and Rakesh 2013).

Microwave hybridization for drying

Microwave hybridization can be considered for various process applications such as drying of moisture content, evaporation, extraction of valuable bioproducts and thermochemical reactions involving high pressure and high temperature conditions.

Microwaves and steam heating—Microwave heating can be combined with steam heating. In steam heating, the (surface) temperatures involved are likely to be lower than in dry heating (with hot air) or the other cooking techniques mentioned above. Heat flux due to steam heating is not readily available (Braud et al. 2001). A range of heat transfer coefficients due to film condensation of water vapor is provided (Michailidis et al. 2009): 5000–6700 W/m²-K. Heat flux is obtained from $q = h_c(T_{sat} - T_s)$ where q is flux, h_c is the heat transfer coefficient due to condensation, T_{sat} is the saturation temperature of water, and T_s is the surface temperature. The heat flux values are reported in the vicinity of 100000 W/m², a value significantly higher than infrared or hot air.

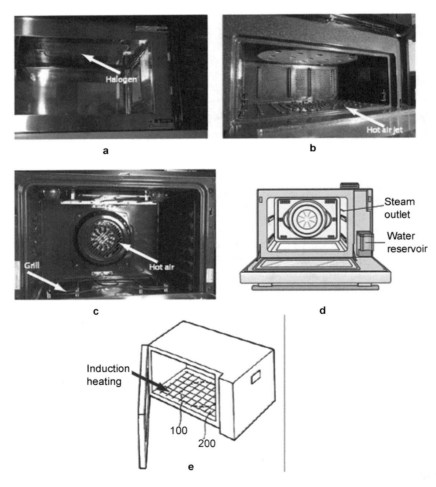

Figure 3. Examples of combination ovens: (a) microwave plus infrared (halogen), GE Advantium oven Model SCA2000BBB 03; (b) microwave plus hot air jet, showing the openings at the bottom for air (Model CJ302UB, technology licensed from Enersyst Development Center, Dallas, TX, USA); (c) microwave plus infrared plus hot air fan (Profile Trivection, Model no. JT930BHBB, General Electric Co., Louisville, KY, USA); (d) microwave oven with steam heating (Sharp 2011); (e) schematic from patent literature of microwave plus induction heating (LG Electronics 1997) (Datta and Rakesh 2013).

Microwave vacuum combined heating

Microwaves have higher penetration and energy transfer efficiencies under vacuum conditions. Therefore it is desirable to develop a microwave combination heating under low pressure (near vacuum) conditions. A microwave unit operating under vacuum is shown in Figure 4. Jindarat et al. (2013) analyzed the energy consumption during the drying of biomaterials using a combined unsymmetrical double-feed microwave and vacuum system (CUMV, Figure 4). Using the first law of thermodynamics, energy efficiency of the system was evaluated and the effect of microwave power, vacuum pressure, and microwave operation mode on energy consumption were considered in detail. In another study, they developed an energy model based on the first law of thermodynamics to evaluate energy efficiency and specific energy consumption in a combined microwave-hot air spouted bed (CMHS) drying of biomaterial (coffee beans). Influencing process parameters such as hot air temperature and initial weight of coffee beans on energy consumption were evaluated. The CMHS drying method with hot air temperature of 60°C, an air velocity of 12 m/s, and initial coffee bean weight of 280 g produced a good quality dehydrated coffee bean with low specific energy consumption in the drying method (Jindarat et al. 2015).

Figure 4. Combined unsymmetrical double-feed microwave and vacuum system (CUMV) (Jindarat et al. 2013).

Microwave Combination Extraction Techniques

Microwave Hydrodiffusion and Gravity (MHG) (Figure 5) (Vian et al. 2008) is relatively a new technique for the extraction of essential oils. This green extraction technique is an original "upside down" microwave distillation combining microwave heating and natural gravity at atmospheric pressure. MHG was conceived for laboratory and industrial scale applications for the extraction of essential oils from different kind of aromatic plants. Based on a relatively simple principle, this method involves placing plant material in a microwave reactor, without adding any solvent or water. The internal heating of the *in situ* water within the plant material distends the plant cells and leads to the rupture of glands and oleiferous receptacles. The heating action of microwaves thus frees essential oil and *in situ* water which are transferred from the inside to the outside of the plant material.

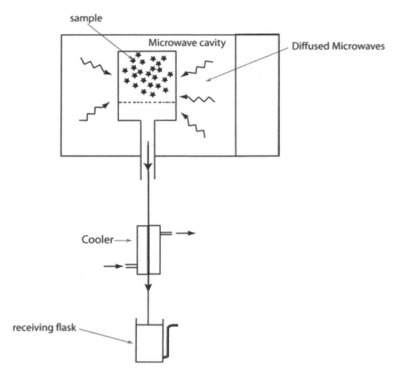

Figure 5. Microwave hydro-diffusion and gravity process for extraction of essential oils and compounds (Zill-E-Huma 2010).

Vacuum Microwave Hydrodistillation (VMHD) is another technique based on selective heating by microwaves combined with sequential application of a vacuum (Mengal and Mompon 1996). The plant material is placed in a microwave cavity with water to refresh the dry material. The plant material is then exposed to microwave irradiation to release the natural extract. Working pressures between 100 and 200 mbar enable evaporation of the azeotropic water–volatile oil mixture from the biological matrix. The procedure is then repeated in a stepwise fashion to extract all the volatile oil from the plant. Up to 30 kg/h material can be treated (Mengal and Mompon 1996). VMHD also provides yields comparable to those obtained by traditional hydrodistillation but with extraction times only one tenth of those required with hydrodistillation. The thermally sensitive crude notes seem to be preserved with VMHD, in contrast to conventional hydrodistillation. VMHD is suggested as an economical and efficient technique to extract high-quality natural products on a large scale (Toursel 1997, Mengal and Mompon 2006).

Microwaves and high shear mixer

Microwave unit can be combined with a mechanical mixing unit to provide uniform distribution of the reactants prior to their exposure to microwave heating. A continuous flow system integrating multi-rotor high-shear mixer and a multimode microwave reactor was developed by Choedkiatsakul et al. (2015). Refined palm oil was converted into biodiesel using methanol and a base catalyst, sodium hydroxide. Significant improvement in heat and mass transfer caused by the integrated high shear mixing and microwave heating achieved a near-complete conversion of palm oil into biodiesel within 5 min (biodiesel yield of 99.8%). Biodiesel met the international specifications and analytical ASTM standards. Figure 6a shows the schematic of this integrated continuous flow high shear mixing and microwave reactor. They also studied other process schemes namely, turbo mixer followed by microwave heating and high shear mix alone as shown in Figures 6a and 6b. Multi-rotor high shear mixing and turbo mixer combined with microwave heating resulted in higher biodiesel yields when compared with shear mix alone.

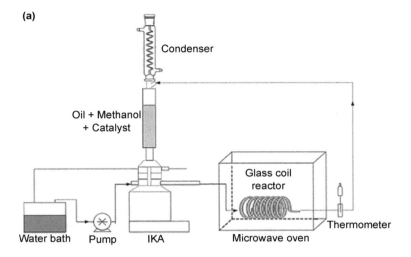

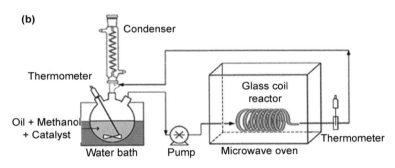

Figure 6. Combination of microwave oven with (a) multi-rotor high-shear mix unit and (b) a turbo mixer (adapted from Choedkiatsakul et al. 2015).

Microwaves combined with ohmic heating

When electric current passes via conduction through any substance with high electrical resistance, heat is generated instantly, increasing the temperature of the substance internally. This heating technology is called ohmic heating or electro-heating. Ohmic heating has been widely applied in the food industry in processes such as aseptic processing and pasteurisation of particulate foods. In comparison with microwave heating, ohmic heating is more efficient because nearly all of the energy enters the food as heat and ohmic heating has no limitation of penetration depth. Ohmic heating also has advantages over conventional heating such as high heating rate, high energy conversion efficiency, volumetric heating, etc. (Li and Sun 2002).

Heating food materials by internal heat generation without the limitation of conventional heat transfer is an advantage of microwave heating. However, some of the non-uniformity commonly associated with microwave heating is associated with its limited dielectric penetration. Heating takes place volumetrically and the product does not experience a large temperature gradient within itself as it heats.

- Higher temperature in particulates than liquid can be achieved, which is impossible for conventional heating.
- Reducing risks of fouling on heat transfer surface and burning of the food product, resulting in minimal mechanical damage and better nutrients and vitamin retention.
- High energy efficiency because 90% of the electrical energy is converted into heat.
- Optimization of capital investment and product safety as a result of high solids loading capacity.

- Ease of process control with instant switch-on and shut-down.
- Reducing maintenance cost (no moving parts).
- Ambient-temperature storage and distribution when combined with an aseptic filling system.
- A quiet environmentally friendly system (Ruan et al. 2002).

A large number of actual and potential applications exist for OH, including blanching, evaporation, dehydration, fermentation, extraction, sterilization, pasteurization and heating of foods to serving temperature, including in the military field or long-duration space missions (Knirsch et al. 2010).

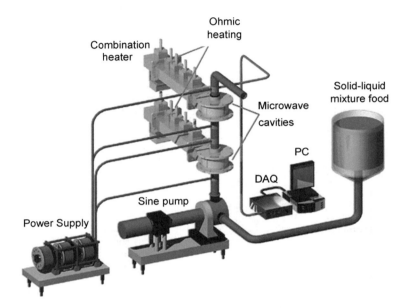

Figure 7. Combined microwave and ohmic heating system (Choi et al. 2015).

Use of Ultrasound in Biofuel Production

Transesterification reaction can be performed under conventional heating methods, which usually employ heated plates (laboratory scale), oil, or sand baths; water heated jacketed reactors combined with mechanical mixing. This process usually takes longer times to complete the reaction. Novel heating techniques like microwaves and ultrasound are known to drastically reduce the reaction time while improving the biodiesel yields significantly. Microwaves provide rapid and convenient heating due to the aforementioned factors. On the other hand, ultrasound activation is entirely different from microwave mechanism. Ultrasound induces cavitations that produce microbubbles to increase the mass transfer rates and generate heat at microscopic levels due to continuous rarefaction and compression cycles of acoustic waves (Gude and Grant 2013). Ultrasonic irradiation is unique from other conventional energy sources in exposure, pressure, and energy per molecule and the duration. The immense local temperatures and pressures, and the extraordinary heating and cooling rates generated by collapse of cavitation bubbles provide an unusual mechanism for generating high-energy chemistry. In ultrasonication, very large amounts of energy are introduced in a short period of time leading to high temperatures but the duration is very short (by $> 10^4$) and the temperatures are even higher (by five- to ten-fold) when compared to pyrolysis (Suslick 1989). Figure 8 shows the overall energy efficiency in ultrasound processing and a biodiesel process set–up. The electric power will be converted into cavitation effect through a generator and transducer to produce acoustic waves into the process liquid. The overall efficiency can be as high as 80–90%. This conversion efficiency is much higher than other techniques. Moreover, these can be scaled up to industrial production just by a simple linear approximation design.

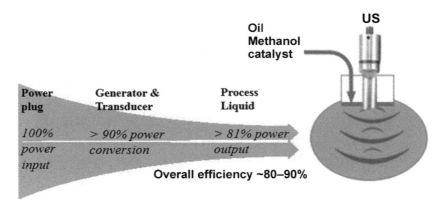

Figure 8. Process intensification by ultrasound (cavitation field), energy conversion efficiency.

Cost of process and equipment

Biodiesel costs from ultrasound processing have not been reported much in literature. Ultrasound processing may result in energy savings because there is no need to heat the reaction mixture and for additional mixing. However, using both heat and ultrasound simultaneously can result in a faster and more complete reaction. A German company, Hielscher Ultrasonics, is a pioneer in industrial ultrasonic processing equipment which produces a wide range of processors with sizes ranging between 500 W and 16,000 W (Figure 9a). According to a recent estimate, 1 kW ultrasonic system consisting of a UIP1000hd ultrasonic processor with a flow cell costs approximately $16,000, and can process approximately 160 to 320 gallons of biodiesel per hour. The UIP2000hd (2 kW) ultrasonic system with a flow cell processes 320 to 640 gal/hr which costs approximately $28,000. A U.S.-based company, Ultrasonic Power Corporation, sells a six-foot-long flow-through reactor and ultrasonic generator for about $12,000 (Figure 9b). The generator provides 2,000 watts of power, and is available in several frequencies, from 40 kHz to 170 kHz. This set-up can produce about 13 gallons of biodiesel per minute (780 gallons per hour).

Beijing Ultrasonic company also manufactures industrial scale ultrasonic processors suitable for biodiesel production (Figure 10). These ultrasound reactors operate at 27 kHz, and a power output range of 500–2000 W. The ultrasound processor/probe has a diameter of 73 mm and a length between 323–1123 mm depending on the process capacity.

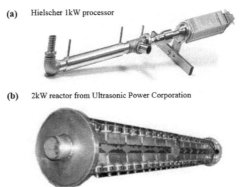

Figure 9. Ultrasound processors suitable for biodiesel production from Hielscher and Ultrasonic Power Corporation: (a) UIP1000 hd ultrasonic processor (1 kW power); (b) six-foot-long flow-through reactor (2 kW power) (http://www.hielscher.com/ultrasonics/index.htm; http://www.upcorp.com/ultrasonic_flowthrough_reactors.html).

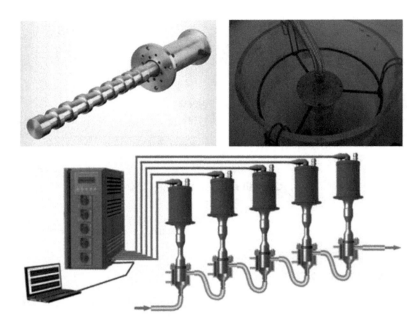

Figure 10. Ultrasound processors suitable for biodiesel production from Beijing Ultrasonic company (https://www.bjultrasonic.com/ultrasonic-biodiesel-reactor).

Process intensification—comparison of heating methods

Mode of heating affects the process chemistry from many aspects. A comparison of the different heating methods is provided in Table 1. In terms of biodiesel production (as an example), the notable differences would be reaction times, solvent requirements, yields, separation times, and specific energy consumption. Shorter reaction times result in significant energy savings, which are attributed to the special effects of microwaves and ultrasound. As shown in Figure 11, increased reaction times result in increased energy expenditures. Conventional heating on a laboratory hot plate requires about 3150 kJ of energy to perform transesterification while microwave and ultrasonic processes required 288 and 60 kJ of energy. This shows that with appropriate reactor design, non-conventional techniques have the potential to reduce the process energy requirements significantly. Another important benefit with microwave process is that it provides high quality biodiesel product when compared with other two methods of biodiesel conversion. Conventional and ultrasound based transesterification involves intense mixing of reaction mixtures thus resulting in increased separation times, and reduced product yield and quality.

A comparison of biodiesel production metrics is shown in Figure 11. Transesterification of waste cooking oil was performed using conventional (laboratory hot plate), microwave, and ultrasound methods. The conventional heating method takes the longest reaction time (105 minutes). Microwaves reduce the reaction time significantly to as low as 6 minutes. The reasons for enhanced reaction rates for non-conventional heating are compared with conventional heating in Table 1 (Gude et al. 2013a,b). When direct sonication was applied, the reaction temperature could increase without any external heat addition similar to microwave conditions. Reaction mixture temperatures as high as 85°C were recorded under 2 minutes of reaction time. This depends on the catalyst ratio and the reaction mixture volume (Gude and Grant 2013).

Synergism of microwaves and ultrasound

Microwave or ultrasound mediated organic synthesis has been the focal point in recent years mainly due to the superior effects in shorter reaction times and high product yields (Gude et al. 2011). While these two process intensification effects have been well utilized in various process chemistry and engineering

Table 1. Comparison of conventional, microwave and ultrasonic heating methods for biodiesel production.

Conventional heating	Microwave heating	Ultrasonic heating
Thermal gradient (outside to inside)	Inverse thermal gradient (inside to outside)	Limited thermal gradient due to mixing
Conduction and convection currents	Molecular level hot spots	Microbubble formation and collapse (compression - rarefaction cycles)
Longer processing times	Very short and instant heating	Relatively very short reaction times, not as quick as microwaves
No or low solvent savings	No or low solvent reactions possible	Solvent savings possible
Product quality and quantity can be affected	Higher product quality and quantity possible	Same as conventional heating
Separation times are longer	Very short separation times	Less than conventional heating
High energy consumption	Moderate to low consumption	Moderate to low consumption
Simple process configuration	Very simple process	Moderate complexity

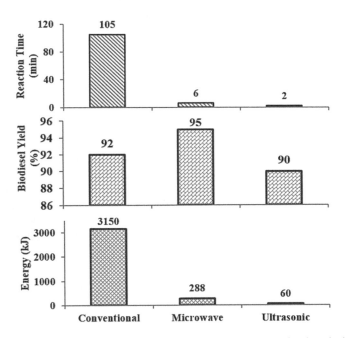

Figure 11. Comparison of waste cooking oil conversion by conventional/non-conventional methods (note: conventional heating was conducted by using a hot plate at 500 W power, microwaves were supplied at 800 W and the ultrasound was applied at 500 W. Other conditions: 10 mL waste cooking oil, 9:1 methanol to oil ratio and 1% NaOH catalyst; Reprinted with permission from (Gude et al. 2013a). Copyright© 2013, Springer.

applications, biofuel industry has yet to explore their beneficial characteristics more extensively. These two non-conventional irradiation processes have been utilized in feedstock preparation, pretreatment, extraction, chemical conversion, and post treatment stages of biofuel production (Gude et al. 2013b). However, the synergism of the two effects has not been explored much.

Microwaves deliver an effect generated by the electromagnetic interaction with reaction materials often resulting in thermal enhancement that produces superior results in chemical synthesis (Gude 2015). Microwaves are capable of providing instant process heat resulting from three major mechanisms in a reaction environment. Ionic conduction, dipolar momentum and interfacial polarization (a combination of ionic conduction and dipolar momentum) are the major causes for this rapid heating (Gude et al. 2013).

Ultrasound are the acoustic cavitations generated by interaction of the sound waves with the reaction compounds resulting in intense mixing which increases the mass and heat transfer among the reaction mixtures leading to higher process efficiency (Gude and Grant 2013, Luo et al. 2014). For example, process intensification by ultrasound can promote mass transfer among gas and liquid components by up to five-fold while the liquid-solid mass transfer can be increased by 20–25 fold and increase product yields significantly (Figure 12). The reaction times can be drastically reduced by microwave heating by up to 1250 times due to rapid heat enhancement (Stankiewicz 2006). These two irradiations can also improve the energy and material efficiencies due to higher product conversion and yields.

Although microwaves provide for rapid heating of the reaction materials, mass transfer of the reaction medium is often compromised in these reactors (Martinez-Guerra, V.G. Gude 2014a). In addition, they interact with reaction materials at a higher rate which results in hot spot formation and thermal runaway. This phenomenon clearly indicates the necessity for a mixing mechanism which can ensure uniform heating of reaction materials and mass transfer promoted by the unusual heating advantage of the microwaves. In the similar context, ultrasound are capable of promoting heat and mass transfer within the reaction medium due to the intense mixing they provide as a result of the acoustic cavitations which form microbubbles with air. The formation-release-collapse of these microbubbles provides cooling and heating cycles at microscales accompanied by high thermal and pressure release. Since this energy release is at micro levels, this energy is not adequate to cause high temperature gains in the reaction medium. This clearly presents a limitation for ultrasound mediated reactions (Martinez-Guerra, V.G. Gude 2014a). These reactions require external heating to enhance the process kinetics.

Considering the aforementioned prospects and limitations for the individual process intensification mechanisms, it is convenient to design a hybrid system that incorporates both non-conventional heating and mixing effects that may lead to enhanced process outcomes. This might lead to greener chemistry since efficient use of chemicals, energy and materials can be anticipated (Cravotto and Cintas 2007, Gude and Martinez-Guerra 2015). Superior benefits gained through the integrated process intensification effects might prove to be economical at large scale applications. It is important to note that this hybrid technology will prove to be ideal for production of high value bioproducts combined with biofuels at present.

Microwave and ultrasound based chemical reactions have been reported to utilize lower amounts of catalysts and solvents along with lower energy consumption in chemical (both organic and inorganic) synthesis. Above all, the reaction times are dramatically decreased (reaction kinetics increased) and the product recovery is greatly enhanced (Gude 2015). These facts clearly support the fundamental principles of green chemistry which refer to atom economy and e-factor (Gude and Martinez-Guerra 2015). Atom

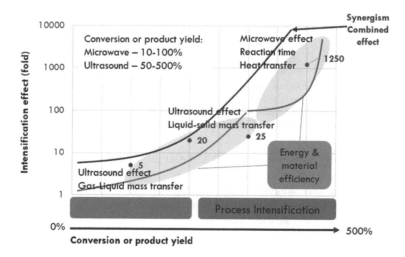

Figure 12. Process intensification effects by microwaves (electromagnetic field) and ultrasound (cavitation field): energy and material efficiency are the potential sustainability effects (printed with permission from Gude 2015).

economy refers to efficient utilization of the raw materials employed in a reaction (maximizing conversion of desired atoms of the reactants into desired products), i.e., converting the raw materials into desired or useful products. The e-factor refers to the amount of waste generated in the process of delivering a desired product. The lower is the e-factor, the greener is the reaction or process. Eliminating waste emissions is the key to accomplishing resource-efficient chemical processes and sustainable biofuel process development.

Microwave and Ultrasound in Biofuel Production

Researchers around the world have investigated the beneficial aspects of microwaves and ultrasonics and accounted for their process intensification benefits in numerous studies. Apart from the pharmaceutical, chemical, and industrial applications, microwaves and ultrasounds have been extensively used and equally investigated for their benefits in biofuel synthesis and bioproduct recovery from various feedstock ever since their discovery. Production of three major energy carriers that microwaves and ultrasound can influence are bioethanol, biodiesel and biogas. In addition to these energy carriers, microwave and ultrasound irradiations can be utilized either in sequential or simultaneous pattern for efficient and selective recovery of various high value bioproducts. Sequential application facilitates process intensification while simultaneous application provides a unique synergy that enhances process chemistry and associated benefits.

Biodiesel production

Among the biofuels, the application of microwaves or ultrasound in biodiesel synthesis or oil extraction and transesterification processes is fairly recent. Conversion of various types of oils (corn, coconut, rice bran, vegetable, rapeseed, sunflower, soybean, cottonseed, safflower, canola, camelina, used or waste vegetable oils, macauba, karanja, *jatropha curcas*, castor bean, castor, palm, yellow horn, animal fats, maize) to biodiesel via esterification and transesterification reactions was evaluated by many researchers (Gude et al. 2013). Similarly ultrasound enhanced extraction, esterification and transesterification reactions of various types of oils (coconut, olive, soybean, palm fatty acid distillate, *jatropha curcas*, *nag champa*, rapeseed, sunflower, and waste cooking oils) were also widely studied (Luo et al. 2014).

The first combined microwave-ultrasound reactor was introduced by Chemat et al. in 1996. In this study, microwave and ultrasound enhanced reactions for pyrolysis of urea and esterification of propanol with acetic acid were evaluated. Conventional, microwave and combined microwave/ultrasound heating mechanisms were evaluated. The yields for urea pyrolysis were 45%, 46% and 57% for a reaction period of 1 hr. for conventional, microwave and combined microwave/ultrasound heating mechanisms, respectively while the yields for esterification reaction in 1 hr reaction time were 80%, 91%, and 99% for the three mechanisms, respectively. Later this group also studied determination of copper in olive oil at a multi-gram scale and ambient conditions (Lagha et al. 1999). Further the application of this hybrid technology in food and total Kjeldahl nitrogen analysis was also reported (Chemat et al. 2004). Cravotto et al. (2008) first investigated the effect of microwave and ultrasound irradiations either simultaneously or individually on extraction of oils from soybean germ and a marine microalga species. Their study reported a reduction in reaction times by up to ten-fold and an increase in oil extraction yields between 50% and 500%. This means higher yields (up to ten-fold can be obtained in a reaction time shortened by 10 times using microwave and ultrasound irradiations. Comparison of individual and sequential applications of microwaves and ultrasound in biodiesel synthesis has been studied recently. Many studies reported shorter reaction times, higher yields for microwave process intensification while reduced solvent volumes and catalyst amounts were reported for ultrasound mediated reactions. Microwaves and ultrasound together have reduced the process reaction time as well as the amount of chemicals required significantly. Microwaves and ultrasound can be used simultaneously in a reaction environment in a single reactor or sequentially in separate reactors (Gude et al. 2013). For example, biodiesel synthesis from a high acid value oil (Nag champa oil) was studied using two-step esterification (to reduce the free fatty acid content that makes the oils suitable for transesterification reaction) followed by transesterification (chemical conversion of triglycerides into fatty acid methyl or ethyl esters) reactions (Gole and Gogate 2013).

Martinez-Guerra and Gude (2014a) studied a novel application of simultaneous microwave and ultrasound (MW/US) irradiations on transesterification of waste vegetable oil (Figure 13). Experiments were conducted in three phases to evaluate: (i) the effect of process parameters on the transesterification reaction mediated by simultaneous MW/US irradiations at a fixed power output rate; (ii) the individual and synergistic effects of the two technologies (by changing the power output rates); and (iii) the synergistic effect on the power density (by changing the sample volume for a fixed MW/US power output). Optimum reaction conditions of 6:1 methanol to oil ratio, 0.75% sodium hydroxide catalyst by wt.%, and 2 min of reaction time at a combined power output rate of 200 W (100/100 MW/US) were reported from the process parametric optimization studies. The biodiesel yields were higher for the simultaneous MW/US mediated reactions (~98%) when compared to MW (87.1%) or US (89.8%) irradiations individually. The power density tests revealed that an optimum power output rate must be determined for energy-efficient biodiesel production. This study concluded that the combined irradiations result in a synergistic effect that enhances the biodiesel process performance and yields significantly (Martinez-Guerra, V.G. Gude 2014a). They also studied the effect of barium oxide as a heterogeneous catalyst in microwave-ultrasound enhanced biodiesel production (Martinez-Guerra, V.G. Gude 2014b). Process parametric optimization studies revealed the optimum process conditions as 6:1 methanol to oil ratio, 0.75% barium oxide catalyst by wt.%, and 2 min of reaction time at a combined power output rate of 200 W (100/100 MW/US). The biodiesel yields were higher for the simultaneous MW/US mediated reactions (93.5%) when compared to MW (91%) and US (83.5%) irradiations individually. A power density of 7.6 W/mL was suggested to be effective for MW, and MW/US irradiated reactions (94.4% and 94.7% biodiesel yields, respectively), while a power density of 5.1 W/mL was appropriate for ultrasound irradiation (93.5%).

As shown in Figure 14, the solvent (alcohol donor) requirements were reduced for both esterification and transesterification reactions when microwave or sequential microwave and ultrasound methods were used (Gole and Gogate 2013). For example, the oil to methanol molar ratios were 1:3 and 1:6 for esterification and transesterification reactions, respectively for a conventional method but the same for the sequential microwave and ultrasound mediated reactions were 1:2 and 1:4 for esterification and transesterification reactions, respectively. In addition, the reaction times were significantly reduced from 20 and 90 minutes to 15 and 6 minutes, respectively. Reductions in these two parameters would also result in energy and cost reductions. As shown in Figure 15, the total energy requirements for both reactions were about 10 times lower for microwave and ultrasound method when compared with the conventional method. Guldhe et al. studied the effects of microwave and ultrasound separately on biodiesel production from algae (*Scenedesmus* sp.). Higher biodiesel conversion was reported for ultrasound (71%) effect when compared with microwave (52%) effect (Guldhe et al. 2014).

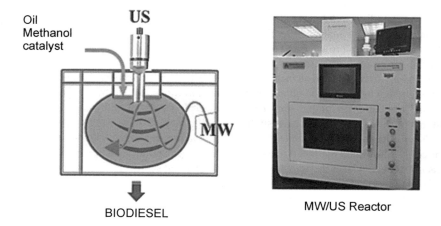

Figure 13. Schematic and photo-image of the microwave and ultrasound combined unit used in biodiesel production by Martinez-Guerra and V.G. Gude (2014a).

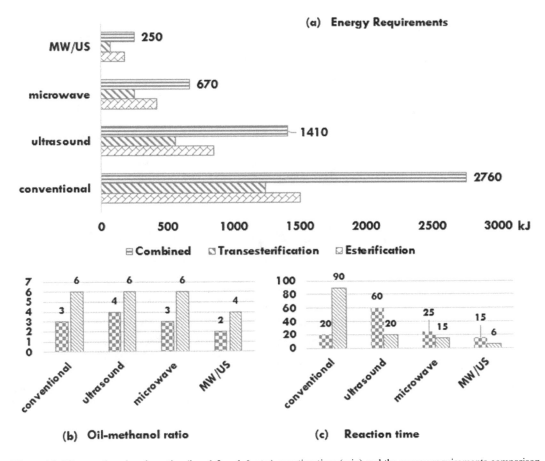

Figure 14. Oil to methanol molar ratios (i.e., 1:3 or 1:6, etc.), reaction time (min) and the energy requirements comparison for conventional, ultrasound, microwave, and sequential microwave and ultrasound based conversion of a non-edible (Nag champa) oil to biodiesel via two-step esterification and transesterification reaction mechanism.

Microwaves and ultrasound can be utilized in other process stages like feedstock pretreatment including harvesting and separation (Gude et al. 2013b). For example, ultrasound can be used to gather (coagulation and flocculation of algal cells facilitated by the cell alignment to ultrasound field in a dilute algal suspension as received from the cultivation stage) the algal cells in the harvesting stage (Fast et al. 2014, Fast and Gude 2015, Gude 2015). Once separated, concentrated algal suspension can be subjected to microwave or microwave-ultrasound irradiations to force-release the cell contents (mainly lipids) from the biological cell matrix and convert them simultaneously into biocrude for further treatment. Supercritical high temperature and high pressure reactions under microwave irradiations are also possible to promote green extraction of lipids and other valuable bioproducts such as proteins (Reddy et al. 2014). In this process, water can be used as a green solvent eliminating toxic releases to the environment.

Bioethanol production

Three major feedstock categories for bioethanol production are: (i) sucrose-containing feedstock (e.g., sugarcane, sugar beet, sweet sorghum, and fruits), (ii) starch-containing feedstock (e.g., corn, milo, wheat, rice, potatoes, cassava and barley), and (iii) lignocellulosic biomass (e.g., wood, straw, grass, wasted crops and crop residues) (Balat and Balat 2009). Starch and sugar based feedstock have proven to be unsustainable due to unfavorable economics and other human interests leaving lignocellulosic and

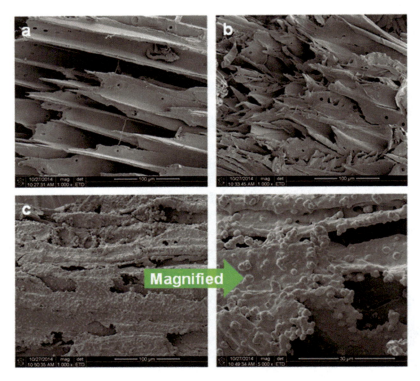

Figure 15. FESEM micrographs of (a) the sawdust, (b) the SR(MW), (c) the SR(MW-UW) and (d) magnified view of the SR(MW-UW) (sawdust, 10 g; mass ratio of solvent/sawdust, 3; mass ratio of PEG 400/glycerol, 4; reaction time, 20 min; and catalyst concentration, 0.3 mol L^{-1}) (Lu et al. 2016).

waste crops and wood residues as attractive feedstock. A variety of valuable products can be recovered from sugars and starch feedstock. These include ethanol, butanol, acetone, lactic acid, and amino acids. But lignocellulosic material is not simple to process. Similar to biodiesel production, bioethanol production from these lignocellulosic and other waste crop feedstock involves three essential steps (Bussemaker and Zhang 2013, Taherzadeh and Karimi 2008): (i) pretreatment of the feedstock; (ii) conversion of cellulose and hemicellulose into fermentable sugars, and (iii) fermentation of sugars into crude biofuels. The pretreatment step is a major hurdle which is required to improve the enzymatic hydrolysis of cellulose by reducing the effect of degree of polymerization, crystallinity of the cellulose, available surface area, lignin content, and moisture content. Complete utilization of available sugars in the biomass is the goal in bioethanol production (Bussemaker and Zhang 2013). Process intensification effects such as microwaves and ultrasound could be useful in enhancing the effectiveness of the pretreatment methods (Nikolić et al. 2011). Microwaves and ultrasound have a long history in the degradation of polysaccharides, water soluble carbohydrates and limited work on starch. Ultrasound was first investigated on degradation of polysaccharides in 1933 by Flosdorf and Chambers (Flosdorf and Chambers 1933). Sonication improves hydrolysis of lignocellulosic materials into sugars and their subsequent fermentation to bioethanol. Degradation of starch from corn meal, maize and potato and other sources was also improved by ultrasound treatment. The effect of ultrasound (at both low and high frequencies) on carbohydrates in sulfuric acid were evaluated in many studies (Choi and Kim 1994, Koda et al. 2011, Portenländer and Heusinger 1997, Czechowska-Biskup et al. 2005). The application of microwaves in starch depolymerization was reported in 1979 by Khan et al., in water, dilute hydrochloric acid and with chloride-based catalyst to enhance hydrolysis (Khan et al. 1979, 1980, Kunlan et al. 2001). Many studies investigated the effect of microwave or ultrasound pretreatment of corn meal, maize sugar, cellulose, switchgrass, rice hull, microcrystalline cellulose, cassava chip, kenaf core fiber, bamboo, waxy rice starch, maize starch, sugarcane, glucose and sugarcane bagasse (Luo et al. 2014). Bioethanol yields from corn meal pretreated by microwave or ultrasound were higher than conventional pretreatment. In a simultaneous saccharification and fermentation process, the yields increased by 6.82% and 8.48% for ultrasound and microwave pretreatments, respectively. The glucose utilization was also increased. Ultrasound and microwave pretreatments increased the maximum ethanol concentration produced in the SSF process by 11.15 and 13.40% (compared to the control sample), respectively (Nikolić et al. 2011).

Simultaneous or sequential use of microwaves and ultrasound has not been explored much in bioethanol production. However, combined use of microwaves and ultrasound in the catalytic conversion of starch-based industrial waste (wet potato sludge) into reducing sugars was evaluated recently (Hernoux-Villière et al. 2013). Two hours of exposure to the combined microwave and ultrasound irradiations at 60°C in sulfuric acid converted 46% of the wet potato sludge to sugars. No significant conversion was observed when ultrasound alone was applied. About 35% of the waste was converted to sugars when microwaves were alone used. For the combined irradiations, dry potato sludge and potato starch yielded 57% and 79% sugars whereas microwave alone produced yields of 87% and 81%, respectively. Microwaves and ultrasound can be used in other process related applications such as preparation of novel catalysts, surface modification of heterogeneous catalysts, and development of ionic liquids for biorefinery applications (Shen 2009). Application of these new materials might bring the benefits of efficient chemical and energy utilization, reduced waste products and alleviation of process reaction conditions.

Biogas production

Digestion of sludge and biomass requires optimum reaction conditions promoted by thermal energy and affected by other important process conditions such as pH, alkalinity, metal concentrations, and presence of micro-pollutants (Khanal et al. 2007). Ultrasound and microwaves can be conveniently used to heat the sludge and then destroy the cell walls to force out the cell components. An alternative would be to apply the ultrasound to weaken the cell walls and then subject the cell components to microwave heating for digestion. Ideally, the two irradiations can be applied simultaneously to increase sludge digestibility. Biogas production remains the most technologically feasible and economically viable process for a variety of sludge and wastewater sources. However, limitations of energy recovery makes the process cost-inefficient. Application of microwaves and ultrasound as process intensification effects may improve the biogas yields and in return, the energy recovery and the overall process economics.

Apart from heating applications, non-thermal effects associated with microwaves align the macromolecules possessing polarization with the electromagnetic field to cause possible breakage of hydrogen bonds (Saha et al. 2011). This may result in enhanced sludge disintegration and hydrolysis which in turn increases the rate of anaerobic digestion, improves dewaterability, and inactivates fecal coliforms to produce Class A sludge. In high and low temperature (above and below boiling points) microwave heating with both batch and continuous flow conditions, the biogas yield and dewaterability were increased by up to 30% and 40%, respectively. Ultrasound was also used frequently to disrupt the cellular matter, although mostly at laboratory scales. Disruption of cellular matter solubilizes the organic matter better and produces higher biogas.

The effects of microwave, ultrasound and combined microwave-ultrasound pretreatment on biogas production, solids removal and dewaterability of anaerobically digested sludge was studied through a comparative study (Yeneneh et al. 2013). The experimental studies revealed that the combined microwave–ultrasonic treatment (2,450 MHz, 800 W and 3 min microwave treatment followed by 0.4 W/mL and 10 min ultrasonication) resulted in better digester performance than ultrasonic or microwave treatment. Mesophilic digestion of combined microwave–ultrasonic-pretreated sludge produced a significantly higher amount of methane (147 mL) after a sludge retention time of 17 days, whereas the ultrasonic- and microwave treated sludge samples produced 30 and 16 mL of methane, respectively. The combined microwave–ultrasonic treatment resulted in total solids reduction of 56.8% and volatile solid removal of 66.8%. Furthermore, this combined treatment improved dewaterability of the digested sludge by reducing the capillary suction time (CST) down to 92 s, as compared to CST of 331 s for microwave-treated and 285 s for ultrasonically treated digested sludge samples. Optimization tests were also carried out to determine the best combination (Yeneneh et al. 2013).

Table 2. Elemental composition of sawdust and of liquefaction products.

Sample[1]	Elementary composition (wt.%)						HHV (MJ/kg)[3]
	C	H	O	N	C/H	C/O	
Sawdust	49	6	44	1	8.2	1.1	17.3
SR(MW)[2]	50	6	43	1	8.3	1.2	17.8
SR(MW-UW)[2]	59	6	33	2	9.8	1.8	22.7
BO(MW)	65	6	28	1	10.8	2.3	25.6
BO(MW-UW)	64	6	29	1	10.7	2.2	25.1

[1]Reaction conditions: sawdust, 10 g; mass ratio of solvent/sawdust, 3; mass ratio of PEG 400/glycerol, 4; reaction time, 20 min; and catalyst concentration, 0.3 mol L^{-1}.
[2]MW and MW-UW represent the microwave-assisted and microwave-ultrasonic-assisted liquefaction, respectively.
[3]Caculated from the Dulong equation: $HHV = 0.3383 \times C + 1.442 \times \left(H + \frac{O}{8}\right)$, where HHV is the Higher Heating Value (MJ Kg^{-1}); C, H and O represent the weight percentage of carbon, hydrogen, and oxygen, respectively.

Microwave-ultrasound liquefaction

Microwave and ultrasound techniques were used in liquefaction of sawdust (woody biomass) in a sulphuric acid/polyethylene glycol 400-glycerol catalytic system (Lu et al. 2016). The results from parametric optimization studies showed that microwave-ultrasound technique reduced the solvent use by 50% and the reaction time from 60 to 20 min when compared with conventional method. The liquefaction yield reached 91% under the optimal conditions. Three main products were obtained which consisted of a solid residue (SR), bio-oil (BO) and water soluble products (WS), the compositions of which were compared for different heating methods as shown in Table 2. The process parametric study also revealed that the influence of microwave and ultrasound irradiations on the yield of some parameters such as catalyst concentration, was similar to that of traditional liquefaction, indicating that the application of microwave and ultrasound possibly only intensified heat and mass transfer rather than altering either the degradation mechanism or pathway.

Under the most suitable liquefaction condition, the sawdust was liquefied with MW-assisted and MW–UW-assisted technology, respectively. The corresponding SR (solid residue) and BO (bio-oil) were marked as MW or MW–UW. Figure 15 shows the FESEM micrographs of the sawdust and the SR. Representative microstructures of wood fiber can be observed in Figure 15a (before treatment). After MW-assisted liquefaction, almost all of the tracheid of sawdust was distorted, but the surface was still as clean and smooth as that of the feedstock (Figure 15b). In Figure 15c and Figure 15d, after the MW–UW assisted liquefaction, the structure of the sawdust was not only completely deconstructed, but also a molten coating with some "granulated slags" appeared and seemed to completely cover the surface of the sawdust to protect it from the corrosion of the acid catalyst and the solvent.

Current Issues with the Hybrid Technology and Future Prospects

The prospects for synergistic effect of microwave and ultrasound enhanced process intensification based on the green chemistry principles are tremendous in any organic or inorganic synthesis or process as shown in Figure 16. These two novel process intensification techniques have the potential to transform the current chemical synthesis and induce a paradigm shift. However, the costs (both capital and operational) of the process development seems to be an immediate concern. Apart from cost issues, other important factors influencing beneficial application of this hybrid technology are discussed next.

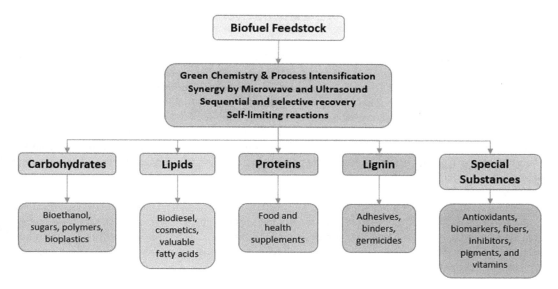

Figure 16. An advanced microwave/ultrasound enhanced biorefinery concept for sequential or selective recovery of various compounds suitable for synthesis of biofuels and bioproducts.

Understanding the biofuel chemistry—green chemistry is the path forward

It is paramount to understand and evaluate the process chemistry prior to determining the use of the MW/US technology. Different chemicals and solvents and materials interact with microwaves and ultrasound in different ways. Some materials absorb microwaves either completely or partially, some reflect and some transmit (let microwaves pass through). Similarly for ultrasound, the effect could be intense in some chemical synthesis and in some, it could be negligible. Microwave enhanced extraction using high dielectric solvents can be performed in closed vessels under high pressures (Mason et al. 2011). Since these solvents are heated rapidly by the microwaves, the temperature differential available in the reaction medium will drive the mass transfer between extractants and the feedstock. This process can be used for production of materials that are not sensitive to high temperatures. However, in the case of biodiesel production, high temperatures may not be favorable due to degradation of the valuable products such as algal biomass. A low dielectric constant possessing solvent can be used to extract the lipids. An example of low dielectric constant solvent is hexane. In a simultaneous extraction and transesterification reaction, addition of hexane helps increase the reaction time to expand the exposure of microwaves to the reaction mixture. Hexane is an excellent solvent but has very low microwave absorption properties. Hexane can be used as a temperature controlling agent while at the same time increasing heat and mass transfer in the extraction and transesterification reactions. Another example is ethanol. It is a good solvent and reactant but its role in biodiesel synthesis is quite different from methanol. Ethanol also has good miscibility into organic solvents when compared with methanol which may increase the mass transfer properties especially under ultrasound irradiation.

The synergism of microwaves and ultrasound has to be properly identified. This requires a mechanistic approach to elucidate the individual effects on the desired reaction. MW/US synergism may not be beneficial in all biofuel production processes. Either microwave or ultrasound could be adequate for simple reactions. This needs to be verified to avoid misuse of resources. For example, in our recent study, we reported that ultrasound irradiation produced the highest biodiesel yield when compared with microwaves and combined microwave and ultrasound effect in a transesterification reaction.

Improving the energy efficiency in the MW/US hybrid reactors

Energy efficiency refers to utilization of energy in the reaction and the conversion efficiency of the electrical energy into microwave or ultrasound energy. The knowledge gap in efficient utilization of energy released through the two process intensification techniques must be addressed. The power density (W/m^3) and energy intensity (W/m^2) have to be analyzed for efficient utilization of energy (Martinez-Guerra, V.G. Gude 2014a, Gude 2015). These two process related factors will aid in process reactor design and product optimization. Power density, which depends on the volume of the reaction mixture, expressed in small scales as W/mL gives an estimate of the optimum energy required for the desired biofuel chemistry while the term energy intensity provides the basis for efficient design of the reactor to induce required energy intensity. The dimensions such as diameter and length of a reactor can be determined based on the optimum energy intensity which would also provide ideal power density for the reaction. Simultaneous or sequential application and the order of sequential application of the two process intensification effects should also be studied through well-designed experimental plan and analytical procedures. Standardization of the experimental results is important considering the present limitations in the literature. The energy density, reaction temperature, frequency (microwave or ultrasound) and amplitude (ultrasound) and other pertinent information should be reported along with reactor specifications for all the studies. This ensures unbiased representation and evaluation of the data.

Increasing recovery of valuable products

Current issues with the microwave or ultrasound based methods is the cost intensive nature of the processes. Potential features that make these processes attractive are the high (cost) value of the product, significant advantage in the processing over conventional processing, limited plant space, and low cost electricity. In biofuel production, some of these requirements cannot be met. For example, the desired product (biofuel) costs are not favorable for implementation of these processes. For this reason, if selective recovery of valuable products from the biomass can be enhanced by manipulating the unique characteristics of these two process intensification effects, such process can be justified from all chemical, energy and cost related aspects.

The capital cost requirements for ultrasound, microwave and sequential mode of operations were reported recently (Gole and Gogate 2013). The capital cost requirements were reported to be 17.8%, 10%, and 7.5% higher when compared to the conventional method for sequential approach, microwave and ultrasound, respectively. That study also reported that despite the capital cost requirement being marginally increased for the sequential mode of operation, there will be considerable savings in terms of methanol requirement (34%) and utility (2% preheating of oil) and processing times. The novel sequential operation approach presented this work can significantly reduce the operating cost requirement for biodiesel synthesis, giving overall favorable economics.

Improving reactor design

Reducing the cost of the construction for the hybrid reactors and improving the reactor designs for energy efficiency may make these process intensifications more relevant to biorefineries. For example, a synergistic effect in small reactor systems such as plug-flow reactors can be produced to promote gradient-less mixing which may lead to improved reaction rates and lower energy consumption. In large batch reactors, ultrasonics may not be able to induce the chemical/physical effects into the entire reactor contents, which depend on the reaction contents and their properties. The same applies to the microwave irradiation (Gude et al. 2013). Ultrasound frequencies between 20 and 40 kHz are reported to be suitable for extraction and chemical conversion (e.g., esterification and transesterification reactions) of biofuel feedstock (Gole and Gogate 2012). About 20–25 kHz frequency is frequently reported in these applications. Higher frequencies can be used in other applications such as algal cell cultivation, harvesting and other applications. The ability of microwaves to penetrate through materials also limits its applications. For large-scale design, two essential strategies can be considered. For simple chemical conversions such as a transesterification reaction involving oils, solvents and catalysts, microwave irradiation frequency of 2450 MHz is adequate with ultrasonic horns in a plug-flow type or contact-type reactor design. For oil extraction and transesterification reactions (such as direct extraction and transesterification of algal lipids), the frequency of the microwaves and ultrasound may need to be altered (Gude et al. 2011). Microwaves at 915 MHz (used industrially) have much higher penetration depths into the material when compared to the higher frequency of 2450 MHz commonly used in laboratory-sized equipment. The higher penetration depths allow for much larger diameter tubes and processing flow rates. Microwave generators can be built for significantly higher power efficiencies when compared to smaller generators. For ultrasound, longitudinally vibrating horns can be beneficial for continuous processing (Martinez-Guerra, V.G. Gude 2014a). Control of microwave-ultrasonic reactions is also subject to similar limitations as any thermal process; however, their intensity and energy supply can be controlled easily to achieve desired reactions (Cintas et al. 2010).

Biorefineries Using Renewable Feedstock

Integrating the production of higher-value chemical/material co-products into the biorefinery's fuel and power output will improve the overall profitability and productivity of all energy-related products. Increased profitability makes it more attractive for new biobased companies to contribute to our domestic fuel and power supply by investing in new biorefineries. Increased productivity and efficiency can also be achieved through operations that lower the overall energy intensity of the biorefinery's unit operations; reduce overall carbon dioxide emissions; maximize the use of all feedstock components, byproducts and waste streams; and use economies of scale, common processing operations, materials, and equipment to drive down production costs (Holladay et al. 2007).

In the most advanced sense, a biorefinery is a facility with integrated, efficient and flexible conversion of biomass feedstocks, through a combination of physical, chemical, biochemical and thermochemical processes, into multiple products (Sadhukhan et al. 2016). The concept was developed by an analogy to the complex crude oil refineries adopting the process engineering principles applied in their designs, such as feedstock fractionation, multiple value-added productions, process flexibility and integration. A biorefinery is a facility that integrates biomass conversion processes and equipment to produce fuels, power, heat and value-added chemicals from biomass. The main technologies available to the biorefinery can be classified as extraction, biochemical and thermochemical processes (Budarin et al. 2011). In this construct, biomass is separated into its component parts: sugars (as cellulose, hemicellulose or starch), lignin, protein and oils. In various current biorefinery concepts, the sugar or oil fractions are used to produce liquid transport fuel or products while lignin is most often relegated to low-value uses of combustion (Figure 16). In fact, in the currently operating biorefineries lignin is either burned to produce process heat and recover pulping chemicals in paper mills or sold as a natural component of animal feeds in wet or dry corn mills.

Algae as feedstock

Microwave and ultrasound irradiations can be utilized for selective recovery of carbohydrates, lipids, proteins, lignin, and special substances to produce a variety of high value bioproducts (see Figure 16). Microwaves can be utilized for selective heating that promotes self-limiting reactions which in turn may enhance the selective recovery of valuable byproducts. For example, proteins can be extracted from algal biomass prior to extracting lipids for biodiesel production (see Figure 17). Microalgae can be used in bioelectrochemical systems such as biocatalysts to produce bioelectricity or bio-hydrogen prior to harvesting and processing for fermentation and anaerobic digestion to produce bioethanol and biogas, respectively. As an alternative, the lipid rich algal residue can be processed to produce bioethanol or biogas, using thermochemical or biochemical processes. Combustion or incineration is another well-established method

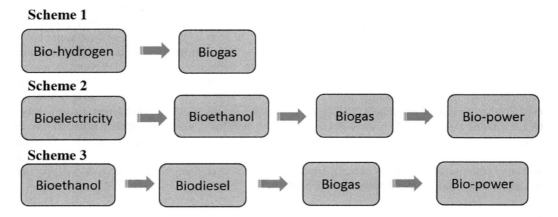

Figure 17. Possible energy recovery schemes from algae and other lignocellulosic feedstock (Bioethanol via fermentation process; Biodiesel via lipid extraction and transesterification process; Biogas via anaerobic decomposition; Bio-power via solids incineration; Bioelectricity and Biohydrogen via microbial fuel cells. The above chemical and energy product recovery schemes can be integrated in a biorefinery concept).

for power generation. These individual process techniques can be combined with complementing schemes through a biorefinery concept to enhance the energy and product recovery benefits. Table 3 shows the application of microwave irradiation for recovery of various valuable byproduct recovery from various microalgae species.

Lignocellulosic feedstock

Lignocellulosic biomass can be treated using biological (hydrolysis–fermentation) and thermochemical (combustion, pyrolysis, gasification) routes which result in different products, including cellulosic ethanol, electricity, pyrolysis oils, charcoal, surplus heat and other transportation fuels (see Figure 18).

Biofuels and bioproducts from catalytic lignin transformations

Lignin offers a significant opportunity for enhancing the operation of a lignocellulosic biorefinery. It is an extremely abundant raw material contributing as much as 30% of the weight and 40% of the energy content of lignocellulosic biomass. Lignin's native structure suggests that it could play a central role as a new chemical feedstock, particularly in the formation of supramolecular materials and aromatic chemicals. Conversion technologies must be developed for value-added applications, especially for the preparation of low-molecular-weight compounds as an alternative to the petrochemical industry.

Lignin-based opportunities could readily be divided into three categories: (1) power, green fuels, and syngas; (2) macromolecules; and (3) aromatics and other chemicals (Holladay et al. 2007). The first category represents the use of lignin purely as a carbon source using aggressive means to break down its polymeric structure. The second category is the opposite extreme and seeks to take advantage of the macromolecular structure imparted by nature in high-molecular weight applications. Somewhere between the two extremes come technologies that would break up lignin's macromolecular structure but maintain the aromatic nature of the building block molecules, the third category. This is intriguing because lignin represents a potential starting source for the roughly 45 billion-pound domestic, non-fuel, aromatic supply chain. Figure 19 shows potential catalytic transformations of lignin into valuable chemical and fuel products which can be conveniently mediated by microwave technology.

Table 3. Microwave based valuable product recovery (Gunerken et al. 2015).

Microalgae	Product	Conditions	Scale	Outcome	Analyses	Reference
Stichococcus sp. *Chlorella* sp.	Chlorophyll a	70 W, 90 s 3 cycles with 5 min breaks	3 ml	Local heat caused degradation of chlorophyll a	Chlorophyll a	Schumann et al. 2005
Scenedesmus dimorphus	Lipid	100 W, 2 min, 2 cycles	15 ml	Lipid recovery 21 wt.% No considerable difference in comparison with methods	Total lipids, dry weight	Shen et al. 2009
Chlorella protothecoides				Lipid recovery 10.7 wt.% Considerable difference in comparison with methods		
Botryococcus sp.	Lipid	10 kHz, 5 min, 0.5% DCW	100 ml	Lipid recovery 8.8 wt.% Considerable difference in comparison with methods	Total lipid, fatty acid composition	Lee et al. 2010
Chlorella vulgaris				Lipid recovery 8 wt.% No considerable difference in comparison with methods		
Scenedesmus sp.				Lipid recovery 9 wt.% No considerable difference in comparison with methods		
Nannochloropsis salina	Anaerobic digestion and Biogas from treated Biomass	200 W, 45 s, 30 kHz	Analytical, volume not given	21% decrease in biogas production in comparison with untreated biomass	Biogas production	Schwede et al. 2011
Chlorella vulgaris	Lipid	600 W, 30 s 34 cycles with 5 second breaks	Laboratory (N50 ml), volume not given	5.11 fold more extraction than untreated cells	Total lipid	Zheng et al. 2011
Chlorella sp.	Lipid	50 kHz, 15 min, 0.5% DCW	100 ml	2.625 fold more extraction than untreated cells	Total lipid	Prabakaran and Ravindran 2011
Nostoc sp.				2.57 fold more extraction than untreated cells		
Tolypothrix sp.				3.625 fold more extraction than untreated cells		
Chlorococcum sp.	Disrupted biomass	130 W, 5 min, 5 cycles, 0.85% DCW	200 ml	Nearly no cell disruption, ≈ 70% of colony diameter reduction after 3rd cycle 36.67 kWh/kg dry biomass energy consumption	Intact cell count, average colony diameter Energy calculations by using the data from Halim et al. (2012b)	Halim et al. 2012 Lee et al. 2012

Scenedesmus obliquus	Fermentable sugars	200 W, 30 s 5 cycles with 10 min breaks, approx. 7–10% DWC	5 ml	Complex sugars were converted to fermentable sugars, yield: 0.025 equal gram of glucose/gram biomass	Total sugars, monosaccharides	Miranda et al. 2012
Synechocystis PCC 6803	Lipid	2 kW, 3 min, 52°C outflow temperature, approx. 0.2% DCW	Analytical, volume not given	27.8% (w/w) Lipid release, SCOD increase as much as 29.8% of total COD of biomass	Total lipid, SCOD analysis	Sheng et al. 2012
		2 kW, 30 s 15 cycles with 30 s breaks, 26°C outflow temperature, approx. 0.2% DCW		14.77% (w/w) lipid release, SCOD increase as much as 6.7% of total COD of biomass		

Microwave Hybridization for Advanced Biorefinery 371

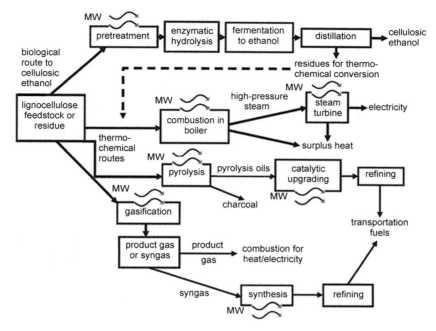

Figure 18. Alternative process routes for conversion of lignocellulose to bioenergy products (Van Zyl et al. 2011).

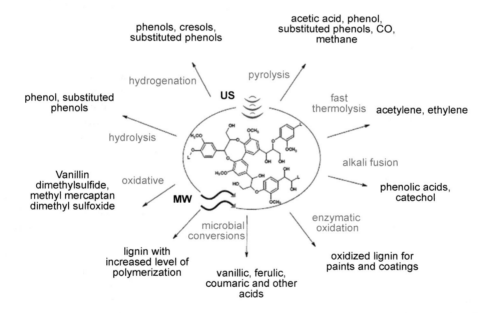

Figure 19. Catalytic lignin transformations (adapted from Holladay et al. 2007).

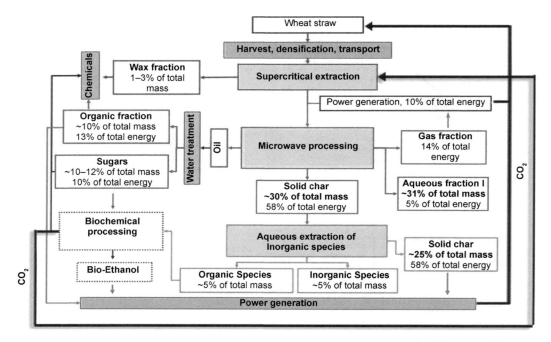

Figure 20. Integrated wheat straw biorefinery utilizing supercritical CO_2 and low temperature microwave pyrolysis to produce a range of products and showing the mass balance of the system (Budarin et al. 2012).

Biorefinery based on agricultural feedstock

Pyrolysis, thermochemical decomposition of organic material at elevated temperatures without the participation of oxygen, is one of the most promising technologies of biorefinery. A new biorefinery concept for wheat straw was proposed by Budarin et al. (2011) as shown in Figure 20. Two novel green technologies, CO_2 extraction and low temperature microwave pyrolysis were integrated, to produce a variety of products, including energy and CO_2 which can be internally recycled to sustain the processes. CO_2 adds value to the process by extracting secondary metabolites including fatty acids, wax esters and fatty alcohols. Low temperature microwave pyrolysis (< 200°C) is shown to use less energy and produce higher quality oils and chars than conventional pyrolysis. The oils can be fractionated to produce either transport fuels or platform chemicals such as levoglucosan and levoglucosenone. The chars are appropriate for co-firing. The quality of the chars was improved by washing to remove the majority of the potassium and chlorine present, lowering their fouling potential. The economic feasibility of a wheat straw biorefinery is enhanced by intergrating these technologies.

The microwave process is very tolerant of water compared to conventional pyrolysis and is suitable for most biomass types without pre-drying. Preliminary energy balance calculations based on the thermodynamic properties of the structural components of wheat straw during the decomposition process indicated an energy requirement of 1.8 kJ g^{-1} for microwave pyrolysis when compared to 2.7 kJ g^{-1} for thermal convectional pyrolysis. The microwave energy requirement decreased from 100 kJ g^{-1} to 2.2 kJ g^{-1} as the sample mass increased from 0.2 g to 200 g wheat straw. Pilot scale studies (30 kg h^{-1}), showed a further increase in energy efficiency, requiring as little as 1.9 kJ g^{-1} biomass. Low temperature microwave pyrolysis therefore produces better quality oils and chars than conventional pyrolysis at 1.5 times the energy efficiency.

Low temperature microwave pyrolysis generates four major product fractions. The main aqueous fraction has possible uses as a disinfectant or antifungal agent. While the solid char and bio-oil can both be upgraded by water treatment, solubilizing sugars and inorganic components, to increase end value. The remaining organic bio-oil may be a suitable starting material for the production of transportation fuels. Gaseous streams from the processing of biomass and combustion of the char can internally generate the CO_2 required for $scCO_2$ extraction and the energy for the whole system. The use of additives is shown to further increase the flexibility of the system.

Bioethanol and biogas from agricultural waste

Palm oil production is a major industry in many countries. For example, Malaysia produced more than 58 million tonnes of Palm Oil Mill Effluent (POME) in 2006. Saifuddin and Fazlili (2009) studied the effects of pre-treatment of palm oil mill effluent by microwave irradiation and ultrasonic on anaerobic digestion. Soluble Chemical Oxygen Demand (soluble COD)/total COD ratio and biodegradability of soluble organic matter increased significantly after both the pre-treatments which indicated an increase in disintegration of the floc structure of the sludge. Three identical bioreactors with working volumes of 5 liters were used as anaerobic digesters at 32–35°C. The reactors were separately fed with pretreated sludge (microwave, ultrasonic and combination of microwave and ultrasonic) and control sludge at different hydraulic residence times to evaluate methane production. The maximum soluble COD/total COD ratio reached almost 29% after 30 min of ultrasonic treatment, while it was 45% after 7 min of microwave irradiation. The BOD_5/soluble COD ratio also increased after the pretreatment with microwave or ultrasound suggesting the biodegradability of the soluble organic material increased during the treatment. Release of total volatile fatty acids increased after both the treatments, with higher yields by microwave treatment. Greatest enhancement in methane production was shown by the 3 min microwave plus 10 min ultrasonic treatment (Saifuddin and Fazlili 2009).

Co-pretreatment of OPF with water and microwave had high cellulose content and a high yield of ethanol when compared with sulphuric acid and microwave, hydrogen peroxide and microwave pretreatment (Srimachai et al. 2014). The co-pretreatment with water and microwave gave maximum ethanol yield of 0.32 g-ethanol/g-glucose which was 62.75% of the ethanol theoretical yield (0.51g-ethanol/g-glucose) and maximum methane yield was 512 ml CH_4/g VS added. Therefore, water and microwave co-pretreatment was helpful because it had low toxicity and low corrosion compared to sulfuric acid and microwave or hydrogen peroxide and microwave pretreatment which this process efficiently increased of enzymatic hydrolysis.

The mass and energy balance were used for assessing the energy output of OPF in the SSF process (Figure 21). The energy output from 1000 g of OPF were 30.31 g ethanol (SAP = 26 g-ethanol, SSF =

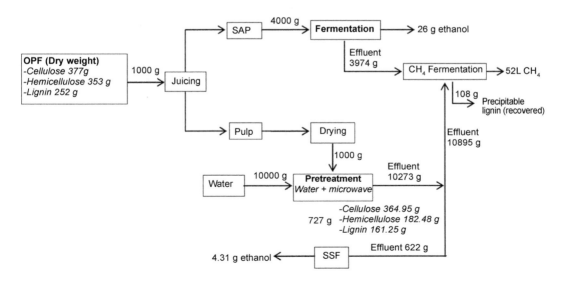

Figure 21. Material balance of the ethanol and methane from palm oil frond by two stage simultaneous saccharification and fermentation, SSF (Saifuddin and Fazlili 2009).

4.31 g-ethanol) and 52 LCH$_4$. The energy value of the ethanol was 30.31 kg-ethanol/ton of OPF or 809 MJ/ton of OPF (lower heating values of 26.7 MJ/kg-ethanol) (Saifuddin and Fazlili 2009). The methane production of effluent from ethanol production was 52 m^3 CH$_4$/ton of OPF. Rabelo et al. 2011 reported that the highest methane production was 72.1 m^3 CH$_4$/ton of bagasse when pretreated with the peroxide. The electricity production of 1 ton of OPF would be 1,872 MJ (1 m^3 CH$_4$ = 36 MJ) or 518 kWh (1 m^3 CH$_4$ = 9.96 kWh) of electricity. Finally, the total energy output from two stage SSF was 2,681 MJ/ton of OPF or 745 kWh/ton of OPF (1 kWh = 3.6 MJ).

Biorefinery using food waste as feedstock

With approximately 100 million cubic tonnes produced per year, citrus fruit is by far the chief fruit crop in the world, with oranges comprising 60% of the total production (Attard et al. 2014). Among the fruit processing industry, orange juicing is a very wasteful process, with typically 50% waste being generated. About 20 million tonnes of unwanted orange peel per year are produced by the juicing industry every year; a potential source of organic molecules of significant value which can be exploited as an alternative natural source of platform molecules. The composition of orange peel in a variety of species has been reported as water (80%); and sugars, cellulose, hemicellulose, pectin and D-limonene which make up the dry matter (20%).

A new biorefinery concept based on orange peels waste as feedstock was proposed by Boukroufa et al. (2015) as shown in Figure 22. Solvent free extraction of essential oil, polyphenols and pectin from orange peel was optimized using microwave and ultrasound irradiations with *in situ* water which was recycled and used as a solvent. The essential oil extraction performed by microwave hydrodiffusion and gravity (MHG) was optimized and compared to steam distillation extraction (SD). MHG and SD produced similar yields which were 4.22 ± 0.03% and 4.16 ± 0.05% for MHG and SD, respectively. Ultrasound assisted extraction (UAE) and conventional extraction (CE) of polyphenols from the MHG residues were conducted using the residual water from MHG process. The optimized conditions of ultrasound power and

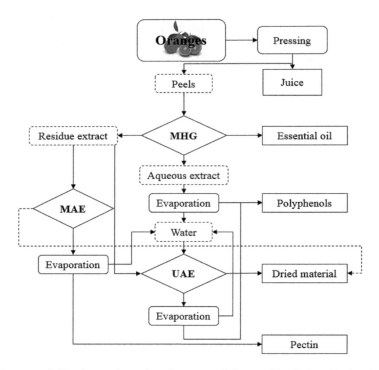

Figure 22. Orange peels biorefinery using under microwave and ultrasound irradiations (Boukroufa et al. 2015).

temperature were reported as 0.956 W/cm² and 59.83°C for a polyphenol yield of 50.02 mg GA/100 g dm. Compared with the conventional extraction (CE), the UAE gave an increase of 30% in TPC yield. Pectin was extracted by conventional and microwave assisted extraction. This technique gives a maximal yield of 24.2% for microwave power of 500 W in only 3 min whereas conventional extraction gives 18.32% in 120 min. Combination of microwave, ultrasound and the recycled *in situ* water of citrus peels allow us to obtain high added values compounds in shorter time and managed to make a closed loop using only natural resources provided by the plant which makes the whole process intensified in term of time and energy saving, cleanliness and reduced waste water.

Concluding Remarks

There are several technologies available for process intensification. Current approach in research is to study these methodologies and technologies individually and promote them as the sole biorefinery technology of the future. However, it is essential that the strengths and weaknesses of all the technologies available are recognized to enable the integration and blending of different technologies and feedstocks to best maximize the diversity of applications and products formed. In this context, superior benefits associated with hybridization of process intensification techniques such as microwaves and ultrasounds come with greater challenges and limitations. These limitations provide opportunities for further research in the near future. Heat and mass transfer limitations have been addressed in the hybrid systems but the power control needs to be improved. One major hurdle that needs to be overcome at present is to make this hybrid system economically viable. The application of longitudinal ultrasound application together with microwaves in continuous plug-flow type reactor design must be developed for efficient utilization of these two process intensification methods. Exploiting the unique characteristics inherent to microwaves and ultrasound such as rapid and selective heating and self-limiting reactions, heat and mass transport and energy dissipation for selective and sequential recovery of multiple high value added bioproducts seems to be an important endeavor to address the cost issues associated with this hybrid technology considering the present biorefinery economics. Integrated approaches to maximize energy and material (resource) utilization efficiencies through a variety of feedstock is essential for sustainable biorefinery development.

References

Attard, T.M., Watterson, B., Budarin, V.L., Clark, J.H. and Hunt, A.J. 2014. Microwave assisted extraction as an important technology for valorising orange waste. *New Journal of Chemistry*. 38(6): pp.2278–2283.
Balat, M. and Balat, H. 2009. Recent trends in global production and utilization of bio-ethanol fuel. *Applied Energy*. 86(11): pp.2273–2282.
Boukroufa, M., Boutekedjiret, C., Petigny, L., Rakotomanomana, N. and Chemat, F. 2015. Bio-refinery of orange peels waste: a new concept based on integrated green and solvent free extraction processes using ultrasound and microwave techniques to obtain essential oil, polyphenols and pectin. *Ultrasonics Sonochemistry*. 24: pp.72–79.
Brunetti, A., Macedonio, F., Barbieri, G. and Drioli, E. 2015. Membrane engineering for environmental protection and sustainable industrial growth: Options for water and gas treatment. *Environmental Engineering Research*. 20(4): pp.307–328.
Braud, L.M., Moreira, R.G. and Castell-Perez, M.E. 2001. Mathematical modeling of impingement drying of corn tortillas. *Journal of Food Engineering*. 50(3): pp.121–128.
Budarin, V.L., Shuttleworth, P.S., Dodson, J.R., Hunt, A.J., Lanigan, B., Marriott, R., Milkowski, K.J., Wilson, A.J., Breeden, S.W., Fan, J. and Sin, E.H. 2011. Use of green chemical technologies in an integrated biorefinery. *Energy and Environmental Science*. 4(2): pp.471–479.
Budarin, V., Ross, A.B., Biller, P., Riley, R., Clark, J.H., Jones, J.M., Gilmour, D.J. and Zimmerman, W. 2012. Microalgae biorefinery concept based on hydrothermal microwave pyrolysis. *Green Chemistry*. 14(12): pp.3251–3254.
Bussemaker, M.J. and Zhang, D. 2013. Effect of ultrasound on lignocellulosic biomass as a pretreatment for biorefinery and biofuel applications. *Industrial & Engineering Chemistry Research*. 52(10): pp.3563–3580.
Chemat, F., Poux, M., Martino, J.D. and Berlan, J. 1996. An original microwave-ultrasound combined reactor suitable for organic synthesis: application to pyrolysis and esterification. *Journal of Microwave Power and Electromagnetic Energy*. 31(1): pp.19–22.
Chemat, S., Lagha, A., Amar, H.A. and Chemat, F. 2004. Ultrasound assisted microwave digestion. *Ultrasonicssonochemistry*. 11(1): pp.5–8.

Choedkiatsakul, I., Ngaosuwan, K., Assabumrungrat, S., Tabasso, S. and Cravotto, G. 2015. Integrated flow reactor that combines high-shear mixing and microwave irradiation for biodiesel production. *Biomass and Bioenergy*. 77: pp.186–191.

Choi, J.H. and Kim, S.B. 1994. Effect of ultrasound on sulfuric acid-catalysed hydrolysis of starch. *Korean Journal of Chemical Engineering*. 11(3): pp.178–184.

Choi, W., Lee, S.H., Kim, C.T. and Jun, S. 2015. A finite element method based flow and heat transfer model of continuous flow microwave and ohmic combination heating for particulate foods. *Journal of Food Engineering*. 149: pp.159–170.

Cintas, P., Mantegna, S., Gaudino, E.C. and Cravotto, G. 2010. A new pilot flow reactor for high-intensity ultrasound irradiation. Application to the synthesis of biodiesel. *Ultrasonics Sonochemistry*. 17(6): pp.985–989.

Cravotto, G. and Cintas, P. 2007. The combined use of microwaves and ultrasound: improved tools in process chemistry and organic synthesis. *Chemistry–A European Journal*. 13(7): pp.1902–1909.

Cravotto, G., Boffa, L., Mantegna, S., Perego, P., Avogadro, M. and Cintas, P. 2008. Improved extraction of vegetable oils under high-intensity ultrasound and/or microwaves. *Ultrasonics Sonochemistry*. 15(5): pp.898–902.

Czechowska-Biskup, R., Rokita, B., Lotfy, S., Ulanski, P. and Rosiak, J.M. 2005. Degradation of chitosan and starch by 360-kHz ultrasound. *Carbohydrate Polymers*. 60(2): pp.175–184.

Datta, A.K. and Rakesh, V. 2013. Principles of microwave combination heating. *Comprehensive Reviews in Food Science and Food Safety*. 12(1): pp.24–39.

Elizabeth Grant, G. and Gnaneswar Gude, V. 2014. Kinetics of ultrasonic transesterification of waste cooking oil. *Environmental Progress & Sustainable Energy*. 33(3): pp.1051–1058.

Energy, C. 2007. European roadmap for process intensification. *Creative Energy*: The Netherlands.

Fast, S.A., Kokabian, B. and Gude, V.G. 2014. Chitosan enhanced coagulation of algal turbid waters–Comparison between rapid mix and ultrasound coagulation methods. *Chemical Engineering Journal*. 244: pp.403–410.

Fast, S.A. and Gude, V.G. 2015. Ultrasound-chitosan enhanced flocculation of low algal turbid waters. *Journal of Industrial and Engineering Chemistry*. 24: pp.153–160.

Flosdorf, E.W. and Chambers, L.A. 1933. The chemical action of audible sound. *Journal of the American Chemical Society*. 55(7): pp.3051–3052.

Gole, V.L. and Gogate, P.R. 2012. A review on intensification of synthesis of biodiesel from sustainable feed stock using sonochemical reactors. *Chemical Engineering and Processing: Process Intensification*. 53: pp.1–9.

Gole, V.L. and Gogate, P.R. 2013. Intensification of synthesis of biodiesel from non-edible oil using sequential combination of microwave and ultrasound. *Fuel Processing Technology*. 106: pp.62–69.

Gude, V.G., Patil, P.D., Deng, S. and Nirmalakhandan, N. 2011. Microwave enhanced methods for biodiesel production and other environmental applications. *Green Chemistry for Environmental Remediation*, New York: Wiley Interscience. pp.209–249.

Gude, V., Grant, G., Patil, P. and Deng, S. 2013a. Biodiesel production from low cost and renewable feedstock. *Open Engineering*. 3(4): pp.595–605.

Gude, V.G., Patil, P., Martinez-Guerra, E., Deng, S. and Nirmalakhandan, N. 2013b. Microwave energy potential for biodiesel production. *Sustainable Chemical Processes*. 1(1): p.5.

Gude, V.G. and Grant, G.E. 2013. Biodiesel from waste cooking oils via direct sonication. *Applied Energy*. 109: pp.135–144.

Gude, V.G. and Martinez-Guerra, E. 2015. Green chemistry of microwave-enhanced biodiesel production. pp.225–250. *In: Production of Biofuels and Chemicals with Microwave*. Springer Netherlands.

Gude, V.G. 2015. Synergism of microwaves and ultrasound for advanced biorefineries. *Resource-Efficient Technologies*. 1(2): pp.116–125.

Guldhe, A., Singh, B., Rawat, I. and Bux, F. 2014. Synthesis of biodiesel from Scenedesmus sp. by microwave and ultrasound assisted *in situ* transesterification using tungstated zirconia as a solid acid catalyst. *Chemical Engineering Research and Design*. 92(8): pp.1503–1511.

Günerken, E., D'Hondt, E., Eppink, M.H.M., Garcia-Gonzalez, L., Elst, K. and Wijffels, R.H. 2015. Cell disruption for microalgae biorefineries. *Biotechnology Advances*. 33(2): pp.243–260.

Halim, R., Danquah, M.K. and Webley, P.A. 2012. Extraction of oil from microalgae for biodiesel production: A review. *Biotechnology Advances*. 30(3): pp.709–732.

Hernoux-Villière, A., Lassi, U., Hu, T., Paquet, A., Rinaldi, L., Cravotto, G., Molina-Boisseau, S., Marais, M.F. and Lévêque, J.M. 2013. Simultaneous microwave/ultrasound-assisted hydrolysis of starch-based industrial waste into reducing sugars. *ACS Sustainable Chemistry & Engineering*. 1(8): pp.995–1002.

Holladay, J.E., White, J.F., Bozell, J.J. and Johnson, D. 2007. Top Value-Added Chemicals from Biomass-Volume II—Results of Screening for Potential Candidates from Biorefinery Lignin (No. PNNL-16983). Pacific Northwest National Laboratory (PNNL), Richland, WA (US).

Huang, K., Wang, S.J., Shan, L., Zhu, Q. and Qian, J. 2007. Seeking synergistic effect—a key principle in process intensification. *Separation and Purification Technology*. 57(1): pp.111–120.

Jindarat, W., Sungsoontorn, S. and Rattanadecho, P. 2013. Analysis of energy consumption in drying process of biomaterials using a combined unsymmetrical double-feed microwave and vacuum system (CUMV)—Case study: Tea leaves. *Drying Technology*. 31(10): pp.1138–1147.

Jindarat, W., Sungsoontorn, S. and Rattanadecho, P. 2015. Analysis of energy consumption in a combined microwave–hot air spouted bed drying of biomaterial: Coffee beans. *Experimental Heat Transfer*. 28(2): pp.107–124.

Khan, A.R., Johnson, J.A. and Robinson, R.J. 1979. Degradation of starch polymers by microwave energy. *Cereal Chemistry*. 56: pp.303–304.

Khan, A.R., Robinson, R.J. and Johnson, J.A. 1980. Starch hydrolysis by acid and microwave energy. *Journal of Food Science*. 45(5): pp.1449–1449.

Knirsch, M.C., Dos Santos, C.A., de Oliveira Soares, A.A.M. and Penna, T.C.V. 2010. Ohmic heating–a review. *Trends in Food Science & Technology*. 21(9): pp.436–441.

Koda, S., Taguchi, K. and Futamura, K. 2011. Effects of frequency and a radical scavenger on ultrasonic degradation of water-soluble polymers. *Ultrasonics Sonochemistry*. 18(1): pp.276–281.

Kumar, V. and Nigam, K.D.P. 2012. Process intensification in green synthesis. *Green Processing and Synthesis*. 1(1): pp.79–107.

Kunlan, L., Lixin, X., Jun, L., Jun, P., Guoying, C. and Zuwei, X. 2001. Salt-assisted acid hydrolysis of starch to D-glucose under microwave irradiation. *Carbohydrate Research*. 331(1): pp.9–12.

Lagha, A., Chemat, S., Bartels, P.V. and Chemat, F. 1999. Microwave-ultrasound combined reactor suitable for atmospheric sample preparation procedure of biological and chemical products. *Analusis*. 27(5): pp.452–457.

Lee, J.Y., Yoo, C., Jun, S.Y., Ahn, C.Y. and Oh, H.M. 2010. Comparison of several methods for effective lipid extraction from microalgae. *Bioresource Technology*. 101(1): pp.S75–S77.

Lee, A.K., Lewis, D.M. and Ashman, P.J. 2012. Disruption of microalgal cells for the extraction of lipids for biofuels: processes and specific energy requirements. *Biomass and Bioenergy*. 46: pp.89–101.

Li, B. and Sun, D.W. 2002. Novel methods for rapid freezing and thawing of foods–a review. *Journal of Food Engineering*. 54(3): pp.175–182.

Lianfu, Z. and Zelong, L. 2008. Optimization and comparison of ultrasound/microwave assisted extraction (UMAE) and ultrasonic assisted extraction (UAE) of lycopene from tomatoes. *Ultrasonics Sonochemistry*. 15(5): pp.731–737.

Lu, Z., Wu, Z., Fan, L., Zhang, H., Liao, Y., Zheng, D. and Wang, S. 2016. Rapid and solvent-saving liquefaction of woody biomass using microwave–ultrasonic assisted technology. *Bioresource Technology*. 199: pp.423–426.

Luo, J., Fang, Z. and Smith, R.L. 2014. Ultrasound-enhanced conversion of biomass to biofuels. *Progress in Energy and Combustion Science*. 41: pp.56–93.

Martinez-Guerra, E. and Gude, V.G. 2014a. Synergistic effect of simultaneous microwave and ultrasound irradiations on transesterification of waste vegetable oil. *Fuel*. 137: pp.100–108.

Martinez-Guerra, E. and Gude, V.G. 2014b. Transesterification of used vegetable oil catalyzed by barium oxide under simultaneous microwave and ultrasound irradiations. *Energy Conversion and Management*. 88: pp.633–640.

Martinez-Guerra, E., Gude, V.G., Mondala, A., Holmes, W. and Hernandez, R. 2014. Microwave and ultrasound enhanced extractive-transesterification of algal lipids. *Applied Energy*. 129: pp.354–363.

Martinez-Guerra, E. and Gude, V.G. 2015. Continuous and pulse sonication effects on transesterification of used vegetable oil. *Energy Conversion and Management*. 96: pp.268–276.

Mason, J.T., Chemat, F. and Vinatoru, M. 2011. The extraction of natural products using ultrasound or microwaves. *Current Organic Chemistry*. 15(2): pp.237–247.

Mengal, P. and Mompon, B. 1996. Method and apparatus for solvent free microwave extraction of natural products. *Eur. Pat. P.* EP, 698(076), p.B1.

Mengal, P. and Mompon, B., Archimex. 2006. *Method and Plant for Solvent-Free Microwave Extraction of Natural Products*. U.S. Patent 7,001,629.

Michailidis, P.A., Krokida, M.K. and Rahman, M.S. 2009. Data and models of density, shrinkage, and porosity. Food properties handbook. CRC Press, Boca Raton, FL, USA, pp.417–500.

Miranda, J.R., Passarinho, P.C. and Gouveia, L. 2012. Pre-treatment optimization of Scenedesmus obliquus microalga for bioethanol production. *Bioresource Technology*. 104: pp.342–348.

Nikolić, S., Mojović, L., Rakin, M., Pejin, D. and Pejin, J. 2011. Utilization of microwave and ultrasound pretreatments in the production of bioethanol from corn. *Clean Technologies and Environmental Policy*. 13(4): pp.587–594.

Patil, P., Gude, V.G., Pinappu, S. and Deng, S. 2011. Transesterification kinetics of Camelina sativa oil on metal oxide catalysts under conventional and microwave heating conditions. *Chemical Engineering Journal*. 168(3): pp.1296–1300.

Portenlänger, G. and Heusinger, H. 1997. The influence of frequency on the mechanical and radical effects for the ultrasonic degradation of dextranes. *Ultrasonics Sonochemistry*. 4(2): pp.127–130.

Prabakaran, P. and Ravindran, A.D. 2011. A comparative study on effective cell disruption methods for lipid extraction from microalgae. *Letters in Applied Microbiology*. 53(2): pp.150–154.

Rabelo, S.C., Carrere, H., Maciel Filho, R. and Costa, A.C. 2011. Production of bioethanol, methane and heat from sugarcane bagasse in a biorefinery concept. *Bioresource Technology*. 102(17): pp.7887–7895.

Rattanadecho, P. and Makul, N. 2016. Microwave-assisted drying: A review of the state-of-the-art. *Drying Technology*. 34(1): pp.1–38.

Reddy, H.K., Muppaneni, T., Patil, P.D., Ponnusamy, S., Cooke, P., Schaub, T. and Deng, S. 2014. Direct conversion of wet algae to crude biodiesel under supercritical ethanol conditions. *Fuel*. 115: pp.720–726.

Rittmann, B.E. 2008. Opportunities for renewable bioenergy using microorganisms. *Biotechnology and Bioengineering*. 100(2): pp.203–212.

Ruan, R., Ye, X., Chen, P., Doona, C., Taub, I. and Center, N.S. 2002. Ohmic heating. *The Nutrition Handbook for Food Processors*. pp.407–422.

Sadhukhan, J., Ng, K.S. and Martinez-Hernandez, E. 2016. Novel integrated mechanical biological chemical treatment (MBCT) systems for the production of levulinic acid from fraction of municipal solid waste: A comprehensive techno-economic analysis. *Bioresource Technology*.

Saifuddin, N. and Fazlili, S.A. 2009. Effect of microwave and ultrasonic pretreatments on biogas production from anaerobic digestion of palm oil mill effleunt. *American Journal of Engineering and Applied Sciences.* 2(1).

Schumann, R., Häubner, N., Klausch, S. and Karsten, U. 2005. Chlorophyll extraction methods for the quantification of green microalgae colonizing building facades. *International Biodeterioration & Biodegradation.* 55(3): pp.213–222.

Schwede, S., Kowalczyk, A., Gerber, M. and Span, R. 2011, May. Influence of different cell disruption techniques on mono digestion of algal biomass. *World Renewable Energy Congress, Linkoping, Sweden.* pp. 8–13.

Shen, X.F. 2009. Combining microwave and ultrasound irradiation for rapid synthesis of nanowires: a case study on Pb (OH) Br. *Journal of Chemical Technology and Biotechnology.* 84(12): pp.1811–1817.

Shen, Y., Pei, Z., Yuan, W. and Mao, E. 2009. Effect of nitrogen and extraction method on algae lipid yield. *International Journal of Agricultural and Biological Engineering.* 2(1): pp.51–57.

Sheng, J., Vannela, R. and Rittmann, B.E. 2012. Disruption of Synechocystis PCC 6803 for lipid extraction. *Water Science & Technology.* 65(3).

Silva, C.M., Ferreira, A.F., Dias, A.P. and Costa, M. 2015. A comparison between microalgae virtual biorefinery arrangements for bio-oil production based on lab-scale results. *Journal of Cleaner Production.*

Srimachai, T., Thonglimp, V. and Sompong, O. 2014. Ethanol and methane production from oil palm frond by two stage SSF. *Energy Procedia.* 52: pp.352–361.

Stankiewicz, A.I. and Moulijn, J.A. 2000. Process intensification: transforming chemical engineering. *Chemical Engineering Progress.* 96(1): pp.22–34.

Stankiewicz, A. 2006. Energy matters: alternative sources and forms of energy for intensification of chemical and biochemical processes. *Chemical Engineering Research and Design.* 84(7): pp.511–521.

Stouten, S.C., Noël, T., Wang, Q. and Hessel, V. 2013. A view through novel process windows. *Australian Journal of Chemistry.* 66(2): pp.121–130.

Suslick, K.S. 1989. The chemical effects of ultrasound. *Scientific American.* 260(2): pp.80–86.

Taherzadeh, M.J. and Karimi, K. 2008. Pretreatment of lignocellulosic wastes to improve ethanol and biogas production: a review. *International Journal of Molecular Sciences.* 9(9): pp.1621–1651.

Toursel, P. 1997. Extraction high-tech pour notes fraîches: Dossier arômessalés. *Process.* 1128: pp.38–41.

Van Gerven, T. and Stankiewicz, A. 2009. Structure, energy, synergy, time. The fundamentals of process intensification. *Industrial & Engineering Chemistry Research.* 48(5): pp.2465–2474.

Van Zyl, W.H., Chimphango, A.F.A., Den Haan, R., Görgens, J.F. and Chirwa, P.W.C. 2011. Next-generation cellulosic ethanol technologies and their contribution to a sustainable Africa. *Interface Focus.* 1(2): pp.196–211.

Vian, M.A., Fernandez, X., Visinoni, F. and Chemat, F. 2008. Microwave hydrodiffusion and gravity, a new technique for extraction of essential oils. *Journal of Chromatography A.* 1190(1): pp.14–17.

Yeneneh, A.M., Chong, S., Sen, T.K., Ang, H.M. and Kayaalp, A. 2013. Effect of ultrasonic, microwave and combined microwave–ultrasonic pretreatment of municipal sludge on anaerobic digester performance. *Water, Air, & Soil Pollution.* 224(5): pp.1–9.

Zheng, H., Yin, J., Gao, Z., Huang, H., Ji, X. and Dou, C. 2011. Disruption of Chlorella vulgaris cells for the release of biodiesel-producing lipids: a comparison of grinding, ultrasonication, bead milling, enzymatic lysis, and microwaves. *Applied Biochemistry and Biotechnology.* 164(7): pp.1215–1224.

Zill-E-Huma, H. 2010. *Microwave Hydro-diffusion and Gravity: A Novel Technique for Antioxidants Extraction* (Doctoral dissertation, Universitéd'Avignon).

Index

A

Absorber 35, 39, 55
Absorption 278, 280, 283, 296
Acetogenesis 206, 207, 212
Acid 160–162, 171–182, 184, 185, 187–191, 194, 195
Acid catalyst 101, 109, 118, 119, 124, 125, 134, 136
Acid pretreatment 209
Acid-microwave 182
Acidogenesis 206, 207, 212
Acoustic 354, 358
Activated carbon 40
Activation energy 47, 49, 111, 112, 118, 278, 279, 293–295, 297, 299–301, 303–305, 307
Agricultural 206, 219, 225, 226, 230, 346, 372, 373
Alcohol 46, 47, 52–54
Alcohol-oil ratio 119, 124
Aldehydes 179, 181
Algae 112, 113, 116, 126–131, 133, 138
Algorithm 284
Alkali-microwave 184
Alkaline 171–174, 177, 178, 182, 188
Alkaline pretreatment 210
Alkylation 76, 77
Amino acids 317
Ammonia 171, 176, 178, 182, 188, 206, 208, 209, 214, 219, 225, 227
Amorphous 171, 174, 179–181
Anaerobic digestion 206–209, 211–215, 218–220, 223, 225–229, 231, 233, 234, 302, 303, 363, 367, 369, 373
Analyte 104, 115, 116, 129
Anions 311–313, 315–320, 327–329, 334, 336, 340
Anton-Paar 66, 67, 75
Applicator 62–67, 70, 71, 75, 81–83, 87, 90, 92
Arrhenius equation 46, 49, 50, 110, 111, 115
Arrhenius law 304
Atom economy 12, 13, 16, 19–21, 23

B

Base catalyst 101, 107, 124
Beating 210
Behavior 165, 166, 182
Benzylation 49
Biochemical 240, 243, 248
Biochemical oxygen demand 226
Biodegradation 212, 213, 227, 229
Biodiesel 7, 9–12, 16–19, 22, 39, 50–53, 58, 283, 286, 291, 293, 294, 296, 308, 317–320, 323–327, 348, 352, 354–357, 359–362, 365–368

Biodiesel composition 137, 139, 140
Biodiesel properties 140, 141
Biodiesel quality 145
Bioethanol 7, 9, 10, 12, 18, 22, 158, 159, 161, 162, 164, 169, 171, 177, 185, 188, 189, 193, 194, 291, 300, 308, 359, 361–363, 367, 368, 373
Biofuels 1, 3, 7, 9, 14, 15, 21–23
Biogas 206–216, 218, 219, 221–234, 302, 308, 317, 318, 334, 335, 359, 363, 367–369, 373
Biological 160, 171, 176–178
Biomass 57, 58, 109, 114–117, 124, 126, 127, 129, 131, 142, 145, 147, 158–163, 167, 171–174, 176–179, 181, 182, 184–192, 194, 346, 361–370, 372, 373
Biomass loading 176, 186, 187, 190, 191, 194
Bio-oil 104, 115, 126, 127, 131, 243, 246–249, 251–257, 259, 260, 262, 264, 268–271
Bioproducts 194
Biorefineries 346, 347, 366, 367
Biotage 71, 72, 74–76
Biphasic 311, 317–319, 327, 336
Blending 210
BOD 217, 218, 231
Boiling point 33, 44–48, 52
Boiling point elevation 45, 46
Boltzmann's Constant 299
Bond energy 167
Bonding 327, 334
Brownian motion 29, 30

C

Carbohydrate 159–161, 165, 167, 171, 173, 177, 184, 194
Carbon dioxide 160, 161, 176
Carbon materials 39, 40
Carbon nanotube 40
Catalysis 3, 12–15, 21
Catalyst 290, 291, 293–296, 298, 299, 301, 305, 307
Cathode 62, 63
Cations 311, 312, 314–319, 327–329, 336, 340
Cavitation 354, 355, 358
Cavity 62, 65–68, 70–74, 80–82, 87, 90, 95
Cell disruption 287, 288
Cellulose 161–164, 167, 170–176, 178–182, 184–187, 190–192, 194, 196, 247, 249–252, 255, 256, 317, 327–329, 332–336, 338, 339
CEM Discover 72, 73
Char 241–247, 249–256, 259, 260, 263–267, 269, 270
Characteristics 28, 29, 35, 40, 57, 58
Charcoal 40

Chemical 100, 102, 104, 107, 114, 116, 118, 131, 134, 136, 142, 143
Chemical bond 163, 167
Chemical Oxygen Demand 214, 216, 217, 221, 229
Chemical synthesis 2, 3, 6, 12–14, 19
Chirality 312
Choke 64, 65
Coal 40
COD 210, 212–215, 220, 221, 226, 227, 231
Codigestion 227, 229, 230
Combination 171, 172, 177, 181, 184–186, 348–351, 353, 354, 357, 363, 367, 373, 375
Combined heating 351
Combustion 240, 241, 243
Complex permittivity 277
Components 62, 64, 73, 81, 90, 92
Composition 160–162, 170, 172, 177, 184, 185, 191, 208, 211, 219, 221, 223, 226, 227, 241–243, 246–248, 251–254, 256, 257, 259, 262, 264–266, 268, 269
Conductor 35, 39
Configurations 62, 246, 248, 258–261, 268
Continuous flow 72, 81, 92
Conventional 278–281, 284–287, 291–295, 298–301, 303, 305, 307
Conventional heating 2, 3, 5, 10, 15, 22
Conversion 293–296, 300–302, 304, 305, 307, 346–348, 352–363, 366–368, 371
Conveyor 88, 90, 93
Co-solvent 101, 116, 118, 126, 129, 134

D

Dairy industry 226, 227
Decomposition 206, 209, 211
Degradation 290, 291, 303, 305
Derivatives 4, 12, 13, 15
Design 62, 64–66, 68–70, 72, 73, 76, 80, 81, 88, 90, 95
Dielectric constant 28, 29, 34, 35, 37, 38, 45, 52–57
Dielectric loss 28, 33, 35–38, 44, 52–56, 58, 216, 217
Dielectric properties 216, 234, 246, 267, 268
Dielectric property 57
Dielectrometer 216, 217
Diels-Alder 48
Diffusion coefficient 279
Digestion 363, 367, 369, 373
Diglyceride 110, 145
Dipolar rotation 33
Dipole 28, 29, 31, 32, 34, 36, 38, 39, 46
Direct heating 43, 44
Direct transesterification 126, 127
Disruption 114–116
Dissolution 327–329, 333, 334
Distillation 92, 93
Distribution 66–68, 71, 73, 80, 88–90
Domestic microwave 41, 50
Drying 241, 250, 254, 259, 280, 350, 351, 372, 373

E

Economic analysis 95
Economics 223, 226, 232, 248, 268, 269, 271
Edible oils 112, 113, 120, 126

E-factor 16–21, 23
Efficiency 62, 65, 67, 68, 80, 82, 85, 87, 92
Electric field 28, 29, 32, 34, 35, 39–42
Electromagnetic 276, 277, 281, 288
Electromagnetic spectrum 29
Energy 319, 324, 329, 330, 334, 340
Energy balance 99, 142
Energy consumption 179, 186–188, 192–194, 290, 291
Energy density 288, 290
Energy efficiency 7, 13, 14, 21, 23, 62, 68, 85, 87, 104, 125, 142, 143, 347, 351, 353, 354, 366, 372
Energy intensity 348, 366, 367
Energy recovery 223, 227, 232
Energy return on investment 143–145
Energy transport 280
Enthalpy 111, 116, 299
Entropy 111, 116, 278, 299
Enzymatic 319, 326, 334
Enzymes 161, 171, 173, 176–178, 186
EPS 212, 213
ESS 212
Esterification 4, 9–11, 44, 47, 48, 50–52, 109, 136
Ethanol 286, 291–295
Exposure 164, 167, 169, 184, 187, 188, 190, 278, 279
Extraction 99–106, 109, 114–117, 125–127, 129, 137, 138, 144, 147, 348–352, 354, 357, 359, 361, 365–369, 372–375
Extraction index 348, 349
Extraction yields 284–286, 290
Extractive-transesterification 126, 127

F

Factors 164, 165, 174, 177, 186, 187
Fatty acid ethyl esters 108, 125, 131
Fatty acid methyl esters 100, 102, 108, 110, 125, 127, 128, 132–134, 288, 319, 320
Feedstock 52, 55, 158–161, 165, 167, 169–171, 174, 178, 181, 182, 186, 187, 189, 193, 194, 241–243, 246, 248–250, 252, 254–256, 259–262, 264–266, 268–271, 285, 286, 290–293, 308, 317–321, 327, 329, 335, 336, 338, 340, 346, 347, 357, 359, 361, 362, 364–368, 371, 372, 374, 375
Fermentation 160, 161, 171, 177, 179, 185, 187–192, 194, 354, 362, 363, 367, 368, 371, 373
Flexibility 311, 319
Focused microwave 73, 81–83
Food waste 215, 229, 374
Fractionation 172, 174, 186
Free energy 278, 279, 299
Free fatty acids 99, 101, 106, 118, 119, 125, 137
Frequency 1, 2, 6, 23, 62–64, 66–71, 80, 81, 87, 88, 92, 164, 165, 190
FTIR 167, 168, 181
Functional groups 320, 323, 324
Furfural 247, 250, 253, 259, 262, 301, 302, 308, 335–337

G

Gas chromatography 138, 139, 145
Gas constant 279, 293, 297, 305
Gas fraction 243, 255

Gasification 159, 160, 240–242, 247, 307
Generator 62, 64, 68, 83, 86, 92
Gibb's energy 278, 279
Glycerol 9, 17–19, 101, 105, 110, 118, 119, 124, 131, 134–138
Granular 165
Green chemistry 1, 6, 9, 12–14, 17, 21–23, 358, 364, 365
Green metrics 16
Green solvent 14
Greenhouse gas 247, 271
Ground state 111

H

Heat and mass transfer 347, 348, 352, 358, 364, 365, 375
Heat loss 103
Heat transfer 277, 281, 283
Heating mechanism 30
Hemicellulose 161–164, 167, 169–174, 176–179, 182, 184–187, 190, 191, 194, 362, 367, 373, 374
Heterogeneous 68, 71, 73, 77, 80, 88, 89, 91
Hexane 104, 105, 115, 127, 129, 130, 139
High shear mixer 349, 352
HMF 161, 173, 179, 187, 188, 194–196, 301, 302, 308, 335–339
Homogenous 73, 79
Hot air 349–351
Hot spot 33, 44, 45, 66, 68, 77
Hybrid reactor 366
Hybrid technology 358, 359, 364, 375
Hybridization 346, 349, 350, 375
Hydraulic retention time 208, 213, 219, 221, 229, 234
Hydrocarbons 102, 103, 134
Hydrodiffusion 351, 374
Hydrodistillation 352
Hydrodynamic 41, 44
Hydrolysis 160, 161, 171–179, 182, 184–186, 188–194, 206, 209–212, 216, 219, 221, 222, 225, 226, 231–233, 300–302, 362, 363, 368, 371, 373
Hydrophilicity 315, 317

I

Imaginary impact 49
in situ transesterification 126, 127
Industrial scale 73, 82, 83, 88, 89, 92
Inedible oils 113
Influent 208, 225, 232
Infrared 349, 350
Inhibition 209, 225, 227, 229
Inhibitors 161, 172–174, 176, 178, 187
Inorganic synthesis 5
Instantaneous 244
Instantaneous heating 43
Insulator 35
Intensity 213, 215, 216, 227
Interaction 28–30, 32–35, 37, 39–41, 43, 52, 55, 58, 59, 164, 180, 181, 244, 258, 261
Interfaces 29, 41, 44, 57, 278, 279, 283
Interfacial polarization 29, 31, 32
Interference 29, 41, 42
Inter-polymer 163

Intra-polymer 163
Ionic conduction 28, 31–34
Ionic liquids 311–330, 332–336, 338, 340
Irradiation 278, 280, 283, 287, 288, 295, 296, 302, 303, 307

K

Kinetics 33, 44, 46, 47, 49

L

Large scale 62, 71, 77, 81, 87, 88, 95
LCA 271
Leakage 65, 70, 80, 90
Levoglucosan 243, 247, 249–251, 257, 259, 262
Levulinic acid 335
Lignin 160–164, 167, 170–174, 176–179, 182, 184–186, 190, 191, 196, 362, 367, 368, 371, 373
Lignocellulose 159, 161, 163, 176–179, 186, 209, 330, 332, 340, 371
Lipid yield 287, 288
Lipids 103, 104, 109, 114–117, 125–129, 131, 133, 137, 138, 140, 144, 145
Liquefaction 100, 102, 127, 364
Liquid fraction 243, 246, 255
Liquid-solid suspension 283

M

Magnetron 62–64, 71, 80, 85–87
Magnitude 29, 35, 40, 41, 57
MARS 72–74
Mass intensity 16, 20, 21, 348
Mass transfer 276, 280, 281, 283, 284, 287, 288
Material 28–36, 38–40, 43–45, 57–59
Mathematical analysis 296
Maxwell 1, 2
Meat industry 226
Mechanical 114–117, 171, 172, 176, 178, 181, 185, 186
Mechanical pretreatment 210
Mechanical-microwave 185
Mechanisms 164, 165, 175, 181, 244, 271, 319, 320, 322, 325, 326, 328, 334, 336, 337
Mesophilic 209, 212, 213, 215, 226, 229
Methane 206–215, 219–221, 223–231, 233
Methanogenesis 206, 212, 229
Methanol 33, 38, 45, 46, 50–55, 288, 292–296, 298
Microalgae 104, 113, 115–117, 126, 131, 137–139, 144, 211, 212, 219–225, 233, 283–285, 287, 303
Microwave effect 6, 14, 15
Microwave energy 29, 32, 33, 35, 36, 43, 47, 52
Microwave heating 1, 3–6, 9, 10, 13–15, 23
Microwave power 246, 248, 249, 251–254, 257, 261, 265, 267, 268
Microwave pyrolysis 149, 156
Microwave technology 1, 2, 23
Microwave-chemical 182, 185, 188
Microwaves 1–6, 9–15, 17, 21–23
Milestone 71–75
Mode 241, 248–250, 255, 258, 260, 261, 263, 268
Model 282, 283, 287–289, 291, 292, 296, 299–303, 306, 308

Modeling 41, 42
Moisture 160, 164–166, 169, 179
Molecules 279, 294, 295, 301
Multimode 62, 65–67, 70, 71, 73, 76, 77, 81, 82, 87, 90

N

Net energy 209, 231, 232
Nicholson-Ross-Weir technique 57
NMR 181
Non-conventional 356–358
Non-thermal 47, 49
Non-thermal effect 211, 214, 216

O

Object size 39
Ohmic heating 349, 353, 354
Oil 31, 48, 50–55, 57
Oil bath 31, 48
Oil Extraction 283–288
Oil-palm 57
Open vessel 31, 48
Operations 62, 64, 65, 67–69, 71, 72, 75–77, 83, 87, 92
Optimization 348, 353, 360, 363, 364, 366
Organic acids 317
Organic matter 208, 211, 213, 214, 216–219, 222, 226, 227, 231–234
Organic solvents 311, 312, 316, 326–328, 340
Organic synthesis 3, 9, 10
Oxidation 240–242
Oxidative 171, 176, 177

P

Parameters 207, 211, 216–218, 226, 227, 232, 234
Partial oxidation 240–242
Penetration 209, 221, 234
Penetration depth 28, 35, 42, 55–57, 67, 68, 71, 80, 82, 87, 276
Permittivity 28, 29, 34–36, 57
Perturbation 57, 58
Phenolic 246, 247, 253–257, 263
Photobioreactor 126
Physical 100, 114, 116
Planck's constant 299
Polarity 104, 105, 110–112, 129, 130, 312, 316, 317, 327
Polarization 28, 29, 31–34, 39
Power 161, 162, 164, 169, 171, 178, 179, 184–188, 190–192, 194, 196, 197
Power density 276, 288, 290
Power effect 33, 36, 44, 46, 50, 52
Power output 63, 69, 87, 92
Pre-exponential 110, 111
Pre-exponential factor 49, 278, 279, 293, 294, 299–301, 305, 307
Pressure 210, 211, 218, 221
Pretreatment 102, 114, 158, 161, 162, 164, 171–174, 176–179, 181, 182, 184–194, 197, 209–219, 221–223, 225–234, 248–254, 256, 257, 262, 268, 280, 291, 300, 302, 303, 357, 361–363, 371, 373

Process development 359, 364
Process equipment 347, 355, 367
Process intensification 23, 346–349, 355, 356, 358, 359, 362–364, 366, 375
Production 99–105, 108, 112–114, 117–120, 124–126, 129, 130, 132, 134–136, 140, 142–147
Productivity 348, 349, 367
Property 312, 327
Psychrophilic 209
Purification 126, 136
p-xylene 49
Pyrolysis 39, 55–58, 160, 180, 240–262, 264–271, 303–308
Pyrolytic gasification 241

R

Radiation 1–3, 10, 30, 34–36, 41
Raman spectroscopy 167
Rate enhancement 3, 4, 278, 279
Reaction 62, 68–81, 83, 85–88, 92
Reaction Kinetics 276, 278, 291, 293, 295, 296, 298, 300, 301, 308
Reaction mass efficiency 16, 21
Reaction mechanism 107–109, 111, 118, 131
Reaction parameters 293, 299
Reaction products 246
Reaction rate 3, 5, 15, 22, 278, 279, 291, 293–298, 302, 304, 305, 307
Reaction temperature 278, 283
Reaction time 43, 48, 50–52, 319–321, 324, 326, 329, 330, 335, 354, 356–362, 364, 365
Reaction yields 288, 294, 295, 298, 303, 304
Reactor design 248, 258
Real-time monitoring 145
Recovery 318–321, 326–330, 340
Reflection 41, 57, 58
Reflux 70, 75, 78, 79
Renewable energy 99, 142
Renewable feedstock 12, 13, 15, 16
Reproducibility 62, 73, 76, 77, 80
Residual 210, 223, 225, 226, 230
Resistance 35
Ring 164, 181

S

Safety 64, 68–70, 73, 80, 81
Scalability 68, 76, 77
Scale-up 68, 69, 71–73, 75, 76, 79–81, 83, 85
Selective heating 33, 43, 44
Separation 101–103, 114, 119, 124–126, 134, 136, 138, 139, 147
Sewage sludge 206, 212, 213, 218, 226, 227, 229
Shaking 210
Single mode 62, 65–67, 70, 72, 73, 75–77, 81, 82, 85
Sludge 206, 208, 212–219, 221, 226–231, 234
Soap 103, 106, 119, 124, 129, 137
Solid fraction 242, 246, 255
Solid interaction 33, 34, 39
Solids 206, 208, 209, 212–215, 217, 218, 221, 223, 227, 229, 232, 234

Solubility 114–116, 119, 134, 213, 214, 216, 217, 221, 227, 232, 233, 311, 316, 317, 328, 329, 332, 333
Solubilization 210, 211, 213–215, 221, 223, 226, 227, 231, 233, 234
Solvent 28, 31–33, 35, 37–39, 43–46, 48, 49, 53, 100–102, 104–106, 109–112, 114–118, 124–127, 129–131, 134, 136–139, 142–144, 147, 311, 312, 316–319, 321, 323, 325–330, 335, 336, 338, 340
Solvent extraction 105, 115–117, 125, 126
Source 62, 65–67, 69, 89
Spectroscopic 165
Stability 311, 316, 320, 323, 327–329
Starch 158, 159, 161, 164–169, 171, 177, 182, 194, 196
Steam 349, 350, 371, 374
Steam explosion 172, 173, 177, 182, 186, 188, 209–211, 233
Steam gasification 241, 247
Steam heating 349, 350
Steam pretreatment 210, 211
Stretching 165, 167
Structural change 165, 166, 176
Structure 209–211, 213, 216, 217, 221–223, 225, 226, 233
Sucrose 158–161, 194, 196
Super-boiling 46
Supercritical 100, 102, 103, 105, 109, 110, 112, 114, 116–119, 124–127, 130–133, 136, 144
Superheating 31, 32, 39, 44, 47
Surface temperature 30
Sustainable chemistry 16
Synergism 346–348, 356, 357, 365
Synergistic effect 347, 348, 360, 364, 366
Syngas 230
Synthesis 104, 109, 124, 125, 127, 132, 134, 136, 142, 143, 311, 312, 315–317, 319, 320, 323, 327, 340
Synthesis gas 241, 243, 247

T

Tangent loss 28
TEM 223
Temperature 28, 30–33, 35–37, 39, 43, 44, 46–58
Temperature control 267, 268
Temperature distribution 245
Thermal effect 30, 46, 49, 50
Thermal energy 277
Thermal gasification 241
Thermal gradient 30, 31, 49
Thermal pretreatment 211, 213, 221, 225, 226, 231–233
Thermal runaway 68, 69, 80, 248, 268, 271
Thermochemical 102, 127, 240, 241, 243, 247
Thermodynamic 110
Thermodynamic parameters 293, 299–301

Thermophilic 209
Thin layer chromatography 127, 138
Toxicity 312, 328, 329, 340
Transesterification 9–11, 17–20, 22, 109, 126, 127, 136, 144, 288, 291–296, 298, 320
Transformation 348, 368, 371
Transmission 57, 58
Transparent 33, 35, 38, 39
Traveling wave 62, 67, 87
Triglycerides 294–296, 298
Triphasic 318, 327
Tubular reactor 88
Turbo mixer 352, 353
Two-stage 185, 187–189

U

Ultrasound 209, 221, 232, 233, 346, 347, 349, 354–367, 373–375
Uniformity 65–67, 71, 83, 87

V

Vacuum 351, 352
Vapor pressure 311, 312, 316
Vapor transport 280
Variables 248, 262, 267, 268, 271
Vegetable oils 99, 100, 102, 106, 109, 110, 112, 113, 115, 119–122, 125, 126, 139
Vendor 71, 81
Vessel 68, 70–76, 78–81, 85, 87, 95
VFA 206, 207, 213, 229
Volatile solids 208, 213, 221, 223, 229, 234
Volatility 311, 328, 329
Voltage breakdown 69, 70, 268, 271
Volumetric 244, 266
Volumetric heating 30, 43
Voyager 71–74, 77–79

W

Waste 346–348, 356, 357, 359, 360, 362, 363, 367, 373–375
Waste intensity 348
Wastewater 212–214, 216–221, 225, 227, 229, 230, 232
Waveguide 62, 64–66, 70, 83, 87, 88, 90
Wavelength 29, 34–37, 39–41, 43

Y

Yield 160, 161, 171, 173, 176–180, 182, 184–194, 196, 347, 352, 354, 356, 358–360, 363–365, 370, 373–375